Ecology and Management of

Cowbirds and Their Hosts

Ecology and Management
of Cowbirds and Their Hosts

Studies in the Conservation of
North American Passerine Birds

EDITED BY
JAMES N. M. SMITH,
TERRY L. COOK,
STEPHEN I. ROTHSTEIN,
SCOTT K. ROBINSON,
AND SPENCER G. SEALY

Foreword by Paul R. Ehrlich

University of Texas Press

Austin

First edition, 2000

Requests for permission to reproduce material from this work should
be sent to Permissions, University of Texas Press, Box 7819, Austin, TX
78713-7819.

(∞) The paper used in this publication meets the minimum requirements of American National Standard for Information Sciences—
Permanence of Paper for Printed Library Materials, ANSI Z39.48-1984.

ISBN 0-292-77738-8

Library of Congress Cataloging-in-Publication Data

Ecology and management of cowbirds and their hosts : studies in the
conservation of North American passerine birds / edited by James
N. M. Smith . . . [et al.] ; foreword by Paul R. Ehrlich. — 1st ed.
    p. cm.
  Includes bibliographical references.
  ISBN 0-292-77738-8 (hardcover : alk. paper)
  1. Cowbirds. 2. Brood parasitism. 3. Wildlife management.
I. Smith, James N. M., 1944– .
QL696.P2475E26 2000
598.8'74—dc21   99-26583

*Title page photograph of a flock of juvenile Brown-headed Cowbirds
by Stephen I. Rothstein.*

Contents

Foreword

Cowbirds and Bull-hockey

PAUL R. EHRLICH

As the loss of biodiversity escalates, a determined effort is being made to convince people that there is nothing to worry about—that there is no extinction crisis (Simon and Wildavsky 1993). This effort is part of a broader campaign to quell environmental concerns, an antiscience outflowing of rhetoric in support of the wise-use movement that has been christened the "brownlash" (Ehrlich and Ehrlich 1996). The depth of ignorance represented in the brownlash is difficult to overestimate. Consider the following statement: "North and west of New York City and London and Chicago, south of Paris and Bonn, east of San Francisco and Moscow, in all directions around Atlanta and Denver and Warsaw and Madrid, and in many similar locations worldwide, extensive tracts of habitat that have known only occasional human intervention abut centers of mechanistic excess" (Easterbrook 1995, p. 9).

Think about going east from San Francisco, through grossly polluted San Francisco Bay with some 90% of its wetlands destroyed, through fully developed and smog-soused East Bay cities and suburbs to the plowed-under Central Valley, virtually devoid of even remnant natural habitat, with most of its native flora and fauna gone, and now populated by cowbirds whose invasion was aided by anthropogenic habitat changes (Rothstein 1994). In fact, one could continue eastward around the world without finding any "extensive tracts of habitat that have known only occasional human intervention," unless one wished to stretch that definition to include some ocean depths.

However, just getting to the cowbirds in California's Central Valley allows me to make the points I want to emphasize here. First, most people are ignorant of the degree to which *Homo sapiens* has modified the biosphere. Second, although biodiversity loss is usually characterized in terms of the rate of species extinction, population extinction is also very important. Third, invasions and subtle community alteration are crucial factors in the extinction epidemic. And finally—as *Ecology and Management of Cowbirds and Their Hosts*

amply demonstrates—the cowbird makes a fascinating case study illustrating these points.

In the past 150 years, humanity has become a global force. Measured by energy use—the best general surrogate for the power to do environmental damage—the human enterprise has expanded roughly 20-fold since 1850 (Holdren 1991). The current state of global overpopulation is evident because humanity is unable to support itself solely on income from its natural capital, but only by also depleting that capital. This is most readily seen in loss of deep, rich agricultural soils, "fossil" groundwater, and, of course, biodiversity (Ehrlich and Ehrlich 1990).

In the last case, emphasis has been placed much too heavily on the extinction of entire species, whereas from the viewpoint of the all-important delivery of ecosystem services (Daily 1997), population extinctions are equally or more important (Ehrlich and Daily 1993). After all, if all remaining nonhuman species could be preserved as minimum viable populations in zoos, aquaria, botanical gardens, and buckets of soil, there would be no loss of species diversity. Nonetheless we would all quickly die from starvation and other consequences of the termination of ecosystem services. Thus, one question of great interest in relationship to cowbirds is how they have (or have not) contributed to the population extinctions that have afflicted a substantial fraction of neotropical migrants.

Continuing population growth and increasing per capita consumption are bound to increase those losses, almost certainly reducing Earth's carrying capacity for *Homo sapiens,* since there is little sign that making technologies more environmentally friendly can or will compensate for expanding human numbers and affluence. Thus whatever the role of cowbirds in the critically important extinction of populations (as well as, of course, species) of their hosts, it will be played out on a stage modified by increasing anthropogenic disruption.

The role of people in causing extinctions is often seen primarily as overhunting (as typified by the extermination of the Passenger Pigeon, Dodo, Eskimo Curlew, and numerous populations of bison, wolves, snow leopards, elephants, rhino, whales, and oceanic fishes) and outright habitat destruction (as in the cases of the Ivory-billed Woodpecker, Bachman's Warbler, and many populations of Northern Spotted Owls, lemurs, black-footed ferrets, prairie plants, and Edith's checkerspot butterfly) (e.g., Ehrlich and Ehrlich 1981, Ehrlich et al. 1992). There is much less public recognition of the role of invasions in causing extinctions (Heywood 1995). Avian extinctions, of course, are most clearly seen in the extirpation of entire species from numerous islands by invading *Homo sapiens* and the animals they transported with them (e.g., King 1984, Steadman 1989). Very often, of course, hunting, habitat destruction, invasions, and the vulnerabilities associated with small population size can all conspire to do in a species, as in the case of the Tasmanian tiger or thylacine (Lunney and Recher 1986).

Sometimes the alterations associated with invasions are so long-standing and widespread that it is very difficult to evaluate their overall impact. For example, it is highly likely that the movement of people into the western hemisphere was a major cause of the Pleistocene megafaunal extinctions (Martin 1984). These, in turn, transformed the landscape of the Americas, creating shifts in community structure and causing other extinctions of populations and species (Guthrie 1984, Owen-Smith 1989), most of which were unrecorded. More recent but probably equally transforming has been the overgrazing and consequent desertification of much of the western United States (e.g., Ferguson and Ferguson 1983, Jacobs 1991), which also certainly led to many population extinctions.

The inevitability of such unsung anthropogenic extinctions has been made even clearer by recent advances in our understanding of the subtlety of habitat requirements. Who, for example, might have imagined that the persistence of Tree and Violet-Green Swallow populations in some subalpine areas would depend on the co-occurence of sapsuckers, aspens, willows, and a fungus (Daily et al. 1993)?

There would seem to be a prima facie case that cowbird populations have a negative impact on the dynamics of their host populations. In several cases, such as those of the Black-capped Vireo and Kirtland's Warbler, remnant populations of endangered species have been threatened by very high rates of cowbird parasitism. In some cases, cowbird removal programs have been part of the operations designed to rescue these species from imperilment. The results, as you will see in the chapters that follow, are variable and sometimes difficult to interpret.

This book underlines several important issues in conservation biology. First of all, in many cases biologists already know more than enough to suggest sensible starting directions for trying to protect biodiversity—both globally and in specific cases. Undertaking cowbird control programs in various cases has been such a sensible direction at the local level. But conservation efforts almost always involve allocation questions, and cowbird control may not be the best place to target scarce available funds. That's where the need for further scientific investigation arises. We need to discover the least expensive steps that will accomplish conservation goals. But that need can, in turn, present difficult trade-offs between the use of funds for continuing conservation efforts such as control programs or purchasing or restoring more habitat, and for determining the relative values of cowbird control and habitat protection and restoration. Resolving such dilemmas is daunting, but conservation biologists should at a minimum always be explicitly aware of them.

Fortunately, the evolutionary and ecological questions surrounding an abundant brood parasite that has recently expanded its range widely are so fascinating, as *Ecology and Management of Cowbirds and Their Hosts* illustrates, that research on cowbird biology would doubtless continue even if cowbirds had negligible impact on threatened or endangered species. The information in this volume would be of great value to biologists even if there were no need to put "management" in the title. Finally, as conservation biologists struggling to protect elements of biodiversity at local and regional levels, we should never forget the global picture. Unless expansion of the scale of the human enterprise can be halted soon, virtually all of the populations and species that may or may not now be threatened with extinction by cowbirds will disappear. They will go extinct from the effects of habitat destruction, climatic change, toxification, or other anthropogenic causes interacting with each other and perhaps with cowbird parasitism. After that, continued increase in human numbers and per capita affluence could make it a question of whether industrial civilization or the cowbirds themselves will disappear first. The kind of bull-hockey produced by the brownlash makes having to face that question ever more likely.

References Cited

Daily, G., P. Ehrlich, and N. Haddad. 1993. Double keystone bird in a keystone species complex. Proc. Natl. Acad. Sci. USA 90: 592–594.

Daily, G. C. (ed). 1997. The value of ecosystem services. Island Press, Washington, DC.

Easterbrook, G. 1995. A moment on the earth: The coming age of environmental optimism. Viking, New York.

Ehrlich, P. R., and G. C. Daily. 1993. Population extinction and saving biodiversity. Ambio 22:64–68.

Ehrlich, P. R., D. S. Dobkin, and D. Wheye. 1992. Birds in jeopardy. Stanford University Press, Stanford, CA.

Ehrlich, P. R., and A. H. Ehrlich. 1981. Extinction: The causes and consequences of the disappearance of species. Random House, New York.

———. 1990. The population explosion. Simon and Schuster, New York.

———. 1996. Betrayal of science and reason: How anti-environmental rhetoric threatens our future. Island Press, Washington, DC.

Ferguson, D., and N. Ferguson. 1983. Sacred cows at the public trough. Maverick Books, Bend, OR.

Guthrie, R. D. 1984. Mosaics, allelochemics, and nutrients: An ecological theory of late Pleistocene megafaunal extinctions. Pp. 259–298 *in* Quaternary extinctions: A prehistoric revolution. (P. S. Martin and R. G. Klein, eds). University of Arizona Press, Tucson.

Heywood, V. H. (ed). 1995. Global biodiversity assessment. Cambridge University Press, Cambridge.

Holdren, J. 1991. Population and the energy problem. Population and Environment 12:231–255.

Jacobs, L. 1991. Waste of the West: Public lands ranching. Lynn Jacobs, Tucson, AZ.

King, C. 1984. Immigrant killers: Introduced predators and the conservation of birds in New Zealand. Oxford University Press, Auckland, New Zealand.

Lunney, D., and H. F. Recher. 1986. The living landscape: An ecological view of national parks and nature conservation. Pp. 294–328 *in* A natural legacy: Ecology in Australia, 2nd ed. (H. F. Recher, D. Lunney, and I. Dunn, eds). Pergamon Press, Sydney.

Martin, P. S. 1984. Prehistoric overkill: The global model. Pp. 354–403 *in* Quaternary extinctions: A prehistoric revolution (P. S. Martin and R. G. Klein, eds). University of Arizona Press, Tucson.

Owen-Smith, N. 1989. Megafaunal extinctions: The conservation message from 11,000 years B.P. Conserv. Biol. 3:404–412.

Rothstein, S. I. 1994. The cowbird's invasion of the far West: History, causes, and consequences experienced by host species. Studies in Avian Biol. 15:301–315.

Simon, J., and A. Wildavsky. 1993. Facts, not species, are endangered. New York Times.

Steadman, D. W. 1989. Extinction of birds in eastern Polynesia: A review of the record and comparisons with other Pacific Island groups. J. Archaeol. Sci. 16:177–205.

Ecology and Management of

Cowbirds and Their Hosts

General Introduction

Brown-headed Cowbirds as a Model System for Studies of Behavior, Ecology, Evolution, and Conservation Biology

JAMES N. M. SMITH AND STEPHEN I. ROTHSTEIN

In this introduction, we present background information on the behavior, ecology, and evolution of brood parasites, with special reference to the Brown-headed Cowbird. Our intent here is to highlight the considerable breadth of biological questions and conservation issues that arise from studying brood parasites. We also describe the genesis of this book and briefly comment on how cowbirds may affect the population levels and even survival of other species. The connection between cowbirds and conservation biology is especially interesting for two reasons. Cowbirds can affect numerous other species, and the anthropogenic changes to landscapes that have benefited the cowbird may simultaneously have harmed other bird species. This intertwining of anthropogenic and natural biotic effects in turn raises the question of whether declines of songbirds are due to parasitism or habitat loss, or some combination of these two potential impacts.

Three species of cowbird now occur in North America, the Brown-headed Cowbird (*Molothrus ater*), the Bronzed Cowbird (*Molothrus aeneus*), and the Shiny Cowbird (*M. bonariensis*). Informative accounts of the general biology of the native Brown-headed and Bronzed Cowbirds are given by Lowther (1993, 1995). The third species, the Shiny Cowbird, is a recent immigrant from the south (see Cruz et al. Chapter 4). A recent and detailed review of the behavior and ecology of cowbirds is given by Robinson et al. (1995a). Much additional information on the biology of avian brood parasites will be elaborated in a forthcoming book (Rothstein and Robinson 1998). That information is complementary with the more applied work that is the main, but not sole, focus of this book. Together, we hope that these two books will help to bring the many unsolved basic and applied questions about brood parasites to a wider audience and into sharper relief.

The Brood Parasitic Life History

Brood parasitism is an unusual and fascinating life history found in some birds, insects, and fishes that share the common habit of rearing their young in a fixed nest (Clutton-Brock 1991). Brood parasites lay their eggs in the nests of other individuals or colonies, thus duping the unsuspecting foster parents into providing parental care. Some brood parasites, like the cowbirds discussed in this book, are obligate, in that they never rear their own young and lay only in the nests of other species. Other animals, like many species of colonial waterfowl, are facultative brood parasites, in which an individual may lay some eggs in the nests of conspecifics or other similar species while incubating the rest of its eggs in its own nest (Lyon and Eadie 1991).

Because they depart so strongly from the usual avian model of strong parental care, brood parasitic birds have always attracted much attention, and numerous authors over the last century have collected data on cowbirds in particular. Much of these data were collected incidentally in the course of nesting studies focused on various host species. There was little attempt to synthesize what was known, with the notable exception of Herbert Friedmann's studies from the 1920s to the 1980s, and the reviews by Lack (1968) and Payne (1977). Indeed, most research that has focused on brood parasitism itself has appeared only in the last two decades. It has recently become apparent that brood parasites and their hosts raise many fascinating problems in five major biological disciplines: behavior (Davies and Brooke 1988, Lotem et al. 1995), population ecology (May and Robinson 1985, Robinson et al. 1995b), evolution (Rothstein 1975a, 1990), neurobiology (Sherry et al. 1993, Clayton et al. 1997), and conservation biology (Robinson et al. 1995a,b). A brief account of the general biology of brood parasites and of cowbirds in particular follows as an introduction to this book.

The principal life history trade-off involved in brood parasitism is that by foregoing parental care, brood parasites greatly increase their annual fecundity. Most workers are agreed that wild Brown-headed Cowbirds lay eggs on 70–80% of days during a two- to four-month breeding season, for a total of at least 40 and perhaps as many as 100 eggs per year (Scott and Ankney 1980, 1983; Fleischer et al. 1987; Smith and Arcese 1994). In captivity, one female laid more than 70 eggs, and fecundity depended strongly on female age (Holford and Roby 1993). Kattan (1997) reports fecundities of 120 eggs per year in wild Shiny Cowbirds. However, recent estimates using molecular methods to track egg laying by individual females (e.g., Gibbs et al. 1997) have revealed that individual fecundities can also be much lower, perhaps 5 to 15 eggs per year (S. G. Sealy pers. comm., D. C. Hahn pers. comm.). In contrast, the average female in two short-lived, open-nesting songbirds, the Song Sparrow (*Melospiza melodia*) and the Meadow Pipit (*Anthus pratensis*), lays 7–11 eggs per year (Smith 1988, Hotker 1989). Thus, the fecundity advantage gained by brood parasitism is at least two- to three-fold and may exceed ten-fold. High fecundity gives cowbirds a high potential intrinsic rate of population growth when they colonize new areas, as they have done almost continuously in the past 200 years (Rothstein 1994).

Reproductive success depends on more than high fecundity, however, and survival of eggs and young in brood parasites is presumably lower than in parental species. In many cases, cowbirds lay eggs in nests of inappropriate host species or lay at inappropriate times (Robinson et al. 1995b, Kattan 1997). A review of studies on a range of host species led Scott and Ankney (1980) to conclude that only about 15% of cowbird eggs result in fledglings. Cowbirds can, however, achieve much higher survival from laying to fledging and may even match the fledging success of especially suitable hosts such as the Song Sparrow (Smith and Arcese 1994). It is not known how the reproductive strategy of cowbirds influences their life span, but there is little reason to believe that they are more long-lived than other similar-sized songbirds.

One noteworthy aspect of the life history of cowbirds is the strongly male-biased sex ratios, which contrast with the female-biased sex ratios typical of nonparasitic blackbirds (Orians 1980). This bias implies that there is a survival cost to the high fecundity of female cowbirds. The survival cost is further indicated by Darley's (1971) discovery that males and females have similar survival rates in their first year of life, when females have not yet laid eggs, but that survivorship among mature females is lower than for males. Keys et al. (1986) documented a possible physiological cost, lowered volumes of red blood cells, that females may incur because of their extended egg laying (but see Ankney and Scott 1980).

The Evolution of Brood Parasitism

The evolution of obligate brood parasitism as a reproductive strategy is not well understood. One possibility is that it evolved via intraspecific nest parasitism (Nolan and Thompson 1975). Two facts, however, suggest that this explanation

is unlikely. First, obligate parasitism is rare in waterfowl, the group of birds with the highest levels of intraspecific parasitism. Second, intraspecific parasitism is rare in blackbirds closely related to cowbirds (Lyon et al. 1992, Rothstein 1993). Another hypothesis is that it arose via the habit of stealing the nests of other birds (Friedmann 1929). Nest stealing by the so-called Bay-winged Cowbird (*Molothrus badius*) is common (Fraga 1998), but this nonparasitic species is actually not closely related to other cowbirds (Lanyon 1992).

Present-day classifications indicate that obligate brood parasitism has evolved seven times, as it occurs in seven different taxa: Old World cuckoos (family Cuculidae, subfamily Cuculinae), New World cuckoos (Cuculidae, subfamily Neomorphinae), honeyguides (Indicatoridae), blackbirds (Emberizidae, subfamily Icterinae), widow birds (Estrildidae, subfamily Viduinae), Old World sparrows (Ploceidae), and ducks (Anatidae). The last two taxa each contain a single obligately parasitic species and numerous nonparasitic ones. Recent DNA studies suggest that the single parasitic Ploceid, the Cuckoo Finch (*Anomalospiza imberbis*), is related to the widow birds rather than to Ploceids (R. Payne pers. comm.), which could mean that obligate parasitism has evolved only six times.

Obligate brood parasites differ greatly in their degree of specialization on host species. Most African widow birds parasitize only a single host species and show remarkable mimicry of the host's song, nestling behavior, and even the gape markings inside the mouth of the nestling (Nicolai 1964, Payne 1982). Other brood parasites, like the well-studied Common Cuckoo (*Cuculus canorus*), usually specialize on a single host species in any particular part of Europe (Brooke and Davies 1987). The degree of local specialization, however, is only moderate (Moksnes and Røskaft 1995), and there seems to be little genetic differentiation between cuckoos with different mimetic egg types (Gibbs et al. 1996). Further flexibility in host use is shown by cuckoos in Japan, where the same cuckoo commonly parasitizes several hosts in a single region and often does so at much higher rates than seen in Europe (Nakamura et al. 1998). When several Old World cuckoo species coexist locally, they tend to parasitize different host species (Friedmann 1948, Higuchi and Sato 1984).

The five species of parasitic cowbirds are remarkable because they include both the most generalized and nearly the most specialized among the world's 80 or so known obligate brood parasites. The most specialized is the Screaming Cowbird (*M. rufoaxillaris*), which was known to parasitize only one host species until two other hosts were noted in the last decade (Fraga 1996, 1998; Rothstein et al. in prep.) The two most generalized cowbirds, the Brown-headed and the Shiny, have by far the largest number of known host species among brood parasites. Each has been recorded laying in the nests of more than 200 other species (Friedmann and Kiff

1985, Friedmann et al. 1977), and individuals are known to lay in the nests of several hosts (Fleischer 1985, S. G. Sealy pers. comm.). Host specialization has the benefit of increasing the chance of choosing a suitable host, but it decreases the number of potential hosts and therefore the growth potential of a parasitic population. Also, there is a risk that a host will evolve a successful defense against a specialized parasite, if the parasite exerts a strong selection pressure on its host. This defense could, in turn, select for a host shift on the part of the parasite (Rothstein 1990).

Reviews of coevolution between parasitic birds and their hosts have suggested that generalized parasites will become more specialized over time as more and more host species evolve defenses such as egg recognition (Davies and Brooke 1989a, Rothstein 1990). According to these arguments, increasing specialization is likely because the most effective adaptation by the parasite to host egg recognition is mimicry of the host egg. Genetic constraints (the need for assortative mating among genotypes) mean that a parasitic population cannot be subdivided to produce many different genetically determined egg types at any one time.

In contrast to previous work, Lanyon (1992) argued that cowbirds have become progressively more specialized over time because the first species to branch off in the cowbird clade is the most specialized, while the two most recent species are the most generalized. Rothstein et al. (in prep.), however, have argued that host number is too labile for one to assume that current host numbers reflect numbers when each cowbird species appeared (see also Freeman and Zink 1995). They have also pointed out that the relationship between branching order and host number is consistent with the hypothesis of increasing specialization, because the potential hosts of the oldest cowbird species have had the greatest amount of time to develop defenses. Neither hypothesis, that of increasing versus decreasing specialization over time, can be invalidated at present.

Behavioral Evolution in Brood Parasites and Their Hosts

The best-studied behavioral traits that affect the success of brood parasites are egg ejection and nest desertion by hosts. In Common Cuckoos, which show strong host specialization, egg ejection is a population-specific trait in at least some hosts, and its expression depends on the historical association of the population with cuckoos (Davies and Brooke 1989a) and even on the present levels of parasitism (Davies et al. 1996). In some populations of cuckoo hosts, egg ejection or nest desertion is maintained at intermediate levels by the costs of egg recognition (Davies and Brooke 1989b, Davies et al. 1996) or by the need for young birds to learn to recognize their own eggs (Lotem et al. 1995).

Hosts eject Brown-headed Cowbird eggs in two ways. In most species, the female host grasps the cowbird egg with her open beak and removes it, something that is easier for larger hosts to do (Rothstein 1975a). In a few species, females spike the cowbird egg with their beak to remove it (e.g., Rothstein 1977, Sealy 1996).

Nearly all actual and potential hosts of Brown-headed and Shiny Cowbirds are apparently made up entirely of acceptor or rejector individuals (Rothstein 1975a,b; Mason 1986; but see Dufty 1994, Sealy and Bazin 1995). That is, close to 100% of individuals eject cowbird eggs placed in their nests, or close to 100% accept them.

Many studies have shown that variable numbers of individuals within acceptor species desert naturally parasitized nests and renest (Rothstein 1975a,b; Graham 1988; Hill and Sealy 1994; Goguen and Matthews 1996). Such desertions seem not to be caused by the cowbird eggs themselves (Rothstein 1975a, 1982, 1986) but rather by the presence of cowbirds near the host nest (Rothstein 1975a, Burhans Chapter 18) or by clutch reduction when eggs are removed by cowbirds (Rothstein 1982, Hill and Sealy 1994, Sealy 1995). Grzybowski and Pease (Chapter 16) use simulation models to explore how the effects of nest desertion interact with varying frequencies of parasitism.

A puzzling fact is that most cowbird host species do not eject cowbird eggs, even when their eggs differ greatly from the nonmimetic cowbird egg. The absence of egg ejection in many frequent cowbird hosts is often explained under the evolutionary lag hypothesis (Rothstein 1990, Ward et al. 1996), where ejection is assumed to be adaptive but has not evolved because of the absence of the appropriate genetic variation in host populations. Alternatively, ejection behavior may not have evolved because it inflicts greater costs than rearing the parasite, the evolutionary equilibrium hypothesis (Davies and Brooke 1989a, Petit 1991, Rohwer et al. 1989, Brooker and Brooker 1996). Finally, rejection of brood parasite eggs is unlikely to evolve in species with low historical or present levels of parasitism (Brooke and Davies 1987). Rejection costs can arise from errors in egg recognition (Davies and Brooke 1988, Lotem et al. 1995), which can lead hosts to reject their own eggs. Most cowbird hosts, however, have eggs that are easily distinguished from the egg of the cowbird. If a host is too small to eject a parasitic egg easily, it might break some of its own eggs during ejection attempts (Rothstein 1976a, Rohwer et al. 1989). This cost of rejection is probably low for most acceptor species, especially given Sealy's (1996) demonstration that a small host can eject cowbird eggs with minimal cost. A host might have to pay the costs of renesting if desertion is its only option (Rohwer and Spaw 1988). Nest site limitation in the Prothonotary Warbler (*Protonotaria citrea*) may make renesting difficult, thus favoring acceptance of cowbird eggs (Petit 1991).

Although egg ejection is lacking in many species in which it would seem to be adaptive, it is still a widespread response to brood parasitism relative to the rarity of discrimination against parasitic chicks by the foster parents. In fact, although there are some cases of clear nestling mimicry among parasitic birds, no bird is known to reject non-mimetic nestlings. Lotem (1993, Lotem et al. 1995) has attributed the absence of nestling rejection to the risk that a host individual may reject its own nestlings and accept only parasites if it learns to recognize its own young from the first nestling that appears in the nest. Such a "first to appear" learning mechanism underlies rejection of cowbird eggs in the Gray Catbird (Rothstein 1974).

Perhaps the most familiar behavioral adaptation of brood parasites is ejection behavior of the cuckoo chick. Within a few hours of hatching, day-old Common Cuckoo chicks eject host eggs or nestlings by pushing themselves under the egg or chick and pushing the luckless victim from the nest (Chance 1922). There is a recent report of ejection behavior by a Brown-headed Cowbird chick (Dearborn 1996), and it remains to be determined whether this is an isolated case of an accidental ejection or an example of a rare but purposeful behavior. Newly hatched honeyguides are equipped with piercing mandibles and use them to kill their foster siblings (Friedmann 1955).

Adult brood parasites also may use destruction of host eggs or young to persuade hosts to accept their eggs. Soler et al. (1995) presented experimental evidence that Great-spotted Cuckoos (*Clamator glandarius*) punish their magpie hosts by destroying the magpie eggs if the magpies eject parasitic cuckoo eggs. Additional work on this system is needed to determine whether such punishment is a regular feature of the cuckoo-magpie interaction, because the same research group had previously shown that the magpies recently increased the rate at which they reject cuckoo eggs despite the possible punishment (Soler et al. 1994). It is unlikely that comparable punishment occurs with cowbird parasitism, because numerous researchers have removed cowbird eggs from parasitized nests without reporting such behavior (Rothstein 1982). However, a related claim has been made for cowbirds by Arcese et al. (1996). They proposed that female cowbirds prey on host eggs and nestlings to force hosts whose nests cannot be parasitized successfully to relay and thus increase the supply of future nests that can be parasitized. Experimental work by Smith and Taitt (unpubl. data) supports a prediction of this idea, that the rate of host nest failure should decline sharply when cowbirds are removed, a result also reported in this volume by Whitfield. Other studies, however, have not found this result (Braden et al. 1997, Stutchbury 1997). If predation by brood parasites occurs commonly, it could depress host breeding success to a greater degree than is evident from the normal costs of parasitism, egg removal, and losses of host young while rearing parasitic young.

Brood parasites face several other interesting behavioral problems. They have to locate appropriate host nests at the best time for laying, during the laying period of the host. They also have to lay in the nests of potentially aggressive hosts, who usually recognize them as enemies (Robertson and Norman 1976, Neudorf and Sealy 1992) and may be capable of injuring them. Brood parasites solve these problems by laying very rapidly (Davies and Brooke 1988, Sealy et al. 1995) and by doing so at dawn, when hosts are least likely to detect them (Scott 1991, Burhans Chapter 18, Sealy et al. Chapter 19).

Finding host nests must require strong skills in nest searching behavior, a measure of caution when approaching nests of large host species, and a specialized memory. Sherry et al. (1993) have shown that eastern female Brown-headed Cowbirds (*M. a. ater*) develop a larger hippocampus, the site of spatial memory in birds and mammals, during the nesting season, presumably to help them keep track of the locations of potential host nests. There may, however, be regional variation in this trait, as Uyehara and Narins (in review) failed to find a hippocampal difference in western cowbirds (*M. a. artemisiae*).

Another problem faced by brood parasites is the development of species recognition and of species-specific behavioral patterns such as songs. Just how cowbirds come to know that they are cowbirds is unclear, but the process appears to be virtually foolproof. There are no reports of cowbirds in nature attempting to court heterospecifics or singing the songs of other species. The lack of such reports led some workers to speculate that the cowbird's development of species recognition and of song is innate and resistant to experience (e.g., Mayr 1979). Cowbirds would thus differ from most other songbirds, whose song development depends heavily on learning.

In fact, we now know that although some cowbird behaviors develop nearly normally in complete isolation from conspecifics (King and West 1977), much of the behavior of cowbirds is learned, and even innate cowbird behaviors can be modified in response to experience (O'Loghlen and Rothstein 1993, West and King 1996). Indeed, some aspects of the cowbird's vocal repertoire, such as the development of flight whistles, show a greater input from learning than do the songs of most nonparasitic songbirds, as described by Rothstein et al. (Chapter 7). In recent years, cowbird studies have become prominent in the vast literature on vocal development and are widely cited in basic animal behavior texts (e.g., Alcock 1993) for two reasons: first, because of their intricate developmental systems, and second, because of the large geographic variation in their vocalizations on both local and continental scales (Rothstein and Fleischer 1987, King and West 1990). Last, cowbirds have been recognized as attractive subjects for controlled learning experiments because they adapt so readily to captivity.

Host Choice by Brood Parasites

There are several parallels between the reproductive decisions of brood parasites and the choices made by foraging animals. Like foragers (Stephens and Krebs 1986), brood parasites exhibit functional responses to host density, i.e., they lay more eggs in denser host populations (Smith and Arcese 1994). Also, like foragers choosing more nutritious foods, generalist parasites ought to select the host species that are most successful at rearing their young. Host species of cowbirds vary in their suitability for several reasons: some are egg ejectors, some desert their nests when parasitized, and others feed their nestlings a mainly vegetable diet, on which young cowbirds cannot survive. We might therefore expect cowbirds to avoid such unsuitable species and to select out the same set of best host species across regions.

In apparent agreement with this general expectation, certain host species do seem to be consistently favored by cowbirds across large areas. The Red-eyed Vireo (*Vireo olivaceus*) stands out as perhaps the most heavily used of all hosts (see, e.g., Thompson et al. Chapter 32, Winslow et al. Chapter 34, Hahn and Hatfield Chapter 13). Other larger hosts, like Song Sparrows, may be used less frequently than Red-eyed Vireos (e.g., Peck and James 1987) but may contribute more cowbirds to the continental population. There are, however, host species like the Red-winged Blackbird (*Agelaius phoeniceus*) that are sometimes used heavily but their frequency of use varies greatly among regions and sites (see Carello and Snyder Chapter 11). Other similar examples are described in this book.

Cowbirds may avoid parasitizing ejector host species (Sealy and Bazin 1995), but it is hard to estimate how often ejector host species are used by brood parasites, because parasitic eggs disappear rapidly from the host nests if they are laid there (Scott 1977). Some studies have shown frequent use of ejectors, even when unparasitized nests of acceptors were available (Rothstein 1976b, Scott 1977, Friedmann et al. 1977:36–37).

While there have been few explicit studies of the fitness consequences of host selection for cowbirds, the evidence to date is far from compelling that they generally select the best hosts available. Weatherhead (1989) found similar frequencies of parasitism in Red-winged Blackbirds and Yellow Warblers (*Dendroica petechia*), even though the blackbirds were far more effective hosts. At the same study site, Briskie et al. (1990) found that Least Flycatchers (*Empidonax minimus*), another good host, were used much less often than the less suitable Yellow Warbler. Although such comparisons of species pairs are a useful start in examining host choice, a major current gap in the cowbird-host literature is community-wide studies that compare the use of hosts by cowbirds with their relative nest abundances (e.g., Hahn and Hatfield Chapter 13). In summary, Brown-headed Cowbirds clearly exhibit host selectivity, but they apparently do not always make the best choices among the local spectrum of potential hosts. Shiny Cowbirds may be even less selective (Kattan 1997).

Another reproductive decision facing a brood parasite is, when it finds a suitable nest, how many eggs to lay in that nest. Since songbird nests often fail completely, and sibling parasites are likely to compete strongly with each other in the nests of smaller hosts, a simple rule would seem to be to lay one egg only per nest. However, if suitable nests are in short supply, and if hosts can rear more than one parasite chick, it should then be adaptive for the parasite to lay more than one egg in the same nest. Multiple parasitism occurs in about one third of all reported cases of parasitism (Friedmann 1963, Lowther 1993), and most students of brood parasitism (e.g., Orians et al. 1989) have assumed that eggs are generally laid at random by cowbird females without regard to whether nests are already parasitized. When several parasitic eggs are laid in a single host nest, it is thus possible that this results from scramble competition between several female parasites for limited hosts and not from adaptive laying behavior (but see Smith and Arcese 1994). In opposition to this argument, laying more than one egg per host nest is commoner, and likely advantageous, in the larger hosts that can often rear more than one cowbird per nest (see Robinson et al. Chapter 33, Trine Chapter 15).

Population Dynamics of Avian Brood Parasites

Cowbirds are abundant in many parts of North America. One conservative estimate of continental numbers is 20 million–40 million (Lowther 1993), but Ortego (Chapter 5) reports 38 million cowbirds at a single roost. It is therefore likely that Brown-headed Cowbirds number 100 million or more across the continent. There have been no detailed studies of populations of cowbirds or other brood parasites because of their high mobility and the large areas over which breeding individuals move (Rothstein et al. 1984, Thompson and Dijak Chapter 10). Thus far, only models of host-parasite interactions have provided much useful insight into the dynamics of brood parasite populations (e.g., May and Robinson 1985, Pease and Grzybowski 1995).

In addition to this major gap, there have been few studies of the effects of brood parasitism on host population dynamics. These may be small in populations where parasitism is infrequent or when hosts can escape in time (e.g., Smith and Arcese 1994). They can be large in landscapes where parasitism is frequent and breeding habitat for cowbirds is limited (Robinson et al. 1995a,b; Brawn and Robinson 1996; Rogers et al. 1997). Cowbird-host dynamics are also likely to vary geographically (e.g., Weatherhead 1989).

It is also not well known which host species do most to promote cowbird numbers. Some of the limited available knowledge on this topic was assembled recently by Scott and Lemon (1996). Scott and Lemon noted that Song Sparrows, Red-winged Blackbirds, and perhaps Chipping Sparrows seem to be particularly successful hosts. Trine (Chapter 15) also found Wood Thrushes to be very effective hosts at rearing cowbirds. In riparian communities in coastal British Columbia, which lack many other abundant and suitable hosts, Song Sparrows care for about 75% of all fledgling cowbirds being fed by foster parents and routinely rear two or more cowbird fledglings from a single nest (J. N. M. Smith, unpubl. data).

Community Effects of Cowbirds

As abundant host generalists, Brown-headed Cowbirds have the potential to generate strong population and community-level interactions among their hosts, i.e., to act as a keystone species (Power et al. 1996). If a widespread host species is used heavily and is vulnerable to the effects of parasitism, it may be driven to become rare locally (Robinson et al. 1995b). If a rare host species is used heavily, it may be extirpated (Robinson et al. 1995a). In areas where many hosts are heavily parasitized, communities may be made up of more resistant host individuals of perhaps fewer species (De Groot et al. in press). On the other hand, cowbirds might increase local host diversity by concentrating their parasitism on especially competitive host species, thus releasing weaker competitors from interspecific competition (Holt 1984). In contrast, in areas where cowbirds are uncommon, they are unlikely to generate strong population or community effects in their hosts.

There is little information available on any of these topics, other than a recent simulation model by Grzybowski and Pease (in press). Several chapters in this book, however, attempt to determine the extent of cowbirds' effects on other species.

The Impacts of Brown-headed Cowbirds on Their Hosts

The notion that cowbirds might have serious impacts on entire host populations was initially raised by the researchers who found some of the first Kirtland's Warbler (*Dendroica kirtlandii*) nests (Leopold 1924, Wood 1926). The first detailed accounts of the effects of cowbirds on hosts came from Nice's (1937) work on Song Sparrows and Mayfield's studies on the Kirtland's Warbler (Mayfield 1960). Nice, Mayfield, and later Walkinshaw (1983) noted that increasing frequencies of cowbird parasitism were associated with low host reproductive success and declines in host numbers. Further examples are discussed by Robinson et al. (1995b).

Echoing an earlier paper by Mayfield (1977), Brittingham and Temple (1983) published an influential article that noted the spectacular range expansion of the cowbird and its heavy use of hosts breeding at forest edges. They argued that cowbirds could depress numbers of many host species in fragmented habitats. Brittingham and Temple's paper helped to propel the cowbird to the front of the conservation stage and stimulated much applied work on cowbirds.

Soon afterward, the popular book by Terborgh (1989) took Brittingham and Temple's findings and used them to fuel the notion of the cowbird as a "conservation villain," a notion that has since taken firm hold among naturalists (Holmes 1993).

Against this backdrop of rising interest in the possible impacts of cowbirds on North American songbirds, a meeting of biologists and managers interested in cowbirds and their ecological effects was held in Austin, Texas, in November 1993. This meeting, which was discussed by Rothstein and Robinson (1994) and Holmes (1993), brought together for the first time numerous researchers with an academic interest in brood parasitism and conservationists and habitat managers who regarded the cowbird as a management problem in need of urgent action. The papers presented at this meeting and the overview chapters discussing them form the backbone of this book.

Acknowledgments

We thank K. De Groot, C. Fonnesbeck, and A. Lindholm for helpful comments.

References Cited

Alcock, J. 1993. Animal behavior, 5th ed. Sinauer Associates, Inc. Sunderland, MA.

Ankney, C. D., and D. M. Scott. 1980. Changes in nutrient reserves and diet of breeding Brown-headed Cowbirds. Auk 97:684–696.

Arcese, P., J. N. M. Smith, and M. I. Hatch. 1996. Nest predation by cowbirds and its consequences for passerine demography. Proc. Natl. Acad. Sci. USA 93:4608–4611.

Braden, G. T., R. L. McKernan, and S. M. Powell. 1997. Effects of nest parasitism by the Brown-headed Cowbird on nesting success of the California Gnatcatcher. Condor 99:858–865.

Brawn, J. D., and S. K. Robinson. 1996. Source-sink population dynamics may complicate the interpretation of long-term census data. Ecology 77:3–12.

Briskie, J. V., S. G. Sealy, and K. A. Hobson. 1990. Differential parasitism of Least Flycatchers and Yellow Warblers by the Brown-headed Cowbird. Behav. Ecol. Sociobiol. 27:403–410.

Brittingham, M. C., and S. A. Temple. 1983. Have cowbirds caused forest songbirds to decline? BioScience 33:31–35.

Brooke, M. de L., and N. B. Davies. 1987. Recent changes in host use by cuckoos *Cuculus canorus* in Britain. J. Anim. Ecol. 56:873–883.

Brooker, M., and Brooker, L. 1996. Acceptance by the Splendid Fairy Wren of parasitism by Horsfield's Bronze Cuckoo: Further evidence for evolutionary equilibrium in brood parasitism. Behav. Ecol. 7:395–407.

Chance, E. P. 1922. The cuckoo's secret. Sedgewick and Jackson, London. 239 pp.

Clayton, N. S., J. C. Reboreda, and A. Kacelnik. 1997. Seasonal changes in hippocampus volume in parasitic cowbirds. Behavioral Processes 41:237–243.

Clutton-Brock, T. H. 1991. The evolution of parental care. Cambridge University Press, Cambridge. 352 pp.

Darley, J. A. 1971. Sex ratio and mortality on the Brown-headed Cowbird. Auk 88:560–566.

Davies, N. B., and M. de L. Brooke. 1988. Cuckoos versus Reed Warblers: Adaptations and counter adaptations. Anim. Behav. 36: 262–284.

———. 1989a. An experimental study of coevolution between the cuckoo, *Cuculus canorus,* and its hosts, 2: Egg marking, chick discrimination, and general discussion. J. Anim. Ecol. 58:225–236.

———. 1989b. An experimental study of coevolution between the cuckoo, *Cuculus canorus,* and its hosts, 1: Host egg discrimination. J. Anim. Ecol. 58:207–224.

Davies, N. B., M. de L. Brooke, and A. Kacelnik. 1996. Recognition errors and probability of parasitism determine whether Reed Warblers should accept or reject cuckoo eggs. Proc. Roy. Soc. Lond. B 263:925–931.

Dearborn, D. C. 1996. Video documentation of a Brown-headed Cowbird nestling ejecting an Indigo Bunting from the nest. Condor 98:645–649.

De Groot, K. L., J. N. M. Smith, and M. J. Taitt. 1999. Cowbird removal programs as ecological experiments: Measuring community-wide impacts of nest parasitism and predation. Studies in Avian Biology 18:229–234.

Dufty, A. M., Jr. 1994. Rejection of foreign eggs by Yellow-headed Blackbirds. Condor 96:799–801.

Fleischer, R. C. 1985. A new technique to identify and assess the dispersion of eggs of individual brood parasites. Behav. Ecol. Sociobiol. 17:91–99.

Fleischer, R. C., A. P. Smyth, and S. I. Rothstein. 1987. Temporal and age-related variation in the laying rate of the Brown-headed Cowbird in the eastern Sierra Nevada. Can. J. Zool. 65:2724–2730.

Fraga, R. M. 1996. Further evidence of parasitism of Chopi Blackbirds (*Gnorimopsar chopi*) by the specialized Screaming Cowbird (*Molothrus rufoaxillaris*). Condor 98:866–867.

———. 1998. Interactions of the parasitic Screaming and Shiny Cowbirds (*Molothrus rufoaxillaris* and *M. bonariensis*) with a shared host, the Bay-winged Cowbird (*M. badius*). Pp. 173–193 *in* Brood parasites and their hosts (S. I. Rothstein and S. K. Robinson, eds). Oxford University Press, New York.

Freeman, S., and R. M. Zink. 1995. A phylogenetic study of blackbirds based on variation in mitochondrial DNA restriction sites. Syst. Biol. 44:409–420.

Friedmann, H. 1929. The cowbirds: A study in the biology of social parasitism. C. C. Thomas, Springfield, IL.

———. 1948. The parasitic cuckoos of Africa. Washington Acad. Sci. Monogr. 1.

———. 1955. The honey-guides. U.S. Natl. Mus. Bull. 208.

———. 1963. Host relations of the parasitic cowbirds. U.S. Natl. Mus. Bull. 233.

Friedmann, H., and L. F. Kiff. 1985. The parasitic cowbirds and their hosts. Proc. Western Found. Vert. Zool. 2:226–304.

Friedmann, H., L. F. Kiff, and S. I. Rothstein. 1977. A further contribution to the knowledge of the host relations of the parasitic cowbirds. Smithsonian Contrib. Zool. 235.

Gibbs, H. L., Brooke, M. de L., and N. B. Davies. 1996. Analysis of genetic differentiation of host races of the Common Cuckoo, *Cuculus canorus,* using mitochondrial and microsatellite DNA variation. Proc. Roy. Soc. B. 273:89–96.

Gibbs, H. L., P. Miller, G. Alderson, and S. G. Sealy. 1997. Genetic analysis of Brown-headed Cowbirds *Molothrus ater* raised by different hosts: Data from mtDNA and microsatellite DNA markers. Mol. Ecol. 6:189–193.

Goguen, C. B., and N. E. Matthews. 1996. Nest desertion by Blue-gray Gnatcatchers in association with cowbird parasitism. Anim. Behav. 52:613–619.

Graham, D. S. 1988. Responses of five host species to cowbird parasitism. Condor 99:588–591.

Grzybowski, J. A., and C. M. Pease. In press. A model of the dynamics of cowbirds and their host communities. Auk.

Hahn, D. C., and J. S. Hatfield. Host selection in the forest interior: Cowbirds target ground-nesting species. This volume.

Higuchi, H., and S. Sato. 1984. An example of character release in host selection and egg colour of cuckoos, *Cuculus* spp., in Japan. Ibis 126:398–404.

Hill, D. P., and S. G. Sealy. 1994. Desertion of nests parasitized by cowbirds: Have Clay-colored Sparrows evolved an anti-parasite defence? Anim. Behav. 48:1063–1070.

Holford, K. C., and D. D. Roby. 1993. Factors limiting fecundity of Brown-headed Cowbirds. Condor 95:536–545.

Holmes, B. 1993. An avian arch villain gets off easy. Science 262:1514–1515.

Holt, R. D. 1984. Spatial heterogeneity, indirect interactions, and the coexistence of prey species. Amer. Natur. 124:377–404.

Hotker, H. 1989. Meadow Pipit. Pp. 119–133 *in* Lifetime reproduction in birds (I. Newton, ed). Academic Press, London. 479 pp.

Kattan, G. H. 1997. Shiny Cowbirds follow the "shotgun" strategy of brood parasitism. Anim. Behav. 53:647–654.

Keys, G. C., R. C. Fleischer, and S. I. Rothstein. 1986. Relationships between elevation, reproduction, and the haematocrit level of Brown-headed Cowbirds. Comp. Biochem. Physiol. 83A:765–769.

King, A. P., and M. J. West. 1977. Species identification in the North American Cowbird: Appropriate responses to abnormal song. Science 195:1002–1004.

———. 1990. Variation in species-typical behavior: A contemporary issue for comparative psychology. Pp. 331–339 *in* Contemporary issues in comparative psychology (D. A. Dewsbury, ed.). Sinauer Associates, Inc., Sunderland, MA.

Lack, D. 1968. Ecological adaptations for breeding in birds. Methuen, London. 409 pp.

Lanyon, S. M. 1992. Interspecific brood parasitism in blackbirds (Icterinae): A phylogenetic perspective. Science 255:77–79.

Leopold, N. F., Jr. 1924. The Kirtland's Warbler in its summer home. Auk 41:44–58.

Lotem, A. 1993. Learning to recognize nestlings is maladaptive for cuckoo *Cuculus canorus* hosts. Nature 362:743–745.

Lotem A., H. Nakamura, and A. Zahavi. 1995. Constraints on egg discrimination and cuckoo-host coevolution. Anim. Behav. 49:1185–1209.

Lowther, P. E. 1993. Brown-headed Cowbird (*Molothrus ater*). *In* The birds of North America 47 (A. Poole and F. Gill, eds.). Academy of Natural Sciences, Philadelphia; American Ornithologists' Union, Washington, DC.

———. 1995. Bronzed Cowbird (*Molothrus aeneus*). *In* The birds of North America 144 (A. Poole and F. Gill, eds). Academy of Natural Sciences, Philadelphia; American Ornithologists' Union, Washington, DC.

Lyon, B. E., and J. M. Eadie. 1991. Mode of development and interspecific brood parasitism. Behav. Ecol. 2:309–318.

Lyon, B. E., L. D. Hamilton, and M. Magrath. 1992. The frequency of conspecific parasitism and the pattern of laying determinacy in Yellow-headed Blackbirds. Condor 94:590–597.

Mason, P. 1986. Brood parasitism in a host generalist, the Shiny Cowbird, 1: The quality of different species as hosts. Auk 103: 52–60.

May, R. M., and S. K. Robinson. 1985. Population dynamics of avian brood parasitism. Amer. Natur. 126:475–494.

Mayfield, H. F. 1960. The Kirtland's Warbler. Cranbrook Institute, Bloomfield Hills, MI.

———. 1977. Brown-headed Cowbird: Agent of extermination? Amer. Birds 31:107–113.

Mayr, E. 1979. Concepts in the study of animal behavior. Pp. 1–18 in Reproductive behavior and evolution (J. S. Rosenblatt and B. R. Komisaruk, eds). Plenum, New York.

Moksnes, A., and E. Røskaft. 1995. Egg morphs and host preference in the Common Cuckoo (Cuculus canorus): An analysis of cuckoo and host eggs from European museum collections. J. Zool. (London) 236:625–648.

Nakamura, H. S., S. Kubota, and R. Suzuki. 1998. Coevolution of the Common Cuckoo and its major hosts in Japan. Pp. 94–112 in Brood parasites and their hosts (S. I. Rothstein and S. K. Robinson, eds). Oxford University Press, New York.

Neudorf, D. L., and S. G. Sealy. 1992. Reactions of four passerine species to threats of predation and cowbird parasitism: Enemy recognition or generalized response? Behaviour 123:84–105.

Nice, M. M. 1937. Studies in the life history of the Song Sparrow, part 1. Trans. Linn. Soc. New York 4. 237 pp.

Nicolai, J. 1964. Der Brutparasitismus der Viduinae als ethologisches Problem. Z. Tierpsychol. 21:129–204.

Nolan, V., Jr., and C. F. Thompson 1975. The occurrence and significance of anomalous reproductive activities in two North American non-parasitic cuckoos, Coccyzus spp. Ibis 117:496–503.

O'Loghlen, A. L., and S. I. Rothstein. 1993. An extreme example of delayed vocal development: Song learning in a population of wild Brown-headed Cowbirds. Anim. Behav. 46:293–304.

Orians, G. H. 1980. Some adaptations in marsh-nesting blackbirds. Princeton University Press, Princeton, NJ.

Orians, G. H., E. Røskaft, and L. D. Beletsky. 1989. Do Brown-headed Cowbirds lay at random in the nests of Red-winged Blackbirds? Wilson Bull. 101:599–605.

Payne, R. B. 1977. The ecology of brood parasitism in birds. Annu. Rev. Ecol. Syst. 8:1–28.

———. 1982. Species limits in the indigobirds (Ploceidae, Vidua) of West Africa: Mouth mimicry, song mimicry, and description of new species. Misc. Publ. Mus. Zool., University of Michigan, Ann Arbor. 162 pp.

Pease, C. M., and J. A. Grzybowski. 1995. Assessing the consequences of brood parasitism and nest predation on seasonal fecundity in passerine birds. Auk 112:343–363.

Peck, G. K., and R. D. James. 1987. Breeding birds of Ontario: Nidiology and distribution, vol. 2, passerines. Royal Ontario Museum, Toronto.

Petit, L. J. 1991. Adaptive tolerance of cowbird parasitism by Prothonotary Warblers: A consequence of nest site limitation? Anim. Behav. 41:425–432.

Power, M. E., D. Tilman, J. A. Estes, B. A. Menge, W. J. Bond, L. S. Mills, G. Daily, J. C. Castillo, J. Lubchenco, and R. T. Paine. 1996. Challenges in the quest for keystones. BioScience 46:609–620.

Robertson, R. J., and R. F. Norman. 1976. Behavioral defenses to brood parasitism by potential hosts of the Brown-headed Cowbird. Condor 78:166–173.

Robinson, S. K., S. I. Rothstein, M. C. Brittingham, L. J. Petit, and J. A. Grzybowski. 1995a. Ecology and behavior of cowbirds and their impact on host populations. Pp. 428–460 in Ecology and management of neotropical migratory birds (T. E. Martin and D. M. Finch, eds). Oxford University Press, New York.

Robinson, S. K., F. R. Thompson III, T. M. Donovan, D. R. Whitehead, and J. Faaborg. 1995b. Forest fragmentation and the regional population dynamics of songbirds. Science 267:1987–1990.

Rogers, C. M., M. J. Taitt, J. N. M. Smith, and G. J. Jongejan. 1997. Nest predation and cowbird parasitism create a population sink in a wetland breeding population of song sparrows. Condor 99: 622–633.

Rohwer, S., and C. D. Spaw. 1988. Evolutionary lag versus bill-size constraints: A comparative study of the acceptance of cowbird eggs by old hosts. Evolutionary Ecology 2:27–36.

Rohwer, S., C. D. Spaw, and E. Røskaft. 1989. Costs to Northern Orioles of puncture-ejecting parasitic cowbird eggs from their nests. Auk 106:734–738.

Rothstein, S. I. 1974. Mechanisms of avian egg recognition: Possible learned and innate factors. Auk 91:796–807.

———. 1975a. An experimental and teleonomic investigation of avian brood parasitism. Condor 77:250–271.

———. 1975b. Evolutionary rates and host defenses against avian brood parasitism. Amer. Natur. 109:161–176.

———. 1976a. Experiments on defenses Cedar Waxwings use against cowbird parasites. Auk 93:675–691.

———. 1976b. Cowbird parasitism of the Cedar Waxwing and its evolutionary implications. Auk 93:498–509.

———. 1977. Cowbird parasitism and egg ejection in the Northern Oriole. Wilson Bull. 89:21–32.

———. 1982. Successes and failures in avian egg and nestling recognition with comments on the utility of optimality reasoning. Amer. Zool. 22:547–560.

———. 1986. A test of optimality: Egg recognition in the Eastern Phoebe. Anim. Behav. 34:1109–1119.

———. 1990. A model system for coevolution: Avian brood parasitism. Annu. Rev. Ecol. Syst. 21:481–508.

———. 1993. An experimental test of the Hamilton-Orians hypothesis for the evolution of brood parasitism. Condor 95: 1000–1005.

———. 1994. The cowbird's invasion of the far West: History, causes, and consequences experienced by host species. Studies in Avian Biol. 15:301–315.

Rothstein, S. I., and R. C. Fleischer. 1987. Vocal dialects and their possible relation to honest status signalling in the Brown-headed Cowbird. Condor 89:1–23.

Rothstein, S. I., and S. K. Robinson. 1994. Conservation and coevolutionary implications of brood parasitism by cowbirds. Trends Ecol. Evol. 9:162–164.

———, eds. 1998. Brood parasites and their hosts. Oxford University Press, New York.

Rothstein, S. I., J. Verner, and E. Stevens. 1984. Radio-tracking confirms a unique diurnal pattern of spatial occurrence in the parasitic Brown-headed Cowbird. Ecology 65:77–88.

Scott, D. M. 1977. Cowbird parasitism on the Gray Catbird at London, Ontario. Auk 94:18.

———. 1991. The time of day of egg-laying by the Brown-headed Cowbird and other icterines. Can. J. Zool. 69:2093–2099.

Scott, D. M., and C. D. Ankney. 1980. Fecundity of the Brown-headed Cowbird in southern Ontario. Auk 97:677–683.

———. 1983. The laying cycle of Brown-headed Cowbirds: Passerine chickens? Auk 100:583–592.

Scott, D. M., and R. E. Lemon. 1996. Differential reproductive success of Brown-headed Cowbirds with Northern Cardinals and three other hosts. Condor 98:259–271.

Sealy, S. G. 1995. Burial of eggs by parasitized Yellow Warblers: An empirical and experimental study. Anim. Behav. 49:877–889.

———. 1996. Evolution of host defenses against brood parasitism: Implications of puncture-ejection by a small passerine. Auk 113:346–355.

Sealy, S. G., and R. C. Bazin. 1995. Low frequency of observed cowbird parasitism on Eastern Kingbirds: Host rejection, effective nest defense, or parasite avoidance? Behav. Ecol. 6:140–145.

Sealy, S. G., D. L. Neudorf, and D. P. Hill. 1995. Rapid laying by Brown-headed Cowbirds *Molothrus ater* and other parasitic birds. Ibis 137:76–84.

Sherry, D. F., M. R. L. Forbes, M. Khurgel, and G. O. Ivy. 1993. Females have a larger hippocampus than males in the brood parasitic Brown-headed Cowbird. Proc. Natl. Acad. Sci. USA 90:7839–7843.

Smith, J. N. M. 1988. Determinants of lifetime reproductive success in the Song Sparrow. Pp. 154–172 in Reproductive success: Studies of individual variation in contrasting breeding systems (T. H. Clutton-Brock, ed). University of Chicago Press, Chicago, IL.

Smith, J. N. M., and P. Arcese. 1994. Brown-headed Cowbirds and an island population of Song Sparrows: A 16-year study. Condor 96:916–934.

Soler, M., J. J. Soler, J. G. Martinez, and A. P. Møller. 1994. Micro-evolutionary change in host response to a brood parasite. Behav. Ecol. Sociobiol. 35:295–301.

———. 1995. Magpie host manipulation by Great Spotted Cuckoos: Evidence for an avian Mafia? Evolution 49:770–775.

Stephens, D. W., and J. R. Krebs. 1986. Foraging theory. Princeton University Press, Princeton, NJ.

Stutchbury, B. M. 1997. Effects of female cowbird removal on reproductive success in Hooded Warblers. Wilson Bull. 109:74–81.

Terborgh, J. 1989. Where have all the birds gone? Princeton University Press, Princeton, NJ. 207 pp.

Walkinshaw, J. H. 1983. Kirtland's Warbler. Cranbrook Institute of Science, Bloomfield Hills, MI.

Ward, D., A. K. Lindholm, and J. N. M. Smith. 1996. Multiple parasitism of the Red-winged Blackbird: Further experimental evidence of evolutionary lag in a common host of the Brown-headed Cowbird. Auk 113:408–413.

Weatherhead, P. J. 1989. Sex-ratios, host-specific reproductive success, and impact of Brown-headed Cowbirds. Auk 106:358–366.

West, M. J., and A. P. King. 1996. Eco-gen-actics: A systems approach to the ontogeny of avian communication. Pp. 20–38 in Ecology and evolution of acoustic communication in birds (D. E. Kroodsma and E. H. Miller, eds). Cornell University Press, Ithaca, NY.

Wood, N. A. 1926. In search of new colonies of Kirtland's Warblers. Wilson Bull. 38:11–13.

Note added in proof: A major new monograph on the biology of cowbirds was published too late for consideration in this volume: Ortega, C. P. 1998. Cowbirds and other brood parasites. University of Arizona Press, Tucson.

PART I

*Population
Trends of Cowbirds
and Hosts and
Relevant
Methodology*

1. Introduction

STEPHEN I. ROTHSTEIN AND SCOTT K. ROBINSON

In addition to its effects on several endangered species with restricted distributions (see Part V), cowbird parasitism may also have caused the decline of numerous common and widespread passerines (Mayfield 1977, Brittingham and Temple 1983, Böhning-Gaese et al. 1993). However, other factors such as habitat loss and degradation (Terborgh 1989, Martin and Finch 1995) also contribute to such declines, and the effects of cowbirds are unclear (Rothstein 1994). Determining geographic and temporal trends in cowbird abundance is therefore vital to elucidating the impact, if any, of cowbird parasitism on the distribution and abundance of widespread passerines.

The six chapters in this section deal with assessments of spatial and temporal variation in cowbird numbers. Two studies report data on changes in cowbird numbers and use this information, plus additional data on population trends of potential hosts, to search for possible effects of parasitism on widespread host species. Several studies deal with cowbird range expansions and the methods used to quantify cowbird abundance, and one addresses the biology of wintering cowbirds. The last chapter describes the acoustic structure, function, and development of cowbird vocalizations and is designed to give researchers a better understanding of the cowbird behaviors (songs and calls) that are used to quantify abundance.

Trends in Cowbird Population Sizes

The Breeding Bird Survey (BBS) is perhaps the largest database, worldwide, dealing with the abundance and distribution of vertebrates. Chapters by Peterjohn et al. and by Wiedenfeld use BBS data to determine whether overall cowbird numbers are changing. These authors describe the basic BBS methodology. Over 100,000 sites are sampled along most of 3,400 routes, each 39.4 km in length. Although each site is visited for only 3 min per year, so many sites are sampled that data on temporal trends for large regions are likely to be reliable.

Both Peterjohn et al. and Wiedenfeld dispel the common assumption that overall cowbird numbers have been increasing in recent years. The authors of Chapter 2 show that cowbird numbers averaged a marginally significant ($.05 < P < .1$) decline of 0.88% per year from 1966 to 1992 averaged over the continental United States and southern Canada. BBS data we have accessed on the World Wide Web show that overall cowbird numbers remained low in 1993 and 1994 and that the decline from 1966 to 1994 is significant at $P < .01$. Additional data available on the Web covering 1966 to 1995 show that relative abundances for the United States and Canada in each of the last six years were below the entire range for the previous 24 years, again indicating an overall decrease. Averaged over the entire BBS survey area, there were about 14 cowbirds per BBS route in the late 1960s compared with about 11 from 1990 to 1995, for an overall decline of roughly 21%. Wiedenfeld's statistical methods for assessing population trends are somewhat different from those of Peterjohn et al. (which methods are best is controversial, Thomas 1996), and his main analyses contrast numbers in 1970–1972 with those in 1986–1988. But he too finds an overall decrease, although in this case it amounts to only 4%.

The two chapters show general agreement in their estimates of geographic variation in the relative abundance of cowbirds. Both indicate that cowbirds are two to four times as common in the central part of North America as in the East and West. Thus cowbirds continue to be most numerous in the Great Plains, their presumed center of abundance under primeval, pre-European conditions (Friedmann 1929, Mayfield 1965, Thompson et al. Chapter 32). But cowbird abundance varies greatly even within the plains, reaching its maximum in the northern plains. For example, the average number of cowbirds per BBS route was 93.9 in North Dakota but only 35.0 and 24.5 in Kansas and Oklahoma, respectively. However, the latter numbers are still much higher than the 10 or fewer cowbirds per route that occur in nearly all states east and west of the plains (Peterjohn et al. Chapter 2).

While cowbirds may well reach their maximum abundances in the northern Great Plains, we believe that caution is needed when using BBS numbers to compare relative cowbird abundances across different regions because cowbirds are likely to be much more detectable in the plains than in other areas. The elevated perches that cowbirds often use for nest finding and social interactions are rare to absent in much of the plains except along roads where barbed-wire fences and power and telephone lines occur. Thus, not only are cowbirds especially visible in the mostly treeless plains but they also may be concentrated along roads, which is where BBS observers count birds. Observations by one of us (SIR) in June 1995 in Jones County, South Dakota, demonstrated this roadside concentration as cowbirds that flew out over grassy areas returned repeatedly to fencerows along roads. Therefore, the true difference in absolute cowbird abundance between the plains and other regions may be lower than the five- to ten-fold differences suggested by BBS data. The moderate levels of parasitism in the northern Great Plains (Davis and Sealy Chapter 26, Koford et al. Chapter 27) provide further indirect evidence that cowbirds may be much less abundant in this region than the BBS data suggest.

Although they do not address the problem of relating relative to absolute abundances, Peterjohn et al. (1995) discuss other potential biases of BBS methodology. Later in this chapter, under "Cowbird Monitoring Techniques," we discuss other biases that may apply primarily to cowbirds. However, none of these potential biases are likely to be serious as regards the primary goal for which the BBS was designed, i.e. assessing regional temporal trends in a species' relative abundance.

BBS data demonstrate that virtually all widely distributed passerines show a mosaic of increasing and decreasing population trends in different regions, and the cowbird is no exception. Both Peterjohn et al. and Wiedenfeld indicate that cowbirds have generally increased in the northern plains where they are already most abundant. However, declines are occurring in the southern plains and in most of the East. Only the last change has a clear explanation, namely the increase in forest cover due to the decline of agriculture. Reforestation appears to be causing widespread declines in most grassland and second-growth species in the Northeast (Askins 1993). Cowbirds in the East tended to be increasing in the southern part of the region as a result of the species' colonization of the southeastern states over the last several decades (Cruz et al. Chapter 4, and see below). Thus, the pattern of increases in the northern plains is not part of an increase throughout the northern part of the cowbird's range. Indeed, cowbirds are decreasing in most of Canada except for the prairie provinces.

Recent cowbird trends in the West are especially interesting because the far West, including the entire Pacific seaboard, is the largest region to be colonized by cowbirds in this century (Rothstein 1994). Thus if cowbirds depress the populations of widespread species shortly after they colonize an area, and with host populations stable at depressed levels thereafter, BBS data from the far West may be especially important in revealing widespread cowbird impacts. The far western colonization began around 1900 in southern California, and cowbirds reached the Seattle area in 1955. Cowbirds increased in the Pacific Northwest from 1966 to 1976 but have declined since then (Peterjohn et al. Chapter 2). Overall, cowbirds have maintained steady numbers in the West throughout the BBS period, although some areas have seen declines and others increases.

One important part of the West, the Central Valley of California, represents the only major discrepancy between the analyses by Peterjohn et al. and Wiedenfeld. This region is particularly significant because it has seen a virtual collapse in the populations of many passerines since cowbirds first colonized it in the early 1900s. Once widespread species such as the Willow Flycatcher (Whitfield Chapter 40) and the Least Bell's Vireo (Griffith and Griffith Chapter 38) are now gone from this region, and many other species have declined significantly since cowbirds colonized the Central Valley (Gaines 1974). It is likely that there is no other region in North America where so many once-common species have declined and where the declines seem so linked to cowbird parasitism. However, the linkage may be more apparent than real, because the cowbird's arrival in the Central Valley coincided with the destruction of about 95% of the riparian habitat (Smith 1977) that was used by most of the hosts that have declined. Indeed, the question of whether habitat loss and degradation or cowbird parasitism is the main cause of the Central Valley declines is a controversial one (see Gaines 1974, Laymon 1987, and Rothstein 1994 for conflicting views). Nevertheless, relevant data on cowbird trends are especially important.

Wiedenfeld reported that the Central Valley has shown the greatest proportionate increase in cowbirds in North America, from a mean of 2.23 per count in 1970–1972 to 12.56 in 1986–1988, for a statistically significant increase of 560%. By contrast, Peterjohn et al. found a nonsignificant 6.98% annual increase in the Central Valley between 1968 and 1992. Reasons for the differences in results may be methodological, as the two studies considered different time periods and used differing analysis methods. Wiedenfeld's methods are sensitive to outlying values near the beginning or end of a time series (Thomas 1997). In addition, Wiedenfeld used only routes run consistently over time so that the same routes were present at the beginning and end of the analysis period. Peterjohn et al. used all routes that met certain minimum standards regardless of whether they were run continuously, and they thus had a larger sample size, 24 routes versus 12 for the Central Valley. Without looking in detail at the individual routes that each study used, it is difficult to determine which has provided the more reliable trend estimate for the Central Valley. But we do note that despite its large overall increase, even Wiedenfeld's analysis did not show an across-the-board increase in the Central Valley as 4 of the 12 routes he used did not show increasing trends. Overall, Central Valley data graphed for each year from 1968 to 1995 show no increase in cowbird abundance (World Wide Web).

Despite different interpretations of Central Valley trends, Peterjohn et al. and Wiedenfeld's chapters are in general agreement that overall cowbird numbers across North America have not increased since the late 1960s. Thus, any overall passerine declines that have occurred during this period are not due to increasing cowbird numbers. Nevertheless, significant host declines before this period could have been due, at least in part, to increased cowbird numbers before BBS data were collected. Cowbirds undoubtedly increased greatly in areas such as the East in the late 1700s to mid-1800s (Mayfield 1965). It is unclear, however, whether they have increased throughout this century except in regions that have been newly colonized such as the far West (Rothstein 1994), the Southeast (Cruz et al. Chapter 4) and the Canadian maritime provinces (Baird et al. 1957). The only evidence for an overall increase is Brittingham and Temple's (1983) data showing an increase between 1900 and 1980 in the proportion of Christmas Bird Counts (CBCs) that have reported cowbirds in southern states. While suggestive, these data need further analysis as the percentage of counts with a species does not control some important variables in CBCs (Root 1988). CBC data available on the Web (Christmas Bird Counts 1998) show that the number of cowbirds per 100 party-hours decreased from 1959 to 1988 in 10 of 11 states south of 37 degrees latitude, contrary to Brittingham and Temple's analysis for the same period and region.

Brittingham and Temple hypothesized that cowbird abundance increased because changes in agriculture in the South increased winter food availability. Indeed, some parts of the Southeast can support enormous winter populations in relatively small areas (Ortego Chapter 5). But if changes in agriculture did increase cowbird numbers, our analysis of CBC data indicates that the changes occurred before the BBS was established. Brittingham and Temple's hypothesis assumes that cowbird populations are limited on a macrogeographic or even continental scale by winter food availability. The extent to which this is true is unknown. There is strong evidence, however, that local and regional cowbird populations are at least partly limited by the availability of feeding areas during the breeding season (see Part IV). The extent to which cowbird populations are limited by winter and/or summer food is an interesting area in need of further study.

Possible Impacts of Cowbirds on Host Numbers

It is clear that cowbirds have the potential to limit population growth in the few host species that are classified as endangered (see Part V), but it is unknown whether they also limit the size or range of breeding populations of widespread and common species. In exploring the latter issue with BBS trend data, Peterjohn et al. and Wiedenfeld make the reasonable assumption that widespread inverse trends between cowbirds and other species are consistent with the hypothesis that cowbirds have large-scale effects on the abundances of other species. That is, decreasing host populations should occur where cowbirds are increasing, with the reverse occurring where cowbirds are decreasing. Although the BBS was not designed to assess the effects that bird species may have on one another, it still seems reasonable to use the massive amount of BBS data available to address such questions. However, it is important to acknowledge that any inverse trends that might be found could demonstrate only correlations and not causation. Furthermore, the general nature of relationships between trends in cowbird and host numbers is difficult to predict, as we discuss below and as Peterjohn et al. and Wiedenfeld also acknowledge.

There are many ways that BBS data could be partitioned to test for inverse relationships between trends for cowbirds and other widespread species. Peterjohn et al. identified five groups of birds on the basis of each species' potential sensitivity to parasitism and on the extent to which species migrate. Except for one group composed of species likely to be unaffected by cowbird parasitism, these workers assumed that species in these groups will show inverse trends relative to cowbirds if parasitism is causing declines. In contrast, direct positive relationships where other species and cowbirds are both decreasing or both increasing on the same routes, indicate that cowbirds do not depress host populations. As it turns out, all five species groups, including neotropical migrants, show more direct than inverse trends. Importantly, this is true for both cowbird hosts and species that are either not parasitized or that avoid major costs of parasitism by removing cowbird eggs from their nests, i.e. rejector species (Rothstein 1975).

In a further analysis, Peterjohn et al. used geographic areas where neotropical migrants as a group are either increasing or decreasing, rather than individual species, to search for patterns consistent with cowbird effects at the population level. Again they assumed that inverse relationships with cowbird trends will prevail if cowbirds affect host species, and again they found that more trends were direct than inverse. For example, both cowbirds and neotropical migrants were increasing or decreasing concordantly in 60.5% of the areas. Thus Peterjohn et al. conclude that there is no evidence for the hypothesis that changes in cowbird numbers influence host populations. Instead they argue that cowbirds and other species do not influence one another's trends and that most species tend to respond to the same large-scale factors, such as changes in weather patterns. Similarly, Thompson et al. (Chapter 32) report that cowbirds may be more abundant in areas where their hosts are more abundant. Instead of concordant responses to the same large-scale extrinsic factors, such as weather patterns, these direct relationships may mean that host population sizes influence cowbird population sizes but that the reverse is not true, i.e. cowbirds have no effects on host population sizes.

Wiedenfeld selected 10 host species and used BBS physiographic strata (see Table 2.1 for a list of these strata) to determine whether cowbird and host populations show inverse relationships. Each species co-occurred with cowbirds in from 6 to 51 strata. Five of the species showed direct relationships and 4 inverse ones, and no correlation coefficients were significant statistically. Wiedenfeld points out that 2 species of neotropical migrants that are heavily parasitized, Yellow Warbler and Common Yellowthroat, do tend to have inverse relationships with cowbird trends. But he acknowledged that these relationships are weak, and he agrees with Peterjohn et al.'s conclusion that there is no evidence for a causal link between cowbird and host populations.

There are a number of possible reasons for the lack of cowbird effects on host populations indicated by Peterjohn et al. and Wiedenfeld's analyses. Even though they decrease the production of host young, cowbirds would limit the abundance and distribution of breeders in widespread species only if hosts are limited by recruitment. But hosts may be limited by winter mortality (Sherry and Holmes 1995) such that the same number of birds survive to breed regardless of the number of young that are produced above the level of the winter carrying capacity. If winter mortality is density dependent in such species, cowbird parasitism would lower breeding populations only if it reduces recruitment so greatly that a host species starts the winter off at a population that is below the carrying capacity.

Similarly, hosts may be limited by the availability of breeding habitat. Indeed a number of studies have shown that passerines have floating populations during the breeding season, i.e. nonbreeding individuals that lack territories or nest sites because their species' habitat is already saturated (Stewart and Aldrich 1951). In some species, large breeding populations may force some individuals to nest in suboptimal habitat where reproductive success is low (possible population sinks: Pulliam 1988). In these cases, cowbird parasitism would only depress the number of nonbreeding birds or birds nesting in suboptimal habitat. Thus, while cowbird parasitism can cause a decline in overall host reproductive output, this decline need not result in a decrease in the size of a host's breeding population. Nest predation presents a parallel situation. Although birds will renest after nest predation, these renests are also subject to predation, which undoubtedly means that predation causes an overall decline

in a population's reproductive output (Peterjohn et al. 1995). Even though predation affects 50% or more of the nests of nearly all passerine species (Ricklefs 1973, Martin 1992), most passerines do not have declining populations. So it is clear that a decrease in recruitment alone does not lead to population declines.

Peterjohn et al. and Wiedenfeld's analyses of cowbird and host populations emphasized population trends because they used BBS methodology, which was designed to detect trends. But the absence of evidence for cowbird effects does not necessarily mean that such effects are absent, because cowbird and host population trends may be unrelated even if cowbirds limit hosts. Brood parasites will affect the breeding population size of a host only if their impact on that host's recruitment exceeds a certain threshold level (May and Robinson 1985). Thus cowbirds could depress a given host's breeding population even if they also have a declining trend where that host occurs, so long as the cowbird decline does not result in its impact going below the threshold level. While the rate of a host's decline might slow down if cowbirds decrease, neither BBS analysis attempted to detect changes in the strength of trends. Alternatively, if cowbirds increase, they could continue to do so without affecting host populations until they increase to the point where they are great enough to exceed the threshold level. Similarly, a decline in cowbirds will have no effect on a host if cowbirds were previously not affecting that host. Additional complications are the dynamic natures of cowbird and host trends and time delays in detecting declines, especially if there are floaters. Because floaters may not be detected on BBS censuses, declining populations may appear to be stable until the number of breeding adults starts to decline. As Gibbs and Faaborg (1990) have argued, some species may even appear to increase as a result of higher singing rates of unmated males. All of these factors limit the potential for comparisons of host and cowbird trends to reveal cause and effect.

A second problem with emphasizing cowbird trends is that identical proportionate increases or decreases in cowbirds can have very different effects on host populations because it is the absolute numbers of cowbirds and the total amount of their parasitism that affect host populations, if effects do indeed occur. In other words, a large proportionate increase in cowbird numbers in an area where cowbirds are uncommon might still leave them far below the level needed to affect local host species negatively, whereas a smaller proportionate increase in an area where cowbirds are common could cause host species to decline.

A final problem with using trend data is that it is not clear that we would always see inverse relationships between changes in cowbirds and the populations of all or most host species if cowbirds affect host populations. If cowbirds depend on hosts for recruitment into their breeding populations, a long-term inverse relationship in which cowbirds are increasing and all hosts are decreasing would be impossible. This would not be a problem if one assumes that cowbird numbers are independent of host numbers, which is possible for the same reasons that host populations may not be limited by recruitment (e.g., if they are winter-limited). Another possibility is that cowbird numbers depend primarily on the abundances of a small number of common species that are especially good hosts. The majority of host species may be much less common than cowbirds, and some, perhaps most, may provide cowbirds with trivial amounts of recruitment even if they are heavily parasitized and are declining as a result of cowbird parasitism. Thus the assumption of treating cowbird numbers or trends as independent variables that affect host populations may be valid for some host species but not for others. This complication has not entered into the analyses presented here, except for Peterjohn et al.'s separate consideration of species that are and are not parasitized.

Peterjohn et al. and Wiedenfeld acknowledge that there are problems with the assumption that cowbirds and hosts will show inverse BBS trends if the former affect the latter. Nevertheless, their analyses are valuable first steps in assessing the effects of cowbirds on widespread host species, and they suggest future analyses that might employ alternative methodologies. In particular, while BBS trend data on host populations are ideal dependent variables, other potential measures of cowbird impacts besides trend data might be better candidates for independent variables. For example, relative abundances of cowbirds, as determined from BBS data, may provide a more sensitive measure of possible cowbird effects than population trends. If cowbirds depress the populations of sensitive host species, the latter should be declining more where cowbirds are abundant than where cowbirds are scarce, regardless of whether cowbirds have declined or increased in those areas. However, as we have noted above, comparisons of relative abundances among regions may introduce some errors.

An even better candidate for an independent variable is the region-specific rate of parasitism on a particular host species. Parasitism rates are more appropriate than cowbird abundance because cowbirds vary geographically in the extent to which they use many host species (Friedmann 1963). Thus similar levels of cowbird abundance in two regions may not translate directly to similar levels of impacts on host species present in both regions (see Part V). Last, analyses could be made more sensitive by increased emphasis on the degree of reproductive loss each host species experiences if parasitism occurs. Some acceptor species, such as Red-winged Blackbirds and Song Sparrows (Freeman et al. 1990, Smith and Arcese 1994), experience little loss from parasitism, whereas others, such as small flycatchers, warblers, and vireos, nearly always lose their entire brood (Friedmann 1963).

We agree with Peterjohn et al. and Wiedenfeld that their analyses provide no justification for large-scale cowbird control programs designed to boost host populations over large regions (see also Rothstein and Robinson 1994). Of course even in the face of considerable uncertainty, managers might be tempted to try to limit cowbirds over large areas simply because it has proven to be possible to do so over limited areas (Griffith and Griffith Chapter 38, DeCapita Chapter 37, Whitfield Chapter 40). Furthermore, cowbird control might seem attractive because it is more easily done than reducing nest predation, even though predation may exert a greater cost than parasitism (Martin 1992). However, it is worth remembering that the same fragmentation of habitats that allowed cowbirds to expand their ranges and gain access to new hosts has also increased nest predation rates (Robinson et al. 1995). Many heavily parasitized host populations experience so much predation on eggs and nestlings that they might not be self-sustaining even if parasitism is eliminated (Rogers et al. 1997, Stutchbury 1997). Indeed, the lack of clear overall declines in the majority of widespread host species, even neotropical migrants (Peterjohn et al. 1995), may be due to dispersal from areas where nesting success is especially high due to low rates of parasitism and/or predation (see Part IV).

Cowbird Range Extensions into the Southeastern United States

It is well known that Brown-headed Cowbirds have extended their range in North America over the last two centuries (Mayfield 1965) and that they reached some regions only in the last 40 years. In Chapter 4, Cruz et al. describe a current range extension that is unique because it involves two cowbird species. They demonstrate that the Brown-headed Cowbird has been colonizing the southeastern states throughout this century and first began to breed in the Carolinas and Georgia between 1929 and 1957. It has moved south in Florida since the late 1960s and is now found throughout the state. While this southward range extension has occurred, the Brown-headed Cowbird's sister species (Friedmann 1929, Lanyon 1992), the Shiny Cowbird, has moved north in the Caribbean from mainland South America and Trinidad. The latter reached the Florida Keys by 1985 and has since spread throughout much of the state. Remarkably, outliers occurred as far away as Oklahoma and Maine by the early 1990s. The Shiny Cowbird will probably continue to spread northward and may eventually colonize much of North America. These two species have expanded their ranges very quickly in the Southeast but perhaps not as quickly as the 70–78 km per year northward advance shown by the Brown-headed Cowbird in the Pacific Northwest in the first half of the twentieth century (Rothstein 1994).

Both cowbirds are generalists in their host use, with each known to have parasitized over 200 species. Although it is difficult to distinguish between the eggs and young of the two species, there is evidence that both species breed in Florida. If the Brown-headed Cowbird continues to spread southward, it could mean that both the West Indies and eastern North America may go from having a single generalist parasite to having two such species. If the two species compete and interfere with one another, it is possible that their combined sympatric populations will be no larger than allopatric populations. But more worrisome is the possibility that one or both regions could end up with combined cowbird populations that exceed the level reached by a single species.

Florida, in particular, poses some potential conservation problems as it has a unique avifauna, including some species that may be especially susceptible to cowbird-induced declines. A number of species or endemic subspecies in Florida occur in small, patchy populations. This pattern is a characteristic of hosts that appear to be endangered by cowbirds (see Part V). Fortunately, parasitism levels so far do not seem to be high, as Cruz et al. found no parasitism in 38 Prairie Warbler and 19 Yellow Warbler nests, two species that are especially prone to being parasitized. Nevertheless, Cruz et al. argue convincingly that the cowbird-host situation in Florida and nearby regions deserves close monitoring. They point out that this monitoring needs to precede any management activities so that the magnitude of any problems can be determined. Elsewhere, Cruz et al. (in press) note that a third cowbird species, the Bronzed Cowbird, is moving into the southeastern states from the Southwest. But this species is spreading slowly and is more specialized in its host use (Friedmann and Kiff 1985).

Winter Biology and Cowbird Roosts

Chapter 5, by Ortego, is the only chapter in this book that focuses on the winter biology of cowbirds. Ortego reports that a single winter roost in Louisiana had from 28,000 to 38,000,000 and a mean of 8,700,000 cowbirds each year between 1974 and 1993. Other blackbirds, especially Red-winged Blackbirds, also used the roost. Birds were tallied as part of a local Christmas Bird Count that typically had more cowbirds than any other in the nation. Numbers at the roost fluctuated wildly from year to year. The minimum of 28,000 occurred between two years with 16,000,000 and 5,950,000 cowbirds, and the roost has had relatively few birds in recent years.

Ortego suggests that large winter roosts offer an opportunity to remove huge numbers of cowbirds and that winter control could accomplish the selective removal of cowbird populations that are having large and widespread impacts on hosts. We think it unlikely that selective removal is possible, because available data indicate that local breeding populations winter over extremely large areas and in the company of birds that breed over a large part of North America (Crase et al. 1972, Dolbeer et al. 1982).

More studies of the winter biology of cowbirds are needed, and these could result in new management strategies. For example, cowbirds color banded during the breeding season in Santa Barbara County, California, and in the Fraser River delta of British Columbia have been seen in the winter within several kilometers of their banding sites (S. I. Rothstein pers. obs., J. N. M. Smith pers. obs.). These birds are usually in small flocks of 100 or fewer cowbirds and are often with other blackbird species, such as Brewer's and Red-winged Blackbirds. This may mean that cowbirds that breed in lowland areas on the Pacific coast winter near their local breeding grounds and do not join the huge winter flocks of cowbirds that occur in some agricultural regions. The latter may be composed primarily of long-distance migrants. These speculations suggest that the effectiveness of local cowbird control programs designed to aid the recovery of the Least Bell's Vireo and the Southwestern Willow Flycatcher (Griffith and Griffith Chapter 38, Whitfield Chapter 40) might be increased by targeting small cowbird flocks that winter near these hosts' breeding grounds.

Cowbird Monitoring Techniques

Chapters by Miles and Buehler and by Rothstein et al. present information on various ways to assess cowbird numbers. Because previous studies (Dufty 1982, Rothstein et al. 1988, Yokel 1989) have shown that both male and female cowbirds are attracted to playback of the female's chatter call (Burnell and Rothstein 1994), these two studies determined whether chatter playbacks increase cowbird detections over those from typical point counts. Increasing detections is useful because female cowbirds are secretive and yet are clearly the more important sex as regards impacts on hosts. Also, boosting the number of detections allows for more statistical power as it reduces the numbers of sites with no cowbirds. This increased statistical power could reduce manpower costs, because fewer sites could be censused to track changes in cowbird numbers.

In both studies, cowbirds were counted at the same sites with and without playback. Miles and Buehler found no differences between these counts, but Rothstein et al. reported significant increases in detections of both sexes when playback was used. As expected, Rothstein et al. found that playback boosted female detections more than male detections. There are several possible reasons for the contrasting results of these two studies. The most likely explanation is the fact that cowbirds were much more common in the Kentucky and Tennessee areas where Miles and Buehler's study was conducted than in the Sierra Nevada, where the other study was done. Without playbacks, Miles and Buehler had an average of 0.89 cowbirds per 10-min count whereas Rothstein et al. had an average of 0.37. As the latter study discusses, cowbirds are often unresponsive to chatter playback if they are already near another cowbird (see also Dufty

1982). So chatter may have been on average a weaker attractant in Miles and Buehler's study and may therefore be most valuable in monitoring cowbirds when they are relatively uncommon. There are two other possible reasons for the increased effectiveness of playbacks in only Rothstein et al.'s study. They played chatter for 10 as opposed to 5 min and used a strong amplifier instead of only a tape recorder.

Another insight that comes out of these two studies is that cowbirds usually interact vocally with one another when they are with conspecifics in the morning and that this affects their detectability. One consequence of this increased detectability is that the per capita detection of cowbirds probably increases with increasing cowbird numbers, i.e. observers will detect a larger proportion of the cowbirds in an area as the actual density of cowbirds increases because proportionately more cowbirds will be with conspecifics. This nonlinear relationship between actual and detected numbers of cowbirds means that comparisons of cowbird numbers between areas may inflate disparities in cowbird numbers, i.e. the actual differences in cowbird abundance among regions or study sites may be less than the differences suggested by monitoring efforts. Similarly, studies that assess the number of cowbirds per host based on point counts or other survey techniques may inflate differences in this ratio if different study sites have very different densities of cowbirds. Similar nonlinear detection problems are probably less serious in other passerines because their vocalization rates are less dependent on the presence of conspecifics.

Nevertheless, Robinson et al. (Chapter 33) and Thompson et al. (Chapter 32) report that ratios of cowbird : host abundances indexed from censuses without playbacks are strongly correlated with community-wide parasitism levels. Cowbird abundance alone was a poor predictor of parasitism level because host density varied so greatly. In the Illinois landscape where the Robinson et al. study was done, cowbirds were abundant everywhere and females frequently interacted vocally. In such landscapes, playbacks may not be necessary; indeed, playbacks may tend to obscure differences in relative cowbird abundance by attracting additional cowbirds from neighboring breeding areas. The strong relationship reported in Thompson et al. depends mainly on the contrast between very low cowbird : host ratios in unfragmented landscapes where there are virtually no cowbirds and very high ratios in fragmented landscapes where cowbirds are very abundant. Thus it is possible that biases inherent in sampling cowbirds affect studies only if comparisons are made between areas with high and moderate cowbird abundances, which did not occur in these two studies of cowbird : host ratios.

In addition to assessing the usefulness of playbacks, Miles and Buehler also address the problem of the optimal duration for point counts. They conducted counts for 60 min, and while the number of cowbird detections continued to increase over the entire period, the rate of additional detections declined after 10 min, leading them to suggest that 10-min periods make the best use of manpower.

Cowbird Vocalizations

In addition to their playback experiment, Rothstein et al. also present an extensive overview of cowbird vocalizations, which have been the subjects of numerous publications in the last 20 years. This extreme interest comes from the fact that song development in birds has become a classic example of learned behavior, in part because of some remarkable parallels between the development of song in birds and speech in humans (Alcock 1993). Biologists initially expected cowbirds to have largely innate vocalizations as a result of their lack of early contact with conspecifics (West et al. 1981). Cowbirds, however, have turned out to display an enormous amount of learning in their vocal development and have become one of the most studied birds in this area of research.

Rothstein et al. describe variation in several types of calls and in both cowbird song types, flight whistles (FW) and perched songs (PS). The former song is undoubtedly one of the most geographically variable songs of any bird in the world. Its acoustic structure can be that of pure-tone whistlelike sounds, buzzes, or trills. In short, just about any sort of sound that any songbird makes can be found among FWs. Despite this extreme geographic variation, most males within an area sing the same type of FW, so there are distinct FW dialects. These can be so different from one another that an observer trained in one area might fail to recognize FWs in nearby areas. To deal with this problem both Rothstein et al. and Miles and Buehler suggest that chatter playbacks may be useful as a training aid because they attract cowbirds in close to an observer, who can then experience local cowbird vocalizations. The cowbird's other song type, PS, is also especially interesting as it has the widest frequency range of any bird song in the world and always follows a particular acoustic pattern that allows observers to attribute it to a cowbird (West et al. 1981). Learning to distinguish among the different types of cowbird vocalizations and the situations in which they are given should allow observers to interpret the behavior of cowbirds more readily.

Acknowledgments

We thank Chris Farmer and an anonymous reviewer for their comments on drafts of this overview.

References Cited

Alcock, J. 1993. Animal Behavior. Sinauer, Sunderland, MA.

Askins, R. A. 1993. Population trends in grassland, shrubland, and forest birds in eastern North America. Curr. Ornithol. 11:1–34.

Baird, J., R. I. Emery, and R. Emery. 1957. Northeastern maritime region. Audubon Field Notes 11:387–390.

Böhning-Gaese, K., M. L. Taper, and J. H. Brown. 1993. Are declines in North American insectivorous songbirds due to causes on the breeding range? Conserv. Biol. 7:76–81.

Brittingham, M. C., and S. A. Temple. 1983. Have cowbirds caused forest songbirds to decline? BioScience 33:31–35.

Burnell, K., and S. I. Rothstein. 1994. Variation in the structure of female Brown-headed Cowbird vocalizations and its relation to vocal function and development. Condor 96:703–715.

Christmas Bird Counts. 1998. www.im.nbs.gov.//bbs/cbclist.html.

Crase, F. T., R. W. De Haven, and P. P. Woronecki. 1972. Movements of Brown-headed Cowbirds banded in the Sacramento Valley, California. Bird-Banding 43:197–204.

Cruz, A., W. Post, J. W. Wiley, C. P. Ortega, T. K. Nakamura, and J. W. Prather. In press. Evolutionary and ecological significance of cowbird range expansion in south Florida. In Parasitic birds and their hosts (S. I. Rothstein and S. K. Robinson, eds). Oxford University Press, New York.

Dolbeer, R. A., P. P. Woronecki, and R. A. Stehn. 1982. Migration patterns for age and sex classes of blackbirds and starlings. J. Field Ornithol. 53:28–46.

Dufty, A. M., Jr. 1982. Response of Brown-headed Cowbirds to simulated conspecific intruders. Anim. Behav. 30:1043–1052.

Freeman, S., D. F. Gori, and S. Rohwer. 1990. Red-winged Blackbirds and Brown-headed Cowbirds: Some aspects of a host-parasite relationship. Condor 92:236–240.

Friedmann, H. 1929. The cowbirds: A study in the biology of social parasitism. C. C. Thomas, Springfield, IL.

———. 1963. Host relations of the parasitic cowbirds. U.S. Natl. Mus. Bull. 233.

Friedmann, H., and L. F. Kiff. 1985. The parasitic cowbirds and their hosts. Proc. Western Found. Vert. Zool. 2:225–304.

Gaines, D. 1974. A new look at the nesting riparian avifauna of the Sacramento Valley, California. Western Birds 5:61–80.

Gibbs, J. P., and J. Faaborg. 1990. Estimating the viability of Ovenbird and Kentucky Warbler populations in forest fragments. Conserv. Biol. 4:193–196.

Lanyon, S. M. 1992. Interspecific brood parasitism in blackbirds (Icterinae): A phylogenetic perspective. Science 255:77–79.

Laymon, S. A. 1987. Brown-headed Cowbirds in California: Historical perspectives and management opportunities in riparian habitats. Western Birds 18:63–70.

Martin, T. E. 1992. Breeding productivity considerations: What are the appropriate habitat features for management? Pp. 455–493 in Ecology and conservation of neotropical migrant landbirds (J. M. Hagan III and D. W. Johnston, eds). Smithsonian Institution Press, Washington, DC.

Martin, T. E., and D. M. Finch. 1995. Ecology and management of neotropical migratory birds. Oxford University Press, New York.

May, R. M., and S. K. Robinson. 1985. Population dynamics of avian brood parasitism. Amer. Natur. 126:475–494.

Mayfield, H. F. 1965. The Brown-headed Cowbird with old and new hosts. Living Bird 4:13–28.

———. 1977. Brown-headed Cowbird: Agent of extermination? Amer. Birds 31:107–113.

Peterjohn, B. G., J. R. Sauer, and C. S. Robbins. 1995. Population trends from the North American breeding bird survey. Pp. 3–39 in Ecology and management of neotropical migratory birds (T. E. Martin and D. M. Finch, eds). Oxford University Press, New York.

Pulliam, H. R. 1988. Sources, sinks, and population regulation. Amer. Natur. 132:652–661.

Ricklefs, R. E. 1973. Fecundity, mortality, and avian demography. Pp. 366–434 in Breeding biology of birds (D. S. Farner, ed). National Academy of Sciences, Washington, DC.

Robinson, S. K., F. Thompson III, T. M. Donovan, D. R. Whitehead, and J. Faaborg. 1995. Regional forest fragmentation and the nesting success of migratory birds. Science 267:1987–1990.

Rogers, C. M., M. J. Taitt, J. N. M. Smith, and G. J. Jongejan. 1997. Nest predation and cowbird parasitism create a demographic sink in wetland-breeding Song Sparrows. Condor 99:622–633.

Rothstein, S. I. 1975. An experimental and teleonomic investigation of avian brood parasitism. Condor 77:250–271.

———. 1994. The cowbird's invasion of the far West: History, causes, and consequences experienced by host species. Studies in Avian Biol. 15:301–315.

Rothstein, S. I, and S. K. Robinson. 1994. Conservation and coevolutionary implications of brood parasitism by cowbirds. Trends Ecol. Evol. 9:162–164.

Rothstein, S. I., D. A. Yokel, and R. C. Fleischer. 1988. The agonistic and sexual functions of vocalizations of male Brown-headed Cowbirds, Molothrus ater. Anim. Behav. 36:73–86.

Sherry, T. W., and R. T. Holmes. 1995. Summer versus winter limitation of populations: What are the issues and what is the evidence? Pp. 85–120 in Ecology and management of neotropical migratory birds (T. E. Martin and D. M. Finch, eds). Oxford University Press, New York.

Smith, F. 1977. A short review of the status of riparian forests in California. Pp. 1–2 in Riparian forests in California: Their ecology and conservation (A. Sands, ed). Institute for Ecology Publications 15. 122 pp.

Smith, J. N. M., and P. Arcese. 1994. Brown-headed Cowbirds and an island population of Song Sparrows: A 16-year study. Condor 96:916–934.

Stewart, R. E., and J. W. Aldrich. 1951. Removal and repopulation of breeding birds in a spruce-fir forest community. Auk 68:471–482.

Stutchbury, B. J. M. 1997. Effects of female cowbird removal on reproductive success of Hooded Warblers. Wilson Bull. 109:74–81.

Terborgh, J. W. 1989. Where have all the birds gone? Princeton University Press, Princeton, NJ.

Thomas, L. 1996. Monitoring long-term population change: Why are there so many analysis methods? Ecology 77:49–58.

———. 1997. Evaluation of statistical methods for estimating long-term population change from extensive wildlife surveys. Ph.D. Thesis, University of British Columbia, Vancouver.

West, M. J., A. P. King, and D. H. Eastzer. 1981. The cowbird: Reflections on development from an unlikely source. Amer. Scientist 69:56–66.

Yokel, D. A. 1989. Intrasexual aggression and the mating behavior of Brown-headed Cowbirds: Their relation to population densities and sex ratios. Condor 91:43–51.

2. Temporal and Geographic Patterns in Population Trends of Brown-headed Cowbirds

BRUCE G. PETERJOHN, JOHN R. SAUER, AND SANDRA SCHWARZ

Abstract

The temporal and geographic patterns in the population trends of Brown-headed Cowbirds are summarized from the North American Breeding Bird Survey. During 1966–1992, the survey-wide population declined significantly, a result of declining populations in the Eastern BBS Region, southern Great Plains, and the Pacific coast states. Increasing populations were most evident in the northern Great Plains. Cowbird populations were generally stable or increasing during 1966–1976, but their trends became more negative after 1976. The trends in cowbird populations were generally directly correlated with the trends of both host and nonhost species, suggesting that large-scale factors such as changing weather patterns, land use practices, or habitat availability were responsible for the observed temporal and geographic patterns in the trends of cowbirds and their hosts.

Introduction

Brown-headed Cowbirds have received considerable attention in recent years, as their nest parasitism has been implicated in the declines of populations of various host species (Brittingham and Temple 1983, Mayfield 1977). In localities where high rates of nest parasitism imperil remnant populations of endangered species, cowbird control measures improve the fledging rates of these species (see chapters in Part V of this volume). Regional control programs have also been proposed, but these programs remain controversial.

While numerous recent studies examined aspects of cowbird breeding biology and geographic variations in rates of nest parasitism (see chapters in this volume), the recent trends of Brown-headed Cowbird populations have received less attention. Potter and Whitehurst (1981) and Rothstein (1994) documented their range expansion in portions of the southeastern and western United States during the past century. Dolbeer and Stehn (1979) and Robbins et al. (1986) utilized data from the North American Breeding Bird Survey (BBS) to summarize continental and regional population trends between 1966 and 1979. They described generally increasing populations for the entire continent, especially in central North America, but some declines were evident in eastern North America. Droege and Sauer (1990) and Peterjohn and Sauer (1993) have summarized more recent continental population trends briefly.

We used BBS data to describe trends in Brown-headed Cowbird populations between 1966 and 1992. Geographic and temporal patterns were analyzed to demonstrate the complexities of cowbird population trends in recent decades. We also examined broad regional associations among population changes of cowbirds and population changes of various groups of bird species. An understanding of these results has important implications for the effectiveness of management strategies to control regional populations of cowbirds.

Methods

BBS Methodology

The BBS consists of approximately 3,700 randomly located routes established along secondary roads throughout the continental United States and southern Canada. Each route is 39.4 km (24.5 miles) long, and consists of 50 stops spaced at 0.8-km (0.5-mile) intervals. Each route is surveyed once annually during the peak of the breeding season, with most surveys being conducted during June. All birds heard or seen within 0.4 km (0.25 mile) of each stop during a 3-min period are recorded. Additional information on the history of the survey and its methodology is provided by Robbins et al. (1986).

Trend and Annual Index Estimation

Regional population trends were estimated using the route-regression method. For each survey route, a trend was estimated using a regression of the natural logarithm of counts (plus a constant of 0.5) against year and covariables that adjusted for differences in counts associated with different observers (Geissler and Sauer 1990). If these observer covariables are not included in the analysis, trends are positively biased for many species (Sauer et al. 1994). The slope estimate, when transformed to a multiplicative scale through exponentiation, was the trend for the route. Regional trends were estimated by averaging route trends, weighting the relative contribution of each route by the mean abundance of the species on the route and the consistency of route coverage. In practice, the abundance weight had little effect on the estimation, but inclusion of the consistency weight greatly increased the precision of the weighted mean relative to an unweighted mean (J. R. Sauer unpubl. analysis). Precision of the weighted regional trend was estimated by bootstrapping, in which the component routes were randomly selected to form 400 subestimates of the regional trend, and a mean and variance calculated among the subestimates were used as the regional trend and its variance. See Geissler and Sauer (1990) for details of the methods.

We estimated trends for the entire survey period (1966–1992), and for this period divided into thirds (1966–1976, 1977–1984, and 1985–1992). Because time periods used in these analyses are arbitrary, we examined patterns in temporal changes in cowbird populations within these intervals by estimating composite annual indices of abundance for BBS regions and the survey area. We fitted observer effects and the regional mean trend to each route in the region, estimated the residual for each year with data, and averaged the residuals by year. The mean residuals were added to the yearly predicted count from the weighted trend estimate (Sauer and Geissler 1990). We used LOESS smooths (James et al. 1990) to examine patterns in the composite annual indices.

Analysis of Geographic Patterns in Trends

Population trends were also estimated for states and provinces; physiographic strata (see Butcher 1990); Eastern, Central, and Western BBS regions (see Bystrak 1981); and the entire survey area. Because trends may not be homogeneous in any region, we demonstrated geographic patterns in population change using an Arc/Info Geographic Information System (Environmental Systems Research Institute 1993). The starting points of survey routes were projected onto a North American map. For each route, we associated the population trends for the different time intervals, weighting each trend by a measure of route consistency to minimize the contribution of routes with missing years and data with large variances. Finally the weighted trends were smoothed using ordinary kriging (Cressie 1991) to create a contour map of Brown-headed Cowbird population changes. We also projected the average count from each route onto the route locations and produced a contour map of cowbird relative abundance for the entire survey period.

Associating Cowbird Population Change with Changes in Other Species

Because Brown-headed Cowbirds parasitize some species of birds that are experiencing regional population declines heavily (e.g., Robbins et al. 1989), we examined associations between cowbird trends and trends of potential host species over the entire survey period. These associations were studied by correlating the route-specific population trends of each bird species with the trends of Brown-headed Cowbirds. In this analysis, we estimated trends on individual survey routes using linear regression as described above, but the first survey of each observer is omitted. These data were omitted to prevent positive bias due to start-up effects, i.e., lower counts associated with the first year an observer surveys a route. These effects have been documented for many species (Kendall et al. 1996) and could introduce spurious direct correlations among trends.

We used these correlations to evaluate several hypotheses about associations among trends of cowbirds and other bird species. We tested the hypothesis that increasing populations of cowbirds result in decreased populations of species that are strongly susceptible to cowbird parasitism, but that species insensitive to parasitism do not have associations with cowbird trends. We defined 18 host species and 12 species that are rejectors or rarely parasitized (see Appendix), using information from Friedmann (1963) and Friedmann et al. (1977). For each group, we presented the percentage of species with trends that were directly correlated with cowbird trends. If no association existed, we expected 50% of the species to have direct associations. Significant deviations from 50%, as shown by binomial tests of the observed percentage, suggested that an association existed between the groups' trends and trends of the cowbird population.

Because cowbird population changes might be associated with changes in neotropical migrant birds, we conducted a similar analysis of correlations for neotropical migrants, short-distance migrants, and permanent resident bird species. Membership within these groups followed Peterjohn and Sauer (1993).

If most bird species in a region tend to increase or decline concordantly, any correlations between species could simply reflect regional consistencies in pattern (Barker and Sauer 1992). To evaluate the relationship between geographic patterns of population change at broad regional scales, we calculated the mean trend for neotropical migrant birds on each survey route and used ordinary kriging to define a surface summarizing areas of mean increase or decline for these species. We then superimposed a map of regions of increase and decline of Brown-headed Cowbirds, to identify areas of (1) cowbird increase and neotropical migrant bird increase, (2) cowbird decline and neotropical migrant bird increase, (3) cowbird increase and neotropical migrant bird decline, and (4) decline in both groups. If no associations exist between the trends of cowbirds and neotropical migrants, these regions should have equal areas. Of course, if associations exist in this geographic analysis, a cause-effect relationship cannot be definitively established.

Results

Relative Abundance

The relative abundance of Brown-headed Cowbirds, expressed as mean numbers of individuals per BBS route, is shown in Figure 2.1. They are most numerous along the Great Plains from southern Canada to Oklahoma and become progressively less numerous towards the eastern and western edges of their range and in the southwestern states.

Population Trends, 1966–1992

The survey-wide Brown-headed Cowbird population declined since 1966 (Table 2.1, Figure 2.2). Similar trends occurred within the Eastern BBS Region (Table 2.1), where declines were widely distributed from the eastern Great Lakes and Mid-Atlantic states north through New England to southern Canada (Figure 2.2). Long-term trends in the Central BBS Region were nonsignificant (Table 2.1), but exhibited some geographic variation with long-term declines in the southern Great Plains, especially portions of Oklahoma and Texas, contrasting with increases farther north in the Dakotas and southern Canada (Figure 2.2). In the Western BBS Region, cowbird populations have nonsignifi-

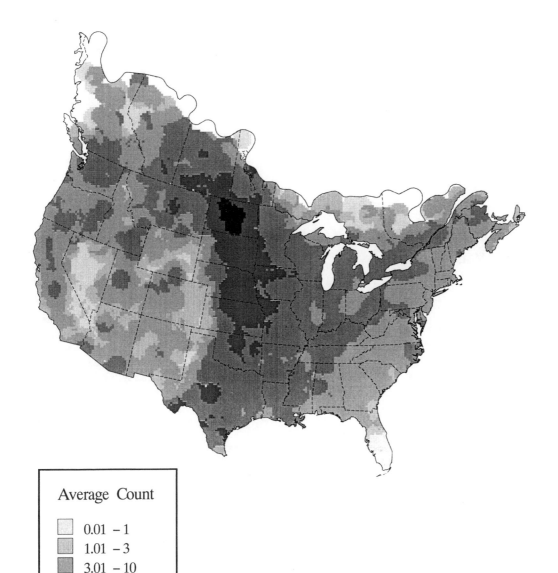

Average Count

- 0.01 – 1
- 1.01 – 3
- 3.01 – 10
- 10.01 – 30
- 30.01 – 100
- > 100

Figure 2.1. The relative abundance of Brown-headed Cowbirds on BBS routes, smoothed into relative abundance categories expressed as mean numbers of individuals per BBS route. The map is constrained to the area covered by the BBS.

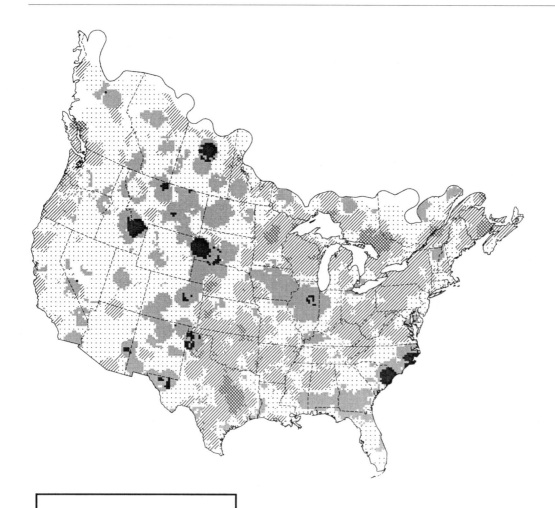

Percent Change per Year

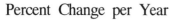 < − 1.5
 − 0.26 − − 1.5
      0.25 − − 0.25
      0.26 − 1.5
      > 1.5

Figure 2.2. Population trends of the Brown-headed Cowbird, 1966–1992. The map presents regions of consistent population change, grouped into categories of declining (trend < -1.5%/year), tending to decline (-0.26 > trend ≥ -1.5), indeterminate (0.25 > trend > -0.25), tending to increase (0.26 < trend ≤ 1.5), and increasing populations (trend > 1.5), based on smoothed weighted, route-specific trends. Unshaded regions represent areas with insufficient BBS data.

cant trends (Table 2.1) with few consistent geographic patterns, except for declines along the Pacific coast (Figure 2.2).

Significant long-term declines occurred in 19 states/provinces and 18 physiographic strata, primarily in eastern North America and the southern Great Plains (Table 2.1). Increases were found in 9 states/provinces and 5 strata, concentrated in the northern Great Plains, the southeastern states, and western North America.

Population Trends, 1966–1976

Brown-headed Cowbird populations were stable or increasing in a large portion of the continent between 1966 and 1976 (Table 2.1). Increasing trends were most consistent within the Central BBS Region (Figure 2.3) and were largely responsible for similar trends in the survey-wide population. In the Eastern BBS Region, cowbirds increased in the southeastern states, but this expansion contrasted with declines in New England, the eastern Great Lakes region, and the southern Appalachians to produce a nonsignificant regional trend (Table 2.1, Figure 2.3). Population trends in the West-

Table 2.1. BBS Population Trend Estimates for the Brown-headed Cowbird

	Interval				N	Abund.
	1966–1976	1977–1984	1985–1992	1966–1992		
State/province						
Alabama	1.72	1.00	−6.77***	−0.15	58	8.92
Alberta	4.21	0.15	0.01	−0.25	54	13.30
Arizona	−11.73	−5.64	−3.30	−1.84	39	7.21
Arkansas	7.69**	−4.12*	−1.61	−1.08	30	13.73
British Columbia	−1.97	−1.00	−4.03*	−1.82*	64	7.72
California	4.14	4.96**	−1.06	−0.41	175	5.03
Colorado	—	−5.98	−2.97	7.21***	50	4.50
Connecticut	−2.50	5.16	6.84*	0.42	17	6.36
Florida	—	—	−1.94	−0.64	52	2.03
Georgia	17.17***	5.44*	1.61	4.35**	60	5.44
Idaho	—	—	−4.45	7.99	47	7.39
Illinois	3.66	1.53	3.01	2.60**	63	12.82
Indiana	1.36	3.29	3.37**	0.50	42	12.97
Iowa	6.70***	−1.31	2.48	1.40	34	26.06
Kansas	0.97	−4.57***	−3.28*	−1.14*	39	35.00
Kentucky	−2.73	−6.85***	3.05	−0.67	45	13.53
Louisiana	2.94	−7.38***	−8.61***	−1.95	45	13.81
Maine	−4.93	−0.50	0.03	−3.29*	52	5.80
Manitoba	—	−8.56***	−1.94	−2.12	29	21.81
Maryland	−0.62	9.25**	1.34	1.43*	58	10.18
Massachusetts	1.22	1.08	6.09**	−1.20	23	6.11
Michigan	0.33	−1.39	−1.27	−1.89**	75	11.78
Minnesota	4.74*	−8.90***	3.39	−3.10***	52	17.81
Mississippi	−4.45	−6.28**	1.20	−0.27	33	11.26
Missouri	0.90	−3.01	−0.42	−0.14	44	25.20
Montana	4.70	−1.90	1.82	2.81**	48	14.15
Nebraska	2.58	−3.42*	−0.06	−1.17	41	29.21
Nevada	—	—	—	−4.41	18	1.29
New Brunswick	−2.68*	−7.13**	−6.26	−5.66***	29	9.18
New Hampshire	−2.33	0.51	−3.59	−0.57	23	6.86
New Jersey	−3.44	12.47**	−3.53	1.81	34	5.42
New Mexico	6.97	7.55	−3.75	3.39**	37	4.39
New York	−2.50**	−1.59	−2.52	−3.08***	117	9.69
North Carolina	7.93**	10.92***	−1.85	4.13***	44	7.77
North Dakota	5.60***	−2.43*	−0.45	3.40***	46	93.88
Nova Scotia	0.68	0.93	—	−4.55***	24	3.14
Ohio	−4.69***	0.87	−3.05	−2.57***	51	10.31
Oklahoma	0.83	−7.49***	−0.13	−2.50***	38	24.54
Ontario	3.06	−7.69***	−7.28***	−5.20***	74	5.86
Oregon	−2.01	−3.51	−2.14	−1.86*	63	9.09
Pennsylvania	−7.12***	−1.70	−2.10	−5.47***	113	8.47
Quebec	1.33	−9.51**	−8.80***	−4.34***	59	3.16
Saskatchewan	4.55**	0.01	0.42	1.76	52	26.19

Table 2.1. Continued

	Interval				N	Abund.
	1966–1976	1977–1984	1985–1992	1966–1992		
State/province						
South Carolina	10.28*	−3.32	−9.53***	0.31	23	4.15
South Dakota	11.55***	2.99	1.90	1.44	39	40.96
Tennessee	−3.57**	−1.35	−0.66	−1.87**	45	10.33
Texas	1.55	−2.76**	−4.57**	−3.30***	119	16.62
Utah	—	—	−7.48*	3.90*	33	5.33
Vermont	1.90	10.10**	7.08*	1.76	24	6.37
Virginia	−3.82***	−1.26	−6.52***	−2.86**	51	6.10
Washington	0.91	2.47	−2.70	−1.66	49	9.99
West Virginia	−6.24***	−0.07	−2.58	−4.37***	41	9.73
Wisconsin	−0.31	−5.73***	0.44	−3.03***	72	18.89
Wyoming	—	14.04**	−7.27	3.60	72	3.38
Stratum (number)						
Coastal flatwoods (03)	9.05	11.31***	−1.83	4.32***	82	6.54
Upper coastal plain (04)	1.87	0.97	−3.58***	0.65	220	10.59
Mississippi alluvial plain (05)	7.89	−9.33***	0.03	−1.83*	33	16.33
Coastal prairies (06)	0.07	—	—	−0.95	19	11.11
South Texas brushlands (07)	0.11	0.42	—	−3.07***	18	21.35
East Texas prairies (08)	2.55	−5.41*	−9.34**	−5.40***	22	19.52
Northern piedmont (10)	−4.98***	4.56*	0.41	−0.63	44	8.08
Southern piedmont (11)	−3.05*	−1.40	−0.65	−0.84	60	5.56
S. New England (12)	−2.56	3.32	4.98**	−0.30	47	5.93
Ridge and valley (13)	−3.74***	−2.00	−3.46*	−2.77***	104	8.17
Highland rim (14)	−4.78***	−2.86*	1.98	−1.06	53	13.82
Lexington plain (15)	−0.16	−8.83**	1.11	−1.99*	24	13.60
Great Lakes plain (16)	−0.72	−1.54	0.56	−1.44**	88	14.02
Driftless area (17)	−4.80*	−4.44	0.53	−2.91**	24	23.53
St. Lawrence River plain (18)	3.00	−7.47***	−2.54	−3.83***	50	17.68
Ozark-Ouachita plateau (19)	1.55	−5.41**	−0.76	−0.29	35	18.73
Great Lakes transition (20)	−0.43	−3.13	−1.72	−3.76***	53	15.68
Cumberland plateau (21)	—	—	—	−0.73	18	9.69
Ohio hills (22)	−7.44***	0.04	−3.02	−3.69***	51	11.54
Blue Ridge Mountains (23)	—	—	—	−4.50**	16	3.40
Allegheny plateau (24)	−4.37***	−1.88	−2.61	−4.98***	113	8.66
Adirondack Mountains (26)	−1.39	−0.13	−6.46**	−2.94**	24	8.95
N. New England (27)	−2.01	−0.07	1.13	−1.40	52	8.04
N. spruce-hardwoods (28)	2.49	−7.75***	−6.71***	−5.34***	218	7.93
Closed boreal forest (29)	—	−16.10***	—	−8.21***	33	1.06
Aspen parklands (30)	4.06*	−2.61*	0.39	0.45	80	16.84
Till plains (31)	2.51	0.90	2.50	1.69	83	11.48
Dissected till plains (32)	5.04***	−1.37	1.02	1.02	61	28.91
Osage Plain–Cross Timbers (33)	0.02	−8.65***	−1.32	−3.81***	37	29.99
High Plains border (34)	2.00	−3.50*	−2.68	−0.59	38	37.73
Rolling red prairies (35)	2.17	−6.51***	0.17	−2.69***	17	34.45

Table 2.1. Continued

	Interval				N	Abund.
	1966–1976	1977–1984	1985–1992	1966–1992		
Stratum (number)						
High plains (36)	—	0.89	8.46**	2.73	43	8.75
Drift prairie (37)	5.96**	0.25	1.59	1.80*	54	56.05
Glaciated Missouri plateau (38)	7.12***	−1.05	−0.79	2.19*	46	40.87
Great Plains roughlands (39)	10.77***	−0.52	−0.64	3.99***	72	25.46
Black prairie (40)	4.60	−6.84***	3.28	−0.60	37	31.03
Edwards plateau (53)	—	−1.77	−2.88	−1.07	18	26.09
Chihuahuan Desert (56)	2.85	6.78	11.79	2.82	23	10.73
Southern Rockies (62)	—	—	−16.71**	1.62	19	4.33
Fraser plateau (63)	—	—	—	0.91	16	5.23
Central Rockies (64)	−2.74	0.17	−2.41	−1.51	76	6.60
Dissected Rockies (65)	2.88	0.96	−1.43	0.52	43	11.93
Sierra Nevada (66)	—	—	—	−2.06	16	2.98
Great Basin deserts (80)	—	—	—	1.88	17	2.78
Sonoran Desert (82)	—	—	—	3.42	18	7.17
Pinyon-juniper woodlands (84)	52.65***	—	−8.20***	4.38	30	5.80
Pitt-Klamath plateau (85)	22.71	−7.92	2.52	0.26	26	13.36
Wyoming basin (86)	—	—	−9.55	2.27	33	4.53
Intermountain grasslands (87)	−1.57	8.99	−7.76	1.36	41	3.95
Basin and range (88)	—	—	−2.55	5.02*	31	4.51
Columbia plateau (89)	4.61	2.90	−2.67	6.87	48	8.17
S. California grasslands (90)	19.40***	—	—	−5.14	15	2.75
Central Valley (91)	6.98	−1.38	2.08	0.96	24	9.47
California foothills (92)	8.06	7.70**	−2.48	−1.20	50	7.31
S. Pacific rainforests (93)	−5.23***	−0.33	−4.75**	−3.65***	47	8.61
N. Pacific rainforests (94)	5.45	—	—	−3.15*	18	8.79
Region						
FWS Region 1	1.68	2.03	−2.01	0.05	352	6.12
FWS Region 2	1.27	−3.89***	−3.15**	−2.88***	233	12.99
FWS Region 3	1.91**	−3.04***	1.43*	−0.56	433	17.66
FWS Region 4	1.41	−1.55	−2.02**	−0.33	435	9.02
FWS Region 5	−4.23***	−0.19	−1.95**	−3.40***	568	7.76
FWS Region 6	5.54***	−2.50**	−0.54	0.94	369	24.54
Eastern BBS Region	−0.71	−3.22***	−1.36*	−2.18***	1438	7.27
Central BBS Region	4.12***	−2.97***	−0.33	−0.21	701	25.55
Western BBS Region	0.30	0.50	−2.54**	−0.68	641	6.60
Canada	2.37*	−4.03***	−2.58**	−1.55**	391	8.75
United States	1.98***	−2.33***	−0.64	−0.74	2389	14.34
Surveywide	2.04***	−2.68***	−0.94*	−0.88*	2780	12.91

Note: See Butcher (1990) for a map of BBS physiographic strata (the numbers in parentheses refer to the strata numbers on the map). Trend estimates are expressed as average percentage change per year; absence of trend estimates indicates inadequate sample size for analysis. N is the number of routes on which cowbirds have been recorded. Relative abundance (Abund.) is expressed as mean number of individuals per BBS route. Statistical significance of the test that the percentage differs from 0.0 is indicated by asterisks: * $.05 < P < .10$; ** $.01 < P < .05$; *** $P < .01$.

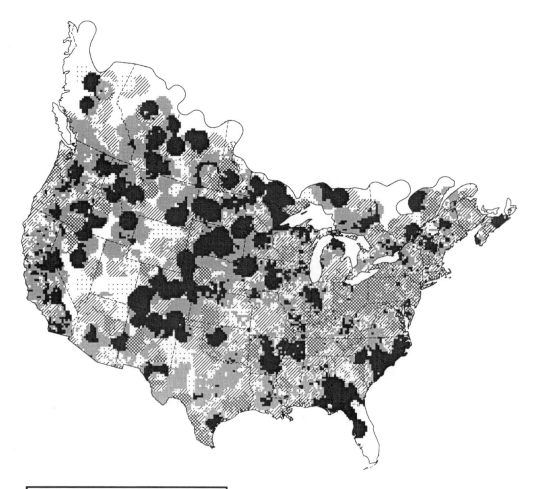

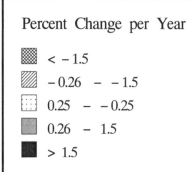

Percent Change per Year

- < − 1.5
- − 0.26 − − 1.5
- 0.25 − − 0.25
- 0.26 − 1.5
- > 1.5

Figure 2.3. Population trends of the Brown-headed Cowbird, 1966–1976. The map presents regions of consistent population change, grouped into categories of declining (trend < -1.5%/year), tending to decline (-0.26 > trend ≥ -1.5), indeterminate (0.25 > trend > -0.25), tending to increase (0.26 < trend ≤ 1.5), and increasing populations (trend > 1.5), based on smoothed weighted, route-specific trends. Unshaded regions represent areas with insufficient BBS data.

ern BBS Region were generally increasing but not significantly so. Significant increases occurred in 9 states/provinces and 7 strata, primarily in the Central and Western regions, while significant decreases in 7 states/provinces and 8 strata were largely confined to eastern North America (Table 2.1). No consistent temporal patterns were apparent in any BBS region in 1966–1976 (Figure 2.4).

Population Trends, 1977–1984

Cowbird trends were generally more negative during this period, with significant declines for the survey-wide, United States, Canada, and Eastern and Central BBS regional populations (Table 2.1). Declines were widespread in central North America, especially from Missouri and Arkansas through the southern Great Plains (Figure 2.5). Populations decreased in most of eastern North America except for the southeastern states. No geographic patterns were apparent in western North America, where regional population trends remained nonsignificant. The tendency towards cowbird population decreases was reinforced by significant declines in 15 states/provinces and 13 strata, as opposed to significant increases in 7 states/provinces and 3 strata (Table 2.1).

In the Eastern BBS Region, declines occurred during 1977–1979, followed by a recovery in 1981 and declines through 1983 (Figure 2.4). The Central BBS Region exhibited a different temporal pattern, with a continuous decline during 1977–1982 (Figure 2.4). In the Western BBS Region, declines during 1977–1981 preceded increases through 1984.

Population Trends, 1985–1992

The survey-wide population continued to decline as did populations in the Eastern and Western BBS regions and Canada (Table 2.1). In eastern North America, these declines extended from New England south along the Appalachian Mountains to include most southeastern states (Figure 2.6). The western declines were widely distributed. Cowbird populations in the Central BBS Region exhibited nonsignificant trends, but decreases in Texas contrasted with increases in the northern Great Plains. Significant decreases in 10 states/provinces and 8 strata exhibited a similar geographic distribution, while significant increases were widely scattered in 4 states/provinces and 2 strata (Table 2.1).

Regional differences in the timing of these declines were also apparent. Cowbird populations peaked during 1987 in the Eastern BBS Region and gradually declined in subsequent years (Figure 2.4). In the Central BBS Region, a sharp decline occurred between 1988 and 1989 (Figure 2.4), with relatively stable populations subsequently. The western population exhibited considerable annual fluctuations with no consistent temporal patterns (Figure 2.4).

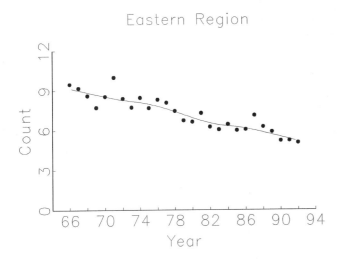

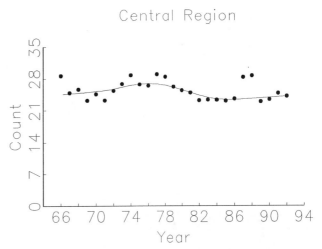

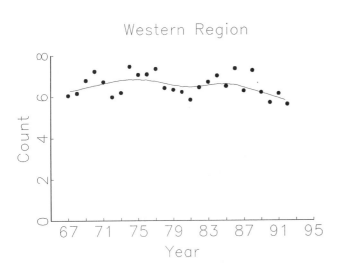

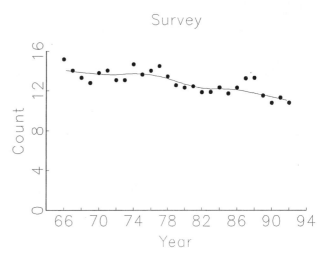

Figure 2.4. Annual indices of abundance for Brown-headed Cowbird population trends in the Eastern, Central, and Western BBS regions, and for the entire survey area (Survey). The count is a measure of mean regional relative abundance, adjusted to incorporate observer and route effects.

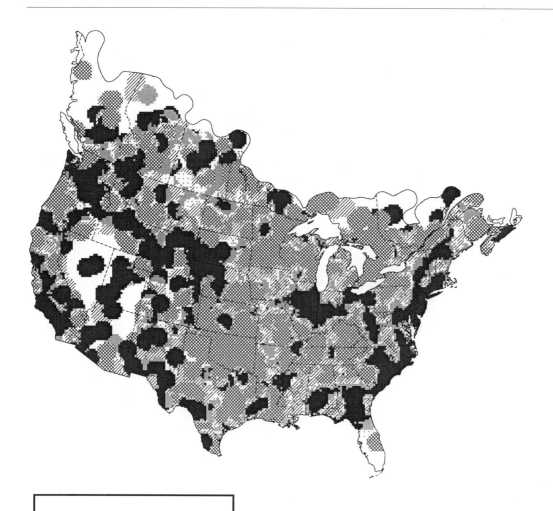

Percent Change per Year

 < − 1.5

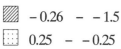 − 0.26 − − 1.5

0.25 − − 0.25

0.26 − 1.5

 > 1.5

Figure 2.5. Population trends of the Brown-headed Cowbird, 1977–1984. The map presents regions of consistent population change, grouped into categories of declining (trend < -1.5%/year), tending to decline (-0.26 > trend ≥ -1.5), indeterminate (0.25 > trend > -0.25), tending to increase (0.26 < trend ≤ 1.5), and increasing populations (trend > 1.5), based on smoothed weighted, route-specific trends. Unshaded regions represent areas with insufficient BBS data.

Correlations of Cowbird Trends with Trends in Other Species

Cowbird trends for the entire survey period were directly correlated with trends for all species groups we examined (cowbird host species 72.2%, $N = 18$, $P < .10$; unparasitized species 83.3%, $N = 12$, $P < .04$; neotropical migrants 75.3%, $N = 130$, $P < .001$; short-distance migrants 64.0%, $N = 100$, $P < .007$; permanent residents 73.4%, $N = 79$, $P < .001$). Thus, cowbirds tended to increase or decrease in numbers in concert with other bird species, regardless of whether the other species are affected by parasitism. However, although

more than 50% of all species groups had trends that were correlated with cowbird trends, the mean correlations for each group were quite small (range 0.145–0.052).

Geographic Analysis of Mean Neotropical Migrant Trends and Cowbird Trends

Results from the geographic analysis were similar to the correlation analyses, with increasing cowbird and mean neotropical migrant trends in 33.4% of the survey area. Cowbirds were increasing and mean neotropical migrant trends declining in only 11.4% of the area, while in 28.5% of the area cowbirds were declining and mean neotropical migrant trends increased. Both groups declined in 27.1% of the area (Figure 2.7).

The direct association between neotropical migrant trends and cowbird trends occurred primarily in the western states and provinces and portions of the southeastern United States (Figure 2.7). Illinois exhibited a mixed pattern with both groups increasing in the northern counties, but southern Illinois showed a concentrated area of cowbird increase and neotropical migrant decline. The urban corridor from New York City to Washington, D.C., also tended to

show mean neotropical migrant decline and cowbird increases, as did portions of Florida, coastal Georgia, and South Carolina. In the southeastern mountains and Arkansas, both groups showed general declines.

Discussion

The current pattern of relative abundance reflects the historic distribution of Brown-headed Cowbirds in North America. They are believed to have been originally concentrated in the Great Plains of central North America but to have

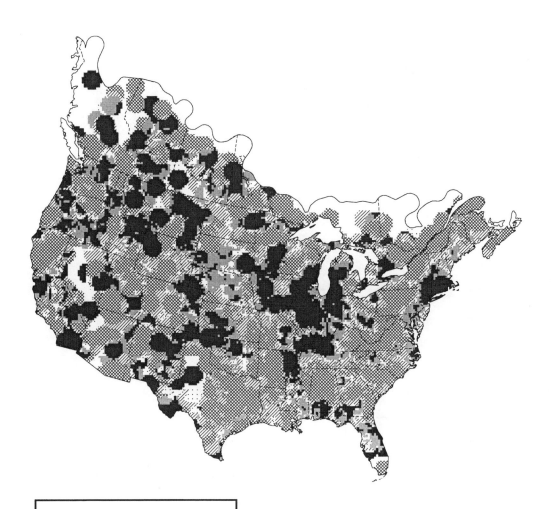

Percent Change per Year

 < −1.5

−0.26 − −1.5

0.25 − −0.25

0.26 − 1.5

> 1.5

Figure 2.6. Population trends of the Brown-headed Cowbird, 1985–1992. The map presents regions of consistent population change, grouped into categories of declining (trend < -1.5%/year), tending to decline (-0.26 > trend ≥ -1.5), indeterminate (0.25 > trend > -0.25), tending to increase (0.26 < trend ≤ 1.5), and increasing populations (trend > 1.5), based on smoothed weighted, route-specific trends. Unshaded regions represent areas with insufficient BBS data.

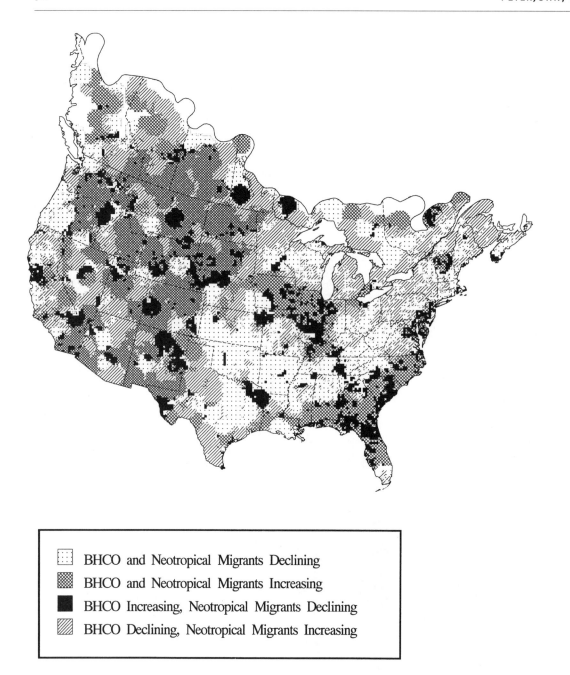

BHCO and Neotropical Migrants Declining

BHCO and Neotropical Migrants Increasing

BHCO Increasing, Neotropical Migrants Declining

BHCO Declining, Neotropical Migrants Increasing

Figure 2.7. Map of population changes of Brown-headed Cowbirds (BHCO) superimposed on a map of composite population trends of neotropical migrant birds for the entire survey period (1966–1992). Regions are summarized to indicate where: (1) neotropical migrant populations were decreasing and Brown-headed Cowbird populations were decreasing, (2) neotropical migrant populations were increasing and Brown-headed Cowbird populations were increasing, (3) neotropical migrant populations were decreasing and Brown-headed Cowbird populations were increasing, and (4) neotropical migrant populations were increasing and Brown-headed Cowbird populations were decreasing.

spread eastward and westward during the 19th and 20th centuries (Mayfield 1965, Rothstein 1994). Despite this extensive range expansion, they remain most numerous throughout the Great Plains.

Previous analyses of Brown-headed Cowbird trends from BBS data showed generally increasing trends through the 1970s, except in eastern North America (Dolbeer and Stehn 1979, Robbins et al. 1986). Our results indicated that the Eastern and Central regional populations generally peaked during the 1970s and subsequently declined, producing a similar trend for the continental population. In contrast, the population in the Western BBS Region remained reasonably stable through the mid-1980s, but subsequently declined.

The importance of these geographic patterns in trends should be viewed within the context of the geographic pattern of relative abundance. Population declines tended to occur along the eastern and western peripheries of the cowbird's range where it is least abundant. These declines may not be as important to the survey-wide population dynamics as the trends in the Great Plains where cowbirds are more numerous, and their populations generally increased in the northern states and Canada but declined in the southern states.

While the BBS documents patterns of population trends, it does not identify the factors responsible for these patterns (Temple and Wiens 1989). Given the considerable temporal and geographic heterogeneity evident in these patterns, the local, regional, and continental trends are probably influenced by a number of factors that defy a simple explanation. For example, severe weather conditions during the winters of 1976–1978 are responsible for significant declines in populations of Carolina Wrens (*Thryothorus ludovicianus*), Eastern Bluebirds (*Sialia sialis*), and other species (Robbins et al. 1986). Declines in some Brown-headed Cowbird populations began during these years, especially in eastern North America. However, inspection of the annual indices for the states and strata of this region did not reveal any geographically consistent declines corresponding with these severe winters, and any relationships between cowbird population trends and unusually severe winter weather may be purely coincidental.

Our analysis provides no evidence that change in cowbird populations differentially influences population change in cowbird host and nonhost species. Trends in both cowbird host species and species rarely parasitized by cowbirds show the same pattern of direct association with cowbird trends, and all of the correlations tend to be low. As summarized by Figure 2.7, the general direct associations documented between cowbird trends and trends of neotropical migrants clearly reflected the broad regional patterns of increasing bird populations in western North America and declines in portions of the southern United States. Although there were some areas where we could speculate that cowbirds may be influencing populations of neotropical migrants, most large-scale patterns of population change in both groups were probably associated with factors that were similarly influencing the trends of both cowbirds and their hosts.

Correlation studies of survey data can provide misleading results and should be viewed with caution. Aside from the well-known difficulties with assigning cause-effect relationships from correlations, analyses over space or time have additional complications. For example, geographic associations exist among trends for species and groups of species (Figure 2.7), and these associations may influence the significance of correlations among routes (Barker and Sauer 1992). Also, each survey route has a different history of changes in observers over time. Although we include observer covariables and start-up effects in our analyses, a lack of independence among species may still exist on individual survey routes. These features should be carefully considered in any correlation analysis of BBS data. We agree with Temple and Wiens (1989) that analyses based on survey data provide weak tests of hypotheses about causes, and that designed experiments at smaller scales should be used to test hypotheses about causal factors.

At the broad regional scale of the BBS, several factors could have contributed to the absence of strong inverse associations between cowbird population trends and the trends of potential host species. Because the BBS was not designed to address these specific associations, it may not be able to assess them adequately. However, given the large sample size for both cowbirds and many host species, some inverse associations should have been apparent if they were actually occurring. A more likely explanation is that large-scale factors such as long-term changes in weather patterns, land use practices, and habitat availability are primarily responsible for the direct associations between the population trends of cowbirds and their hosts. Additionally, since some host species are able to raise their own young at the same time as a fledgling cowbird, cowbird parasitism may not cause substantive reductions in their populations. These latter factors suggest that a national cowbird control program would produce few measurable benefits for the regional populations of potential hosts. However, control programs may still be necessary at the local level in order to maintain viable populations of endangered species experiencing high levels of nest parasitism.

Acknowledgments

We gratefully acknowledge the thousands of volunteers who have conducted Breeding bird surveys over the years and provided the data used in these analyses. We thank D. Dawson, C. Hahn, S. Rothstein, and an anonymous reviewer for their comments on earlier drafts of the manuscript.

APPENDIX
Species Used in the Correlation Analysis of Cowbird Trends

Host Species

Acadian Flycatcher, *Empidonax virescens*
Willow Flycatcher, *E. traillii*
Wood Thrush, *Hylocichla mustelina*
White-eyed Vireo, *Vireo griseus*
Bell's Vireo, *V. bellii*
Red-eyed Vireo, *V. olivaceus*
Yellow Warbler, *Dendroica petechia*
Prairie Warbler, *D. discolor*
Ovenbird, *Seiurus aurocapillus*

Common Yellowthroat, *Geothlypis trichas*
Summer Tanager, *Piranga rubra*
Northern Cardinal, *Cardinalis cardinalis*
Dickcissel, *Spiza americana*
Clay-colored Sparrow, *Spizella pallida*
Lark Sparrow, *Chondestes grammacus*
Grasshopper Sparrow, *Ammodramus savannarum*
Song Sparrow, *Melospiza melodia*
Brewer's Blackbird, *Euphagus cyanocephalus*

Neutral/Rejectors

Black-billed Cuckoo, *Coccyzus erythropthalmus*
Yellow-billed Cuckoo, *C. americanus*
Great Crested Flycatcher, *Myiarchus crinitus*
Barn Swallow, *Hirundo rustica*
Blue Jay, *Cyanocitta cristata*
White-breasted Nuthatch, *Sitta carolinensis*
American Robin, *Turdus migratorius*
Gray Catbird, *Dumetella carolinensis*
Brown Thrasher, *Toxostoma rufum*
Cedar Waxwing, *Bombycilla cedrorum*
Bullock's Oriole, *Icterus bullockii*
Baltimore Oriole, *I. galbula*

References Cited

Barker, R. J., and J. R. Sauer. 1992. Modelling population change from time series data. Pp. 182–194 *in* Wildlife 2001: Populations (D. R. McCullough, and R. H. Barrett, eds). Elsevier Applied Science, New York.

Brittingham, M. C., and S. A. Temple. 1983. Have cowbirds caused forest songbirds to decline? BioScience 33:31–35.

Butcher, G. S. 1990. Audubon Christmas bird counts. Pp. 5–13 *in* Survey designs and statistical methods for the estimation of avian population trends (J. R. Sauer and S. Droege, eds). USFWS Biol. Rep. 90(1).

Bystrak, D. 1981. The North American breeding bird survey. Studies in Avian Biol. 6:34–41.

Cressie, N. 1991. Statistics for spatial data. Wiley, New York. 900 pp.

Dolbeer, R. A., and R. A. Stehn. 1979. Population trends of blackbirds and starlings in North America, 1966–76. USFWS Special Scientific Rep. 214. 99 pp.

Droege, S., and J. R. Sauer. 1990. North American breeding bird survey annual summary, 1989. USFWS Biol. Rep. 90(8).

Environmental Systems Research Institute. 1993. Understanding GIS: The Arc/Info Method. Environmental Systems Research Institute, Redlands, CA.

Friedmann, H. 1963. Host relations of the parasitic cowbirds. U.S. Natl. Mus. Bull. 233.

Friedmann, H., L. F. Kiff, and S. I. Rothstein. 1977. A further contribution to knowledge of host relations of the parasitic cowbirds. Smithsonian Contr. Zool. 235.

Geissler, P. H., and J. R. Sauer. 1990. Topics in route-regression analyses. Pp. 54–57 *in* Survey designs and statistical methods for the estimation of avian population trends (J. R. Sauer and S. Droege, eds). USFWS Biol. Rep. 90(1).

James, F. C., C. E. McCulloch, and L. E. Wolfe. 1990. Methodological issues in the estimation of trends in bird populations with an example: The Pine Warbler. Pp. 84–97 *in* Survey designs and statistical methods for the estimation of avian population trends (J. R. Sauer and S. Droege, eds). USFWS Biol. Rep. 90(1).

Kendall, W. L., B. G. Peterjohn, and J. R. Sauer. 1996. First-time observer effects in the North American breeding bird survey. Auk 113:823–829.

Mayfield, H. F. 1965. The Brown-headed Cowbird with old and new hosts. Living Bird 4:13–28.

———. 1977. Brown-headed Cowbird: Agent of extermination? Amer. Birds 31:107–113.

Peterjohn, B. G., and J. R. Sauer. 1993. North American breeding bird survey annual summary 1990–1991. Bird Populations 1: 1–15.

Potter, E. F., and G. T. Whitehurst. 1981. Cowbirds in the Carolinas. Chat 45:57–68.

Robbins, C. S., D. Bystrak, and P. Geissler. 1986. The breeding bird survey: Its first fifteen years, 1965–1979. USFWS Resource Publ. 157. 196 pp.

Robbins, C. S., J. R. Sauer, R. S. Greenberg, and S. Droege. 1989. Population declines in North American birds that migrate to the neotropics. Proc. Natl. Acad. Sci. USA 86:7658–7662.

Rothstein, S. I. 1994. The cowbird's invasion of the far west: History, causes, and consequences experienced by host species. Studies in Avian Biol. 15:301–315.

Sauer, J. R., and P. H. Geissler. 1990. Estimation of annual indices from roadside surveys. Pp. 58–62 *in* Survey designs and statistical methods for the estimation of avian population trends (J. R. Sauer and S. Droege, eds). USFWS Biol. Rep. 90(1).

Sauer, J. R., B. G. Peterjohn, and W. A. Link. 1994. Observer differences in the North American breeding bird survey. Auk 111: 50–62.

Temple, S. A., and J. A. Wiens. 1989. Bird populations and environmental changes: Can birds be bio-indicators? Amer. Birds 43: 260–270.

3. Cowbird Population Changes and Their Relationship to Changes in Some Host Species

DAVID A. WIEDENFELD

Abstract

Analysis of Breeding Bird Survey (BBS) data using nonlinear semiparametric route-regression shows that the Brown-headed Cowbird is decreasing over much of the eastern part of its range in the northeastern United States and southeastern Canada. The species is increasing strongly in the northern Great Plains and Central Valley of California. Although its numbers show an overall decline in the central midwestern United States over the period of the BBS, its recent trend in that area is increasing. In the southeastern U.S. coastal plains, the cowbird has increased overall during the BBS period, but in recent years the numbers have stabilized and may be slightly declining. Cowbird parasitism does not appear to have a close linkage with population changes in the host species examined in this analysis, including nine species of neotropical migrants. Because the cowbird is already declining in much of the eastern part of its range, and because it does not seem to have much direct impact on its hosts' overall populations, habitat improvement may be the most effective way to control cowbird populations. Direct control (killing cowbirds) is probably unnecessary except in some specific cases, such as for Kirtland's Warbler (*Dendroica kirtlandii*) or Black-capped and Least Bell's Vireos (*Vireo atricapillus* and *V. bellii pusillus*).

Introduction

To understand the effects of Brown-headed Cowbirds on the populations of bird species that they parasitize, it is first necessary to know and understand the cowbirds themselves. An important first step is to know the cowbird's current geographic distribution and the changes in that distribution and abundance through time. This information is necessary before making decisions on management, so that management efforts can be concentrated where they will make the greatest impact.

The study of populations is a large-scale problem that can only be addressed using a large-scale data set. The only data set suitable for analyzing cowbird populations in this way is the North American Breeding Bird Survey (BBS). Although the BBS is not ideal, it is the only data set capable of sampling breeding cowbird populations on a regional or continental scale through time.

Cowbirds have been implicated as important factors contributing to the decline of some rare species, including Kirtland's Warbler (DeCapita, Chapter 37, this volume) and Black-capped Vireo (Hayden et al., Chapter 39, this volume), and local populations of common species (Mayfield 1977), especially in the western United States (Schram 1994). For most host species, however, it is clear that there are many other limiting factors affecting host populations besides cowbird parasitism. It does seem logical, however, that Brown-headed Cowbirds, because they can limit recruitment by a host species, should have some effect on their hosts' population sizes. It has been shown that, at least in some areas, cowbirds parasitize some of their hosts so heavily that the hosts produce too few recruits to sustain their local populations (Robinson 1992).

I analyzed the Brown-headed Cowbird BBS data with two objectives. My first objective was to examine cowbird populations alone: to determine where cowbird populations are most dense, where the species has increased most during the BBS period, and whether some areas are showing a greater rate of recent increase, or a slowdown or reversal. My second objective was to determine if there are large-scale relationships between cowbird populations and those of other birds. To do this I also used the BBS data and compared the overall change in population numbers in several neotropical migrant host species and the Red-winged Blackbird (*Agelaius phoeniceus*), which is not a neotropical migrant, with the change in cowbird numbers in the same area.

Methods

The Breeding Bird Survey is a program of standardized roadside censuses cosponsored by the U.S. National Biological Service (NBS; formerly part of the Fish and Wildlife Service) and the Canadian Wildlife Service, and conducted by knowledgeable volunteers (Droege 1990). Birds seen and heard are recorded at 50 stops 0.8 km apart. The BBS consists of data from about 2,000 routes each year and has been under way since 1966, with 27 years of data available at the time of writing (1966–1992). To understand the continent-wide distribution and population changes in the Brown-headed Cowbird, I analyzed the BBS data using Nonlinear Semiparametric Route Regression (NSRR), described in James et al. (1996). The advantage of a nonlinear technique is that it can show not only the overall population trend but

also changes in rate or reversals of that trend. In addition, NSRR uses only routes that have been run consistently so that the routes in the analysis at the beginnings and ends of the time period are the same.

The geographical units of analysis are "strata," physiographic units that have been defined by the NBS. The technique developed by James et al. (1996) uses some slight modifications from the system of strata as defined by the NBS, combining some strata with small numbers of routes. NSRR uses 54 strata in the conterminous United States and southern Canada (Figure 3.1). NSRR also includes an observer-effect term in its semiparametric model, to compensate for differences among observers. To obtain an estimate of the overall change in numbers and the change in each stratum for each species analyzed, the NSRR technique compares estimates for the three-year periods 1970–1972 and 1986–1988.

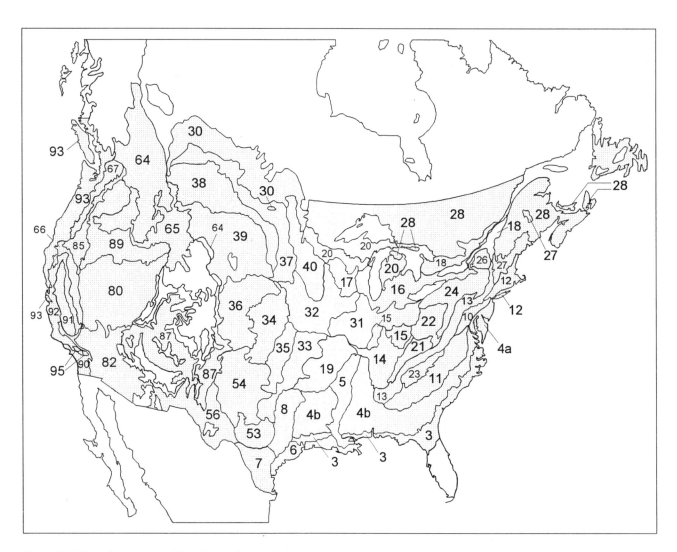

Figure 3.1. Map of locations and boundaries of strata. Strata are based on those used by the U.S. National Biological Service, but some NBS strata have been combined.

To examine the linkage between Brown-headed Cowbird population changes and changes in their hosts, I calculated linear regressions of the change in host numbers in each stratum where it occurs against the change in cowbird numbers for that same stratum. For this part of the analysis I used linear regression because the response of the host population should be linear if the very stringent assumptions are met. To visualize variation from linearity, I also plotted the LOESS smooth (smoothing parameter $f = 0.75$) of the data. I used only strata from the eastern United States and Canada (east of the Rocky Mountains), because BBS data for the host species I chose are sparse from the west, except for the very abundant Red-winged Blackbird, for which I used all strata in which it and the cowbird both occur.

It was not feasible to analyze data for all hosts of the Brown-headed Cowbird. Therefore, because of the current concern about population changes in neotropical migrants, I selected eight species of wood-warblers from the eastern United States and Canada that are neotropical migrants. I chose four open/edge species (Yellow Warbler *Dendroica petechia*, Prairie Warbler *D. discolor*, Common Yellowthroat *Geothlypis trichas*, and Yellow-breasted Chat *Icteria virens*) and four forest-interior species (Prothonotary Warbler *Protonotaria citrea*, Black-throated Blue Warbler *Dendroica caerulescens*, Canada Warbler *Wilsonia canadensis*, and Hooded Warbler *W. citrina*). The set included species with increasing populations (Black-throated Blue and Hooded Warblers; see James et al. 1996) as well as decreasing species (Prairie and Canada Warblers, the yellowthroat, and the chat; see James et al. 1996). I also selected a nonparuline neotropical migrant, the Wood Thrush (*Hylocichla mustelina*), because it experiences high rates of parasitism in some areas (e.g., Robinson 1992), and a resident and short-range migrant, the Red-winged Blackbird, because it is in the same subfamily (Icterinae) as the cowbird and is approximately the same size. These species have varying levels of cowbird parasitism, even within a species (e.g., Hoover and Brittingham 1993 for Wood Thrush or Friedmann and Kiff 1985 for the yellowthroat). General levels of parasitism vary from high (Yellow Warbler) to low (Prothonotary Warbler, generally a cavity nester).

Results and Discussion

Trends in Cowbird Numbers

The number of cowbirds per route for each stratum for the period 1986–1988 can be seen in Table 3.1 and is also shown categorized into four levels on the map in Figure 3.2. The cowbird is most abundant in the central and northern Great Plains. In the western United States, abundance is generally low, although strata 91 and 85 show relatively high numbers.

Overall, Brown-headed Cowbirds declined slightly and nonsignificantly during the BBS period (Table 3.1). Cowbird numbers have been stable or decreasing over most of the eastern United States and southeastern Canada, except for the tallgrass prairie in Ohio through Illinois (strata 31 and 32, Figure 3.3). In most of the southern Great Plains, the cowbird is decreasing, but its increase is greatest in the northern Great Plains. It is also increasing in numbers very strongly in the Central Valley of California.

The areas showing greatest proportional declines are around the eastern Great Lakes, southern Great Plains, and in Idaho, eastern Oregon, and Washington (Figure 3.4). Note that in an area with low numbers of cowbirds a small change in absolute numbers can produce a large proportional change. The area showing the greatest proportional increase is Stratum 91, the Central Valley of California, but some areas in the Great Plains are also showing strong proportional increases (cf. Table 3.1).

The overall trend in Brown-headed Cowbird numbers has been fairly stable through time (Figure 3.5), although the Eastern Region shows a slight but steady decline since about 1974, and the Western Region has shown an accelerating rate of increase since the mid-1980s. In southeastern Canada cowbird numbers show parallel declines in strata 18 and 28, both beginning about 1975 (Figure 3.6). In Stratum 18 the cowbird has declined to less than half of its 1975 numbers. The decline in general of agriculture and commensurate reforestation in the northeastern United States (Brooks and Birch 1988, Dickson and McAfee 1988) may be a major factor in the decline of the cowbird. However, it would be interesting to know exactly why the species began its precipitous decline in the late 1970s in that area.

Adjacent strata in the central midwestern United States all show similar patterns, especially strata 14, 15, and 17 (Figure 3.7). All of these strata except 19 show a decline until about 1980. After that year all five show a trend toward rapidly increasing numbers of cowbirds. Although most of these strata show an overall decline through the BBS period, they are presently increasing and could be at their previous levels or higher by the end of the century if the trend continues.

In the southeastern coastal plains of the United States, the cowbird has increased fairly strongly in overall numbers (Figure 3.4). In the more southerly part of this region (strata 3 and 4B), however, the increase leveled off during the early 1980s and numbers declined in the late 1980s and early 1990s (Figure 3.8). In the northern part of the Upper Coastal Plain (Stratum 4A), however, the cowbird has increased strongly through the entire BBS period.

In the northern Great Plains from South Dakota and Montana to southern Canada (Figure 3.9), the cowbird has strongly increased in the two strata where it was already most abundant, strata 37 and 38. In addition, these two strata show an almost identical pattern of humps and troughs, which because of their scale may be weather-related. Adjacent strata to the north and south (strata 30 and 39) show neither the very strong increase nor the same trend pattern.

Table 3.1. Density and Change in Brown-headed Cowbird Numbers by Stratum

Stratum	No. Birds per Route	Change	Percent Change	Percent Increasing Routes	SE	t	df	Sig.	Increasing or Decreasing
3	3.25	1.27	65	62	1.57	1.16	20		
4A[a]	9.48	5.35	129	66	1.80	3.12	28	*	+
4B[a]	9.31	2.20	31	63	1.52	2.01	62	*	+
5	11.59	2.17	23	45	2.24	0.49	10		
6	5.62	−2.95	−34	33	1.05	−1.29	5		
7	21.24	2.60	14	30	2.97	0.31	9		
8	13.05	−9.12	−41	29	2.87	−1.43	13		
10	7.36	−0.75	−9	38	1.72	−0.38	23		
11	5.06	−2.28	−31	36	1.92	−0.90	13		
12[b]	5.07	−0.21	−4	38	1.43	−0.16	23		
13	6.88	−2.70	−28	28	1.21	−2.56	42	*	−
14	13.97	1.00	8	44	1.52	0.52	33		
15	13.00	−4.04	−24	18	0.87	−2.00	10		
16	11.19	−6.60	−37	29	1.63	−3.42	40	*	−
17	17.85	−4.64	−21	30	1.60	−1.44	19		
18	11.89	−14.02	−54	10	1.98	−3.70	19	*	−
19	14.68	−1.05	−7	38	1.36	−0.45	20		
20	11.71	−10.83	−48	24	1.52	−3.99	20	*	−
21	4.67	−4.39	−48	33	1.34	−1.55	5		
22	8.61	−5.90	−41	21	1.29	−2.95	18	*	−
23	0.31	−1.90	−86	50	1.20	−1.55	3		
24	4.10	−7.98	−66	10	1.83	−5.01	39	*	−
26	9.51	−4.89	−34	20	1.02	−1.56	4		
27	6.92	1.04	18	60	1.57	0.58	19		
28	3.92	−3.57	−48	27	1.71	−3.70	69	*	−
30	13.46	0.21	2	52	2.02	0.06	20		
31	8.88	3.26	58	58	2.01	1.87	37		
32	27.75	8.90	47	73	1.98	2.96	39	*	+
33	18.75	−14.53	−44	14	1.79	−3.76	21	*	−
34	27.73	−5.95	−18	35	2.15	−1.27	25		
35	28.53	−8.90	−24	30	1.96	−1.25	9		
36	3.04	1.59	109	60	1.73	0.99	9		
37	39.60	13.52	52	77	1.63	2.63	12	*	+
38	50.87	11.38	29	71	3.15	0.71	6		
39	19.59	5.76	42	63	1.80	1.57	15		
40	28.85	2.29	9	55	1.77	0.55	19		
53	31.82	6.86	27	63	1.07	1.71	7		
54[c]	5.12	−5.68	−53	33	1.40	−2.19	8		
56	5.75	3.07	115	71	1.43	1.41	6		
64	7.88	2.53	47	60	2.37	0.66	9		
65	9.45	0.66	7	56	1.63	0.20	8		
66	3.23	1.08	50	50	3.62	0.22	5		

Table 3.1. Continued

Stratum	No. Birds per Route	Change	Percent Change	Percent Increasing Routes	SE	t	df	Sig.	Increasing or Decreasing
67	8.36	2.89	53	50	1.89	0.72	5		
80[d]	0.24	0.02	7	50	0.30	0.10	3		
82[e]	1.99	−1.96	−50	50	1.53	−0.75	3		
85	11.27	−0.24	−2	63	3.93	−0.03	7		
87	1.60	−0.07	−4	60	0.77	−0.08	4		
89	2.17	−3.62	−63	55	4.62	−0.67	10		
90	0.54	0.27	101	67	1.37	0.39	5		
91	12.56	10.33	464	67	2.56	2.78	11	*	+
92	5.17	−1.19	−19	50	2.04	−0.49	15		
93[f]	4.77	−2.96	−38	38	1.55	−1.97	25		
95	4.29	1.58	59	67	0.43	1.72	2		
Overall	9.59	−0.42	−4	43	0.09	−0.73			

Note: Change and percent change were calculated using the period 1970–1972 as beginning, 1986–1988 as ending. The percent increasing routes is the proportion of routes showing an increase over the same periods. The df (degrees of freedom) is the number of routes in the stratum used in the paired t-test minus 1. Asterisks in the significance column indicate *P* = .05.

[a] Stratum 4 was divided into northern (4A) and southern (4B) sections.
[b] Stratum 9 was joined to Stratum 12.
[c] Stratum 55 was joined to Stratum 54.
[d] Stratum 88 was joined to Stratum 80.
[e] Stratum 83 was joined to Stratum 82.
[f] Vancouver Island and Stratum 98 were joined to Stratum 93.

In the far western United States, cowbird populations in the central-northern California strata all increased strongly in the late 1980s through early 1990s (Figure 3.10), although the number of BBS routes (sample size) in each of these strata is small. All three strata in Figure 3.10 also show a brief decline in the late 1980s, but quickly resumed their steep upward trend.

Areas of Potential Management Concerns

From this analysis, what areas can we determine to be of the greatest concern to conservationists? After examining both the map and the graphs, it is apparent that the cowbird is increasing greatly in the northern Great Plains, and the trend through 1992 shows no signs of slowing down. This is also true of the northern part of the Lower Coastal Plain (Stratum 4A). Several areas in the western United States are also showing strong population increases, proportionally the greatest on the continent. The several strata in the midwestern United States that show an overall decline during the BBS period, but with an increase in recent years (strata 14, 15, 17, 19, and 31; Table 3.1, Figure 3.7), also suggest a possibly deleterious trend in that region. The areas showing a *current* increase in cowbird numbers should be the areas receiving greatest attention by conservationists.

Although the cowbird has increased in numbers in the Upper Coastal Plain and southern part of the Lower Coastal Plain (strata 3 and 4B), the increase has leveled off since the mid-1980s and may currently be showing some signs of reversal. Therefore, although the southern coastal plains have had an overall increase in cowbird numbers during the period of the BBS, it is probably not necessary to focus cowbird research there, although monitoring of population levels is still necessary. In the northeastern United States and southeastern Canada, cowbird populations are almost all declining. Scarce resources should probably not be used in this area for cowbird control except for specific problems such as maintaining Kirtland's Warbler. Continued monitoring of cowbird population changes is necessary, not only for this area but also for the entire range of the cowbird.

Relationships between Cowbird and Host Population Sizes

Under the assumptions that (1) the Brown-headed Cowbird affects its host's population size, (2) the cowbird is the *only* factor affecting the host's population, and (3) the cowbird's population is regulated by recruitment, one would expect to see a straight-line, negative correlation between changes in the host's and cowbird's populations. Obviously, in the real world cowbirds are not the only factor affecting host popu-

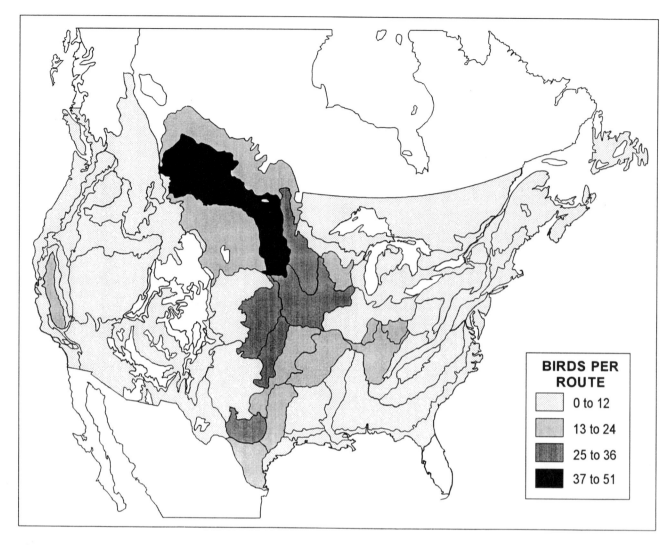

Figure 3.2. Relative abundance (as expressed by number of birds per BBS route) of Brown-headed Cowbirds by stratum, from Breeding Bird Survey data averaged for the period 1986–1988. Compare with Table 3.1. In white areas, either cowbirds do not occur or there are too few data to analyze.

lations; the factors are myriad. The relationship between the cowbird and any particular host is complicated, because the cowbird is not restricted to a single host and may shift its parasitism among a set of hosts without relying on any one. The host may also affect the cowbird's population; as the host population declines, the cowbird population may decline too as it loses its host, or (more likely) the cowbird may shift to another host. In addition, host effects may interact among species if there is a differential impact of parasitism on different hosts. The whole is complicated by the fact that cowbird mortality may not be directly related to cowbird reproductive success or lack thereof on the breeding ground. Winter mortality or density independent mortality of the cowbird can affect cowbird populations independently of the cowbird's effects on its hosts. Finally, any apparent cor-

relations between host and cowbird populations may be not the result of any interaction between the two species but rather a response of both to some other factor, such as habitat change.

With this web of interacting factors acting on the host directly, on the cowbird directly, and interconnecting the cowbird and host, it is unlikely that the relationship between overall host population and overall cowbird population would be very close. However, because of the concern in the conservation community over cowbird parasitism, it is important to examine the possible relationship between cowbird and host numbers. If cowbird parasitism has a very weak effect or none on the *overall* population of a host, then efforts to control cowbirds might be better spent elsewhere.

None of the correlation coefficients between Brown-

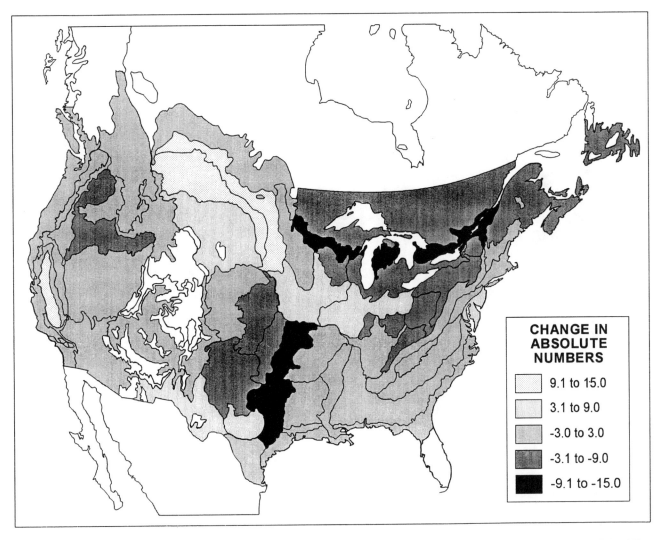

Figure 3.3. Change in numbers of cowbirds between 1970–1972 and 1986–1988, by stratum.

headed Cowbird population changes and host population changes within individual strata were significant or very high (Table 3.2). Only the Prairie Warbler reached an r^2 of greater than 0.1. The slope of the regression line, expected to be negative if cowbirds were a major factor in the host population changes, was negative in only four of ten comparisons.

Because the Brown-headed Cowbird is primarily an edge habitat species, I have divided the host species into two groups, open/edge versus forest-interior nesting. Cowbirds would be expected to have greater impact on the edge-nesting species. Three of the four negative relationships are indeed in the edge species, and the edge species do tend to have slightly higher coefficients of correlation.

To visualize the relationship between host and cowbird, I plotted the change in population for cowbird versus host in each stratum for five of the ten host species and overlaid the points with their linear regression and LOESS smooth. This technique can help to show if some nonlinear relationship exists between the host and cowbird population changes, a relationship that linear regression would miss. The five species were chosen to include open/edge species and forest-interior species, and parulines as well as nonparulines. I present only five of the ten species to reduce the number of graphs to be published. Graphs of the other five species show similarly chaotic patterns.

In areas where the Brown-headed Cowbird is increasing (right half of the graph), the Yellow Warbler is not usually decreasing, counter to the expectation that the cowbird would be affecting the warbler's populations (Figure 3.11). One possible explanation is that Yellow Warblers are not de-

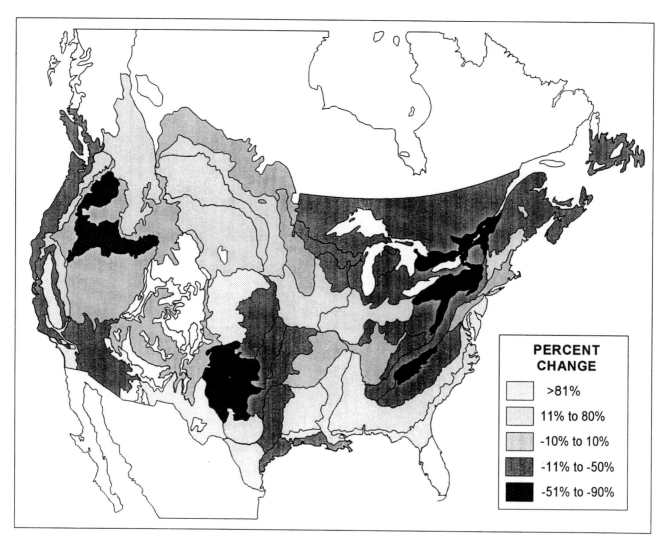

Figure 3.4. Proportional change in numbers of cowbirds between 1970–1972 and 1986–1988, by stratum.

creasing because the cowbird has already extinguished the warbler in areas where the cowbird is increasing strongly. In most instances, however, the warbler's populations could have declined further, but did not. For example, the right-most point represents changes in Stratum 37, where the Yellow Warbler population has a density of 3.26 birds per route.

There may, however, still be some effect of cowbirds on the warbler's populations. Where the cowbird is decreasing (left half of the graph), the Yellow Warbler tends to be increasing. This suggests there is a saturation level for the effect of cowbirds on the Yellow Warbler. That is, an increase in cowbird numbers does cause the warbler to decline, but there is a threshold below which the warbler population does not go, perhaps as the cowbird shifts to another host. Both the cowbird and the warbler are edge species, so it seems unlikely that habitat changes such as loss of edge habitat would adversely affect one and not the other.

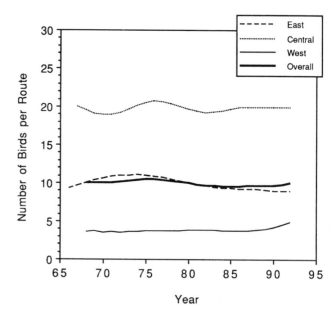

Figure 3.5. Trends in cowbird numbers through the period of the BBS. The thick "Overall" line is a weighted combination of the other three lines.

Table 3.2. Linear Correlation Coefficients and Direction of Slopes for Host Species Population Changes Versus Cowbird Population Changes

Species	N	r^2
Open/edge species		
Yellow Warbler (*Dendroica petechia*)	30	−0.071
Prairie Warbler (*Dendroica discolor*)	15	+0.105
Common Yellowthroat (*Geothlypis trichas*)	35	−0.063
Yellow-breasted Chat (*Icteria virens*)	21	+0.056
Red-winged Blackbird (*Agelaius phoeniceus*)	51	−0.006
Forest-interior species		
Wood Thrush (*Hylocichla mustelina*)	24	+0.005
Prothonotary Warbler (*Protonotaria citrea*)	10	+0.024
Black-throated Blue Warbler (*Dendroica caerulescens*)	7	0.000
Canada Warbler (*Wilsonia canadensis*)	6	−0.007
Hooded Warbler (*Wilsonia citrina*)	13	+0.043

Note: N is the number of strata in which the host and cowbird co-occur. No correlation coefficients were significant at the *P* = .05 level.

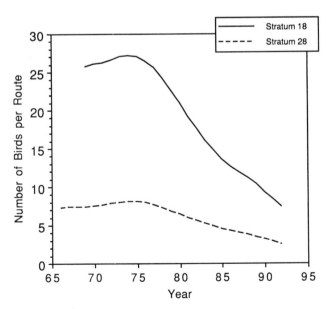

Figure 3.6. Trends in cowbird numbers in strata 18 and 28. (For locations of these strata, see Figure 3.1.)

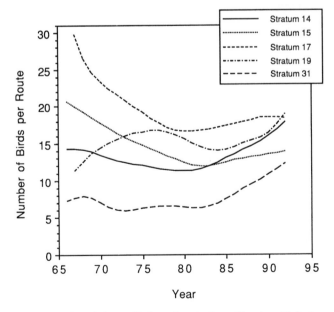

Figure 3.7. Trends in cowbird numbers in five midwestern United States strata.

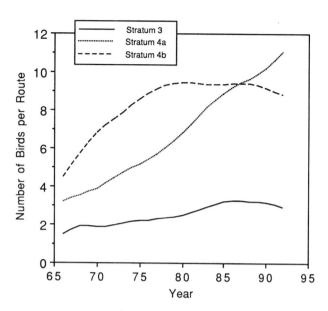

Figure 3.8. Trends in cowbird numbers in three southeastern United States strata.

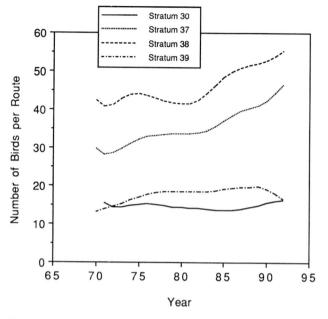

Figure 3.9. Trends in cowbird numbers in four strata in the northern Great Plains.

The Common Yellowthroat (Figure 3.12) is also an edge species and is declining over much of its range, especially in the eastern United States (James et al. 1996). Compared with the Yellow Warbler, the yellowthroat shows a greater tendency to decrease where the cowbird has increased. Almost all areas where the cowbird is increasing show a decline in yellowthroat numbers. Where the cowbird is decreasing, the yellowthroat tends to be increasing. The relationship, however, is not tight, and the greatest declines in

yellowthroat numbers are in strata where the cowbird is also declining or increasing only slightly. Again, both birds are edge species, so it seems unlikely that a habitat change affecting one would not affect the other.

There seems to be little or no relationship between cowbird and Hooded Warbler population changes (Figure 3.13). The warblers seem to be increasing regardless of cowbird population changes, and in the four strata where the warbler is decreasing, the cowbird is too. The warbler is a forest-interior species, and the cowbird is an edge species, so it seems likely that habitat changes favoring one (for example, an increase in forest cover and forest fragment size, which would favor the warbler's populations) would disfavor the other.

The Wood Thrush is also a forest-interior species, but one that is declining over most of its range. As in the case of the Hooded Warbler, the thrush's population changes show no relationship to changes in the cowbird (Figure 3.14). The thrush is declining where the cowbird is declining also, and the thrush's greatest declines, in fact, are in strata where the cowbird is also declining. This result is strongly counter to expectation. Robinson (1992) and others have shown that the Brown-headed Cowbird parasitizes Wood Thrush nests very heavily, at least in some areas. However, the results I have presented show that even with heavy parasitism, the cowbird is probably not a controlling factor on *overall* Wood Thrush populations.

The Red-winged Blackbird shares the cowbird's use of open habitats. In some areas, particularly the Great Plains, the blackbird is fairly heavily parasitized by Brown-headed Cowbirds. Often, around 50% of nests are parasitized

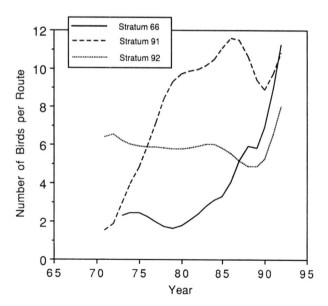

Figure 3.10. Trends in cowbird numbers in three California strata.

Cowbird parasitism does not appear to have a close linkage with population changes in some of its hosts, including neotropical migrants.

Habitat improvement is probably a better way to control cowbird populations than direct control (killing cowbirds) except for some specific situations (for example, Kirtland's Warbler or Black-capped Vireo).

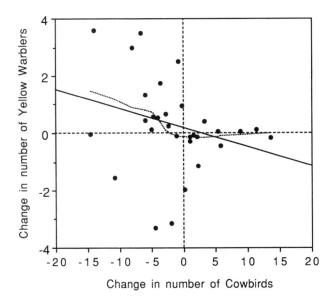

Figure 3.11. Change in Brown-headed Cowbird numbers plotted against the change in Yellow Warbler numbers in the same strata. Dashed lines divide areas where both the cowbird and warbler are increasing (upper right), where the warbler is increasing but the cowbird decreasing (upper left), etc. The solid line is the linear regression line; see Table 3.2 for r^2 and sample size. The LOESS smooth ($f = 0.75$) of the points is the dotted line.

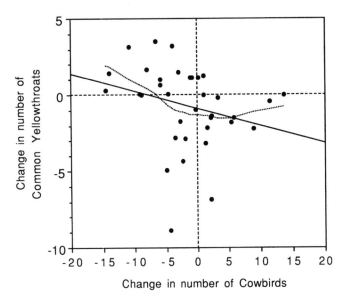

Figure 3.12. Change in Brown-headed Cowbird numbers plotted against the change in Common Yellowthroat numbers in the same strata. Conventions as in Figure 3.11.

(Friedmann and Kiff 1985), although parasitism does not drastically lower reproductive output (Røskaft et al. 1990). There does, however, seem to be some effect of cowbird population increases on blackbird numbers (Figure 3.15). Indeed, the blackbird is decreasing most where the cowbird is increasing most, and the blackbird's greatest increases are in strata where the cowbird is decreasing. The relationship, however, is still not very tight, because the blackbird is also decreasing in strata where the cowbird is also decreasing.

None of the species considered in this chapter showed a strong relationship to cowbird population changes. In three of the edge species (Yellow Warbler, Common Yellowthroat, Red-winged Blackbird) there may be some effect, but it is not consistent or strong. In the Wood Thrush, a species of concern to conservationists because of its declining numbers and intense parasitism by cowbirds in some areas, little effect of parasitism on the thrush's population can be detected; the thrush seems to be declining for reasons not related to the cowbird.

Conclusions

The Brown-headed Cowbird is declining in the eastern part of its range, perhaps as a result of reforestation. It is increasing strongly in the northern Great Plains and Central Valley of California. Although its numbers show an overall decline in the central Midwest over the period of the BBS, its recent trend in that area is sharply increasing.

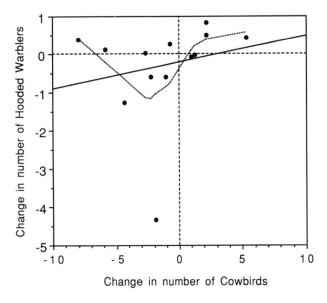

Figure 3.13. Change in Brown-headed Cowbird numbers plotted against the change in Hooded Warbler numbers in the same strata. Conventions as in Figure 3.11.

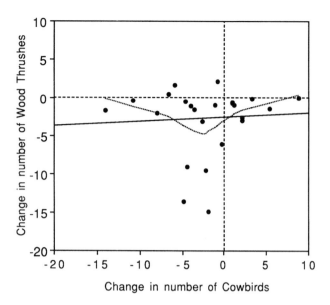

Figure 3.14. Change in Brown-headed Cowbird numbers plotted against the change in Wood Thrush numbers in the same strata. Conventions as in Figure 3.11.

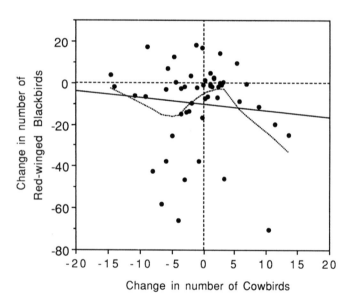

Figure 3.15. Change in Brown-headed Cowbird numbers plotted against the change in Red-winged Blackbird numbers in the same strata. Conventions as in Figure 3.11.

Acknowledgments

I thank Frances C. James and Charles E. McCulloch for collaboration in the development of the methods used to analyze the Breeding Bird Survey data. R. Terry Chesser, Mario Cohn-Haft, Andrew Kratter, J. Van Remsen, Stephen I. Rothstein, and Frederick H. Sheldon made many valuable suggestions that improved the manuscript.

References Cited

Brooks, R. T., and T. W. Birch. 1988. Changes in New England forests and forest owners: Implications for wildlife habitat resources and management. Trans. N. Amer. Wildl. Nat. Resour. Conf. 53:78–87.

Dickson, D. R., and C. L. McAfee. 1988. Forest statistics for Connecticut, 1972 and 1985. USDA Forest Service Northeastern Forest Experiment Station Resources Bull. NE-105.

Droege, S. 1990. The North American breeding bird survey. Pp. 1–4 in Survey designs and statistical methods for the estimation of avian population trends (J. R. Sauer and S. Droege, eds). USFWS Biol. Rep. 90(1).

Friedmann, H., and L. F. Kiff. 1985. The parasitic cowbirds and their hosts. Proc. Western Found. Vert. Zool. 2:225–304.

Hoover, J. P., and M. C. Brittingham. 1993. Regional variation in cowbird parasitism of Wood Thrushes. Wilson Bull. 105:228–238.

James, F. C., C. E. McCulloch, and D. A. Wiedenfeld. 1996. New approaches to the analysis of population trends in land birds. Ecology 77:13–27.

Mayfield, H. 1977. Brown-headed Cowbird: Agent of extinction? Amer. Birds 31:107–113.

Robinson, S. K. 1992. Population dynamics of breeding neotropical migrants in a fragmented Illinois landscape. Pp. 408–418 in Ecology and conservation of neotropical migrant landbirds (J. M. Hagan III and D. W. Johnston, eds). Smithsonian Institution Press, Washington, DC.

Røskaft, E., G. H. Orians, and L. D. Beletsky. 1990. Why do Redwinged Blackbirds accept eggs of Brown-headed Cowbirds? Evol. Ecol. 4:35–42.

Schram, B. 1994. An open solicitation for cowbird recipes. Birding (August), pp. 254–257.

4. The Spread of Shiny and Brown-headed Cowbirds into the Florida Region

ALEXANDER CRUZ, JOHN W. PRATHER,

WILLIAM POST, AND JAMES W. WILEY

Abstract

Recent expansions of the ranges of the Shiny Cowbird and the Brown-headed Cowbird have brought them into contact with avian communities in south Florida that have never experienced brood parasitism. The Shiny Cowbird, which was originally confined to South America, Trinidad and Tobago, has spread dramatically into the West Indies during the past century, and since 1985 it has been recorded in Florida. From the opposite direction, the North American Brown-headed Cowbird has spread rapidly through peninsular Florida since the 1950s. The spread of the Shiny Cowbird into Florida has been mainly through coastal areas, and although we have not found parasitized nests, we characterize south Florida populations as potentially colonizing propagules outside the contiguous range of the species in the West Indies. In contrast to the Shiny Cowbird, host nests parasitized by Brown-headed Cowbirds have been found in Florida. The spread of Brown-headed Cowbirds has been a broad-front advance in range, especially in north Florida. In south Florida, there are colonizing propagules ahead of the spreading contiguous range. In addition to providing a unique opportunity to study parasitism at an early interfacing of host and parasite populations, these formerly allopatric species are expected to have important negative consequences for south Florida passerines. While there are a few records of parasitism of Prairie Warblers by cowbirds in south Florida, we found no parasitism in 38 Prairie and 19 Yellow Warbler nests in Florida Bay and on the Florida Keys. We feel that it will be only a matter of time before parasitism is observed, as both Brown-headed and Shiny Cowbirds have been recorded during the breeding season and, in some instances, in the nesting habitats of these two host species.

Introduction and Background

Historically, the Shiny Cowbird was confined to South America, and the Brown-headed Cowbird to central North America. With the reduction of forest and large-scale alterations associated with agriculture and animal husbandry, cowbirds have expanded their geographical ranges, bringing them into contact with new hosts that lack defenses against brood parasitism (Mayfield 1977, Cruz et al. 1985). Since about 1900, the Shiny Cowbird has been expanding its breeding range northward into the West Indies from the Guianas and Trinidad (Post and Wiley 1977; Cruz et al. 1985, 1989), and it is now spreading into south Florida (Figure 4.1). From the opposite direction, the Brown-headed Cowbird has expanded its range in eastern North America (Figure 4.1) and is colonizing south Florida (Hoffman and Woolfenden 1986).

The spread of these cowbirds is particularly interesting for three reasons: (1) It provides a unique opportunity to study brood parasitism during early contact of host and parasite populations—colonization began within the last 40 years, and both cowbird species are increasing in abundance (Robertson and Woolfenden 1992). (2) Some potential hosts are patchily distributed in south Florida and breed in small populations that may be vulnerable to extirpation by cowbirds. (3) Shiny Cowbirds may spread to other regions in North America from south Florida, and Brown-headed Cowbirds may spread from south Florida to the West Indies. Both cowbird species are obligate brood parasites and are host generalists; more than 200 host species have been recorded for each species (Cruz et al. 1989, Friedmann and Kiff 1985).

Because of the small population size and high degree of isolation of the south Florida avifauna (Robertson and Kushlan 1984), contact with cowbirds is potentially more detrimental than in other North American areas where new contact between species occurs more gradually and over a wider area. Not only are many of the Florida species and subspecies (e.g., vireos and warblers) restricted in range, but they may also be confined to specific habitats, factors that make them particularly susceptible to the detrimental effects of parasitism.

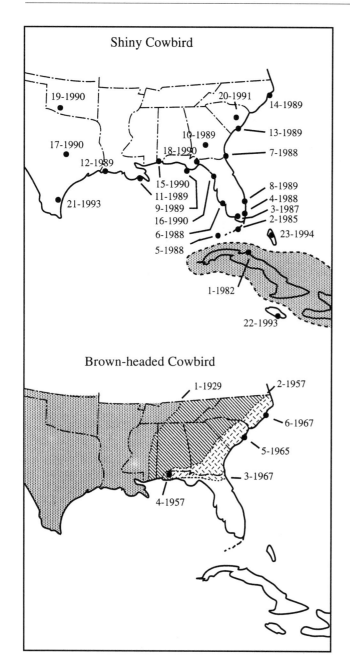

Figure 4.1. *Top:* Shiny Cowbird sightings for the western Caribbean Basin and southeastern North America. In North America, only spring and summer records of range extensions of 100 km or more are shown. Shaded area represents West Indian breeding range. 1. Cuba, 1982, 2. Lower Matecumbe Key, Florida, 1985, 3. Flamingo, Florida, 1987, 4. Homestead, Florida, 1988, 5. Dry Tortugas, Florida, 1988, 6. Fort Desoto, Florida, 1988, 7. Jacksonville, Florida, 1988, 8. Delray Beach, Florida, 1989, 9. Cape San Blas, Florida, 1989, 10. Warner Robbins, Georgia, 1989, 11. Fort Fourchon, Louisiana, 1989, 12. Cameron, Louisiana, 1989, 13. Sullivan's Island, South Carolina, 1989, 14. Aurora, North Carolina, 1989, 15. Bon Secour, Alabama, 1990, 16. Levy County, Florida, 1990, 17. Fort Hood, Texas, 1990, 18. Tallahassee, Florida, 1990, 19. Winborn Spring, Oklahoma, 1990, 20. Kingstree, South Carolina, 1991, 21. Goliad, Texas, 1993, 22. Yallahs Pond, Jamaica, 1993, and 23. Staniard Creek, Andros Island, 1994. Data based on field observations, personal communications, Florida Breeding Bird Atlas Project, and literature review, e.g., *American Birds, Florida Naturalist, Florida Field Naturalist,* and *Tropical Audubon Bulletin.*

Bottom: Brown-headed Cowbird breeding range expansion in southeastern United States. 1. 1929 breeding range limit (Friedmann 1929), 2. 1957 breeding range limit (Webb and Wetherbee 1960, Potter and Whitehurst 1981), 3. 1967 breeding range limit (Newman 1957, Potter and Whitehurst 1981), 4. Escambia County—first Florida breeding record, 1957, 5. Charleston, first coastal South Carolina breeding record, 1965, and 6. Wilmington—first coastal breeding record for North Carolina.

In 1989, we began an ongoing study of Shiny and Brown-headed Cowbirds and potential host species in south Florida, a region that has a unique faunal assemblage with affinities to both the West Indies and North America (Robertson and Kushlan 1984). Our work in Florida is designed to obtain data on the (1) geographical and habitat distribution of Brown-headed and Shiny Cowbirds, (2) breeding biology of cowbirds and potential host species, (3) cowbirds' impact on host nesting success in comparison to nonparasitized species, (4) host resource partitioning among the previously nonsympatric cowbird species. In this chapter, we document the spread of cowbirds into the region, discuss the implications of this colonization for passerines breeding in south Florida, and suggest management strategies.

Study Areas and Methods

South Florida is taken to be the peninsula south of the Caloosahatchee (Okeechobee Waterway) and St. Lucie outlets of Lake Okeechobee and comprises the following counties: Palm Beach, Broward, Dade, Monroe, Collier, Lee, Hendry, southern Charlotte, and southern Glades (shaded area in Figures 4.2 and 4.3). Our work is being conducted in mangroves, freshwater marshes, upland hammocks, and human-modified areas. Documentation of the distribution of cowbirds is based on extensive review of the literature, communications with individuals knowledgeable of the avifauna, and field sampling. At each visit to the study areas (usually at 2–4 day intervals), we inspected potential host nests to determine the number of hosts and, if present, parasite eggs and chicks.

Cowbirds in the South Florida Region

Shiny Cowbird

In 1985, the first Shiny Cowbird was discovered in the middle Florida Keys (Smith and Sprunt 1987) (Figure 4.2). During 1988, a minimum of 35 individuals were seen in eight Florida localities. Two sightings of males were outside of south Florida: at Fort Desoto on the west coast (Langridge 1988) and at Jacksonville on the east coast (Paul 1988).

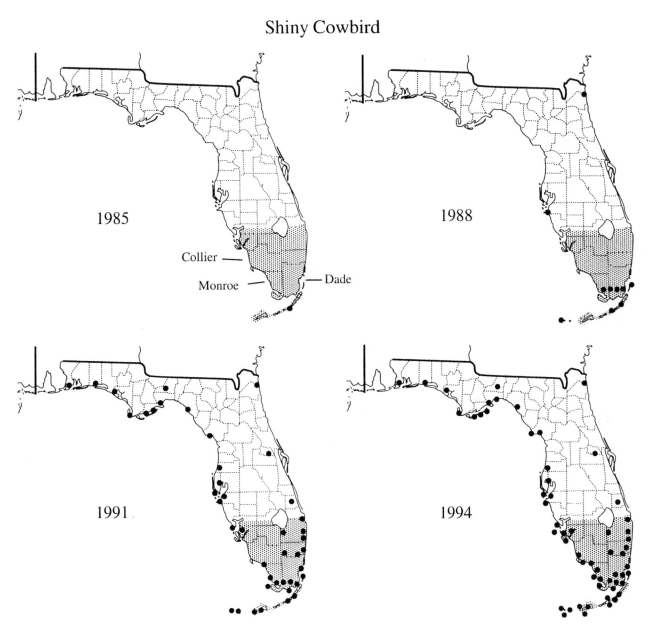

Figure 4.2. The spread of the Shiny Cowbird into the Florida region, 1985 to 1994. Only spring and summer records are shown. The shaded area represents the south Florida region with potential breeding propagules in Dade, Collier, and Monroe counties. Data from sources listed for Figure 4.1.

Brown-headed Cowbird

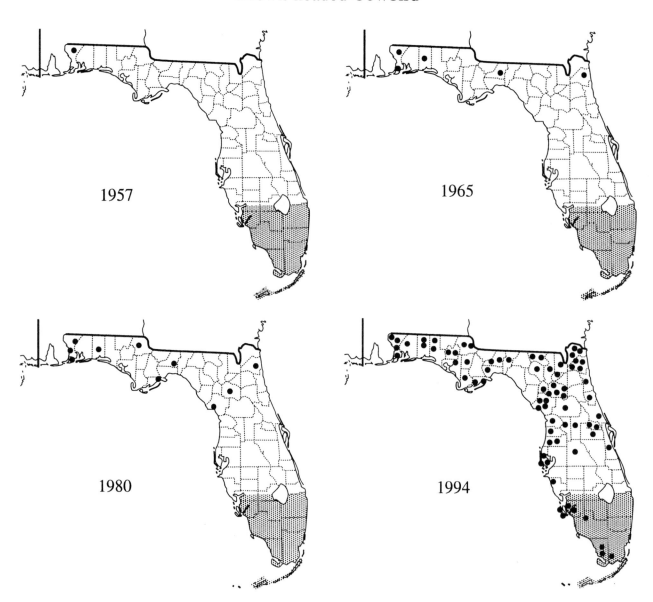

Figure 4.3. Brown-headed Cowbird breeding records for the Florida region, 1957 to 1994. The shaded area represents the south Florida region. Confirmed breeding records are based on the criteria used by the Florida Breeding Bird Atlas Project and include, e.g., female with egg in the oviduct, host nest with cowbird egg or nestling, recently fledged young, and host feeding cowbird young. Data from sources listed for Figure 4.1.

Since then, cowbirds have also been reported in other Florida localities (Figure 4.2), and as far north as the Carolinas and Maine and as far west as Texas and Oklahoma (Figure 4.1) (e.g., Grzybowski and Fazio 1991, Post et al. 1993). In addition, new distributional records continue to accumulate for the West Indies (Figure 4.1). In 1993, Fletcher (1993) observed 4 Shiny Cowbirds (1 male and 3 females) in Yallahs Salt Pond, southeastern Jamaica, and P. Marra (pers. comm.) observed a pair of Shiny Cowbirds in southwestern Jamaica. In the Bahamas, Baltz (1995) observed a group of 2 males and 4 females on Andros during July 1994.

Specific localities visited by us further confirm the range extension of Shiny Cowbirds into south Florida, where cowbirds are now observed year-round (Figure 4.2). We found cowbirds in residential areas, agricultural areas, and modified habitats in the Keys and on the mainland. In Everglades

National Park, cowbirds were present in several localities, but they were most commonly observed on lawns and mangroves around the visitors' center at Flamingo, where groups of up to 13 birds were observed (8 males and 5 females). As in the West Indian region, where Shiny Cowbirds are more common in coastal areas (Cruz et al. 1985), the majority of Florida records have been from coastal sites (Figure 4.2).

Although cowbirds have been observed year-round in south Florida, numbers are higher in spring and early summer, suggesting a migratory population (Post et al. 1993). During 1989–1991, for example, 148 male Shiny Cowbirds were recorded from March to June, and 102 male Shiny Cowbirds from July to February. Because of the similarity between female Shiny Cowbirds and female Brown-headed Cowbirds, we include only reports of males during these periods.

Although we have not found nests parasitized by Shiny Cowbirds in south Florida, we characterize the south Florida populations as potentially colonizing propagules outside the contiguous range in the West Indies. In February 1992, for example, 4 males and 3 females were collected from a flock of at least 10 birds by National Audubon personnel in Collier County, southwestern Florida, suggesting that potentially breeding birds are present in late winter–early spring. Six males collected in southeastern Florida between 30 April and 25 July 1991 had enlarged testes (Post et al. 1993). In 1992, a flock of 5 males and 5 females was present in the Flamingo area of Everglades National Park from mid-May to late June. While we did not observe any evidence of parasitism, we have observed males displaying to females on the mainland and on the Keys. In addition, juvenile cowbirds have been observed in the park and in nearby Homestead (Smith and Sprunt 1987, pers. obs.). Outside the south Florida region, Shiny Cowbird records in North America can be considered accidental, especially the extralimital records far from the main West Indian breeding range.

Brown-headed Cowbird

Before 1954, Brown-headed Cowbirds were not known to breed in Florida (Sprunt 1954), but since the mid-1950s they have spread rapidly through Florida (Hoffman and Woolfenden 1986) and now breed there (Figure 4.3). Breeding in Florida first occurred in 1957 in the northwest, near Pensacola (Weston 1965). By 1965, cowbirds were breeding in the Tallahassee and Jacksonville regions (Ogden 1966), and by 1980, fledglings were reported in north Florida from Pensacola to Jacksonville and south to Gainesville (Edscorn 1980). In 1985, a fledgling cowbird was found in Pinellas County, halfway down the peninsula on the Gulf coast (Hoffman and Woolfenden 1986).

Brown-headed Cowbirds are now observed year-round in south Florida (Figure 4.3). The numbers are augmented from mid-summer on by the arrival of postbreeding flocks from the north. In July–August 1988, at least 1,000 cowbirds were observed in Dade County (Paul 1988). In August 1991,

we observed a flock of approximately 170 Brown-headed Cowbirds in Fort Pierce, just north of Palm Beach County, and in 1992–1994, flocks of 50–75 cowbirds were observed near Homestead, Dade County. During mid-April to late June, we observed a flock of cowbirds in Flamingo, Everglades National Park, the numbers varying from 6 males and 14 females in 1991 to 6 females and 10 males in 1992. On several occasions, male cowbirds were observed singing accompanied by display (song spread), involving the ruffling of feathers and spreading of wings. Although copulations were observed at Flamingo, no parasitized nests were located. In 1993 and 1994, we located Red-winged Blackbird nests that were parasitized by Brown-headed Cowbirds on Sanibel Island (see below).

In contrast to the Shiny Cowbird, the spread of the Brown-headed Cowbird into Florida may be characterized as a broad-front advance in range, especially in the north and central Florida regions. In the south Florida region, there are colonizing propagules ahead of the spreading contiguous range.

Impacts of Brood Parasitism on Host Species

Although parasitism reduces host breeding success (Friedmann et al. 1977, Cruz et al. 1989, Robinson et al. 1993), in most instances hosts have evolved with cowbirds over centuries and thus can survive with them (e.g., Rothstein 1975). Brood parasitism has also constituted a selective pressure favoring the evolution of antiparasite defenses by some hosts (Payne 1977, Cruz and Wiley 1989, Rothstein 1990). In areas where cowbirds have recently become established, there has not been sufficient time for host species to evolve defenses against cowbirds (Cruz et al. 1989). In the West Indian region, Shiny Cowbird parasitism is considered the most important factor in the reduced reproductive output of the endangered Yellow-shouldered Blackbird (*Agelaius xanthomus*), a Puerto Rican endemic (Post 1981, Cruz et al. 1985, Wiley et al. 1991).

Brown-headed Cowbirds have been implicated in the decline of several host species (see chapters in Part V, this volume), including Kirtland's Warbler, *Dendroica kirtlandii* (Walkinshaw 1983), Black-capped Vireo, *V. atricapillus* (Grzybowski et al. 1986), and Least Bell's Vireo, *V. bellii pusillus* (Franzreb 1989). We thus expect potential host species in south Florida to be negatively affected by cowbirds, if cowbirds of one or both species eventually establish breeding populations in the area. In particular, those host species that have small population sizes and restricted ranges are likely to be vulnerable.

Potential Hosts in the South Florida Region

Information on cowbird-host interactions in the West Indies and North America (including Florida) allows us to estimate the potential host species among south Florida passerines (Table 4.1). Some passerines, especially in the Vireonidae, Parulinae, and Emberizidae, are heavily parasitized outside Florida and thus will likely be preferred hosts for both species of cowbirds.

Differences in nesting cycles and habitat preferences between Shiny and Brown-headed Cowbirds, however, may minimize overlap in host use. The breeding season of the Shiny Cowbird in the West Indian region extends from mid-March to September (Cruz et al. 1985, 1989; Wiley 1985), whereas the Brown-headed Cowbird breeds from late April to late July in captivity (Holford and Roby 1993). Thus, late- and early-nesting species in Florida may not escape parasitism by Shiny Cowbirds. In Florida, as in the West Indian re-

gion (Cruz et al. 1989), the Shiny Cowbird has been recorded from coastal habitats (Figure 4.2). In contrast, Brown-headed Cowbirds have been found in both coastal and interior regions (Figure 4.3).

In our studies, we have focused on three subspecies that we believe should be monitored, the Florida Prairie Warbler (*Dendroica discolor paludicola*), the Cuban Yellow Warbler (*Dendroica petechia gundlachi*), and the Red-winged Blackbird (*Agelaius phoeniceus*) (Nolan 1978, Cruz et al. 1989, Ortega and Cruz 1991).

The Prairie Warbler breeds in the Keys and northward along the coast to central Florida (Figure 4.4). The species is primarily restricted to mangrove habitats and is of special concern for this reason (Kale 1978). In North America,

Table 4.1. Abundance, Distribution, and Suitability of Potential Cowbird Hosts in the South Florida Region

Species		Abundance[a]	Range[b]	Suitability[c]
Great Crested Flycatcher	*Myiarchus crinitus*	FC	W	L
Carolina Wren	*Thryothorus ludovicianus*[d]	UC	W	L
Blue-gray Gnatcatcher	*Polioptila caerulea*[e]	UC	P	M
White-eyed Vireo	*Vireo griseus*[d,e]	C	W	H
Black-whiskered Vireo	*V. altiloquus*[e]	R	R	H
Red-eyed Vireo	*V. olivaceus*[e]	UC	P	H
Northern Parula Warbler	*Parula americana*[e]	FC	P	M
Yellow Warbler	*Dendroica petechia*	R	R	H
Pine Warbler	*D. pinus*[d,e]	UC	P	L
Prairie Warbler	*D. discolor*[d,e]	FC	R	H
Prothonotary Warbler	*Protonotaria citrea*	R	P	M
Common Yellowthroat	*Geothlypis trichas*[e]	C	W	H
Summer Tanager	*Piranga rubra*[e]	R	R	H
Northern Cardinal	*Cardinalis cardinalis*[d,e]	C	W	H
Rufous-sided Towhee	*Pipilo erythrophthalmus*[d,e]	C	W	H
Bachman's Sparrow	*Aimophila aestivalis*[d]	UC	P	L
Grasshopper Sparrow	*Ammodramus savannarum*[d]	R	R	H
Cape Sable Seaside Sparrow	*A. maritimus*	UC	R	NR
Red-winged Blackbird	*Agelaius phoeniceus*[d,e]	VC	W	H
Eastern Meadowlark	*Sturnella magna*	FC	W	H
Common Grackle	*Quiscalus quiscula*	VC	W	L
Spot-breasted Oriole	*Icterus pectoralis*	R	R	M

Note: Host suitability is based on frequency of parasitism elsewhere, including Florida. Data based on field survey, Florida Breeding Bird Atlas Project, and review of the literature (e.g., Sprunt 1954, Robertson 1955, Friedmann et al. 1977, Cruz et al. 1985 and 1989, Friedmann and Kiff 1985, Wiley 1985, Kale and Maehr 1990, Robertson and Woolfenden 1992).

[a]VC = very common, C = common, FC = fairly common, UC = uncommon, R = rare.

[b]W = widespread, P = patchy, R = restricted to small areas.

[c]L = low, M = moderate, H = high, NR = not recorded as host.

[d]Species that have races restricted to Florida.

[e]Species recorded as cowbird hosts in Florida.

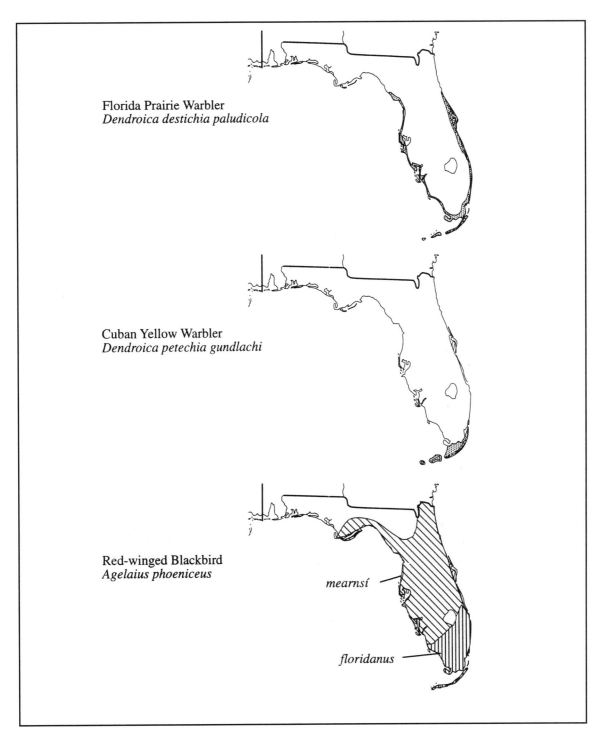

Florida Prairie Warbler
Dendroica destichia paludicola

Cuban Yellow Warbler
Dendroica petechia gundlachi

Red-winged Blackbird
Agelaius phoeniceus

mearnsi

floridanus

Figure 4.4. Breeding distributions of the Florida Prairie Warbler, Cuban Yellow Warbler, Florida Red-winged Blackbird, and Maynard's Red-winged Blackbird. Data based on field observations, personal communications, Sprunt (1954), Kale (1978), Kale and Maehr (1990), and Robertson and Woolfenden (1992).

mangroves are primarily restricted to south Florida, and this habitat is under constant pressure as a result of the increased human population in the region. Thousands of hectares have been destroyed and replaced with filled and developed land (Myers and Ewel 1990). Where mangroves have been decimated, as along the Atlantic coast from Palm Beach to Miami, the Prairie Warbler rarely breeds (Kale 1978). In a study of the nominate race of the Prairie Warbler in Indiana, Nolan (1978) estimated that cowbird parasitism caused about 18% of the nest failures of this species. His study was conducted in what has been generally considered the original habitat of the cowbird, where cowbirds and Prairie Warblers have coexisted for many years. However, this is not the case in Florida. In 1988, parasitism of Prairie Warblers was confirmed when breeding pairs in Pinellas and Sarasota counties, southwest Florida, were observed feeding fledgling cowbirds (Atherton and Atherton 1988). In Everglades National Park, a cowbird fledgling was seen being fed by a Prairie Warbler (Kale 1989). In 1990 and again in 1997, a Prairie Warbler was observed feeding a cowbird young on Captiva Island, south Florida (McGrath pers. comm.).

Despite the wide distribution of Yellow Warblers in North America (Bent 1953), the only breeding population in Florida is the Cuban Yellow Warbler (Prather and Cruz 1995). This population is of great interest because of its recent successful colonization of North America from the West Indies (Robertson 1978). The Yellow Warbler was first found in the lower Florida Keys in 1941 (Greene 1942). This subspecies has spread rapidly through the Keys, and it is now a permanent resident from Key West to Virginia Key off Miami (Figure 4.4). Its present distribution is patchy; the birds occur more commonly in the low outlying keys than on larger islands or along the mainland shore (pers. obs.). Robertson (1978) listed the Cuban Yellow Warbler as rare because its limited range in Florida makes it vulnerable to natural catastrophes such as severe hurricanes. The destruction of mangrove has also reduced potential habitat for Yellow Warblers.

Cowbirds pose an additional threat to the Yellow Warbler in south Florida, where we expect that it will be used as a host by cowbirds. Brood parasitism by cowbirds may be the most important threat to these populations. In the West Indies, this species is an important Shiny Cowbird host, with frequencies of parasitism of 55%, 63%, and 39% for St. Lucia, Puerto Rico, and Hispaniola, respectively (Cruz et al. 1989). Bond considered the Yellow Warbler to be a common bird on Barbados but noted that it had become very rare by 1950 (Bond 1930, 1950). By 1984, Bond wrote that it had declined to near extinction (Bond 1984). Bond (1950, 1984) blamed the Shiny Cowbird for the decline. Other Yellow Warbler races are also heavily parasitized by Brownheaded Cowbirds in North America (Friedmann et al. 1977, Clark and Robertson 1981, Sealy 1992). In Illinois, Eifrig (1937) blamed the decline of the Yellow Warbler population on cowbird parasitism, and the incidence of parasitism of Yellow Warbler nests increased from about 6% before 1900 to 40.5% after 1970 (Graber et al. 1983).

While there are a few records of parasitism of Prairie Warblers by cowbirds for south Florida, we found no evidence of parasitism in 38 Prairie Warbler and 19 Yellow Warbler nests that we examined in Florida Bay and on the Florida Keys. We feel that it is only a matter of time before parasitism will be observed, as both Brown-headed and Shiny Cowbirds have been recorded during the breeding season and, in some instances, in the nesting habitats of these species.

Two subspecies of Red-winged Blackbird, Maynard's Redwing (*A. phoeniceus floridanus*) and the Florida Redwing (*A. p. mearnsi*), occur in south Florida (Figure 4.4). Both subspecies breed primarily in freshwater marshes and mangroves. Yellow-shouldered Blackbirds in Puerto Rico also use the latter habitat. Parasitism by Shiny Cowbirds is considered to be the main cause of endangerment of the Yellowshouldered Blackbird (Wiley et al. 1991).

Of particular significance was the discovery in 1993 of a parasitized Red-winged Blackbird nest on Sanibel Island, southwest Florida. The nest contained 1 cowbird egg and 2 blackbird eggs. In 1994, we found a Red-winged Blackbird nest containing one cowbird young and two host nestlings on Sanibel. Both nests were subsequently predated. Since both cowbird species have been reported in the area and their eggs are very similar, we are not sure of the identity. However, Brown-headed Cowbirds of both sexes (at least 8 males and 5 females) were observed more frequently in the study area, and only a pair of Shiny Cowbirds was observed in the summer. In 1991, a Red-winged Blackbird near Florida City, south Florida, was recorded feeding an immature cowbird, identified as a Shiny Cowbird by its vocalization (Kale and Pranty pers. comm.).

Management Strategies and Conservation Implications

Monitoring the spread of cowbirds in Florida and undertaking studies on the host use by cowbirds will enhance our understanding of the ecology of brood parasitism. In 1994, we began monitoring other potential host species such as the Black-whiskered Vireo (*Vireo altiloquus*), White-eyed Vireo (*Vireo griseus*), Northern Cardinal (*Cardinalis cardinalis*), Rufous-sided Towhee (*Pipilo erythrophthalmus*), and Common Yellowthroat (*Geothlypis trichas*) (Friedmann and Kiff 1985, Cruz et al. 1989).

In a few species (e.g., Kirtland's Warbler, Black-capped Vireo, Yellow-shouldered Blackbird) with small populations that are already threatened or endangered and are severely affected by cowbird parasitism, cowbird populations are being managed intensively (Robinson et al. 1993, Wiley et al. 1991, chapters in Part V of this volume). In south Florida, any management plan (e.g., cowbird trapping) needs to be preceded by additional information on cowbird abundance, distribution, and levels of nest parasitism to determine if cowbird management is necessary. At this point, levels of parasitism are very low.

In our studies of cowbird-host interactions in the West Indies, we have tested several management techniques to reduce the negative effects of brood parasitism (Wiley et al. 1991). The most effective management tool is direct cowbird control. However, it is vital that management attempts are begun as soon as the level of impact is ascertained. Efforts begun when host populations are in serious decline will be much more expensive and less likely to succeed. An effective method is trapping cowbirds. This method, in which live cowbirds in the traps serve as decoys, has halted the decline of the Kirtland's Warbler (Kelley and DeCapita 1982; De-Capita, Chapter 37, this volume; but see introduction to Part V) and has proved successful in managing Shiny Cowbirds in Yellow-shouldered Blackbird nesting areas in Puerto Rico. Cowbird removal has resulted in fewer parasitized blackbird nests (93% before removal vs. 8% with removal) and higher fledging rates (0.3 vs. 2.3 blackbirds/nest) (Wiley et al. 1991, Febles pers. comm.).

While trapping reduces the effects of cowbird parasitism on local populations (e.g., Wiley et al. 1991), it will be difficult to eliminate cowbirds from entire regions such as south Florida, because of the large area involved and continuing colonization by cowbirds from areas outside the region. Nevertheless, by intensively trapping cowbirds in nesting areas of targeted species, we believe that cowbird removal programs have excellent potential to reduce the adverse impact of cowbird parasitism. During the breeding season, cowbirds are known to leave the hosts' nesting area in the late morning or early afternoon to travel to feeding areas (Rothstein et al. 1980). Cowbird removal should be undertaken in the morning, before cowbirds move out to the feeding areas (Sealy pers. comm.). Cowbird control through trapping can be accomplished with little risk to host species, but daily monitoring is necessary to prevent deaths of non-target species in traps.

Another method of control is shooting cowbirds near the breeding grounds of host species. Both male and female cowbirds can be attracted within shotgun range during the breeding season by tape recordings (pers. obs.). Luring cowbirds with recorded calls uses behavior related to breeding, so adult birds are probably targeted. Shooting has been used in conjunction with trapping on Fort Hood, Texas (Hayden et al., Chapter 39, this volume), but the effort did not isolate specific effects of shooting from those of trapping. Cowbird shooting may be more cost-effective in areas with small or scattered groupings of sensitive species (Robinson et al. 1993).

Concluding Remarks

Not only will our ongoing studies on potential hosts in south Florida provide data on the ecology and breeding biology of these species, but the results will provide a baseline to compare to future effects of cowbird parasitism on these populations. It is of the utmost importance to continue to monitor the distribution, status, age, and sex ratio for Shiny and Brown-headed Cowbirds and to gather evidence of breeding. These studies may also provide information for managing cowbirds and host populations.

While this chapter has focused on south Florida, Shiny Cowbirds have been reported in other areas in North America, in particular, the southeast (Figure 4.1). We believe that Shiny Cowbirds will continue to expand their range, especially along the Atlantic and Gulf coasts, possibly developing an annual distribution pattern similar to that of the Cattle Egret (*Bubulcus ibis*). This species has also responded to clearing of forests for animal husbandry and has rapidly colonized North America from the West Indies within the last century (Arendt 1988). We expect to see a northern dispersing population of Shiny Cowbirds that returns to south Florida, and possibly the West Indies, during the winter (Post et al. 1993). The fat reserves of cowbirds collected in Florida in late summer, for example, were sufficient for nonstop flights greater than 500 km, and it is possible that some of the population currently returns to the Greater Antilles in the winter (Post et al. 1993).

The northern boundary of Shiny Cowbird wintering distribution may be determined by metabolic constraints. Root (1988) found that 60% of the bird species wintering in North America have northern range boundaries that coincide with isotherms of minimal daily January temperature. The passerines within that set of species were calculated to have metabolism rates at their northern limits that average 2.5 times the basal metabolic rate. Repasky (1991), however, in a reanalysis of Root's data, failed to find support for Root's prediction. Repasky suggested that temperature probably plays a pervasive, although less acute, role in determining distributions through interactions with biotic factors such as food, competition, predation, and habitat.

In contrast to the gradual expansion of Brown-headed Cowbirds in North America, extending eventually over an enormous area (Mayfield 1977), the expansion of this species into Florida has been more sudden and dramatic. We predict that Brown-headed Cowbirds will continue expanding their range in south Florida, including the Keys. The Keys will likely serve as a stepping stone to the colonization of Cuba. In North America, Brown-headed Cowbirds breed as far south as Veracruz, Mexico (AOU 1983), suggesting that they are adapted to the tropical conditions of the West Indian region. With the continued expansion of these two cowbirds, resource managers should be aware of the potential problems that they may cause for vulnerable species or for populations of special concern.

Acknowledgments

We thank the National Audubon Society for logistical help, in particular A. Sprunt IV, G. T. Bancroft, W. Hoffman, R. J. Sawicki, and A. M. Strong. H. W. Kale II and W. Pranty provided Florida cowbird records. We are grateful to the National Park Service and the United States Fish and Wildlife Service for allowing us to work within lands under their jurisdiction. S. G. Sealy and J. N. M. Smith provided helpful editorial comments.

References Cited

American Ornithologists' Union. 1983. Check-list of North American birds, 6th edition. American Ornithologists' Union, Washington, DC.

Arendt, W. J. 1988. Range expansion of the Cattle Egret (*Bubulcus ibis*) in the Greater Caribbean basin. Colonial Waterbirds 11:252–262.

Atherton, L. S., and B. H. Atherton. 1988. Florida region. Amer. Birds. 42:60–63.

Baltz, M. E. 1995. First records of Shiny Cowbird (*Molothrus bonariensis*) in the Bahama Archipelago. Auk 112:1039–1041.

Bent, A. C. 1953. Life histories of North American wood warblers. U.S. Natl. Mus. Bull. 203.

Bond, J. 1930. The resident West Indian warblers of the genus *Dendroica*. Proc. Acad. Natural Sciences of Philadelphia. 82:329–337.

———. 1950. Check-list of birds of the West Indies, 3rd ed. Wickersham Printing Co., Lancaster, PA.

———. 1984. Twenty-fifth supplement to the Check-list of the Birds of the West Indies (1956). Acad. Natural Sciences of Philadelphia. 16 pp.

Clark, K. L., and R. J. Robertson. 1981. Cowbird parasitism and evolution of anti-parasite strategies in the Yellow Warbler. Wilson Bull. 93:249–258.

Cruz, A., T. Manolis, and J. W. Wiley. 1985. The Shiny Cowbird: A brood parasite expanding its range in the Caribbean region. Ornithol. Monogr. 36:607–619.

Cruz, A., and J. W. Wiley. 1989. The decline of an adaptation in the absence of a presumed selection pressure. Evolution 43:55–62.

Cruz, A., J. W. Wiley, T. K. Nakamura, and W. Post. 1989. The Shiny Cowbird in the Caribbean region: Biogeographical and ecological implications. Pp. 519–540 *in* Biogeography of the West Indies: Past, present, and future. Sandhill Crane Press, Gainesville, FL.

Edscorn, J. B. 1980. The nesting season: Florida region. Amer. Birds 34:887–889.

Eifrig, C. W. G. 1937. The changing status of birds as regards their abundance. Trans. Illinois State Acad. Sci. 30:295–297.

Fletcher, J. 1993. Is the Shiny Cowbird in Jamaica? Gosse Bird Club Broadsheet 61:5–7.

Franzreb, K. E. 1989. Ecology and conservation of the endangered Least Bell's Vireo. USFWS Biol. Rep. 89. 17 pp.

Friedmann, H. 1929. The cowbirds: A study in the biology of social parasitism. C. C. Thomas, Springfield, IL.

Friedmann, H., and L. F. Kiff. 1985. The parasitic cowbirds and their hosts. Proc. Western Found. Vert. Zool. 2:225–302.

Friedmann, H., L. F. Kiff, and S. I. Rothstein. 1977. A further contribution to knowledge of the host relations of the parasitic cowbirds. Smithsonian Contr. Zool. 235:1–75.

Graber, J. W., R. R. Graber, and E. L. Kirk. 1983. Illinois birds: Wood warblers. Biological Notes, Illinois Natural History Survey 188:2–144.

Greene, E. R. 1942. Golden warbler nesting in Lower Florida Keys. Auk 59:114.

Grzybowski, J. A., R. B. Clapp, and J. T. Marshall, Jr. 1986. History and current status of the Black-capped Vireo in Oklahoma. Amer. Birds 40:151–161.

Grzybowski, J. A., and V. W. Fazio. 1991. Shiny Cowbird reaches Oklahoma. Amer. Birds 45:50–52.

Hoffman, W., and G. E. Woolfenden. 1986. A fledgling Brown-headed Cowbird from Pinellas County. Florida Field Natur. 14:18–20.

Holford, K. C., and D. D. Roby. 1993. Factors limiting fecundity of Brown-headed Cowbirds. Condor 95:536–545.

Kale, H. W. II. 1978. Rare and Endangered Plants and Animals, Vol. 2. University of Florida Press, Gainesville, FL.

———. 1989. Florida birds. Florida Natur. 62:14.

Kale, H. W. II, and D. S. Maehr. 1990. Florida's birds: A handbook reference. Pineapple Press, Sarasota, FL.

Kelley, S. T., and M. E. DeCapita. 1982. Cowbird control and its effect on Kirtland's Warbler reproductive success. Wilson Bull. 94:363–365.

Langridge, H. P. 1988. The Florida region. Amer. Birds 42:424–426.

Mayfield, H. 1977. Brown-headed Cowbirds: Agent of extermination? Amer. Birds 31:107–113.

Myers, R. L., and J. J. Ewel. 1990. Ecosystems of Florida. University of Central Florida Press, Orlando, FL.

Newman, G. 1957. The Florida region. Audubon Field Notes 11:412–413.

Nolan, V., Jr. 1978. Ecology and behavior of the Prairie Warbler, *Dendroica discolor*. Ornithol. Monogr. 26:1–595.

Ogden, J. C. 1966. Florida region. Audubon Field Notes 19:534–537.

Ortega, C. P., and A. Cruz. 1991. A comparative study of cowbird parasitism in Yellow-headed Blackbirds and Red-winged Blackbirds. Auk 108:16–24.

Paul, R. T. 1988. Florida region. Amer. Birds 42:1278–1281.

Payne, R. B. 1977. The ecology of brood parasitism in birds. Ann. Rev. Ecol. Syst. 8:1–28.

Post, W. 1981. Biology of the Yellow-shouldered Blackbird: *Agelaius* on a tropical island. Bull. Florida State Mus., Biol. Sci. 25:125–202.

Post, W., A. Cruz, and D. B. McNair. 1993. The North American invasion pattern of the Shiny Cowbird. J. Field Ornithol. 64:32–41.

Post, W., and J. W. Wiley. 1977. The Shiny Cowbird in the West Indies. Condor 79:119–121.

Potter, E. F., and G. T. Whitehurst. 1981. Cowbirds in the Carolinas. Chat 45:57–68.

Prather, J. W., and A. Cruz. 1995. Breeding biology of Florida Prairie Warblers and Cuban Yellow Warblers. Wilson Bull. 107:475–484.

Repasky, R. R. 1991. Temperature and the northern distributions of wintering birds. Ecology 72:2274–2285.

Robertson, W. B., Jr. 1955. An analysis of the breeding bird populations of tropical Florida in relation to the vegetation. Ph.D. Thesis, University of Illinois, Urbana.

———. 1978. Cuban Yellow Warbler. Pp. 61–62 *in* Rare and endangered biota of Florida, Vol. 2, Birds (H. W. Kale II, ed). University Presses of Florida. 121 pp.

Robertson, W. B., Jr., and J. A. Kushlan. 1984. The southern Florida avifauna. Pp. 219–257 *in* Environments of south Florida: Present and past (P. J. Gleason, ed). Miami Geological Society, Coral Gables, FL.

Robertson, W. B., Jr., and G. E. Woolfenden. 1992. Florida bird species: An annotated list. Florida Ornithological Society Special Publication 6.

Robinson, S. K., J. A. Grzybowski, S. I. Rothstein, M. C. Brittingham, L. J. Petit, and F. R. Thompson. 1993. Management implications of cowbird parasitism on neotropical migrant songbirds. Pp. 93–102 in Status and management of neotropical migratory birds (D. M. Finch and P. W. Stangel, eds). USDA Gen. Tech. Rep. RM-229.

Root, T. 1988. Energy constraints on avian distributions and abundances. Ecology 69:330–339.

Rothstein, S. I. 1975. An experimental and teleonomic investigation of avian brood parasitism. Condor 77:250–271.

———. 1990. A model system for coevolution: Avian brood parasitism. Ann. Rev. Ecol. Syst. 21:481–508.

Rothstein, S. I., J. Verner, and E. Stevens. 1980. Range expansion and diurnal changes in dispersion of the Brown-headed Cowbird in the Sierra Nevada. Auk 97:253–267.

Sealy, S. G. 1992. Removal of Yellow Warbler eggs in association with cowbird parasitism. Condor 94:40–54.

Smith, P. W., and A. Sprunt IV. 1987. The Shiny Cowbird reaches the United States. Amer. Birds 41:370–371.

Sprunt, A., Jr. 1954. Florida bird life. Coward-McCaan, New York.

Walkinshaw, L. H. 1983. Kirtland's Warbler: The natural history of an endangered species. Cranbrook Inst. Sci., Bloomfield Hills, MI.

Webb, J. S., and D. K. Wetherbee. 1960. Southeastern breeding range of the Brown-headed Cowbird. Wilson Bull. 31:83–87.

Weston, F. M. 1965. A survey of the birdlife of northwestern Florida. Bull. Tall Timbers Research Station 5:1–147.

Wiley, J. W. 1985. Shiny Cowbird parasitism in two avian communities in Puerto Rico. Condor 87:165–176.

Wiley, J. W., W. Post, and A. Cruz. 1991. Conservation of the Yellow-shouldered Blackbird: An endangered West Indian species. Biol. Conserv. 55:119–138.

5. Brown-headed Cowbird Population Trends at a Large Winter Roost in Southwest Louisiana, 1974–1992

BRENT ORTEGO

Abstract

Brown-headed Cowbirds were monitored in a large winter roost of blackbirds at Millers Lake in Evangeline Parish, Louisiana, from 1974 to 1993. Estimated cowbird numbers ranged from 28,025 in 1980 to 38,000,000 in 1986, with an annual average of 8,743,169. Cowbird numbers at this roost dropped during the winters of 1990–1993 to about 1% of the peak numbers during the winters of 1982-1987. Part, but not all, of this decline may have been due to poisoning of cowbirds and Red-winged Blackbirds (*Agelaius phoeniceus*) during the springs of 1989 and 1990. Large winter roosts, like the one at Millers Lake, provide an opportunity to control large numbers of cowbirds and other blackbirds that impact songbirds and agricultural crops over large geographic areas. Control measures at large winter roosts should be more cost-effective than controls on the breeding grounds, when cowbirds are more widespread in distribution and thus less vulnerable.

Introduction

Brown-headed Cowbirds and Red-winged Blackbirds breed throughout the United States and southern Canada, and northern populations typically migrate south to the Gulf coastal region after breeding (Dolbeer 1982). These northern populations comprise as much as 84% of all blackbirds found in the southern rice states during winter (Meanley et al. 1966). Multispecies blackbird populations, in which the Brown-headed Cowbird is one of seven common species, have been reported to contain 398 million birds (Meanley 1976).

One of 63 major blackbird roost sites along the Louisiana-Texas Gulf Coast in this region is at Millers Lake, Louisiana (Meanley 1971). The impacts that cowbirds using this roost have on songbird populations is unknown, but I suspect it is significant. Economic impacts of all blackbirds at this site on local agricultural crops have been estimated at several million dollars (Wilson et al. 1989). Monitoring of the blackbirds roosting at this site was started in 1974 as part of a broader study of the avifauna of Evangeline Parish. Surveys conducted during Christmas Bird Counts were used to track this large concentration of blackbirds for cooperating landowners and to compare the magnitude of this concentration to other blackbird roosts being monitored elsewhere in the USA.

Methods
Study Area

Millers Lake is a 2,500-ha impoundment surrounding a natural depression used primarily for irrigation of rice (*Oryza sativa*) and for outdoor recreation, mostly hunting and fishing (Ortego 1976). The lake is located in the northeast corner of the rice belt in Louisiana. Bottomland hardwood forests of the Red River are to the east, lower coastal plain pine woods are to the north and west, and to the south and southwest are the former coastal prairies that make up the Louisiana rice belt. The lake is nearly square, and two thirds of it is covered with shrubs and tree swamps. The northern half of the lake borders bottomland and upland forests intermixed with improved pastures. The southern half of the lake borders agricultural fields primarily used for rice production.

Survey Methods

Blackbird surveys in 1974 were part of a special study of the birds at Millers Lake (Ortego 1976). From 1976 to 1993, participants of the Pine Prairie Christmas Bird Count (CBC) counted blackbirds departing the roost at Millers Lake and other areas within a 12-km radius circle centered 2 km north of Millers Lake. Data from individual tally sheets are used here because of minor discrepancies between results published in *American Birds* CBC issues 30–48 and actual field data. Counts from Millers Lake and from other sites are treated separately to distinguish estimates from a consistent sampling program from estimates with more variability (Table 5.1).

Counts at Millers Lake

Major departure flights of blackbirds from the roost in 1974 were counted once during mid-December along an east-west highway 0.8 km south of and parallel to the lake during the early morning (Ortego 1976). Whenever the morning flight consisted of scattered flocks of blackbirds, the species composition and number of individuals of each species within each flock were estimated. When the flight became a continuous sheet of birds (a flock about 50 m deep and with individuals varying from a few centimeters to 3 m apart), the average flock width, speed, and duration of passing were used to estimate the size and density of the flock.

This approach was changed in 1976 because large portions of the flock did not cross the highway and the accuracy of density estimates (birds per m²) was uncertain. From 1976 to 1993, one primary observer was stationed at the boat ramp on the southwest corner of Millers Lake. This observer had a view of most of the morning flight traveling south, southwest, and west. In 1977, an additional primary observer was added along the highway at a point south of the southeast corner of the lake. This second observer counted blackbirds flying southeast and east (Figure 5.1). Additional observers were added near the lake from 1989 to 1993 to count birds on minor flight lines traveling west, northeast, and northwest. These observers added only a few hundred thousand blackbirds to the total each year.

Conspicuous lines of trees were used as reference marks to partition the flight from Millers Lake and to minimize double counting. Morning flights were used instead of evening flights because they were shorter, visibility was better, and because the entire roost departed during the observation period. Unknown portions of birds returned to the roost after dark during winter.

Table 5.1. Brown-headed Cowbird and Blackbird Roost Population Estimates at Millers Lake and the Pine Prairie Christmas Bird Count (CBC) at Evangeline Parish, Louisiana, from 1974 to 1993

Date	Brown-headed Cowbirds		Total Blackbirds		% Cowbirds in Roosts at CBC
	Millers Lake	Remainder of CBC	Millers Lake	Remainder of CBC	
12–18–74	650,000	No count	1,000,000	No count	65
01–04–76	5,007,611	12,084	9,230,802	347,139	52
12–19–76	3,818,000	204,125	4,925,475	243,127	78
12–18–77	322,000	41,700	430,433	74,061	72
12–17–78	2,300,000	3,538	2,613,042	13,767	88
12–23–79	16,000,000	610,000	21,191,799	1,128,455	74
12–21–80	28,025	10,420	109,415	22,893	29
12–20–81	5,950,000	192	8,284,114	39,319	71
12–19–82	19,800,000	6,656	37,577,341	22,788	53
12–18–83	30,000,000	11,080	47,302,276	45,675	63
12–22–84	13,600,000	6,005,000	37,100,036	26,036,354	31
12–28–85	10,100,000	13,200	34,974,286	68,272	29
12–27–86	38,000,000	201,300	108,021,079	627,670	35
12–26–87	20,100,000	4,900	111,103,765	1,560,082	18
12–18–88	2,000,000	195	11,803,257	7,763	17
12–26–89	5,000,000	0	80,000,331	1,061,519	6
12–16–90	150,225	280	6,266,299	1,277,132	2
12–15–91	1,330,020	108,338	17,096,530	1,266,540	8
12–20–92	607,001	1,300	7,650,875	37,579	8
12–19–93	200,407	906,883	1,504,441	28,550,715	4
Total	174,863,389	8,141,191	548,185,596	62,430,850	30
Annual mean	8,743,169	428,484	27,409,280	3,305,223	32

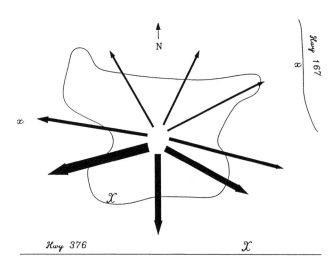

Figure 5.1. Major blackbird flight lines from Millers Lake, Evangeline Parish, Louisiana. *X,* major observer location; *x,* minor observer location. Thickness of arrows corresponds to relative abundance of blackbirds.

Blackbirds leaving the lake were tallied in increments of 1,000, 10,000, and 100,000 by scanning the horizon with binoculars and scanning to the north with spotting scopes. Increments of 1 million were sometimes used in zones of heaviest bird densities (flights to the southwest 2–4 km north of the southwestern boat during 1986, 1987, and 1989).

In normal years, the most abundant blackbirds were tallied as a unit, and ratios of abundance for each species were estimated every 5 minutes during morning flights. Ratios were not used for species representing less than 5% of the flock. These species were counted individually. During return flights in late afternoon, blackbirds flying to the roost frequently landed in agricultural fields adjacent to the lake. Redwing : cowbird ratios were calculated from at least 10 of these flocks at different sites from the morning locations used by primary observers. I averaged these ratios with those for the same two species from the morning flight to increase the number of estimates. This was necessary because it was difficult to differentiate cowbirds from female redwings in flight at a distance. Additional flocks of blackbirds that were not part of the morning departure flight were counted during the day along the lake.

The primary observers covering the southwest and southeast corners of the lake were the same for 18 years. The primary observer covering the southwest corner of Millers Lake was tested repeatedly using standard techniques for training waterfowl surveyors. These included estimating numbers in trays of beans, photographs of flocks, drawn replicas of flocks (Art Brazo pers. comm.), and computer flocks generated by the program Wildlife Counts (1986). On all tests, the observer averaged within 10% of the correct answer. The observer covering the southeast corner was also very experienced.

Counts Outside of Millers Lake

From 10 to 20 observers counted birds at other sites in the Pine Prairie CBC circle. These observers counted blackbirds opportunistically as they occurred within their sector. Most of these blackbirds undoubtedly came from Millers Lake, but because of their reported flight directions, they probably originated from unsurveyed minor flight lines departing the major roost. They are therefore included in the counts reported here.

Results and Discussion

Numbers of Cowbirds at Millers Lake

Early winter population estimates of blackbirds (redwings, cowbirds, and other species) departing Millers Lake averaged 27.4 million and averaged about 8.7 million Brown-headed Cowbirds (Table 5.1). Cowbirds consistently made up more than 5 million birds from 1976 to 1989. Cowbird abundance dropped considerably from 1990, after poisoning programs were employed locally to control numbers of breeding blackbirds (see below).

Observers conducting morning departure flights from Millers Lake gradually got used to estimating the awesome numbers of blackbirds. During a typical flight, flocks containing a few hundred birds started departing the roost at 6:45 a.m. By about 7:00 a.m. departing waves started to form groups of 1,000–100,000 birds. By about 7:10 a.m. observers at major flight lines could not distinguish individual flocks because the departure flight became one continuous sheet of birds that was 4 km wide at the edge of the lake and grew wider as birds traveled away. The uninterrupted flow of blackbirds usually lasted until about 7:50 a.m., when waves reappeared. Waves broke up into small flocks by 8:00 a.m. To visualize the magnitude of the larger departure flights, if the blackbirds did not spread out or land during their departure, the single flock would cover 170 km². Labisky and Brugger (1989) estimated that this concentration of birds foraged over a 5,000–6,000 km² area.

Cowbird concentrations at the Pine Prairie CBC were typically larger than in any other CBC in the nation from 1975 to 1992 (Monroe 1976–1993). There were 1,000–1,600 CBCs conducted annually during the study period. CBCs are distributed through most states and Canadian provinces and are frequently placed where greatest concentrations of birds occur.

Accuracy of Counts

The accuracy of even experienced observers counting large flocks of birds can be low, and the estimates here are therefore open to question. I am confident, however, that these estimates are reliable, because the two primary observers in the study were very experienced at counting large numbers of birds. Furthermore, they were stationed at the same site for 18 of 20 years, and they accounted for about 85% of the birds reported. Trends between years by the two primary observers were very similar. Most other observers participated in the count for at least 5 years.

Fluctuations in Blackbird Numbers

Blackbird numbers roosting at Millers Lake consistently remained high over the 20-year study, and estimates exceeded 10 million during 10 years. However, there was considerable variation among years, even when there were no poisoning programs. Variation in numbers may have had several causes, which I now explore.

In the decade when huge numbers of blackbirds were counted at Pine Prairie, numbers at other major roosts (e.g., at Little Rock) declined dramatically (Monroe 1976–1993). This result suggests that the locations of major cowbird roosts fluctuate on a north-south axis. However, before the surveys described here, blackbirds had been abundant at Pine Prairie for many years (Curry McCauley pers. comm.). The high numbers reported from 1979 to 1989 may therefore not have been due to a permanent shift from abandoned northern roosts.

I now consider if a poison bait program contributed to the variation of blackbird numbers at Millers Lake. During the springs of 1989 and 1990, USDA Animal Damage Control staff applied 112 kg/ha of a 2% DRC-1339 treated brown rice bait at sites under blackbird flight lines near Millers Lake, after most northern migrants had left (Glahn and Wilson 1992). The poisoning program killed between 1.3 and 2.7 million birds annually, mainly Red-winged Blackbirds and Brown-headed Cowbirds. This effort reduced breeding redwing populations by 80–85% in treated areas, with minimal impact on nontarget birds.

At least 90% of the blackbirds counted during the Pine Prairie CBCs originated from the Millers Lake roost. However, in 1984, well before the poisoning program, satellite roosts started to appear, with more than 1 million birds roosting outside Millers Lake during 1984 and 1987. Poisoning, however, could have contributed to reduced numbers of roosting cowbirds from 1989 to 1993, particularly if roosting sites shifted in response to poisoning. Many blackbirds counted near Millers Lake during 1992 and 1993 originated from roosts to the north, judging by their flight directions in the morning.

The initial major drop in cowbird numbers occurred in 1988, one year prior to the poison baiting, and the numbers killed in 1989 and 1990 are insufficient to explain the major drop in roosting numbers. There were no apparent changes in agricultural practices during the years of low numbers in the major foraging areas to the south and southwest. There has been a steady increase in commercial crawfish (*Procambarus* spp.) production in rested rice fields since the early 1980s, but this is unlikely to have caused shifting of roosts or population declines. The nationwide decline of Brown-headed Cowbirds since 1966 (0.9% per year, Peterjohn et al., Chapter 2, this volume) is also not nearly precipitous enough to have caused this local change.

Timing of winter blackbird counts is also known to affect estimates (White 1980, Greenleaf 1982). In general, mild early winters (as in 1977 and 1980) resulted in low counts, and the severe early winter of 1989 led to a high count. Peak numbers at this site usually occurred during January and February with weekly increases of as much as 3 million from December on (Ortego 1976, Wilson 1986). The five counts conducted after Christmas averaged 57 million more blackbirds than the counts before Christmas. Labisky and Brugger (1989) and Wilson (1986) reported that birds at the Millers Lake roost were mostly migrants, and that the species composition and density depended on migration patterns. There were no obvious long-term shifts in winter weather sufficient to explain the decline in cowbird numbers after 1987.

In conclusion, while annual variation in early winter weather and poisoning programs may explain part of the fluctuations in cowbird numbers at Millers Lake, the causes of the sustained decline in cowbird numbers after 1987 are unknown. The factors contributing to regional shifts of cowbirds among Gulf coast roosts merit further study.

Management Implications

The adverse effects of the Brown-headed Cowbird on individual hosts of North American songbirds have been well documented (Southern 1958, Walkinshaw 1961, Friedmann et al. 1977, Mayfield 1977a, Elliott 1978). Large winter roosts, like the one at Millers Lake, provide an opportunity to monitor and control large numbers of cowbirds and other blackbirds that impact songbirds and agricultural crops in large geographic areas.

The U.S. Fish and Wildlife Service conducted seven National Roost Surveys from 1958 to 1980 to map the locations and estimate the sizes of winter blackbird roosts. These data are difficult to compare among years because of inconsistent observer efforts, but they provide information on location, size, and species composition for eastern U.S. roosts (Labisky and Brugger 1989). The data indicate that very large numbers of blackbirds winter in the South. Other monitoring activity, such as CBC counts (see above), tracks early winter populations at more than 1,000 sites in the United States.

Nowadays, the general public in developed countries often views killing birds to protect agricultural crops as unacceptable (Dolbeer 1986). However, if the economic or biological impact of these birds is large enough, population reduction may be justified. Many conservationists have enthusiastically endorsed killing cowbirds to protect populations of endangered hosts (see Part V, this volume).

Brown-headed Cowbirds do not show strong site fidelity to breeding grounds and do not travel to particular wintering grounds from specific breeding grounds (Dolbeer 1982).

However, if some breeding populations of cowbirds use specific roosting areas during late winter, controls could be focused on any problem populations where cowbird parasitism threatens the existence of host populations. Reduction of selected cowbird winter populations would then be easier to justify on a large scale. Otherwise, justification for reducing blackbird (including cowbirds) numbers at large winter roosts is limited to any benefits gained from continent-wide protection of populations of songbirds and agricultural crops. Control measures applied to large populations at winter roosts should be cost-effective, compared to controls on the breeding grounds when individuals are dispersed and large concentrations of birds are uncommon.

Successful population control methods have used poison baits, decoy traps, and surfactants. The application of brown rice treated with 2% DRC-1339 near roosts was very effective at killing large numbers of redwings and cowbirds during spring in Louisiana (Glahn and Wilson 1992). This technique could also be used to reduce large concentrations of cowbirds on winter roosts. A benefit here is that impacts on nontarget species are likely to be small. Baited decoy traps (Meanley 1971) and, under special conditions, floodlight traps (Mitchell 1963) are also effective at catching large numbers of cowbirds on the wintering grounds. Use of surfactants and water to kill wintering blackbirds (Lefebvre and Seubert 1970) has been successful at large roosts and has caused some roosts to shift (Glahn et al. 1991).

Control methods used on the breeding grounds include playback of female chatter calls and selective shooting (Laymon 1987; Hayden et al., Chapter 39, this volume), and decoy traps (Mayfield 1977b; DeCapita, Chapter 37, Griffith and Griffith, Chapter 38, and Hayden et al., Chapter 39, this volume). These methods result in many fewer birds being killed per day compared to the tens of thousands achieved by the winter control methods mentioned above.

Acknowledgments

The efforts of co-compilers—Bill Fontenot, Harland Guillory, and Dwight Leblanc—and participants of the Pine Prairie Christmas Bird Count are greatly appreciated, along with the interest and access provided by the Millers Lake Estate, primarily Curry and Benny McCauley, and Gus Miller.

References Cited

Dolbeer, R. A. 1982. Migration patterns for age and sex classes of blackbirds and starlings. J. Field Ornithol. 53:28–46.
———. 1986. Current status and potential of lethal means of reducing bird damage in agriculture. Acta Congr. Inter. Ornithol. 19:474–483.
Elliott, P. F. 1978. Cowbird parasitism in the Kansas tallgrass prairie. Auk 95:161–167.
Friedmann, H., L. Kiff, and S. I. Rothstein. 1977. A further contribution to knowledge of the host relations of the parasitic cowbirds. Smithsonian Contr. Zool. 235. 75 pp.
Glahn, J. F., A. R. Stickley, Jr., J. F. Heisterberg, and D. F. Mott. 1991. Impact of roost control on local urban and agricultural blackbird problems. Wildl. Soc. Bull. 19:511–522.
Glahn, J. F., and E. A. Wilson. 1992. Effectiveness of DRC-1339 baiting for reducing blackbird damage to sprouting rice. Proc. East. Wildl. Damage Control. Conf. 5:117–123.
Greenleaf, P. A. 1982. Nocturnal roosting behavior of blackbirds and starlings in northwestern Arkansas. M.S. Thesis, University of Arkansas, Fayetteville. 29 pp.
Labisky, R. F., and K. E. Brugger. 1989. Population analysis and roosting- and feeding-flock behavior of blackbirds damaging sprouting rice in southwestern Louisiana. Florida Coop. Fish Wildl. Res. Unit, Tech. Rep. 36. 77 pp.
Laymon, S. A. 1987. Brown-headed Cowbirds in California: Historical perspectives and management opportunities in riparian habitats. Western Birds 18:63–70.
Lefebvre, P. W., and J. L. Seubert. 1970. Surfactants as blackbird stressing agents. Proc. Vertebr. Pest Conf. 4:156–161.
Mayfield, H. 1977a. Brown-headed Cowbird: Agent of extermination? Amer. Birds 31:107–113.
———. 1977b. Brood parasitism: Reducing interaction between Kirtland's Warbler and Brown-headed Cowbirds. Pp. 85–91 in Endangered birds: Management techniques for preserving threatened species (S. A. Temple, ed). University of Wisconsin Press, Madison.
Meanley, B. 1971. Blackbirds and the southern rice crop. USFWS Resource Publ. 100. 64 pp.
———. 1976. Distribution and ecology of blackbird and starling roosts in the United States. USFWS Patuxent Wildl. Res. Cent. Annu. Prog. Rep. 27 pp.
Meanley, B., J. S. Webb, and D. P. Fankhauser. 1966. Migration and movements of blackbirds and starlings. USFWS Patuxent Wildlife Research Center, Laurel, MD. 95 pp.
Mitchell, R. T. 1963. The floodlight trap: A device for capturing large numbers of blackbirds and starlings at roosts. USFWS Special Scientific Rep., Wildl. 77. 14 pp.
Monroe, B. A., Jr. 1976–1993. Summary of highest counts of individuals for Canada and the United States. Amer. Birds 30–47.
Ortego, J. B. 1976. Bird usage by habitat types in a large freshwater lake. M.S. Thesis, Louisiana State University, Baton Rouge. 189 pp.
Southern, W. M. E. 1958. Nesting of the Red-eyed Vireo in the Douglas Lake region, Michigan. Jack-Pine Warbler 36:104–130.
Walkinshaw, L. H. 1961. The effect of parasitism by the Brown-headed Cowbird on Empidonax flycatchers in Michigan. Auk 78:266–268.
White, S. B. 1980. Bioenergetics of large winter-roosting populations of blackbirds and starlings. Ph.D. Thesis, Ohio State University, Columbus. 150 pp.
Wildlife Counts. 1986. Counting wildlife. Wildlife Counts (computer software). Juneau, AK.
Wilson, E. A. 1986. Blackbird depredation on rice in southwestern Louisiana. M.S. Thesis, Louisiana State University, Baton Rouge. 91 pp.
Wilson, E. A., E. A. LeBoeuf, K. M. Weaver, and D. J. LeBlanc. 1989. Delayed seeding for reducing blackbird damage to sprouting rice in southwestern Louisiana. Wildl. Soc. Bull. 17:165–171.

6. An Evaluation of Point Counts and Playbacks as Techniques for Censusing Brown-headed Cowbirds

R. KIRK MILES AND DAVID A. BUEHLER

Abstract

Techniques to monitor the distribution and abundance of Brown-headed Cowbirds accurately are urgently needed. We used both point counts and playbacks of recorded cowbird vocalizations to compare cowbird detection frequency on two study areas in Kentucky and western Tennessee. A total of 148 point counts and playbacks were conducted in both forest openings and in the forest interior from 17 May to 15 July 1993. Ten-min point counts, broken into intervals of 0–3, 3–5, and 5–10 min, are suggested to be the best option for managers monitoring cowbird populations. Point counts and playbacks yielded similar results ($P > .05$) for detecting cowbird presence and estimating cowbird abundance. Finally, emphasis should be placed on monitoring female cowbirds, as including male cowbirds may obscure differences in cowbird distribution and inflate abundance estimates.

Introduction

The distribution and abundance of Brown-headed Cowbirds must be known to estimate the potential threat of cowbird parasitism to songbird populations. Parasitism levels vary geographically for most host species (Robinson et al. 1993). Therefore, it is important that land managers collect local data to assess a potential cowbird problem. To make this assessment, researchers and managers require techniques that monitor cowbird populations efficiently and accurately.

Point counts of various types are widely used for censusing avian populations (Ralph and Scott 1981). These counts are primarily designed to detect singing males on relatively small, breeding territories. Although general avian censusing techniques have been reviewed (Ralph and Scott 1981), no one has evaluated techniques specifically designed for censusing Brown-headed Cowbirds (but see Rothstein et al., Chapter 7, this volume).

Cowbirds maintain relatively large home ranges (Dufty 1981; Darley 1982, 1983; Teather and Robertson 1985) and are highly mobile during the breeding season (Rothstein et. al. 1984, Robinson et. al. 1993; Raim, Chapter 9, and Thompson and Dijak, Chapter 10, this volume). Counts of long duration (e.g., 20 min) may overestimate densities of highly mobile avian species like the cowbird (Scott and Ramsey 1981). Thus, the optimal duration for point counts may be particularly important when censusing cowbirds.

Response of both sexes of Brown-headed Cowbirds to playbacks of the recorded female chatter call is well documented (Dufty 1981, Yokel 1989). However, little information is available on the potential use of playbacks for censusing cowbirds. Playbacks may be especially effective for enhancing female cowbird detection because females are secretive while searching for nests. Information on female cowbirds is especially important because females are directly responsible for nest parasitism. Playbacks also allow positive identification of male and female cowbirds, as both sexes will approach playback equipment closely (Dufty 1981, Yokel 1989).

We assessed the effects of varying the duration of point counts on cowbird detection and abundance estimates. We also compared the efficiency of point counts and playbacks for estimating cowbird presence and abundance.

Methods

Study Areas

We conducted point counts and playbacks on two study areas, Land Between the Lakes (LBL) in Kentucky, and Natchez Trace Wildlife Management Area in western Tennessee. LBL is a 70,000-ha inland peninsula separating Kentucky and Barkley lakes, developed and managed by the Tennessee Valley Authority (TVA) as a national outdoor recreation and environmental education area. TVA maintains 10% of the total LBL area in forest openings. LBL is surrounded on three sides by water. Predominant land use patterns beyond the reservoirs are agricultural.

Natchez Trace is a 16,000-ha wildlife management area,

state forest, and state park with 2% of the area maintained as forest openings. Tennessee Wildlife Resources Agency (TWRA) maintains both linear forest openings and smaller (ca. 1 ha), polygonal forest openings on Natchez Trace. Surrounding land use is primarily agricultural.

Most forest openings on both areas were managed for use by wildlife; however, some campgrounds and recreational areas were censused on LBL. All forest openings were permanently maintained, usually by mowing. Forest types on both areas included oak (*Quercus* spp.), hickory (*Carya* spp.), pine (*Pinus* spp.), or mixed.

Avian Censuses

Fifty-meter fixed-radius point counts were conducted at 296 paired, randomly selected points in forest openings and forest interior from 17 May–15 July 1993. Counts were conducted between sunrise and 10:30 EST. Forest opening counts were conducted at the forest edge. Interior points were located on a randomly selected compass bearing 200 m from the edge of the opening.

Counts were broken into intervals of 0–3, 3–5, 5–10, and 10–20 min. All avian species seen or heard were recorded as being within 50 m of point center, outside the 50-m radius, or as flyovers. Special attention was given to cowbirds seen or heard, specifically noting the sex of the individual, song heard, and activity. Census protocols were the same for forest opening and interior points, although differences in visibility existed.

Recorded Playbacks

Recordings of the female chatter call were used to detect both male and female cowbirds. Free-ranging female cowbirds were recorded in Boise, Idaho, in 1991 with a Marantz PMD 221 cassette recorder and a Sennheiser K3-U power unit and condenser microphone (A. Dufty, Jr., pers. comm.). Playbacks were broadcast with a Sony CFS-208 radio cassette-recorder. The playback emitted two chatters every 10 sec. We broadcast playbacks at a volume clearly audible to humans at up to 100 m.

Playback sessions were conducted immediately after standard point counts. We conducted playbacks at either the forest opening or the paired interior forest point to avoid the possibility that tapes were audible to cowbirds from both points. Thus, both techniques were applied at 148 points. Tapes were played for 5 min, followed by a 5-min observation period. All cowbirds observed during the 10-min period were recorded, including sex and activity (i.e., flying, perched, etc.) of the bird.

Data Analyses

Duration of Point Counts

To predict results for point counts longer than 20 min, a power-function regression curve (Smith et al. 1993) was fitted to cowbird presence (cumulative number of points

with cowbirds) and mean cowbird abundance (cumulative individual cowbirds per count), as a function of point-count duration for both forest opening and interior forest points. This approach allowed us to compare cowbird presence, mean abundance, and habitat effects across a wide range of point-count durations.

Comparison of Point-count and Playback Techniques

We compared the first 10-min interval of point counts with the 10-min playback period. A chi-square test for independence was performed to determine if detection patterns differed between the two techniques. We also compared estimated mean cowbird abundance for each technique and performed a three-way analysis of variance (ANOVA), where we modeled estimated mean abundance as a function of technique (point count or playback), sex (male or female cowbird), and habitat type (opening or interior points).

Habitat Effects

We used data from our 20-min point counts to determine if point counts of different durations led to differences in cowbird detection frequency between forest openings and interior points. A logistic regression was performed, modeling cowbird presence as a function of the duration of point counts (3-, 5-, 10-, and 20-min counts) and habitat type (forest opening or interior). Finally, we performed a two-way ANOVA, where we modeled mean cowbird abundance as a function of the duration of point counts and habitat type.

Results

Duration of Point Counts

Cowbird Presence

Male and female cowbird presence, predicted by the power-function regression equation, continued to increase throughout the 60-min count interval (Figure 6.1). Considering the curve for both sexes combined (points with males and/or females present), 60-min point counts were predicted to have cowbirds present at every point (N = 296).

A marked difference in male and female detections was evident (Figure 6.1). The male presence curve tracked the curve for both sexes combined, but the female curve was at a lower level than the other two curves. In all three cases, the slope of the line began decreasing after 10 min.

Cowbird Abundance

A sharp increase in predicted mean abundance is evident in all three curves between 5- and 10-min intervals (Figure 6.1). After 10 min, predicted mean abundance continued to increase throughout 60-min counts, although at a decreasing rate.

A distinct difference in male and female mean abundance curves was evident (Figure 6.1). The male curve closely

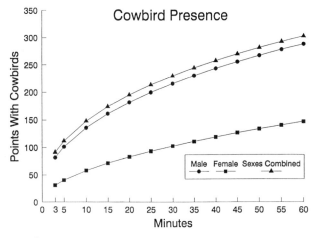

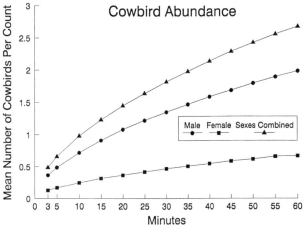

Figure 6.1. *Top:* Cumulative number of points where cowbirds were detected with varying sampling times. *Bottom:* Cumulative mean numbers of cowbirds detected with varying sampling times. Separate curves derived from power-function regressions are plotted for males, females, and both sexes combined.

tracked the curve for both sexes combined (males plus females), but the female curve was at a much lower level, with less slope than the other two curves.

Habitat Effects

Female cowbirds were present in forest openings more often than in the forest interior (Figure 6.2). However, there was little difference in presence of males between the two habitat types (Figure 6.2). A logistic regression model detected no significant interaction between point-count duration and habitat type for all three curves ($P > .05$). Habitat type explained little variation for the curves for males and for the sexes combined ($P > .05$), whereas it did explain some variation in the presence of females ($P = .005$). Point-count duration was significant for all three curves ($P < .05$).

Female and male cowbird mean abundance was greater in forest openings than in the forest interior (Figure 6.3). A two-way ANOVA detected no significant interaction be-

tween point-count duration and habitat type for all three curves ($P > .05$). Main effects of point-count duration and habitat type were significant ($P < .05$) for all three curves.

Comparison of Point-count and Playback Techniques
Cowbird Presence

There was no difference in the distributions of points at which point counts and playbacks detected cowbirds ($P < .05$, Table 6.1). Point counts and playbacks detected male cowbirds at 62 and 56 points, respectively. Point counts detected female cowbirds at 27 points, whereas playbacks detected females at 33 points. Point counts and playbacks detected birds of either sex at 68 and 65 points, respectively.

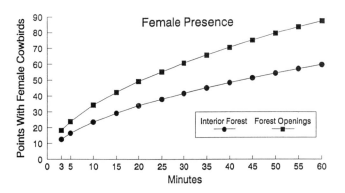

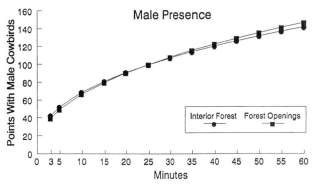

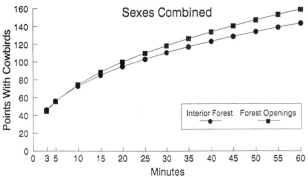

Figure 6.2. Cumulative numbers of sample points in which cowbirds were detected in forest openings and interior forest for females, males, and cowbirds of both sexes. All relationships predicted from power-function regression equations.

Each method contributed additional points to the "present" category that were missed by the other method for males, females, and both sexes combined. For example, playbacks detected males at an additional 23 points that point counts classified as "absent," and point counts detected males at an additional 29 points that were classified as "absent" by playbacks (Table 6.1).

Cowbird Abundance

Point counts detected males more frequently than playbacks (0.54 vs. 0.40 birds per count), while the reverse was true for females (0.19 vs. 0.23, Table 6.2). However, abundance estimates did not differ significantly between the two methods ($P > .05$).

Based on a three-way ANOVA, we detected no significant interactions between technique, sex, or habitat variables ($P > .05$, Table 6.2). Sex and habitat type were the only vari-

Table 6.1. Number of Point Counts and Playbacks in Which Cowbirds Were Present or Absent

Males

		Playback		
		Absent	Present	Total
	Absent	63	23	86
Point count	Present	29	33	62
	Total	92	56	148

$\chi^2 = 10.74$, df = 1, $P = 0.001$

Females

		Playback		
		Absent	Present	Total
	Absent	98	23	121
Point count	Present	17	10	27
	Total	115	33	148

$\chi^2 = 4.14$, df = 1, $P = 0.042$

Sexes combined (males and/or females)

		Playback		
		Absent	Present	Total
	Absent	54	26	80
Point count	Present	29	39	68
	Total	83	65	148

$\chi^2 = 9.22$, df = 1, $P = 0.002$

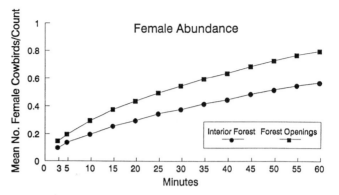

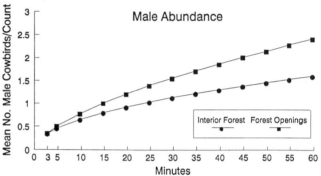

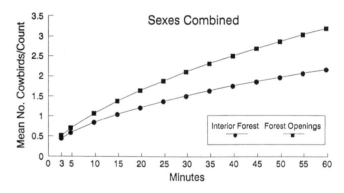

Figure 6.3. Cumulative mean numbers of individual females, males, and cowbirds of both sexes (males plus females) detected in forest interior and forest opening points per unit time, as predicted from power-function regression equations.

ables that explained any significant variation in mean abundance ($P < .05$). Estimated male abundance was consistently greater than estimated female abundance ($P < .001$). Estimated abundance for both sexes combined was consistently higher in forest openings than in the forest interior ($P = .009$).

Discussion

Duration of Point Counts

The estimates of male and female cowbird presence and abundance predicted by the power-function regression equation did not approach asymptotes during 60-min point counts (Figure 6.1). Thus, increasing point-count duration

Table 6.2. Mean Cowbird Abundance per Point Count and Playback[a] by Sex and Habitat Type

Technique	Sex	Interior Forest		Forest Openings		All Points	
		Mean	SE	Mean	SE	Mean	SE
Playback	Male	0.40	0.08	0.68	0.12	0.54	0.07
Point count	Male	0.54	0.09	0.76	0.14	0.65	0.08
Playback	Female	0.23	0.06	0.32	0.08	0.28	0.05
Point count	Female	0.19	0.06	0.28	0.07	0.24	0.05
Playback	Sexes combined[b]	0.63	0.12	1.01	0.16	0.83	0.10
Point count	Sexes combined[b]	0.73	0.13	1.05	0.19	0.89	0.12

Source	Three-way Analysis of Variance		
	df	F-Value	Pr > F
Technique	1	0.27	0.6037
Sex	1	26.02	0.0001
Habitat type	1	6.89	0.0089
Technique–sex	1	1.29	0.2565
Technique–habitat type	1	0.05	0.8264
Sex–habitat type	1	1.43	0.2318
Technique–sex–habitat type	1	0.06	0.8135

[a] $N = 148$ for both point counts and playbacks.
[b] Males plus females.

does not lead to a fixed estimate of the number of points with cowbirds present. In addition, prolonged censuses are not feasible because of constraints on personnel, time, and other responsibilities (Smith et al. 1993). Prolonged censuses also greatly increase the likelihood of multiple counts of the same birds (Scott and Ramsey 1981). Thus, census methods must be based on measures of relative cowbird abundance. The ultimate question, then, is what count duration provides the best estimates for these relative measures?

Cowbird Presence

We feel that the optimal count duration should balance the increasing likelihood of at least one detection at a point with increasing time versus the decreasing efficiency after the first bird has been detected. Ten-min point counts appear to provide a reasonable balance between these competing needs. For example, 10-min counts accounted for 70% and 40% of the total female presence for 20- and 60-min counts, respectively. In contrast, 5-min counts accounted for only 49% and 28% of the total female presence for 20- and 60-min counts, respectively. Thus, a substantial increase in female cowbird presence was recorded from 5 to 10 min. After the 10-min period, however, the relative gain in female detection decreased.

Cowbird Abundance

Ten-min point counts also appear to be suitable for estimating mean cowbird abundance. For example, 10-min counts accounted for 68% and 36% of the estimated female mean abundance for 20- and 60-min counts, respectively. Five-min counts accounted for 46% and 25% of the estimated female mean abundance for 20- and 60-min counts, respectively. Thus, a substantial increase in estimated female abundance was recorded between 5- and 10-min intervals.

Habitat Effects

Cowbirds are likely to be more visible in forest openings than in the forest interior. This difference in detectability could have obscured real differences in cowbird presence or abundance as the duration of point counts increased. However, there was no significant interaction between count duration and habitat type when both were used to estimate cowbird presence or mean abundance for counts up to 20 min.

Comparison of Point-count and Playback Techniques

We strongly suspect that cowbird mobility accounted for the differences in cowbird detection by the two techniques. Because we used only the first 10 min of point counts to compare with the 10-min playback period, a 10-min window existed in which cowbirds could move in or out of a point. This time interval appeared to be sufficient to allow birds detected by the point counts to move out of the area by the time playbacks began. Similarly, cowbirds detected by the playbacks were not necessarily present during the earlier point count.

Perhaps as important as where birds were detected, however, was the frequency of detection. Both male and female presence and abundance estimates were similar for both techniques.

Differences in Male and Female Detections

Male cowbird abundance on our study areas was consistently greater than female cowbird abundance, regardless of detection technique. This is consistent with other studies reporting a strongly male-biased sex ratio (Rothstein et al. 1986, Yokel 1989).

Of the 198 points at which males were observed on our study areas, females were absent 65% of the time (128/198). Conversely, of the 85 points at which females were observed, males were absent only 21% of the time (15/85). Robinson et al. (1993) suggested that many males in an area might be unmated and searching for mates, whereas females are more likely to be searching for nests.

Presumably, unmated males on our study areas were highly mobile (Yokel 1989) and searching for mates (Robinson et al. 1993). Unmated males may provide little insight into the actual breeding population of cowbirds and consequently may be of little value for predicting parasitism frequencies. Including males in abundance estimates, because of their high mobility, may also overestimate cowbird densities. Thus, our results suggest that female cowbird distribution and abundance are the best index of local parasitism frequencies.

Management Implications

We suggest that 10-min counts, broken into intervals of 0–3, 3–5, and 5–10 min, are best for monitoring cowbird populations. After reviewing initial results, managers can decide which count duration works best for their particular situation. For high-density cowbird populations, shorter counts may be more efficient.

Point-count and playback techniques yielded similar results in our study. Playbacks may be useful to managers as a training tool for cowbird identification (including distinguishing cowbird sex) or as a check to ensure that cowbirds are not being overlooked during point counts. Playbacks may also be particularly useful on areas where cowbird densities are low.

Finally, our results indicate that particular emphasis should be placed on monitoring female cowbirds. Including male cowbirds in estimates may hide differences in cowbird distribution or inflate estimates of abundance.

Acknowledgments

We thank A. Dufty, Jr., for providing both advice and the recordings of female cowbird vocalizations that we used in this study. We are grateful to S. Sealy, S. Robinson, S. Rothstein, and C. Nicholson for their advice and input. We also thank M. Muller and S. Strickland for their excellent assistance in field work. The Tennessee Wildlife Resources Agency, Tennessee Valley Authority, U.S. Fish and Wildlife Service, and the University of Tennessee Institute of Agriculture supported this research.

References Cited

Darley, J. A. 1982. Territoriality and mating behavior of the male Brown-headed Cowbird. Condor 84:15–21.

———. 1983. Territorial behavior of the female Brown-headed Cowbird (*Molothrus ater*). Can. J. Zool. 61:65–69.

Dufty, A. M., Jr. 1981. Social organization of the Brown-headed Cowbird, *Molothrus ater*, in New York State. Ph.D. Thesis, State University of New York, Binghamton. 109 pp.

Ralph, C. J., and J. M. Scott (eds). 1981. Estimating numbers of terrestrial birds. Studies in Avian Biol. 6:1–630.

Robinson, S. K., J. A. Grzybowski, S. I. Rothstein, M. C. Brittingham, L. J. Petit, and F. R. Thompson. 1993. Management implications of cowbird parasitism on neotropical migrant songbirds. Pp. 93–102 *in* Status and management of neotropical migratory birds (D. M. Finch and P. W. Stangel, eds). U.S. For. Serv. Gen. Tech. Rep. RM-229.

Rothstein, S. I., J. Verner, and E. Stevens. 1984. Radio-tracking confirms a unique diurnal pattern of spatial occurrence in the parasitic Brown-headed Cowbird. Ecology 65:77–88.

Rothstein, S. I., D. A. Yokel, and R. C. Fleischer. 1986. Social dominance, mating and spacing systems, female fecundity, and vocal dialects in captive and free-ranging Brown-headed Cowbirds. Curr. Ornithol. 5:127–185.

Scott, J. M., and F. L. Ramsey. 1981. Length of count period as a possible source of bias in estimating bird densities. Studies in Avian Biol. 6:409–413.

Smith, W. P., D. J. Twedt, D. A. Wiedenfeld, P. B. Hamel, R. P. Ford, and R. J. Cooper. 1993. Point-counts of birds in bottomland hardwood forests of the Mississippi alluvial valley: Duration, minimum sample size, and points versus visits. U.S. For. Serv. Res. Pap. SO-274.

Teather, K. L., and R. J. Robertson. 1985. Female spacing patterns in Brown-headed Cowbirds. Can. J. Zool. 63:218–222.

Yokel, D. A. 1989. Intrasexual aggression and the mating behavior of Brown-headed Cowbirds: Their relation to population densities and sex ratios. Condor 91:43–51.

7. The Structure and Function of Cowbird Vocalizations and the Use of Playbacks to Enhance Cowbird Detectability: Relations to Potential Censusing Biases

STEPHEN I. ROTHSTEIN,

CHRIS FARMER, AND JARED VERNER

Abstract

The Brown-headed Cowbird is one of the most frequent subjects of research involving bird song. Its vocalizations are highly variable and sometimes confusing, yet proper identification of this species' vocalizations and of their uses can be essential to cowbird research. To provide an overview of cowbird vocalizations, we describe the structure, function, and ontogeny of all five distinct types of cowbird songs and calls. One male song type, the flight whistle, has such variable and distinct dialects that field workers trained in one area may fail to recognize the local whistle in other areas. Another song type, perched song, is also variable but is always recognizable as a cowbird vocalization as its variation never violates species-specific structural rules. The three other types of vocalizations (single-syllable, chatter, and kek calls) show less geographic variation. Cowbirds are highly interactive vocally and respond to playback of the female chatter call with vocalizations and approach responses. Chatter playbacks have potential uses in management efforts concerned with the removal or censusing of cowbirds. A field playback experiment in the Sierra Nevada of California resulted in twice as many male and almost four times as many female cowbirds being detected during playbacks of chatter than during standard 10-min point counts without playbacks. Standard survey techniques may not give completely reliable data on absolute or even relative parasitism burdens experienced by local avifaunas.

Introduction

Bird song has long been a classic example of the function and ontogeny of behavior, especially the ways in which genetic programming and experience interact in the development of complex behavior (Manning and Dawkins 1992, Alcock 1993, Campbell 1993). It was initially assumed that Brown-headed Cowbirds must have genetically programmed vocalizations because they develop species-specific songs even though brood parasitism deprives them of early contact with conspecific adults (Mayr 1979). Whereas extensive studies of cowbird vocalizations have shown that some characteristics are innate (King and West 1977), most aspects of vocal development are modified in response to experience (West et al. 1981, King and West 1983, Rothstein and Fleischer 1987a). Indeed, the cowbird's vocal development is so complex that it has become one of the most prominent species in the bird song literature in the last 15 years. One vocalization, the flight whistle, has an especially large component of learning in its development (O'Loghlen and Rothstein 1993, O'Loghlen 1995) and is among the most variable of all songbird vocalizations (Rothstein and Fleischer 1987b, Rothstein et al. 1988). This vocalization has such extensive geographic variation and well-defined dialects that researchers trained in one area may fail to recognize flight whistles as cowbird vocalizations in another area as close as 20–50 km away.

Knowledge of the vocal repertoire of cowbirds, and of the ways in which each type of vocalization is used, is essential to understanding many facets of cowbird behavior. This information can aid those working with cowbirds either because of management goals or because they are studying basic ecological, evolutionary, and behavioral questions raised by cowbirds and brood parasitism. Furthermore, such knowledge can enable researchers and land managers to maximize the benefits of playbacks of cowbird vocalizations. For example, because cowbirds are highly responsive to some conspecific vocalizations, playbacks can be used in censusing to reveal individuals that might otherwise go undetected and in control efforts to attract and shoot female cowbirds whose removal might be difficult or too costly with other methods.

Thus, this chapter has three goals. First, we present a brief overview of the structure, function, and ontogeny of cowbird vocalizations. Second, we discuss the special problems inherent in using vocalizations to monitor the relative or absolute numbers of cowbirds in an area. Last, we present the results of a playback experiment designed to assess the extent to which playbacks make cowbirds more detectable. The vocalization used in the playback experiment, the chatter or rattle call, is the only loud one given by females (Friedmann 1929, Burnell and Rothstein 1994). Although previous studies have shown that both male and female cowbirds approach chatter playbacks (Dufty 1982a, Rothstein et al. 1988, Yokel 1989), no controlled experiment has compared numbers of birds responding to playbacks to numbers counted by traditional censusing techniques, such as standard point counts (see also Miles and Buehler, Chapter 6, this volume).

The Structure, Function, and Development of Cowbird Vocalizations

Friedmann (1929) provided the first detailed categorization of cowbird vocalizations in his classic monograph covering all cowbird species. Although he based his descriptions on what the unaided human ear could perceive, Friedmann's categories are virtually identical to those identified by modern sound analysis procedures. The only major difference between Friedmann's categories and ours is that he did not distinguish between flight whistles and single-syllable calls. Instead, he lumped these two categories under the former name and remarked (p. 167) that "The whistle varies considerably." We recognize five types of cowbird vocalizations that are distinct in terms of acoustic structure. Most are also distinct as regards function and ontogeny. We now describe these vocalizations in turn.

Flight Whistle (FW)

The FW is a high-amplitude song type given only by males and is by far the most spatially variable of cowbird vocalizations (Figure 7.1). Easily defined dialects occur in eastern (Brewster in Friedmann 1929, Tyler 1920, Rothstein unpubl. data) and western (Rothstein and Fleischer 1987b) North America. In the West, dialect boundaries usually correspond to gaps in cowbird distribution or to areas where cowbird abundance is low, i.e., dialects correspond to patches of suitable cowbird habitat separated by arid or high-elevation areas of unsuitable habitat (Rothstein and Fleischer 1987b). Cowbirds are more evenly distributed in the East, and dialect borders there can occur even where cowbirds are continuously distributed (Rothstein unpubl. data). Most males have only one type of FW, although a large proportion of individuals found along dialect borders are bilingual (Rothstein and Fleischer 1987b).

FWs usually consist of modulated pure whistlelike tones in the 3–9-kHz range lasting 0.5–1.5 sec, but many dialects contain buzzes or trills (Figure 7.1). People perceive puretone FWs to have two to four distinct syllables. On occasion, males in three- or four-syllable FW dialects (such as A, B, D, and E in Figure 7.1) give only the first two syllables of their

complete FW (Rothstein and Fleischer 1987b), and less than 10% may give only these abbreviated FWs (O'Loghlen and Rothstein 1993). Workers unfamiliar with vocal variation in cowbirds may perceive abbreviated FWs as a different type of vocalization.

The degree of structural difference among dialects varies regionally. In much of California west of the crest of the Sierra Nevada, where *M. a. obscurus* breeds, most FW dialects show the same basic coastal form (D and E in Figure 7.1), and dialect differences are usually minor and of a phonetic (i.e., pronunciation) nature (Rothstein et al. 1986). But dialect differences are much more pronounced where cowbird distribution is more disjunct, as in *obscurus* that breed in the deserts of southeastern California (F in Figure 7.1), and where *M. a. artemisiae* breeds east of the crests of the Sierra Nevada and Cascades ranges (A, B, C, K, and L in Figure 7.1, Rothstein and Fleischer 1987b). In these regions and in the East, where *M. a. ater* breeds, dialects often have one or more syllables with no apparent homologue in neighboring dialects and thus have lexical differences analogous to different languages in humans. When people are unfamiliar with a local dialect, FWs may confound attempts to assess the distribution and abundance of cowbirds. This is a significant problem because FWs and the somewhat similar single-syllable calls (SSs, see below) are more commonly heard than the other frequent male vocalization, perched song (PS), in both eastern (Laskey 1950) and western North America (Rothstein unpubl. data). One exception, however, is that males may give long series of PSs near particular favored perching sites (Friedmann 1929). It is unclear whether the more frequent detection of FWs and SSs occurs because these vocalizations are given more commonly than PSs (which are more easily recognizable to most people) or because the higher amplitude of FWs and SSs simply allows them to be heard over a greater distance. Nevertheless, FWs and SSs provide the most common auditory evidence for the presence of cowbirds, which makes it essential for field workers to learn to recognize the dialects in their study areas.

Unfortunately, the circumstances in which FWs are given make it difficult for people to learn them. Stationary males often give long series of PSs but relatively few FWs. Instead, FWs are commonly given in flight. If a male gives an FW while perched, he is likely to fly off within a few seconds or to have just landed (Friedmann 1929, pers. obs.). Thus observers have few opportunities to become familiar with FWs by tracking down their source. FWs are used primarily for long-distance communication. Males respond to distant FWs by giving FWs and SSs and by approaching the source of the FWs (Rothstein et al. 1988, Dufty and Pugh 1994). Females sometimes show similar behavior, except that they give chatter calls in response to FWs. When males come within several meters of other males or females, they switch from FWs to PSs, and males usually do not give FWs when

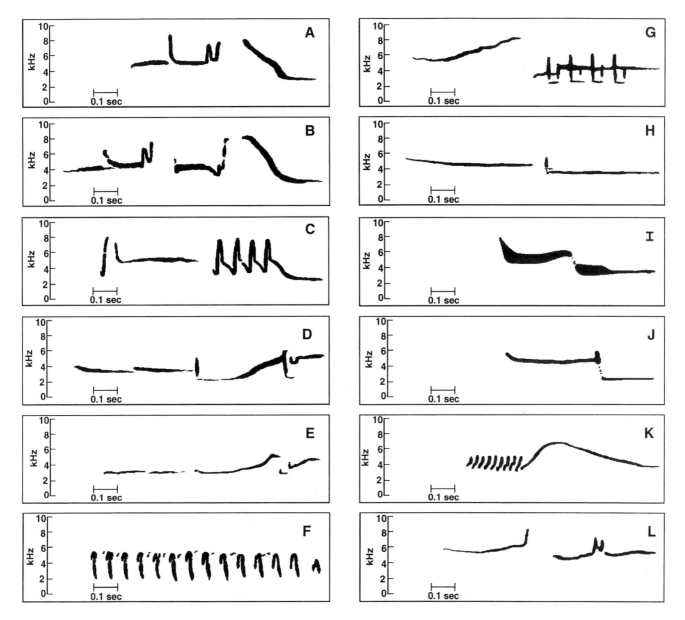

Figure 7.1. Flight whistles from 12 dialects. This figure, and figures 7.2–7.4, present digitized images of original audiospectrograms made with a Kay 7800 Digital Sonograph machine. Panels A, B, and C show FWs of three dialects studied by Rothstein and Fleischer (1987a,b) in Mono County, CA. D and E show two versions of the coastal form of the FW widespread in California west of the crest of the Sierra Nevada (Rothstein et al. 1986). They are from sites 45 km apart in Santa Barbara County, CA, and their relatively slight differences represent stable dialects. These coastal FWs have three perceptible syllables, the first two being the monotonic elements and the third being the remainder of the FW. People do not hear the very short element at the end as a separate syllable. F shows the most divergent FW dialect we have found; this trill-like dialect sounds like the song of a Chipping Sparrow (*Spi-*

zella passerina) and occurs in the Imperial Valley, Imperial County, CA. G and H show FWs from Anne Arundel County, MD, and Orange County, NC, respectively. The former is perceived as a whistle followed by a short trill. I and J show vocalizations categorized as FWs because the abrupt drop in frequency near the midpoint makes them sound like two syllables. Most birds in Canadian County, OK, did both of these vocalizations and no other FW or SS-like vocalization. I differs from J by being higher in frequency and by having a buzzy quality. K and L show FWs from Yakima County, WA, and Douglas County, NV, respectively. The former is perceived as a buzz followed by a whistle. Subspecies responsible for these FWs are as follows: *M. a. artemisiae*—A, B, C, K, L; *M. a. ater*— G, H, I, J; *M. a. obscurus*—D, E, F. The exact localities for the recordings in this and later figures can be obtained from S. I. Rothstein.

another male is nearby. There are two exceptions to the tendency for FWs to be given only when males are alone. FWs are given in an alarm call context when a person, or some other threat, approaches a flock (Rothstein and Fleischer 1987b, Rothstein et al. 1988). Second, FWs are given close to females in the 5 sec immediately preceding nearly all copulations in nature (Rothstein et al. 1988, Dufty and McChrystal 1992). In captivity, females implanted with estradiol respond to FW playbacks with a copulatory, lordosis posture and are more responsive to local FWs than to ones from other dialects (O'Loghlen and Rothstein 1995).

FWs are learned from conspecifics (Rothstein and Fleischer 1987a). Their development has been studied most fully in the High Sierra, where an extreme climate results in a short breeding season providing juvenile cowbirds with little opportunity to hear adult cowbirds (Rothstein et al. 1980, Verner and Ritter 1983). In that region, most yearling males produce foreign FWs or incomplete versions of local FWs. However, nearly all males ≥2 years old give complete versions of the local FW dialect after apparently acquiring the complete template for the local dialect during their first breeding season (Rothstein and Fleischer 1987b, O'Loghlen and Rothstein 1993, O'Loghlen 1995). This is one of the most protracted periods of song learning known among birds (O'Loghlen and Rothstein 1993). However, most cowbird populations breed at lower elevations, and the longer breeding seasons they experience may allow many to most juvenile males to hear enough FWs to enable them to develop complete FWs as yearlings.

Single-syllable Call (SS)

This high-amplitude vocalization consists of a single pure-tone syllable given in the 2–8 kHz range and lasts for less than 0.75 sec. It is more similar to a call note than a song, although most male cowbird vocalizations do not fit easily into the dichotomy between call notes and songs (Rothstein et al. 1988). The acoustic structure of an SS (Figure 7.2) is usually different from any of the syllables that make up the local FW. SSs are often monotonic for most of their duration, but many gradually ascend or descend in frequency and some begin or end with extreme frequency sweeps lasting ≤ 0.1 sec. These extreme frequency sweeps are usually difficult to detect with the unaided ear, and SSs are typically perceived as one-syllable calls in contrast to FWs, which always have two or more distinct syllables and/or buzzes or trills. SSs show a moderate amount of geographic variation, and this variation is concordant with FW variation in the Sierra (Figure 7.2) and possibly elsewhere. Males give only one type of SS in most FW-SS dialects, but two types of SSs are the rule in some dialects. SSs are commonly given both alone and immediately preceding FWs in one of three intensely studied adjacent dialects in the Sierra, although this latter combination is rare in most dialects (Rothstein and Fleischer 1987b; but see Dufty and McChrystal 1992 and

Dufty and Pugh 1994, whose FW1 would be called an SS under our categorization). Although it is usually clear, the structural distinction between FWs and SSs (more than one versus one syllable) is difficult to make in some dialects in certain regions (F in Figure 7.2) and some dialects may lack an SS altogether (see I and J in Figure 7.1). The occasional lack of a clear structural distinction between SSs and FWs and the tendency for males in some areas to combine FWs and SSs in one vocalization has resulted in much confusion. Some previous workers have not distinguished between SSs and FWs (Friedmann 1929, Laskey 1950), whereas others have done so (Darley 1968). However, the distinction is clear and useful in most cases.

When the distinction between local FWs and SSs is clear, their functions and uses (Rothstein et al. 1988) are identical except that: (1) males are more likely to give SSs when conspecifics are nearby, (2) SSs are less likely to precede copulations, and (3) SS playbacks elicit weaker vocal and approach responses. It is unknown whether these differences apply in areas where it is difficult to distinguish between FWs and SSs. Although there are no clear data as to whether males develop an SS from an innate template, males sometimes change their SS between their yearling and subsequent years, indicating that SS development is responsive to social experiences (A. O'Loghlen pers. comm.).

Perched Song (PS)

Unlike FWs and SSs, this moderate-amplitude song type always conforms to certain structural rules defining the sequence and nature of syllables (West et al. 1979). All PSs begin with one or more clusters of brief (<0.1 sec), low-frequency (<1 to 4 kHz) notes (Figure 7.3) that never occur in FWs or SSs. These introductory "glug" notes sound like bubbling water and sometimes go unheard in the field because they are given at relatively low amplitudes and degrade over distances of a few meters (King et al. 1981). After the introductory notes, which last less than a second, PSs immediately jump in frequency to 8 to 12 kHz, giving them the most extreme frequency range of any bird song (Greenewalt 1968). The first high-frequency element is nearly always a brief (<0.1 sec) note known as the interphrase unit, or IPU (West et al. 1981), which is then followed by a longer (0.5 to 0.8 sec) tone that is usually highly modulated and whistle-like in quality. The terminal whistlelike component is often multisyllabic, and these syllables are nearly always unlike those in the local FW, although one case is known in which a local PS and FW shared a similar syllable (Dufty 1988). The terminal component is the loudest part of the song and shows seemingly endless variation (Figure 7.3), sometimes even having a brief trill or buzz embedded within it. Nevertheless, PSs always conform to the species-specific structural rules described here and the low-frequency introductory notes nearly always make PSs recognizable as cowbird vocalizations, unlike FWs.

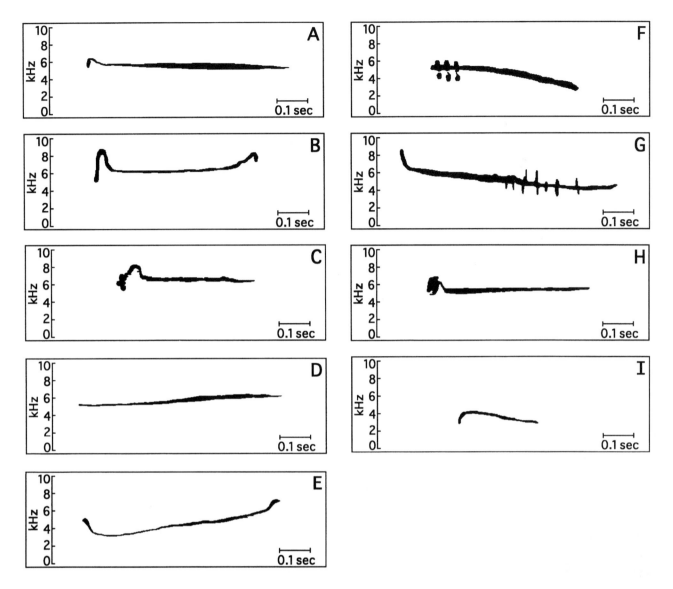

Figure 7.2. Single-syllable calls (SS) from nine dialects. A through D are from the same localities as FWs A–D (Figure 7.1), i.e., most birds in the dialect where FW A occurs give SS A, etc. E is from Nez Perce County, ID. F is from Spokane County, WA, and shows a fairly complex SS, but the three spikes near its beginning are not numerous enough to give it a buzzy or trilled quality or to give it a multisyllabic sound. G was done by the same male in Anne Arundel County, MD, that did FW G in Figure 7.1. H is from Lyon County, KS, where birds did vocalizations easily separable into FWs and SSs, unlike the case further south in the Great Plains in Oklahoma where birds did two types of FWs but no SS (I and J in Figure 7.1). I is from Grant County, WA, and shows a brief low-frequency SS. When such SSs are given, which occurs locally throughout the species' range, individual males often also give a more typical higher-frequency SS, such as those in panels A–E and H. Subspecies responsible for these SSs are as follows: *M. a. artemisiae*—A, B, C, E, F, I; *M. a. ater*—G, H; *M. a. obscurus*—D.

While giving PSs, males commonly do an elaborate and simultaneous song-spread display in which body feathers are fluffed up, the wings spread out in full lateral position, and the entire body tilted downwards (Friedmann 1929). This graded display is done at high intensity only when a PS is directed towards a female or another male within a meter or so of the singer. When lone males give PSs, the only indication of the display is usually a small amount of feather fluffing and a slight spread of the wings with each song.

Males never do any elements of the wing-spread display when giving FWs or SSs or when they are silent.

Each male has a repertoire (Figure 7.3) of one (but usually at least two) to eight distinct PS variants (Dufty 1985, Rothstein unpubl. data). Yearling males in the Sierra averaged 3.1 songs in their PS repertoires, whereas adult males averaged 5.5 song types (O'Loghlen and Rothstein 1993). Adult and yearling males in New York did not differ from each other and had a mean repertoire size of 3.9 (Dufty 1985). Males

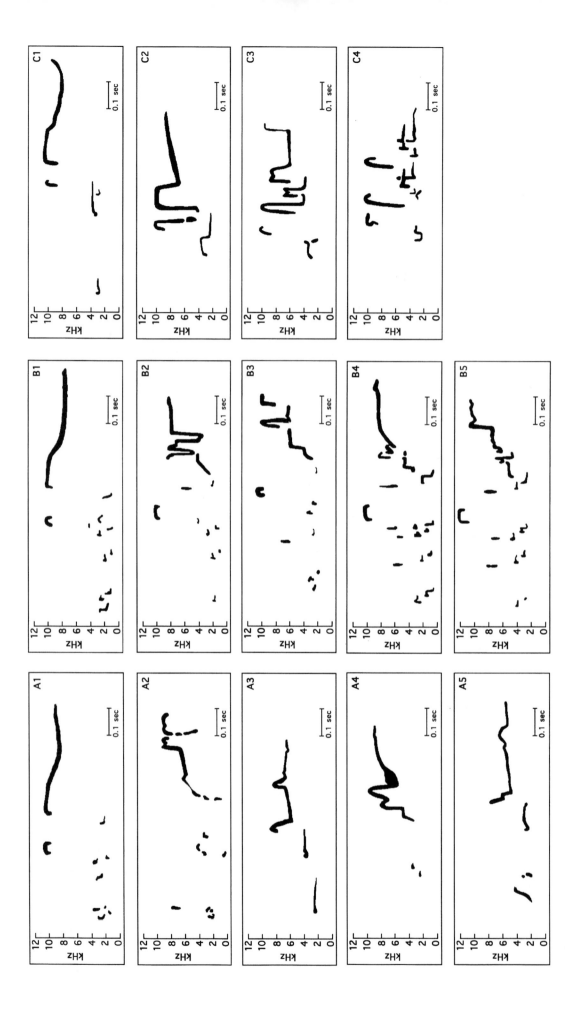

within an area commonly show local song sharing, and the degree of sharing decreases with distance. Dufty (1985) found no well-defined PS dialects with discrete borders in New York, but PS dialects may occur where cowbirds are patchily distributed, as in the Sierra Nevada (Rothstein unpubl. data).

PSs are almost always given while a male is perched and are either directed to a male or female less than 1 m away or nondirected with no conspecifics nearby (Rothstein et al. 1988). On rare occasions, PSs are given in flight as one or more males rapidly pursue a female, which often circles back and forth over the same area (Friedmann 1929, pers. obs.). These chases can go on for 10 min or more and are most likely to occur at the beginning of the breeding season. As with FWs, PSs precede most copulations. However, nearly all copulations occur within 5 sec after a male and female come together after being separated. For this reason, the conspicuous, long sequences of PSs that males often direct at females rarely culminate in copulations (Rothstein et al. 1988, Dufty and McChrystal 1992).

Two field playback studies involving PSs found that cowbirds showed little response (Dufty 1982a, Yokel 1989). But these studies used repertoire sizes of one or three song types and a more extensive experiment that used playbacks with larger and more typical repertoire sizes showed that PS playback elicits male approaches and vocalizations (Zelaya and Rothstein unpubl. data).

There are no obvious differences between the PSs that males direct to other males, which seem to function as threat gestures, and those directed to females, which seem clearly to represent courtship (Rothstein unpubl. data). The function of seemingly undirected PSs given when no nearby conspecifics are visible is unclear. These PSs could be comparable to the broadcast song given by most passerines, or they

could actually be directed to inconspicuous females in the general area.

West and King and their associates (West et al. 1981, Eastzer et al. 1985, West and King 1986, King and West 1988) have studied development of PS primarily in captive birds. Unlike the songs of nearly all songbirds, males develop normal or nearly normal PSs if raised in acoustic isolation. However, they learn to modify these innate PSs in response to the songs other males sing and in response to nonvocal feedback provided by females (King and West 1983). Modification of PS repertoires has also been shown in nature, as males in the Sierra delete unique songs (types shared with no other male) sung as yearlings and replace them with a larger number of shared songs by the age of two years (O'Loghlen and Rothstein 1993). If young captive cowbirds are deprived of all conspecific contact and kept only with a heterospecific in small cages for most of their first year, they may develop some vocalizations learned from other species and may even court them (Freeberg et al. 1995, West et al. 1996). But such extreme conditions never occur in nature, and cowbirds in the wild appear never to learn heterospecific vocalizations despite being reared by numerous host species.

Kek or Chuck

This is a low-amplitude note lasting less than 0.2 sec (Figure 7.4D). It is given commonly by flying or perched males, and less commonly by females, when conspecifics are nearby or when there is reason to behave as if conspecifics are nearby, e.g., when males are attracted to a speaker playing chatter calls. Unlike other cowbird vocalizations, this may be a graded signal, as its structure seems to vary. Hence it includes both the "kek" or "tek" and the "chuck" or "kuk" calls recognized by Friedmann (1929) and Dufty and McChrystal (1992). Alternatively, further study might show that its variation is not continuous but occurs in discrete categories. It is usually not detectable beyond 5 m (Friedmann 1929), so many researchers are unaware of its existence. There is some degree of geographic variation, at least in amplitude, as these notes can be heard from 50 m or more on the west (but not the east) slope of the Sierra (Rothstein pers. obs.). We know of no data on the ontogeny of this call.

Chatter (CH) or Rattle Call

CH is a high-amplitude call (Figure 7.4A–C) commonly given by females and occasionally by males and is the only female vocalization likely to be heard in the field. Its duration is usually 1.5–2.5 sec, and it consists of a rapid series of 10–25 (mean = 18) brief (<0.1 sec) chevron-shaped (triangular) notes in the 2–6-kHz range (Burnell and Rothstein 1994). Univariate and multivariate analyses of 39 frequency and duration parameters of calls from 45 females from all three subspecies showed little or no geographic variation (Burnell and Rothstein 1994). But these analyses also

Facing page

Figure 7.3. Perched song repertoires of three male cowbirds. Male A is an *M. a. artemisiae* from Grant County, WA. Male B is an *M. a. obscurus* from Merced County, CA. Male C is an *M. a. ater* from Prince Georges County, MD. Male B had two other song types in his repertoire in addition to the five shown here. Interphrase units, or IPUs (see text), are missing from the audiospectrograms of songs 2–5 of male A. It is unclear whether males sometimes fail to include this element or whether field recordings without IPUs were made from too far away for the IPU to register. Some panels show few or no low-frequency (<3 kHz) introductory notes (especially for male C) because these are given at low amplitudes and are often not picked up in field recordings. Note that the high-frequency component of the first song for each male (A1, B1, and C1) is shared whereas none of the other song types is shared.

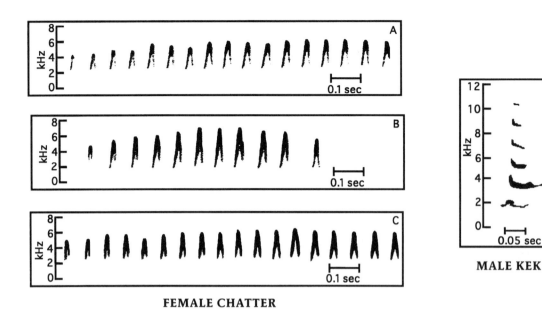

FEMALE CHATTER

MALE KEK

Figure 7.4. Female chatter calls (A–C) and a male kek note (D). Panels A and B show calls recorded from two *M. a. obscurus* females, and C is an *M. a. ater* female. D shows a kek note recorded from an *M. a. obscurus* male in Santa Barbara County, CA.

showed minor quantitative variation among females that may allow for individual recognition.

CH is given in flight or while stationary, and in response to either FWs by distant males or CH calls by distant females (Rothstein et al. 1988, Yokel 1989). This call is also used when another cowbird is close by, either another female or a male giving PSs. Females often begin their CH call before a male finishes a PS, and the two vocalizations can run together so that they sound like one bird (pers. obs.). CH playbacks elicit strong approach and vocal responses by both sexes (Dufty 1982a, Yokel 1989) even if initiated when no cowbirds are known to be nearby (Rothstein et al. 1988). The development of CH appears to be under strong genetic control (Burnell and Rothstein 1994).

Male CH is apparently identical to Friedmann's (1929: 167) "call to the flock" and is sufficiently rare that 95% or more of the CH calls heard in the field are from females (Rothstein pers. obs.). But census takers should be aware that a cowbird detection that is based only on CH may in fact be a male. In our experience, males rarely give more than one or two CH calls in succession, so a long series of CH calls is almost certain to be from a female. Unlike female CH (Figure 7.4), the individual notes in male CH usually show little variation in amplitude or peak frequency (Rothstein unpubl. data). Thus observers can learn to differentiate most male and female CH calls.

The Use of Cowbird Vocalizations in Monitoring and Censusing Studies

Cowbirds present special problems in studies that attempt to gather data on the absolute or relative numbers of breeding birds in an area. For most species of passerines, singing males provide the most common type of reliable censusing data because males give frequent broadcast song on relatively small home ranges. In other words, an observer has a high likelihood of hearing a resident male songbird if a site is visited at the right time of day and year. However, male cowbirds have relatively large and overlapping home ranges (Dufty 1982a,b; Rothstein et al. 1984) and broadcast song much less often than most male passerines. Even when cowbirds have favorite singing perches (Friedmann 1929), these are often far apart and each is used for a relatively small part of each morning (Rothstein pers. obs.). Furthermore, although it is widely applied to most passerines, the assumption that nearly all singing males detected are territorial males with a mate may not apply to cowbirds because male cowbirds are not territorial and because they greatly outnumber females (Dufty 1982a,b; Rothstein et al. 1986; Yokel 1989).

In light of these problems, how should one quantify cowbird abundance? This question is especially acute for females since they are the more important sex in management considerations yet are much more secretive than males. We believe that it is best to count all cowbird detections but that male and female detections should be noted separately. This will allow separate statistical analyses of numbers of males, of females, and of total cowbirds (although it is unlikely that these separate analyses will be independent statistically). The advantage of counting all cowbirds, even when females are the sex of primary interest, is that this results in larger numbers of birds and of sites with cowbirds and adds therefore to statistical power. Furthermore, even though males typically account for most detections, their numbers are likely to be correlated with the numbers of females since males are strongly attracted to females (Dufty 1982a, Rothstein et al. 1986). Also, some studies have shown that unmated subordinate males are relatively silent and otherwise secretive (Rothstein et al. 1986). Thus the males that are detected in censusing efforts may be primarily mated males and therefore a good guide to the numbers of females using an area.

Some types of male detections may indicate cowbird use of a site more clearly than other types. For example, because PSs given in flight are so rare, detection of PSs can be assumed to mean that cowbirds are using a site. On the other hand, some detections of FWs and SSs could be due to males that are merely flying over a site. Nevertheless, we suggest that all types of detections be counted. Both males and females respond to FWs (Rothstein et al. 1988), so a male flying overhead and giving FWs and SSs is probably attempting to solicit vocalizations from conspecifics on the site. Once females or other males respond, it is likely that the flying male will approach them and therefore use the site. On many occasions, we have begun to broadcast CH playbacks while watching a male cowbird give FWs as he rapidly flew past a playback site. These males instantly changed direction and flew towards the playback site as soon as the CH was heard. We suspect that a male flying over an area and making his presence known with FWs is visiting areas where his mate is likely to be found because cowbirds are monogamous in most regions but the members of a pair are often apart (Yokel and Rothstein 1991). Although the areas frequented by a mated male may be those that his mate is likely to use, males will respond to CH from any female if they are not with their mate (Dufty 1982a).

The Chatter Playback Experiment

Because of the interactive nature of cowbirds described above, wild cowbirds are responsive to playback of CH and FWs (Rothstein et al. 1988). Such playbacks provide a rich source of data relevant to basic questions regarding communication and vocal function (Dufty 1982a, Dufty and Pugh 1994, Rothstein et al. 1988) and also have potential uses in censusing and managing cowbird and host populations. If censusing and management efforts employ playbacks, they should use female CH, as it produces the strongest responses (Yokel 1989, pers. obs.) and one need not

secure recordings of local birds (unlike the case for FWs). Playback of CH recorded in California was equally effective in attracting cowbirds in the West and along the east coast (Rothstein et al. 1988).

In CH playbacks, a live or mounted decoy female placed near the speaker results in the closest approaches, but males and females often come within a meter of a speaker even if no decoy is present. However, if no decoy bird is used, cowbirds will approach CH playbacks more closely if the speaker is concealed in vegetation.

Chatter playbacks can aid censusing and management efforts in four ways:

(1) By undergoing a brief training period during which they attract males with CH playbacks, field workers can learn to recognize local FWs and SSs because males will give these vocalizations at distances of 20 m or less. This can result in more accurate censuses. But playbacks should not be done on the same day as formal censusing because playbacks seem to give cowbirds a prolonged interest in a site (pers. obs.) and could affect their movement patterns.

(2) CH playbacks can aid cowbird control programs because females approach playback to within 20 m or less and can then be easily shot. Depending on the size of the local cowbird population, this method may be more economical than the use of decoy traps.

(3) Cowbirds can be captured in breeding habitat for behavioral studies (radio tracking, etc.) or blood samples with mist nets or drop traps after being attracted by CH playbacks (Rothstein et al. 1984). In these cases, it is usually necessary to place a mounted or live decoy female near the net or trap. This method may be more economical than the use of a decoy trap if only small numbers need to be caught. Also, decoy traps provide a source of social attraction and of food and may change local patterns of spatial use. Therefore, decoy traps are clearly unsuitable for some studies.

(4) Female cowbirds are often secretive and therefore difficult to detect in breeding habitat. The use of CH playbacks may result in more accurate censuses by eliciting responses from females that would otherwise go undetected. Playbacks are also likely to increase male detections.

We next present the results of a field experiment designed to assess the extent to which CH playbacks can boost detection of cowbirds in a censusing study.

Methods and Research Area

We performed 10-min counts of cowbirds and all other birds at 99 sites over a 310-km, north-south span along the eastern Sierra Nevada and the adjacent high-elevation Owens Valley in Inyo, Kern, and Mono counties, California. Sites were at least 1.6 km apart and were a subset of sites involved in a 1978 study (Rothstein et al. 1980). Counts were made twice at each site, once without and once with CH playback. The latter were always done after the former because prior observations suggested that cowbirds might return to a site after hearing CH played there even if the play-back is no longer on. In most cases, counts with playback were done the day after those with no playback. But in some cases, up to 3 days elapsed between the counts, and in a few cases playbacks were done immediately after the first count. All counts were done during June and July 1993, and all but a few were initiated before 1000 PST. Our playback tape had CH sequences from two females recorded on 13 July 1985 in Mono County. It consisted of 4- to 10-sec bursts of CH, was silent 40% of the time, and was played continuously for 10 min. A peak playback level of 95 db at one meter was achieved via use of an Ampli-Vox S702 amplifier powered by ten D cells, and a Realistic Minimus 7 speaker.

Results and Discussion

Across all 99 sites, male cowbirds were detected during 30% and 17% of counts with and without playback, respectively. The comparable data for females were 18% and 6%. Cowbirds were detected during significantly more 10-min counts with than without playback ($N = 198$ counts; $\chi^2 = 4.71$, $P = .03$, and $\chi^2 = 6.83$, $P = .01$, for males and females respectively). Consideration of the numbers of cowbirds detected at each site gives these data increased power. Significantly more cowbirds were detected during counts with playback than with no playback regardless of whether analyses considered males only, females only, or the total number of males plus females (all P values $\leq .002$, Wilcoxon tests in Table 7.1).

Many sites chosen for this experiment were places we expected to have low cowbird numbers based on habitat and on results obtained in 1978, and no cowbirds were detected during either count at 59 of the 99 sites. The best measure of the potential of playbacks to increase cowbird detections is to consider only those sites at which at least one cowbird was detected during either count, i.e., sites that definitely had cowbirds. Among these sites, playback increased mean male detections by a factor of 2.34 (1.842/0.789, from Table 7.1) and female detections by 3.71 (1.30/0.35). Thus, playbacks were 1.6 times (3.71/2.34) more effective at increasing female than male detections. This male-female difference was expected because females are more secretive than males (Friedmann 1929). In addition, a radio-tracking study in this region found that females were "generally silent," whereas males were "highly vocal" (Rothstein et al. 1984). The greater effect of playbacks on females can also be seen by the fact that the male : female ratio was 2.25 with no playback (0.789/0.350, from Table 7.1) but 1.42 (1.842/1.300) with playback.

Observations from this and other playback studies indicate that cowbirds may be attracted from as far away as 100–200 m or more. While such large distances could be a censusing problem for most passerine species because they exceed the home range sizes of those species, this is unlikely to be a difficulty with cowbirds because of their huge home ranges. Even the 12.6-ha area covered by a 200-m attraction zone is much smaller than the areas over which cowbirds range and search for nests in the morning, 20 ha in New York (Dufty 1982b) and 68 ha in the Sierra Nevada (Rothstein et al. 1984).

Because CH playbacks increase the number of cowbirds detected, they are likely to add statistical power to a study. This increased statistical power may in turn reduce the costs

Table 7.1. Numbers of Cowbirds Detected during 10-min Counts with and without Playback (pb) of Female Chatter Calls

	Males		Females		Total	
	no pb	pb	no pb	pb	no pb	pb
Mean, all 99 sites	0.30	0.71	0.07	0.26	0.37	0.97
Median, all 99 sites	0.0	0.0	0.0	0.0	0.0	0.0
Mean, known cowbird sites[a]	0.79	1.84	0.35	1.30	0.93	2.40
Median, known cowbird sites	0.0	1.0	0.0	1.0	0.0	2.0
Number of sites with more cowbirds[b]	10	26	3	16	10	29
Wilcoxon test	$P = 0.002$		$P = 0.002$		$P = 0.0004$	

[a]Means for sites where at least one cowbird in a category was detected during at least one of the two count periods at the site. The results of Wilcoxon tests on these sites are identical to the tests for all sites.

[b]Number of sites where one of the counts, no pb or with pb, had a greater number of cowbirds in each category (males, females, or total) than the other count at the same site. Sites with ties, nearly all of which had zero cowbirds on both counts, are not included.

of certain research programs. For example, compared to studies without playback, the use of playback may enable one to monitor a smaller number of sites in a study assessing temporal changes in cowbird numbers. Thus researchers beginning long-term studies may benefit from using playbacks at the initiation of their project. However, cowbirds habituate to CH playbacks, so these should not be used at the same site more often than once every week or two.

Although CH playbacks clearly increased the detection of cowbirds in our study, we suspect that they are likely to be less effective in boosting detections when cowbirds are more common. When cowbirds are relatively abundant, they will often be with conspecifics, and responsiveness to playbacks decreases if cowbirds are already interacting with another cowbird (Dufty 1982a, pers. obs.). In addition, when cowbirds are in the vicinity of conspecifics, they are likely to reveal their presence via auditory cues because cowbirds almost invariably vocalize when together in the morning (pers. obs.); i.e., when cowbirds are abundant, their interactive nature may result in most individuals being detected even without playbacks.

Because cowbirds of both sexes almost invariably interact vocally when they meet conspecifics in morning breeding habitat, the relation between cowbird detections and actual cowbird numbers is probably not linear, i.e., the per capita detection rate without playbacks may increase as cowbird density increases. Thus, workers should use caution in choosing the types of statistical analysis they use for data on cowbird numbers. If there are large differences in cowbird numbers among samples, an analysis using linear regression may be inappropriate. In such cases, nonparametric correlation tests are likely to be more reliable because they require no assumptions about a fixed relationship between some measure of cowbird numbers and other variables. Importantly, it may also be incorrect to assume that there is a consistent relationship between numbers of cowbirds detected and the overall impact cowbirds impose on local host populations, i.e., as *apparent numbers* of cowbirds increase, the impact per detected female may decline.

Acknowledgments

Field work involving cowbird vocalizations was supported by NSF grants BNS 82-16778 and BNS 86-16922 to SIR. The playback experiment was supported by a grant from the Pacific Southwest Research Station, USDA Forest Service, Albany, California. Susan Carter provided able assistance with the playback experiment. We thank Alfred M. Dufty, Adrian O'Loghlen, Scott Robinson, and Spencer Sealy for their helpful comments on this manuscript.

References Cited

Alcock, J. 1993. Animal Behavior. Sinauer Associates, Sunderland, MA.

Burnell, K., and S. I. Rothstein. 1994. Variation in the structure of female Brown-headed Cowbird vocalizations and its relation to vocal function and development. Condor 96:703–715.

Campbell, N. A. 1993. Biology. Benjamin/Cummings: Redwood City, CA.

Darley, J. A. 1968. The social organization of breeding Brown-headed Cowbirds. Ph.D. Thesis, University of Western Ontario, London, Ontario.

Dufty, A. M., Jr. 1982a. Response of Brown-headed Cowbirds to simulated conspecific intruders. Anim. Behav. 30:1043–1052.

———. 1982b. Movements and activities of radio-tracked Brown-headed Cowbirds. Auk 99:316–327.

———. 1985. Song sharing in the Brown-headed Cowbird (*Molothrus ater*). Z. Tierpsychol. 69:177–190.

———. 1988. Flight whistle incorporated in Brown-headed Cowbird song. Condor 90:508–510.

Dufty, A. M., Jr., and R. McChrystal. 1992. Vocalizations and copulatory attempts in free-living Brown-headed Cowbirds. J. Field. Ornithol. 63:16–25.

Dufty, A. M., Jr., and J. K. Pugh. 1994. Response of male Brown-headed Cowbirds to broadcast of complete or partial flight whistles. Auk 111:734–739.

Eastzer, D. H., A. P. King, and M. J. West. 1985. Patterns of courtship between cowbird subspecies: Evidence for positive assortment. Anim. Behav. 33:30–39.

Freeberg, T. M., A. P. King, and M. J. West. 1995. Social malleability in cowbirds Molothrus ater artemisiae: Species and mate recognition in the first year of life. J. Comp. Psychol. 109:357–367.

Friedmann, H. 1929. The cowbirds: A study in the biology of social parasitism. C. C. Thomas, Springfield, IL.

Greenewalt, C. H. 1968. Bird song: Acoustics and physiology. Appleton-Century-Crofts: New York.

King, A. P., and M. J. West. 1977. Species identification in the North American cowbird: Appropriate responses to abnormal song. Science 195:1002–1004.

———. 1983. Epigenesis of cowbird song: A joint endeavour of males and females. Nature 305:704–706.

———. 1988. Searching for the functional origins of song in eastern Brown-headed Cowbirds, Molothrus ater ater. Anim. Behav. 36:1575–1588.

King, A. P., M. J. West, D. H. Eastzer, and J. E. R. Staddon. 1981. An experimental investigation of the bioacoustics of cowbird song. Behav. Ecol. Sociobiol. 9:211–217.

Laskey, A. R. 1950. Cowbird behavior. Wilson Bull. 62:157–174.

Manning, A., and M. S. Dawkins. 1992. An introduction to animal behavior. Cambridge University Press: Cambridge.

Mayr, E. 1979. Concepts in the study of animal behavior. Pp. 1–16 in Reproductive behavior and evolution (J. S. Rosenblatt and B. R. Komisaruk, eds). Plenum: New York.

O'Loghlen, A. L. 1995. Delayed access to local songs prolongs vocal development in dialect populations of Brown-headed Cowbirds. Condor 97:402–414.

O'Loghlen, A. L., and S. I. Rothstein. 1993. An extreme example of delayed vocal development: Song learning in a population of wild Brown-headed Cowbirds. Anim. Behav. 46:293–304.

———. 1995. Culturally correct song dialects are correlated with male age and female song preferences in wild populations of Brown-headed Cowbirds. Behav. Ecol. Sociobiol. 36:251–259.

Rothstein, S. I., and R. C. Fleischer. 1987a. Brown-headed Cowbirds learn flight whistles after the juvenile period. Auk 104:512–516.

———. 1987b. Vocal dialects and their possible relation to honest status signalling in the Brown-headed Cowbird. Condor 89:1–23.

Rothstein, S. I., J. Verner, and E. Stevens. 1980. Range expansion and diurnal changes in dispersion of the Brown-headed Cowbird in the Sierra Nevada. Auk 97:253–267.

———. 1984. Radio-tracking confirms a unique diurnal pattern of spatial occurrence in the parasitic Brown-headed Cowbird. Ecology 65:77–88.

Rothstein, S. I., D. A. Yokel, and R. C. Fleischer. 1986. Social dominance, mating and spacing systems, female fecundity, and vocal dialects in captive and free-ranging Brown-headed Cowbirds. Curr. Ornithol. 3:127–185.

———. 1988. The agonistic and sexual functions of vocalizations of male Brown-headed Cowbirds, *Molothrus ater*. Anim. Behav. 36: 73–86.

Tyler, W. M. 1920. The cowbird's whistle. Auk 37:584.

Verner, J., and L. V. Ritter. 1983. Current status of the Brown-headed Cowbird in the Sierra National Forest. Auk 100:355–368.

West, M. J., and A. P. King. 1986. Song repertoire development in male cowbirds (*Molothrus ater*): Its relation to female assessment of song potency. J. Comp. Psychol. 100:296–303.

West, M. J., A. P. King, and D. H. Eastzer. 1981. The cowbird: Reflections on development from an unlikely source. Amer. Sci. 69:56–66.

West, M. J., A. P. King, D. H. Eastzer, and J. E. R. Staddon. 1979. A bioassay of cowbird song. J. Comp. Physiol. Psychol. 93:124–133.

West, M. J., A. P. King, and T. M. Freeberg. 1996. Social malleability in cowbirds: New measures reveal new evidence of plasticity in the eastern subspecies (*Molothrus ater ater*). J. Comp. Psychol. 110:15–26.

Yokel, D. A. 1989. Intrasexual aggression and the mating behavior of Brown-headed Cowbirds: Their relation to population densities and sex ratios. Condor 91:43–51.

Yokel, D. A., and S. I. Rothstein. 1991. The basis for female choice in an avian brood parasite. Behav. Ecol. Sociobiol. 29:39–45.

PART II

Cowbird Spacing Behavior,
Host Selection, and
Negative Consequences
of Parasitism for
Commonly Used Hosts

8. Introduction

JAMES N. M. SMITH,

SPENCER G. SEALY, AND TERRY L. COOK

The chapters in this part deal with three topics likely to be of interest to both academic biologists and habitat managers. Two of these concern related aspects of cowbird behavior: how individual cowbirds select and use breeding, feeding, and roosting areas and how host species are chosen within such local areas. The third topic is a consequence of host selection, how the frequent use by cowbirds lowers the reproductive success of individual host species. We now discuss each topic in turn.

Cowbird Spacing Behavior

Early students of cowbird behavior (e.g., Friedmann 1929, Nice 1937, Laskey 1950) realized that brood parasitism frees cowbirds from a major constraint common to most songbirds, the need to own a single all-purpose breeding and feeding area. Cowbirds can be much more mobile than most other passerines, because they do not need to incubate their eggs and feed and brood their young. They thus can use different areas for each need at different times of day, usually searching for nests on the breeding area in the morning and foraging in flocks in the afternoon (e.g., Rothstein et al. 1980, 1984). Darley (1968, 1983) conducted the first detailed study of spacing behavior by color-banded cowbirds in an urban riparian area. He found that female Brown-headed Cowbirds defended barely overlapping breeding territories, to which they returned in more than one year.

It was not, however, until cowbirds were fitted with miniature radio transmitters that the full extent of cowbirds' mobility was revealed. Three papers published in the early 1980s showed that female cowbirds used morning breeding ranges that varied in size from 5 to 150 ha, and that the degree of overlap in use of space between neighbors also varied strongly (Dufty 1982, Rothstein et al. 1984, Teather and Robertson 1985). While Dufty found apparent territoriality in an urban habitat similar to the one studied by Darley, the other two studies found much overlap in the ranges of individuals in more natural habitats. One striking result cap-

tured considerable attention. Cowbirds of both sexes were shown to move daily up to 7 km from their morning breeding ranges to afternoon feeding sites in the Sierra Nevada mountains (Rothstein et al. 1984). In these pioneering studies, only a few individuals were fitted with radios, and it was therefore difficult to judge how representative the results were.

The first telemetry study of cowbirds was begun by Raim (1978) well before the above studies, but Raim's detailed results are published here for the first time (Chapter 9). By radiotagging 16 females, and following over 100 color-banded females, Raim established several new results and confirmed others. First, at any one time, most females in an Illinois state park defended nearly exclusive areas, but the sizes of these areas varied greatly over time for some individuals, and some females exhibited breeding dispersal within a season. Two of Raim's marked females used the same or neighboring breeding ranges for at least five years, and other birds relocated to ranges adjacent to ones used in previous years. Thus, fidelity to breeding sites in cowbirds is strong, as noted by others (Smith and Arcese 1994, J. N. M. Smith and M. J. Taitt unpubl. data).

Unlike the cowbirds studied by Rothstein et al. (1984) in California, some of Raim's birds had feeding areas very close to their breeding sites and were thus able to search for nests throughout the day. Other females commuted considerable distances, like the Sierra Nevada birds. Recent studies by B. E. Woolfenden, H. L. Gibbs, and S. G. Sealy (unpubl. data) have also found that most cowbirds breeding at Delta Marsh in Manitoba remain on their breeding ranges to feed.

The second chapter on the spacing behavior of cowbirds, Chapter 10 by Thompson and Dijak (also Thompson 1994) is noteworthy for its large samples of radiomarked birds, its geographic extent, and an illuminating analysis of the selection of feeding habitat. More than 20 cowbirds were tracked in detail using standard methods at three different sites in the Midwest. Despite considerable variation in the extent of forest cover and cowbird density among the sites, most cowbirds traveled less than 1 km from breeding sites to feeding sites, although travel distances over 5 km were noted. Feeding sites on short-grass pastures were preferred, and cattle feedlots were used extensively when they were available. These results suggest that cowbird numbers might be managed locally by minimizing these habitats. Many female cowbirds also returned to their breeding sites in the afternoons. A further new finding was that roost sites tended to be farther from both breeding and feeding sites than breeding and feeding sites were from each other. Raim also noted movements to distant roost sites and that cowbirds use habitats selectively when feeding, again apparently preferring short-grass pastures.

The picture emerging from these two studies of cowbird movements and the previous work referred to above is that cowbird females exhibit variable territoriality on their breeding sites. Why this variability exists is not yet well understood. Feeding sites are undefended, and cowbirds probably travel only as far from their breeding areas as necessary to find food. The long-range commuting first noted by Rothstein et al. (1984) is not unique to the Sierra Nevada, but it is not typical of most individual cowbirds in the parts of North America where cowbirds are a conservation threat. In these areas, cowbirds often feed near, or even on, their breeding ranges.

Host Selection

Generalist brood parasites differ fundamentally from nearly all other birds in selecting not just one but many types of hosts and nest sites (Colias and Colias 1984, Ehrlich et al. 1988). Although they are extreme host generalists as species, Brown-headed and Shiny Cowbirds tend to use only a few hosts heavily at any one site (e.g., Hahn and Hatfield Chapter 13, Nakamura and Cruz Chapter 20, Wiley 1988), and individuals may specialize on particular habitats. If there were simple and stable patterns of host use across regions, it would be possible to predict the suite of local hosts likely to be used at any site, and thus the need for local cowbird management, from information gathered elsewhere.

There are general patterns in host use. Among the three North American cowbird species, there are sharp differences in host selection. The Bronzed is thought to specialize on other icterines as hosts (Lowther 1995), although the host relations of this species have yet to be studied in detail. The Shiny prefers cavity-nesting species (Wiley 1988), while the Brown-headed prefers open-cup nesters (Friedmann and Kiff 1985). Forest-nesting species are usually parasitized by Brown-headed Cowbirds more often than grassland-nesting species in the same region (Robinson et al. Chapter 33), a puzzling result, given the Brown-headed Cowbird's presumed ancestral range in the prairies. Several of the most commonly used hosts of the cowbird, the Yellow Warbler (*Dendroica petechia*), Red-eyed Vireo (*Vireo olivaceus*), Song

Sparrow (*Melospiza melodia*), Wood Thrush (*Hylocichla mustelina*), and Common Yellowthroat (*Geothlypis trichas*), are heavily used across the continent within their ranges (Friedmann et al. 1977, Friedmann and Kiff 1985). The frequency of use among such commonly used hosts is roughly proportional to the regional population density of cowbirds (Hoover and Brittingham 1993 and see below, Smith and Myers-Smith in press). There is also some evidence that unsuitable host species are avoided by cowbirds (see Chapter 1). Some studies have demonstrated clear local host selection (e.g., Briskie et al. 1990) but have failed to find an adaptive explanation for it.

However, it is also evident from the natural history of host relations (e.g., Friedmann et al. 1977, Friedmann and Kiff 1985) that use of particular host species varies geographically to such an extent that patterns of local host use cannot always be generalized readily. In addition, the mechanisms underlying this variation in host use are far from clear.

Three chapters in this part bear on these topics. Dufty provides an additional and clear example of host selectivity (Chapter 12). Yellow-headed Blackbirds (*Xanthocephalus xanthocephalus*) are rarely used as hosts (Friedmann and Kiff 1985), although they can rear cowbird young when parasitized experimentally (Ortega and Cruz 1988). Dufty, however, shows that cowbirds use yellowheads as hosts frequently when nests of Red-winged Blackbirds (*Agelaius phoeniceus*), a preferred local host that almost always shares the marsh habitat of the yellowhead (Orians 1980), are unavailable locally. Carello and Snyder show that redwings at neighboring sites in similar riparian habitat can experience markedly different frequencies of parasitism (Chapter 11), perhaps because of differences in marsh structure and in defensive aggression by redwings and co-occurring Common Grackles (*Quiscalus quiscula*). The final chapter on host selection, Chapter 13 by Hahn and Hatfield, provides an example of selection based on host nesting microhabitat. They found that cowbirds breeding in deciduous forest in New York State used several low- and ground-nesting hosts, while parasitizing some commonly used off-ground nesting species, including the Wood Thrush, less frequently. However, species identity, not nest height per se, explained host selection in Hahn and Hatfield's study.

A general difficulty with studies of host selection under field conditions is illustrated by these studies. In nature, many unmeasured factors may influence host choice, and experimental tests of host selectivity are difficult and perhaps ethically questionable. In particular, whether a moderately preferred host is used heavily or not is likely to depend on the local abundance of other, more preferred hosts. Conclusions on host selectivity can therefore not be generalized uncritically across regions. Local studies on host use are needed when habitat managers are considering active cowbird management.

Impacts of Cowbird Parasitism on Hosts

Cowbird parasitism reduces host reproductive success mainly through two mechanisms, removal of host eggs by the female cowbird and the results of competition between parasite and host nestlings for food and space in the nest (Payne 1977). Cowbirds use a wide range of host sizes, from warblers and vireos that weigh 8–15 g to blackbirds weighing over 100 g (Friedmann et al. 1977, Friedmann and Kiff 1985, Weatherhead 1989, Sealy 1996; chapters by DeCapita, Hayden et al., and Peer and Bollinger in this volume). Small hosts suffer most from parasitism, particularly when their incubation period is long relative to that of the cowbird (Grzybowski 1995). Large hosts generally lose fewer young when parasitized and can often rear one or more cowbirds successfully, in addition to some of their own young (Weatherhead 1989, Robinson et al. 1995a).

When parasitism is particularly frequent, however, as often seen in forest fragments in the Midwest (Robinson et al. 1995b and Chapter 33), even large and usually resistant hosts can struggle to rear many of their own young. Studies of the Wood Thrush have been particularly illuminating in this regard. Using data from nest records in the Cornell University Laboratory for Ornithology, Hoover and Brittingham (1993) found high percentages of thrush nests parasitized and frequent multiple parasitism in the Midwest, in association with high regional cowbird densities.

In perhaps the most striking example of this phenomenon yet reported, Trine describes in Chapter 15 the consequences of extreme parasitism on Wood Thrushes in Illinois, where virtually every thrush nest is multiply parasitized. She demonstrates that host reproductive success declines sharply as the number of cowbird eggs rises. Thrush nests receiving five or more cowbird eggs produce fewer than 0.7 thrush fledglings, but over 1.8 cowbird fledglings, on average. Parasitism at these levels results in host populations declining unless, as seen for Wood Thrushes in Illinois and for Song Sparrows in the Pacific Northwest, immigrant birds colonize from more productive sites elsewhere (Robinson et al. 1995b and Chapter 33, Brawn and Robinson 1996, Rogers et al. 1997). Trine's work also suggests that female cowbirds compete strongly with each other to get their eggs into nests of large and effective hosts and that this scramble competition reduces the reproductive success of individual cowbird females.

A general pattern in studies of cowbird-host relations is for host populations in heavily forested and montane areas to experience very low levels of parasitism in comparison to hosts in agricultural and urban landscapes (e.g., Peck and James 1987, Smith and Myers-Smith in press). Exceptions to this pattern can occur in open mountain forests near urban centers. Chace et al. show in Chapter 14 that Plumbeous Vireos (*V. plumbeus*) suffered fairly frequent parasitism in pine forests in Colorado and consequent greatly reduced nesting success. Chace et al.'s study also reinforces the increasing evidence that vireos are among the most vulnerable of all passerines to parasitism by Brown-headed Cowbirds (Friedmann and Kiff 1985, Grzybowski 1995, Sealy 1996, Barber and Martin 1997, Woodworth 1997, Ward and Smith Chapter 25).

The final chapter in this part raises a general issue concerning the interpretation of much of the data in this volume. Most of the chapters in this book report only data on nest success, i.e., on independent nesting attempts by an unknown number of host females. It is, however, the reduction in seasonal reproductive success of the average host female that determines the total impact of parasitism on host population dynamics (Smith 1981, May and Robinson 1985). Estimating seasonal reproductive success in the many species that nest more than once a year generally requires close study of individually marked host females (e.g., Smith 1981, Woodworth 1997). Population modeling, however, offers an alternative to such intensive field studies.

Grzybowski and Pease use a simulation model in Chapter 16 to explore how varying levels of cowbird parasitism, nest desertion by hosts, and predation on eggs and nestlings combine to influence seasonal reproductive success (their "seasonal fecundity") of hosts. Their results demonstrate several interesting patterns. First, the impact of parasitism depends strongly on the frequency of nest predation. Less obviously, even frequent nest failures induced by parasitism may not reduce seasonal reproductive success greatly, provided the host eventually succeeds in rearing a brood. This occurs commonly because many species of hosts have nesting seasons longer than that of the cowbird (e.g., Braden et al. 1997, Smith and Arcese 1994). Two further factors not considered by Grzybowski and Pease here, but which can be addressed by their models (Pease and Grzybowski 1995), are the quality of host young resulting from late nests and seasonally varying patterns of parasitism and predation. Seasonal trends in cowbird parasitism and nest failure rate are common in open-nesting songbirds (e.g., Dowell et al. Chapter 29, Rogers et al. 1997, J. N. M. Smith and M. J. Taitt unpubl. data). Late nests may suffer more frequent failure, and late-hatched young often survive poorly (e.g., Hochachka 1990), but this result may not apply in host populations where few early young are produced because of cowbird parasitism.

The important implication of Grzybowski and Pease's work for managers contemplating predator or cowbird removal is that it is necessary to estimate seasonal reproductive success of host species that are believed to be threatened, not merely nest success. This sobering conclusion reduces the force of several studies of the ecological impact of parasitism included in this book, and of other studies reviewed elsewhere (Robinson et al. 1995a). However, the demography of small open-nesting songbirds may be sufficiently predictable (e.g., Newton 1989) that models can be used to estimate seasonal production of young with some confidence.

References Cited

Barber, D. R., and T. E. Martin. 1997. Influence of alternate host densities on Brown-headed Cowbird parasitism rates in Black-capped Vireos. Condor 99:595–604.

Braden, G. T., R. L. McKernan, and S. M. Powell. 1997. Effects of nest parasitism by the Brown-headed Cowbird on nesting success of the California Gnatcatcher. Condor 99:858–865.

Brawn, J. D., and S. K. Robinson. 1996. Source-sink population dynamics may complicate the interpretation of long term census data. Ecology 77:3–12.

Briskie, J. V., S. G. Sealy, and K. A. Hobson. 1990. Differential parasitism of Least Flycatchers and Yellow Warblers by the Brown-headed Cowbird. Behav. Ecol. Sociobiol. 27:403–410.

Colias, N. E., and E. C. Colias. 1984. Nest building and bird behavior. Princeton University Press, Princeton, NJ.

Darley, J. A. 1968. The social organization of breeding Brown-headed Cowbirds. Ph.D. Thesis, University of Western Ontario, London, Ontario.

———. 1983. Territorial behavior of the female Brown-headed Cowbird (*Molothrus ater*). Can. J. Zool. 61:65–69.

Dufty, A. M., Jr. 1982. Movements and activities of radio-tracked Brown-headed Cowbirds. Auk 99:316–327.

Ehrlich, P. R., D. S. Dobkin, and D. Wheye. 1988. The birder's handbook. Simon and Schuster, New York.

Friedmann, H. 1929. The cowbirds: A study in the biology of social parasitism. C. C. Thomas, Springfield, IL.

Friedmann, H., and L. F. Kiff. 1985. The parasitic cowbirds and their hosts. Proc. Western Found. Vert. Zool. 2:226–304.

Friedmann, H., L. F. Kiff, and S. I. Rothstein. 1977. A further contribution to the knowledge of the host relations of the parasitic cowbirds. Smithsonian Contrib. Zool. 235.

Grzybowski, J. A. 1995. The Black-capped Vireo (*Vireo atricapillus*). *In* The birds of North America 181 (A. Poole and F. Gill, eds). Academy of Natural Sciences, Philadelphia; American Ornithologists' Union, Washington, DC.

Hochachka, W. 1990. Seasonal decline in reproductive performance of Song Sparrows. Ecology 71:1279–1288.

Hoover. J. P., and M. C. Brittingham. 1993. Regional variation in cowbird parasitism of Wood Thrushes. Wilson Bull. 105:228–238.

Laskey, A. R. 1950. Cowbird behavior. Wilson Bull. 62:157–174.

Lowther, P. E. 1995. Bronzed Cowbird (*Molothrus aeneus*). *In* The birds of North America 144 (A. Poole and F. Gill, eds). Academy of Natural Sciences, Philadelphia; American Ornithologists' Union, Washington, DC.

May, R. M., and S. K. Robinson 1985. Population dynamics of avian brood parasitism. Amer. Natur. 126:475–494.

Newton, I. (ed). 1989. Lifetime reproduction in birds. Academic Press, London.

Nice, M. M. 1937. Studies in the life history of the Song Sparrow. Trans. Linn. Soc. New York 4:1–237.

Orians, G. H. 1980. Some adaptations of marsh-nesting blackbirds. Princeton University Press, Princeton, NJ.

Ortega, C. P., and A. Cruz. 1988. Mechanisms of egg acceptance by marsh-dwelling blackbirds. Condor 90:349–358.

Payne, R. B. 1977. The ecology of brood parasitism in birds. Annu. Rev. Ecol. Syst. 8:1–28.

Pease, C. M., and J. A. Grzybowski. 1995. Assessing the consequences of brood parasitism and nest predation on seasonal fecundity in passerine birds. Auk 112:343–363.

Peck, G. K., and R. D. James. 1987. Breeding birds of Ontario. Nidiology and distribution, vol. 2: Passerines. Royal Ontario Museum, Toronto.

Raim, A. 1978. A radio transmitter attachment for small passerine birds. Bird-Banding 49:326–332.

Robinson, S. K., S. I. Rothstein, M. C. Brittingham, L. J. Petit, and J. A. Grzybowski. 1995a. Ecology and behavior of cowbirds and their impact on host populations. Pp. 428–460 *in* Ecology and management of neotropical migratory birds (T. E. Martin and D. M. Finch, eds). Oxford University Press, New York.

Robinson, S. K., F. R. Thompson III, T. M. Donovan, D. R. Whitehead, and J. Faaborg. 1995b. Forest fragmentation and the regional population dynamics of songbirds. Science 267:1987–1990.

Rogers, C. M., M. J. Taitt, J. N. M. Smith, and G. Jongejan. 1997. Nest predation and cowbird parasitism create a demographic sink in wetland-breeding Song Sparrows. Condor 99:622–633.

Rothstein, S. I., J. Verner, and E. Stevens. 1980. Range expansion and diurnal changes in dispersion of Brown-headed Cowbirds in the Sierra Nevada. Auk 97:253–267.

———. 1984. Radio-tracking confirms a unique diurnal pattern of spatial occurrence in the parasitic Brown-headed Cowbird. Ecology 65:77–88.

Sealy, S. G. 1996. Evolution of host defenses against brood parasitism: Implications of puncture-ejection by a small passerine. Auk 113:346–355.

Smith, J. N. M. 1981. Cowbird parasitism, host fitness, and age of the host female in an island Song Sparrow population. Condor 83:152–161.

Smith, J. N. M., and P. Arcese. 1994. Brown-headed Cowbirds and an island population of Song Sparrows: A 16-year study. Condor 96:916–934.

Smith, J. N. M., and I. H. Myers-Smith. In press. Spatial variation in parasitism of Song Sparrows by Brown-headed Cowbirds. *In* Parasitic birds and their hosts (S. I. Rothstein and S. K. Robinson, eds). Oxford University Press, New York.

Teather, K. L., and R. J. Robertson. 1985. Female spacing patterns in Brown-headed Cowbirds. Can. J. Zool. 63:218–222.

Thompson, F. R. III. 1994. Temporal and spatial patterns of breeding Brown-headed Cowbirds in the midwestern United States. Auk 111:979–990.

Weatherhead, P. J. 1989. Sex-ratios, host-specific reproductive success, and impact of Brown-headed Cowbirds. Auk 106:358–366.

Wiley, J. W. 1988. Host selection in the Shiny Cowbird. Condor 90:289–303.

Woodworth, B. L. 1997. Brood parasitism, nest predation, and season-long reproductive success of a tropical island endemic. Condor 99:605–621.

9. Spatial Patterns of Breeding Female Brown-headed Cowbirds on an Illinois Site

ARLO RAIM

Abstract

I used radiotelemetry and observations of color-marked birds to study spacing patterns of 102 female Brown-headed Cowbirds over seven years in a heterogeneous parklike habitat (180 ha) in east central Illinois. Cowbirds fed both in short grass and leaf litter within their breeding areas and commuted up to 4 km to feed in pastures and row crops. During the study, local cowbird numbers remained stable at 13–16 females. Except for two large areas of open, mowed grass, the entire study area was included within cowbird breeding areas. During one period when 6 adjacent females were tracked in May, adjacent breeding areas overlapped by 12% on average. Breeding areas shifted frequently within a season, and increased from 9.2 ± 5.9 (SD) ha in May and June to 21.5 ± 11.4 ha in late June and July. This increase was because remaining females expanded into areas vacated by territorial females that left the area early in the year. Most females were strongly site-faithful between years, with an annual return rate of 47% ($N = 79$) to identical breeding ranges. Shifts to adjacent areas were observed, but no long movements were seen within the study area. These data are consistent with a site-based dominance interpretation of cowbird spacing.

Introduction

Although female Brown-headed Cowbirds parasitize over 200 species of North American passerines (Friedmann 1963, 1971; Friedmann et al. 1977), their spacing system is still not well understood (Rothstein et al. 1986). Friedmann (1929) believed that cowbirds had a definite breeding area with defense restricted to intimidation displays rather than physical interactions. Nice (1937, 1941) reported that color-banded cowbirds did not defend an area although they were strongly attached to particular areas. Laskey (1950) found no evidence of territorial behavior in color-banded cowbirds

around her yard; instead, pairs gained dominance in an area and they alone used it for pair formation and mating. In a campus setting, Darley (1968, 1983) reported that marked female cowbirds defended territories that overlapped extensively. Elliott (1978, 1980), on the other hand, observed little cowbird aggression and little evidence of territoriality in a grassland habitat. Dufty (1982b) found that female cowbirds responded aggressively to female models, which, in combination with his data from marked individuals and radiotracking (1982a), suggested that female cowbirds defended an exclusive area. Teather and Robertson (1985) found that individually marked and radiotracked cowbirds defended nonexclusive areas with extensive territorial overlap. Yokel (1989) found that females displayed increased aggression in more dense populations but did not defend exclusive areas.

These results suggest that territorial defense in female cowbirds ranges from little aggression with extensive overlap between ranges of female neighbors, to significant aggressive behavior with little or no overlap. Some flexibility may be expected in a species that does not need to defend all-purpose territories because they can feed and breed in separate areas (Rothstein et al. 1986). The secretive nature of much of cowbird breeding behavior (Young 1963), their extensive movements and lack of a single nest to stabilize the location of breeding females over the season, have made it difficult to gather detailed data on cowbird spacing systems.

The purpose of this chapter is to examine patterns of use of breeding areas by color-marked and radiotagged female cowbirds during the breeding season. Birds were studied over a seven-year period on a site in east central Illinois. During this period, 102 female cowbirds were individually color-marked, and 16 of them were also radiotracked. Specifically, I examined: (1) size and change in breeding areas throughout the season, (2) the return of females between years, (3) the overlap of breeding areas, and (4) evidence for territorial and nonterritorial behavior.

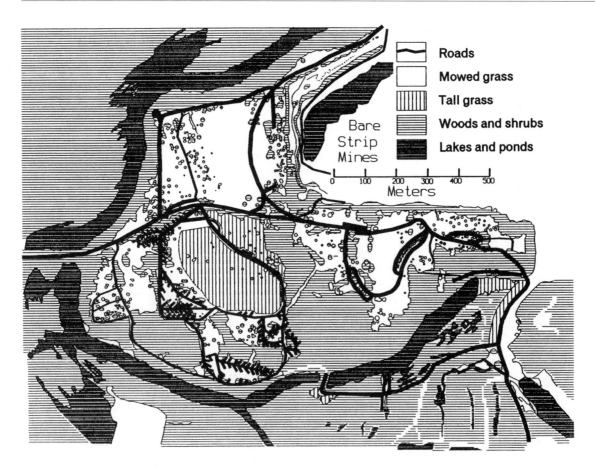

Figure 9.1. Map of the Kickapoo State Park main study area showing roads, lakes and ponds, shrubby and wooded areas, and areas of tall and short grass.

Methods

Study Area

The study area (180 ha) was in east central Illinois in Kickapoo State Park (87.7° W, 40.1° N, Figure 9.1). One third of the study area was composed of gently rolling mowed picnic areas and campgrounds with scattered trees. Hilly land with beech/maple and oak/hickory woods made up 27% of the area. Shrubs and tree borders (18%), old coal strip mines with vegetation ranging from steep bare slopes to second-growth woods (15%), and ponds (9%) made up the remainder of the area. Parts of some picnic and campground areas were left unmowed after the second year. The park is surrounded by cropland (mostly corn and soybeans), pastures, a wooded river valley, and deeply contoured strip-mined land, which varied from densely vegetated in the east to sparsely vegetated spoil piles north of the ponds in the northeast of the study area (Figure 9.1).

Trapping and Radiotelemetry

To identify cowbirds during the breeding season and in successive years, cowbirds were captured in portable decoy traps (1.5 × 1.5 × 1 m) placed in openings throughout the area from late March until mid-July. Between 16 and 31 cowbirds were color banded and marked each year between 1972 and 1977. Approximately 90% of the female and 80% of the male cowbirds observed in the study area were marked by mid-May. Most unmarked birds appeared to be nonresidents that were only present in the area for a short time. Cowbirds were banded with numbered metal and celluloid color bands. Tail feathers were marked with a unique combination of one or two of six colors of acrylic paint on the exposed distal 15–25 mm upper and lower surfaces. The terminal 2 or 3 mm of the tail was left unmarked to absorb ground abrasion. Marked birds were identified with 7 × 35 binoculars and a 20x spotting scope.

Individual cowbirds and their behaviors were recorded on maps (1 cm = 24 m) during surveys from before arrival in the breeding area (late February to late March) until about 1 August and occasionally thereafter into October. The study area was surveyed between 35 and 55 times each year in 1972–1975 and in 1977. There were 95 surveys in 1976 and 12 in 1978. All surveys took place between sunrise and noon. Most surveys in 1976 were completed by 0900 hr.

A tape recording of a chatter call of a resident breeding female was played periodically on about half of the surveys from 1973 to 1978 to increase numbers of observations and to aid in individual identification. Chatter call recordings frequently evoked chatter responses from previously unobserved females (Dufty 1982b; Yokel 1989; Miles and Buehler, Chapter 6, and Rothstein et al., Chapter 7, this volume).

Continuous and pulsed radio transmitters (1.8 g, Cochran 1967) were attached to female cowbirds periodically throughout the breeding season from early April until mid-July in 1972–1976. Continuous transmitters (15 cm long, 0.011-inch spring antenna) lasted for a shorter time but were easier and faster to locate, with some types of behavior being identifiable by the change in audio pitch of the receiver. Pulsed transmitters (with a 22.5-cm combination 0.006 antenna on a 0.011 spring antenna base) were used in early April and late July when greater range and battery life were needed. An LA 12 AVM receiver was used with a handheld, 5-element yagi receiving antenna. A vehicle-mounted horizontal and vertical switchable 9-element antenna was used to locate cowbirds outside the main study area. Range varied from 0.3 to 0.7 km with the handheld antenna, when the cowbird was feeding on the ground, up to several kilometers, depending on the height of the bird or receiving antenna.

The radio transmitters varied from 3 to 5% of the body weight of a cowbird and were attached with threads to a cloth that was glued with an eyelash skin adhesive to the back of the bird, providing a flexible attachment (Raim 1978). Radiotagged birds were watched immediately to look for any behavioral effects of the transmitter. Data from the first day of capture were not used, to eliminate any initial behavioral effects and because of the interruption to the bird's normal activities caused by tagging.

Cowbirds frequenting or breeding in the study area were radiotracked periodically from 29 March to 21 July. From 1 to 7 females were tracked during each of five years. One female was tracked during three different years and three others were tracked during two different years. Sixteen individual female cowbirds were radiotracked for a total of 747 hours over 83 days. Tracking data were recorded on maps of the area and on a portable tape recorder. With few exceptions, each female was tracked continuously from when she left her roost in the morning to her return. Each time the cowbird was observed, the time, general and specific locations, behavior, other birds present, and interactions between them were recorded. Visual observation was attempted whenever it did not appear it would disturb the bird.

Data Analysis

Census data were compiled on maps, with the locations of different breeding activities being plotted separately. Cumulative maps were made for each day and for each female over several days. Because females moved widely when being courted by males and often occurred in multifemale groups, locations of male-female interactions were not used in estimating breeding areas.

Locations of chatter calls, aggressive interactions, and probable nest-searching behaviors were combined by connecting outer observations to form convex polygons describing breeding areas (Mohr 1947). The circumscribed area of these observations often varied from day to day, particularly in large mowed and tall grass areas with scattered trees, although breeding activities during each day of radiotracking generally enclosed the central part of a female's breeding area. A second day, on average, doubled (102.7%, $N = 9$) the circumscribed area, whereas the third and fourth days increased it only by 26.2% and 2% respectively. After mid-June, when the birds' ranges were less stable, an extra day of tracking data was needed to reach the same level of increase. Compilations of 3 or 4 days of tracking (Table 9.1) were used when possible to estimate breeding area sizes during stable periods. Breeding area size and overlap were estimated from radiotracking data by weighing paper cutouts of the mapped breeding areas and measuring them with a compensating planimeter.

Results

The breeding season of cowbirds extended from early April (when banded females returned) through mid-July, by which time most females had abandoned their breeding areas. This analysis is restricted to the period from mid-April (after pair formation of returning females and breeding-area establishment) through mid-July.

Cowbird Breeding Behavior
Aggressive Interactions

The most common interactions included head-up displays and bill wiping. In head-up displays, both birds pointed their bills vertically when two females flew or ran toward each other, or perched beside each other on a branch or on the ground. Bill wiping involved bending forward and wiping the bill from base to tip on the branch or some other object on which the female was perched (Laskey 1950, Darley 1968). Head-up displays were interspersed with bill wiping, with the females alternately following each other through the vegetation (Darley 1968). In some contexts (see below) both females displayed nearly simultaneously or alternately with no clear dominant. In other contexts, one female did most or all of the aggressive displays and following, which suggested behavioral dominance (see also Darley, 1968). Occasionally, more intense interactions were observed, involving fighting and high-speed chases that lasted for many minutes. Fighting and chasing sometimes began in the trees and ended on the ground. These intense physical encounters were interspersed with the less intense aggressive displays

Table 9.1. Territory Sizes of Female Cowbirds Based on Radio Tracking

Year	Individual	Dates	Territory Size (ha)
1972	w	18–23 May	7.8
1973	y/-	23–24 May	4.3
	w	24–25 April	4.9
		16–18 June	14.1
		20–24 June	26.1
		28–30 June	36.1
		3–8 July	38.0
		10–11 July	42.0
	go	10–11 May	3.7
1974	w	10–11 May	10.9
		27–29 June	15.1
		18–21 July	23.7
	ww	2–3 May	6.3
	r-	5 May	2.7[a]
	y/-	7, 31 May, 1 June	5.6
	g-	8 May, 20–22 June	12.4
1975	p	24–26 April	21.6
		3–5 June	24.1

[a]Based on only one day; not used in calculating averages.

described above. In this study, fighting was observed on only 2 occasions, whereas high-speed chases were observed 12 times. Darley (1968) also observed only 2 fights, while none were observed by Dufty (1982b).

Chatter Calls

Females gave the chatter call in a variety of contexts: while responding to males, during aggression directed at other females, after finding and laying in nests, and in response to playbacks. Occasionally, chatter calls were also given by female cowbirds when being handled. Females on adjacent breeding areas often chattered back and forth. Alternating chatter bouts generally occurred when female cowbirds separated by 5–10 m were about to engage in fights, chases, head-up displays, bill wiping, and following. Alternating chatter calls were also given when females were returning to the central part of their respective breeding areas after an aggressive interaction (see below).

Early in the morning, some but not all females seemed to travel around their breeding area chattering from conspicuous perches as if patrolling and proclaiming their ownership. Dufty (1982a) also observed such patrolling. Chatter calling was also observed at certain times in conjunction with nest searching and egg laying.

Searching

Females spent much of their time searching, presumably for hosts or host nests. Female cowbird searching seemed to take several forms. Active searching involved constantly moving through trees, shrubs, grasses, and forest litter while peering up, down, and to the side and making frequent short hops, flights, and/or walks or runs. This behavior was easily spotted and identified while radiotracking by the constant movement and by the particular fluctuating pitch of the audible radio signal. Some females frequently searched in this manner, whereas others did so infrequently (A. Raim, unpubl. data).

I distinguished active searching from a perch from inactive searching. In active searching, perched females frequently turned their heads while staring down or to the side and shifted back and forth on the perch, occasionally stretching up and bending low. In inactive searching, females perched quietly on a branch or wire while only occasionally looking back and forth, turning around, and/or shifting slightly on their perches. Inactive searching was sometimes interspersed with preening and attendance by the mate or other cowbirds. Because active searching was observed only in the breeding area during the laying period (A. Raim, unpubl. data), I believe its main function is searching for nests or hosts rather than for predators, mates, or other cowbirds. On a few occasions, inactive searching from a perch appeared to be directed toward a host building its nest, as reported by others (Hann 1941; Norris 1947; Mayfield 1961; Norman and Robertson 1975; Darley 1983).

Rothstein (1972, Rothstein et al. 1984, 1986) noted that most or all nest searching was in the morning. In my study, females that fed around their breeding area all day (Figure 9.2c,d) and those that returned to their breeding areas in the afternoon on at least some days (Figure 9.2a,e) searched for nests periodically throughout the day.

Feeding Locations

During the breeding season, females fed both in and around their breeding areas when short grass or open-ground habitat was available, and in distant (1–3 km away) pastures and row crops (Figure 9.2). Feeding areas of individuals varied greatly. Some females fed almost entirely in the open understory among leaf litter (Figure 9.2d), while others spent most of the afternoon at distant feeding sites in large flocks (Figure 9.2e). Aggressive interactions between feeding females were usually observed only when feeding areas fell within or bordered breeding areas (see below).

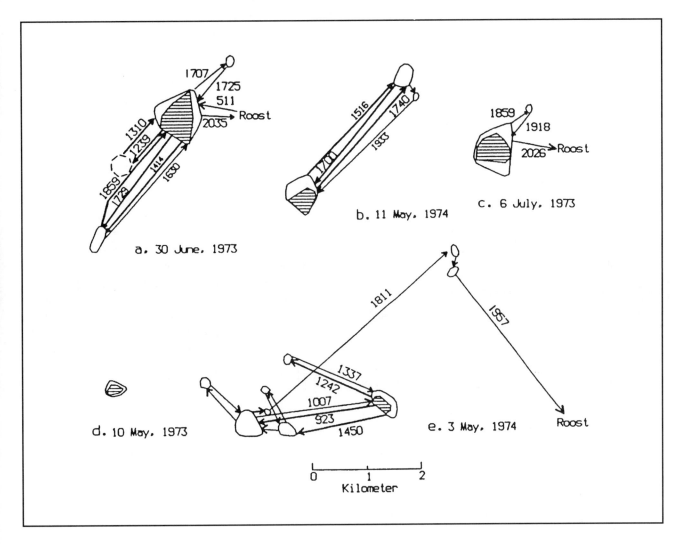

Figure 9.2. Examples of the range of daily feeding patterns observed while radiotracking individual female cowbirds. Patterns a, b, and c are the same female on different days. The horizontal cross-hatched areas represent the breeding areas bounded by plotting chatter, defensive interactions, and nest search locations for that day. Enclosed areas are where feeding was observed that day. Roost locations are shown when they lay outside the breeding area. Arrows indicate flights between areas, and numbers, the times of day when long flights occurred.

Overall Spacing of Females

During the seven years of this study, the number of females occupying the study area during the peak of the breeding season (mid-May to mid-June) varied from 13 (1972) to 16 (1974). Usually there were 14 (1973, 1976, 1977) or 15 territories (1975). In most years, the entire plot was included within breeding areas (Figure 9.3), except for areas of short mowed grass in the northwest central and east central part of the area (see Figure 9.1) where there were only a few host nests in scattered trees.

Spatial Distribution of Breeding Behavior

Most breeding behavior occurred within well-defined areas. Convex polygons generated by mapping male-female interactions, aggressive interactions, chatter locations, and searching activities of radiotracked females were similar on most days (Figure 9.4). Spatial segregation of behaviors observed on some days (e.g., the female observed on 20 June 1974 in Figure 9.4) disappeared when further records were added (A. Raim, unpubl. data).

The sizes of breeding areas varied among individuals and seasons (Table 9.1, see below). Overall, breeding areas averaged 9.2 ± 5.9 (SD) ha (range 3.7–24.1 ha). After the third week of June, breeding areas expanded to average 21.5 ± 11.4 ha (range 15.1–42.0 ha) as the number of breeders decreased. The smallest defended areas may represent a lower limit because all of the females that originally occupied breeding areas of less than 4.3 ha vacated them during 1973 and 1974.

Within-Season Changes in Spatial Patterns

Some breeding areas remained relatively stable throughout the breeding season, whereas others moved, disappeared in mid-season, or expanded to fill vacated breeding areas. Because censusing was most intense then, 1976 best represents the changes that occurred throughout the season (Figure 9.5). Of the 14 females in the core of the study area, 4 occupied the study area for only part of the season (p, w-, bo, and rwb). Female p, for example, disappeared early (she moved west in late April before disappearing on May 3). She was replaced by female bo, which was then replaced in 3 weeks by b-, which moved into the area. Six females shifted their breeding areas from one to three times between April and early July (wy, wr, r-, g-, gg, and y-). Females g- and r- expanded into areas that had been vacated by other females by mid-June. A new female, rwb, arrived late in June about the same time that female g- disappeared, resulting in a partial replacement; rwb returned to this area in the next two years.

Changes in the breeding area of one intensively radiotracked female in 1973 and 1974 further illustrate seasonal changes (Figures 9.6 and 9.7). Female w expanded her breeding area from late April (4.9 ha) through early July

(42.0 ha) in 1973 (see Table 9.1). She first moved west into open areas with scattered trees and then south into an area abandoned by a breeding female in mid-June. In 1974 the same female w again expanded into the same open areas to the west, although she did not move south, as that area remained occupied (Figure 9.7). Active nest searching seemed to occur in the expanded breeding area before any significant aggressive interactions were observed there (Figure 9.6). Thus, females may expand nest-searching activities into the breeding areas of other females as long as there is no aggressive interference.

Spatial Overlap of Breeding Areas

Interpreting spatial overlap was complicated by the seasonal shifts described above. For this reason, I tracked six females consecutively in May 1974 in the central part of the study area. May was the month when space was used most stably. Overlap based on these tracking data averaged 12.0% (range 1.3–24.1%, SD = 6.7, N = 12) between any two breeding areas, with up to three breeding areas overlapping simultaneously (Figure 9.8). Females appeared to tolerate greater overlap toward the end of the breeding season.

Return Rates of Females between Years

About half (47%) of the 79 breeding areas occupied by color-banded females during June were reoccupied by the same female the following year (Figure 9.9). From 1972 to 1973, 42% of 12 territories were reoccupied. Corresponding numbers in subsequent pairs of years were 54% of 11, 25% of 12, 50% of 14, 63% of 16, and 43% of 14 territories. During 1972–1978, 15 females had similar breeding areas for at least two years, 7 for at least three years, and 2 for at least five consecutive years. Only 4 females shifted territories significantly between years (2, 7, 8, 9, in Figure 9.9) and all four occupied adjacent areas.

Nonbreeding and/or Floating Females

Although all females that were radiotracked extensively within the main study area confined their breeding activities to a compact breeding area, two radiotracked females found at the edge of the study area seemed to travel over much larger areas. Female rr, which was observed exclusively in the east end of the study area during 1972 and in May 1973, traveled extensively over the northeast half of the main study area when tracked on 8, 9, and 10 June. She showed little evidence of breeding on 8–9 June, although she may have done some nest searching nearly 1 km from where she was observed in May. Because of the lack of breeding behavior, I believe that her territory broke down.

Female b in 1974 also traveled over a large area. She had a territory in the sparsely vegetated and deeply contoured strip-mine area. She showed low to moderate amounts of breeding behavior, although she was seen to find a Lark

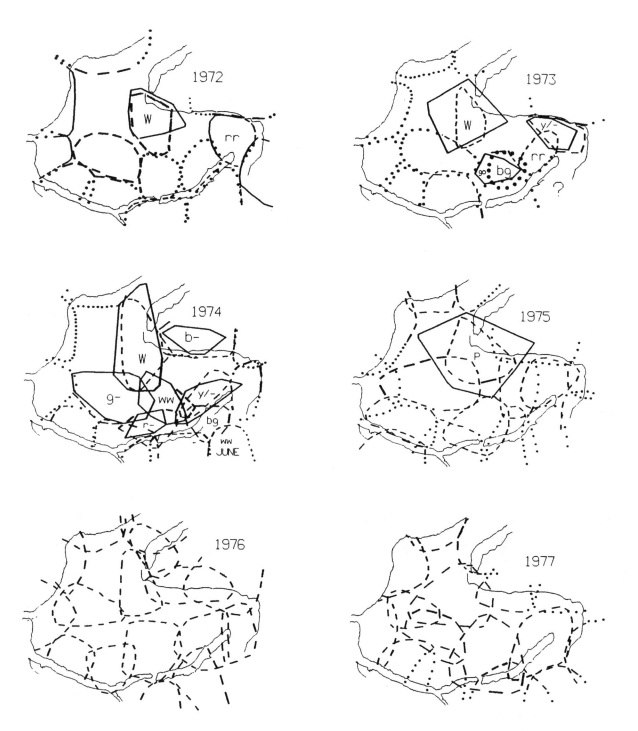

Figure 9.3. Maps of breeding areas in 1972–1977 during the most stable part of the breeding season (May–mid-June). Solid lines outline breeding areas determined by plotting locations of chatter calls, defensive interactions, and nest searching activity obtained while radiotracking over 1–5 days. The dashed lines represent breeding areas estimated using census data from the whole season. Dotted lines indicate that there were insufficient data to draw an accurate boundary.

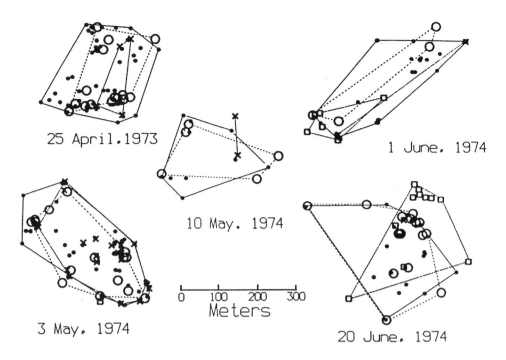

Figure 9.4. Convex polygons drawn for individual days using the locations of: male-female interactions (x), chatter calls (dot), aggressive interactions (circles), and nest searching behavior (squares). The data were recorded while radiotracking five different cowbirds.

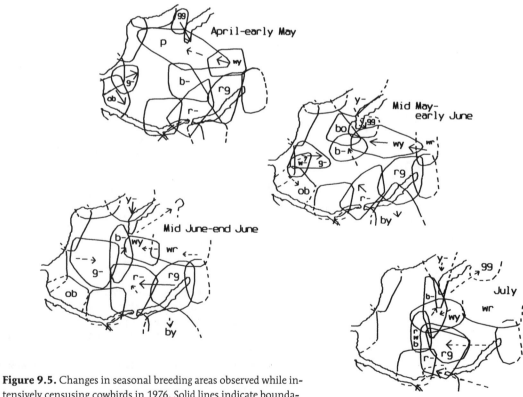

Figure 9.5. Changes in seasonal breeding areas observed while intensively censusing cowbirds in 1976. Solid lines indicate boundaries of breeding areas, and dashed lines indicate approximate boundaries. Dashed arrows indicate shifts from another area, and solid arrows indicate shifts to another area. The dotted line indicates that bird p (tracked in April) shifted west just prior to disappearing in early May; bo and w- disappeared after about 3 weeks.

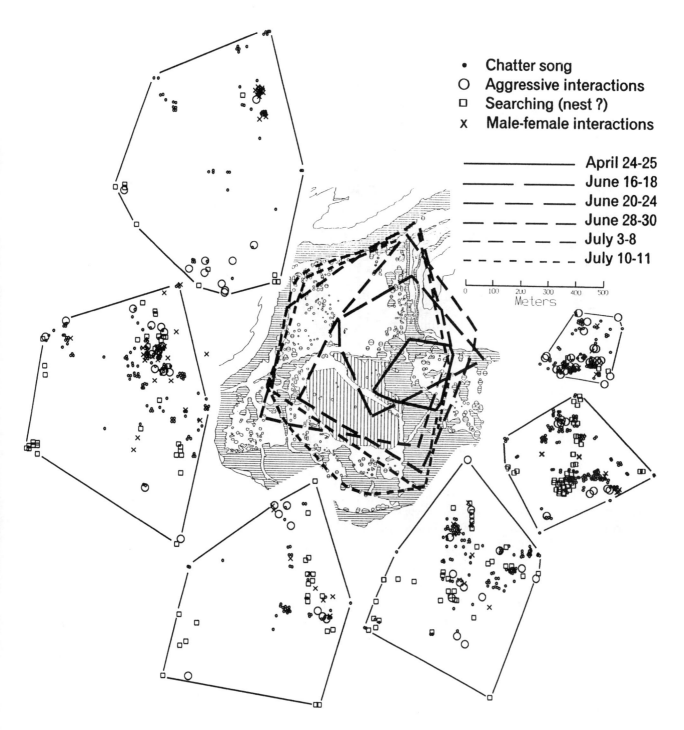

Figure 9.6. Expansion of the breeding area of female w in 1973 based on tracking data. The outlines show where the various breeding behaviors occurred during each period.

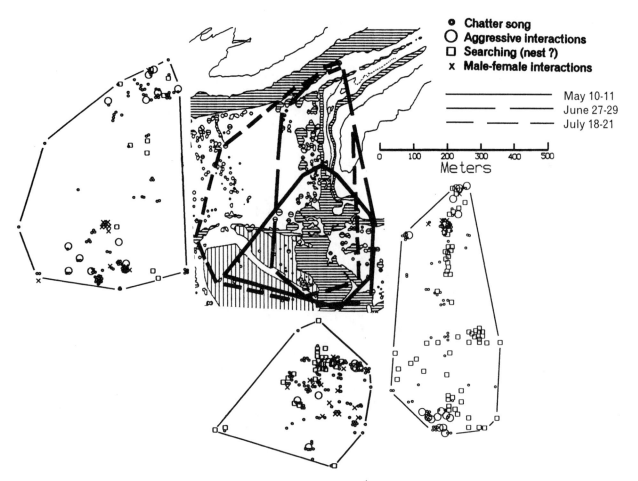

Figure 9.7. Female w's breeding-area expansion in 1974 based on radiotracking data. The outlines show where the various breeding behaviors occurred during that period.

Sparrow nest. However, she spent most of the day at distant (>2 km) feeding sites. She may have bred over a larger area and spent more time at distant feeding areas because the sparse vegetation on her territory provided poor breeding and feeding opportunities.

Discussion

Cowbirds occupied all available woody and shrubby habitat in the study area. The only unoccupied areas were of mowed grass where there were few or no host nests. These areas, however, served as feeding areas for some local females, which thus did not need to leave their breeding areas to forage in agricultural areas, as is commonly found in other landscapes (Rothstein 1972; Rothstein et al. 1984, 1986; Smith and Arcese 1994; Thompson 1994). Areas with mowed grass interspersed with wooded and shrubby areas may represent optimal cowbird habitat.

Within the study area, female cowbirds conformed most closely to a site-based dominance model of territoriality

(Rothstein et al. 1986). Cowbirds were evenly spaced throughout suitable habitat. These results conform closely to those of Dufty (1982b) and Darley (1968, 1983), who thought that cowbirds defend their breeding area to preserve a supply of host nests. The moderate overlap between females, however, may help to explain the high incidence of multiple parasitism in many hosts nesting in agricultural landscapes in the Midwest (e.g., Robinson et al., Chapter 33, this volume).

The sizes of breeding areas in this study are similar to those observed by others. Darley (1968, 1983) found that nonfeeding areas averaged 4.5 ha (range 0.9–13.4), and Dufty (1982a) found an average of 20.4 ha (9.9–33.2 ha). Teather and Robertson (1985) reported an average of 9.9 ha (5.5–13.7 ha). Although each of these studies used slightly different criteria, the methods are similar enough to allow general comparisons. Some of Darley's estimates may be small due to the low numbers of observations of marked birds.

Between-year fidelity to breeding areas was high and similar to that reported in other studies. Friedmann (1929)

noted evidence that females returned to breeding sites, and Nice (1937), Laskey (1950), and Hunt (1977) observed marked females returning for up to six years. Darley (1968) found that 43% of 14 females returned, although only 4 bred in the same area. Dufty (1982a) found that 4 of 8 females returned to either the same or adjacent areas.

The extensive within-season shifts in breeding areas documented here for some females mean that season-long compilations of breeding areas can be poor predictors of spatial distribution at any one time. Cowbirds appear to use space very flexibly. Some females opportunistically shifted breeding areas in response to nearby vacancies. Others, which had small breeding areas with extensive overlap with neighbors, shifted completely to other breeding areas. Breeding area size may also reflect host availability; breeding areas near large openings where hosts were scarce tended to

be larger than those in more wooded areas (e.g., Figures 9.1 and 9.3). Understanding the determinants of breeding area size, however, requires data on host density, which was beyond the scope of this study. Nevertheless, such flexibility in spacing patterns may only be possible in a species that does not need to defend an all-purpose territory.

Acknowledgments

I wish to acknowledge Bill Cochran for his help in most aspects of this study, including suggestions on gathering the data, help in building, use of his equipment, and reviewing early manuscripts. I also thank Illinois Natural History Survey personnel for their loan of equipment and for allowing me the use of a radiotracking vehicle during the study. I also thank Richard Graber and Glen Sanderson for reading an

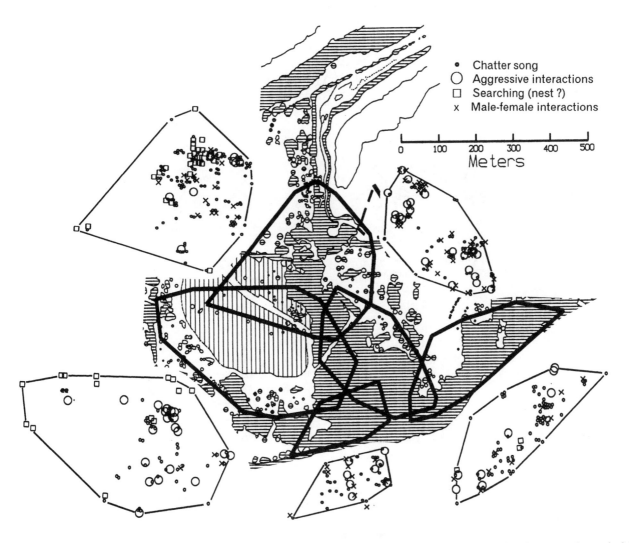

Figure 9.8. Map showing overlap in breeding areas observed while radiotracking several adjacent breeding females in 1974. Outlines bound areas with breeding activities.

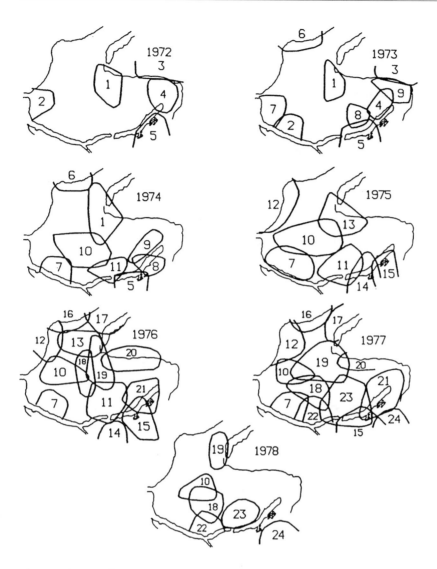

Figure 9.9. Maps showing 24 females that returned to breed for at least one further year after breeding in at least the last half of the previous breeding season from 1972 to 1978. Numbers identify individual females.

early draft of the manuscript, and Scott Robinson for later assistance. This study was in partial fulfillment of Ph.D. degree requirements in the Department of Zoology at the University of Illinois.

References Cited

Cochran, W. W. 1967. 145–160 Mhz beacon (tag) transmitter for small animals. AIBS BIAC Inf. Module M-15. 12 pp.

Darley, J. A. 1968. The social organization of breeding Brown-headed Cowbirds. Ph.D. Thesis, University of Western Ontario, London, Ontario. 88 pp.

———. 1983. Territorial behaviour of the female Brown-headed Cowbird (*Molothrus ater*). Can. J. Zool. 61:65–69.

Dufty, A. M., Jr. 1982a. Movements and activities of radio-tracked Brown-headed Cowbirds. Auk 99:316–327.

———. 1982b. Response of Brown-headed Cowbirds to simulated conspecific intruders. Anim. Behav. 30:1043–1052.

Elliott, P. F. 1978. Cowbird parasitism in Kansas Tallgrass Prairie. Auk 95:161–167.

———. 1980. Evolution of promiscuity in the Brown-headed Cowbird. Condor 82:138–141.

Friedmann, H. 1929. The cowbirds: A study in the biology of social parasitism. C. C. Thomas, Springfield, IL.

———. 1963. Host relations of the parasitic cowbirds. U.S. Natl. Mus. Bull. 233. 273 pp.

———. 1971. Further information on the host relations of the parasitic cowbird. Auk 88:239–255.

Friedmann, H., L. F. Kiff, and S. I. Rothstein. 1977. A further contribution to knowledge of the host relations of the parasitic cowbirds. Smithsonian Contrib. Zool. 235 pp.

Hann, H. W. 1941. The cowbird at the nest. Wilson Bull. 53:210–221.

Hunt, L. B. 1977. Brown-headed Cowbirds form a pair bond for three seasons. IBBA News 49:223–224.

Laskey, A. 1950. Cowbird behavior. Wilson Bull. 62:157–174.

Mayfield, H. 1961. Vestiges of a proprietary interest in nests by the Brown-headed Cowbird parasitizing the Kirtland's Warbler. Auk 78:162–166.

Mohr, C. O. 1947. A table of equivalent populations of North American small mammals. Am. Midl. Nat. 37:223–249.

Nice, M. M. 1937. Studies in the life history of the Song Sparrow. Trans. Linn. Soc. N.Y. 4:1–246.

———. 1941. The role of territory in bird life. Amer. Midl. Natur. 26:441–487.

Norman, R. F., and R. J. Robertson. 1975. Nest searching behavior in the Brown-headed Cowbird. Auk 92:610–611.

Norris, R. T. 1947. The cowbirds of Preston Firth. Wilson Bull. 59:83–103.

Raim, A. 1978. A radio transmitter attachment for small passerine birds. Bird-Banding 49:326–332.

Rothstein, S. I. 1972. Territoriality and mating system in the parasitic Brown-headed Cowbird (*Molothrus ater*) as determined from captive birds. Amer. Zool. 12:659 (Abstract).

Rothstein, S. I., J. Verner, and E. Stevens. 1984. Radio-tracking confirms a unique diurnal pattern of spatial occurrence in the parasitic Brown-headed Cowbird. Ecology 65:77–88.

Rothstein, S. I., D. A. Yokel, and R. C. Fleischer. 1986. Social dominance, mating and spacing system, female fecundity, and vocal dialects in captive and free-ranging Brown-headed Cowbirds. Curr. Ornithol. 3:127–185.

Smith, J. N. M., and P. Arcese. 1994. Brown-headed Cowbirds and an island population of Song Sparrows: A 16-year study. Condor 96:916–934.

Teather, K. L., and R. J. Robertson. 1985. Female spacing patterns in Brown-headed Cowbirds. Can. J. Zool. 63:218–222.

Thompson, F. R. III. 1994. Temporal and spatial patterns of breeding Brown-headed Cowbirds in the midwestern U.S. Auk 111:979–990.

Yokel, D. A. 1989. Intrasexual aggression and the mating behavior of Brown-headed Cowbirds: Their relation to population densities and sex ratios. Condor 91:43–51.

Young, H. 1963. Breeding success of the cowbird. Wilson Bull. 75:115–122.

10. Differences in Movements, Home Range, and Habitat Preferences of Female Brown-headed Cowbirds in Three Midwestern Landscapes

FRANK R. THOMPSON III AND WILLIAM D. DIJAK

Abstract

We radiotracked breeding female cowbirds in three land-scapes in Missouri and Illinois that ranged from 50 to 93% forest. We determined differences in cowbird home range size, movements, habitat use and preference, and spatial relations of breeding, feeding, and roosting areas among these landscapes. We obtained a mean of 42 locations/female of 84 cowbirds. Most cowbirds used spatially distinct areas for morning nonfeeding ranges, afternoon ranges, and evening roosting areas. Median home-range size ranges were 43–253 ha, 89–242 ha, and 261–845 ha, based on morning nonfeeding locations, afternoon locations, and all locations, respectively. Cowbird movements between breeding and feeding locations were similar for all study areas, but movements between feeding and roosting areas and roosting and breeding areas differed among study areas. Forest and shrubland habitats were generally the most preferred habitats in the morning; grass, feedlot, and developed habitats were the most preferred in the afternoon. Overall, cowbird behavior was similar among study areas. Significant differences among areas were all related to the use of a distant, large communal roost on the Jonesboro study area. While cowbird behavior was similar among study areas, cowbird population size varied greatly. We suggest cowbirds have similar behavior over a range of landscape composition in midwestern landscapes but that their abundance is regulated by the availability of suitable feeding habitat.

Introduction

Brown-headed Cowbird breeding and feeding activities can be separated spatially and temporally because cowbirds provide no parental care. This uncoupling of breeding and feeding allows cowbirds to select separate areas appropriate for each activity (Rothstein et al. 1986). In the Midwest, most cowbirds breed, feed, and roost in spatially distinct areas and on average move 3.6 km between evening roosting areas and morning breeding areas, 1.2 km between morning breeding areas and afternoon feeding areas, and 2.6 km between afternoon feeding areas and evening roosting areas (Thompson 1994). Cowbirds may move as far as 7 km between breeding and feeding areas (Rothstein et al. 1984, Thompson 1994). Spatial segregation of activities and the extent of movements between breeding, feeding, and roosting areas appear to vary across the cowbird's range. Breeding and feeding ranges may overlap more in eastern North America than in California (Rothstein et al. 1986, Lowther 1993).

Cowbirds usually feed in short grass habitats or with large grazing mammals (Friedmann 1929, Mayfield 1965, Dufty 1982, Rothstein et al. 1986, Thompson 1994). They breed in a wide variety of habitats from prairie to forest, but may select habitats with high host densities (Rothstein et al. 1986; Thompson et al., Chapter 32, this volume). Habitat use by cowbirds has been described, but habitat selection or preferences have not been analyzed for morning breeding areas and afternoon feeding areas. Habitat selection or preference infers that habitat use differs from that expected by chance (Johnson 1980).

We identified differences in behavior of Brown-headed Cowbirds in three midwestern landscapes. We determined spatial and temporal patterns in behavior in these land-scapes by radiotracking female cowbirds. We determined the size of morning nonfeeding ranges, afternoon ranges, and entire home ranges. We also examined habitat use and selection on morning nonfeeding areas and afternoon feeding ranges on three distinct study areas. We also extend the analysis of Thompson (1994) to compare movements and segregation of breeding, feeding, and roosting among study areas.

We expected differences in cowbird behavior among the study areas because the distributions of breeding resources (hosts) and food resources differed among the three areas. Cowbird numbers, cowbird/host ratios, and brood parasitism levels vary greatly in relation to forest cover in these three landscapes (Thompson et al., Chapter 32, this volume). As it was difficult to predict a priori what differences might occur, this study was largely exploratory. For instance, cowbird movements and home ranges could be smaller in more fragmented landscapes because resources are interspersed. Alternatively, movements and home ranges could be larger because of increased host competition resulting from greater numbers of cowbirds in more fragmented forests (Thompson et al., Chapter 32, this volume). Or, cowbird behavior could be inflexible and movements and home ranges similar among study areas. We expected habitat selection to be similar in all three areas except for differences related to habitat availability.

Study Areas and Methods

Study Areas

We selected three study sites in Illinois and Missouri. The Jonesboro, Illinois, site was in Union County and included private lands, portions of the Jonesboro District of Shawnee National Forest, and Trail of Tears State Forest. The area is predominately forest and cropland, with limited amounts of pasture. The Carr Creek, Missouri, site was in Shannon, Reynolds, and Carter counties and included private lands and Carr Creek and Deer Run state forests. The area is predominately forested (>90%) with some pasture. The Ashland, Missouri, site was in southern Boone County and included the Thomas S. Basket Wildlife Education and Research Center, adjacent private lands, and portions of the Cedar Creek District of Mark Twain National Forest. The Ashland study area was predominately forest and pasture (habitat composition of these landscapes is reported below).

Brown-headed Cowbird numbers and cowbird/host ratios vary greatly among these study areas. The Jonesboro, Carr Creek, and Ashland areas are within the southwest Illinois, south-central Missouri, and north-central Missouri study areas for which Thompson et al. (Chapter 32) report cowbird and host abundance and parasitism levels. The percentages of nests parasitized were 59, 67, and 5% for the areas that included our Ashland, Jonesboro, and Carr Creek study areas, respectively. The number of female cowbirds/point count and number of female cowbirds/host were 0.21 and 0.052, 0.54 and 0.079, and 0.03 and 0.006 for the regions that included our Ashland, Jonesboro, and Carr Creek study areas, respectively. This represents a 10-fold difference in cowbird abundance and parasitism level among these study areas.

Field Methods

We trapped cowbirds from 10 May to 5 June in 1991 and 1992 at the Carr Creek and Ashland sites, and in 1992 at the Jonesboro site. We used walk-in funnel traps (Stoddard 1931) baited with millet and placed near the center of the study sites in both forested and agricultural habitats. We selected trapping sites in areas used by cowbirds and limited

trapping to 1 or 2 trapping periods of 2 to 3 days to reduce the chance of altering cowbird behavior. Each female cowbird was banded and fitted with a radio transmitter. Transmitters had a mass of 2 g and a battery life of 30 to 40 days. We attached the transmitter to the cowbird's back with a harness made from elastic cord using a technique developed for Mourning Doves (*Zenaida macroura;* Fuemmeler 1992). We tied elastic loops around the bird's body behind and in front of the wings; these loops were then tied together on the bird's breast. The transmitter was centered on the bird's back between its wings with a 15-cm antenna extending down the bird's back and slightly past the end of its tail.

Cowbirds were radiotracked by four to six field assistants each year. We located individual cowbirds between one and three times a day and stratified our searching to obtain nearly equal numbers of locations for each cowbird in 3-hr periods from 0500 to 2000 CST and a nocturnal period from 2000 to 0500. Cowbirds were never located more than once in a 3-hr period on the same day. This procedure was designed to improve the independence of locations and to ensure that locations were representative of female cowbird activity throughout the day (White and Garrott 1990). Cowbirds were located primarily by homing on a transmitter's signal with a portable receiver and four-element yagi antenna until the bird could be seen, and occasionally by triangulating from two locations within 500 m of the bird. Locations were recorded in the field on aerial photographs (scale 1 : 12000) or USGS topographic maps (scale 1 : 24000). At each location, we recorded the date, time, habitat, the bird's behavior, and the number of male and female cowbirds present.

Daytime cowbird behavior was classified as feeding, nonfeeding, or unknown if the bird was not sighted. Birds that were observed actively foraging for or gleaning insects or seeds were defined as feeding. All other daytime behaviors were classified as nonfeeding. Birds located after dark at roosts were classified as roosting.

Data Analysis

We determined universal transverse Mercator coordinates of each cowbird location by comparing locations of cowbirds on field maps with a georeferenced aerial photographic image on a geographic information system. Scanning aerial photographs of the study areas and assigning coordinates to them from USGS topographic maps created the georeferenced aerial photographic image. Coordinates of each cowbird location were combined with the date, time, behavior, and habitat of each location for statistical analysis.

We analyzed only data from individuals for which we had 15 or more locations. This ensured a reasonable number of observations in morning and afternoon analysis periods. In addition, preliminary analyses suggested that home range size was not significantly related to the number of observations for $N \geq 15$.

We classified observations as being on morning nonfeeding ranges, afternoon ranges, or evening roosts. Females observed in nonfeeding behavior before 1100 CST were classified as on morning nonfeeding ranges. We excluded feeding observations in the morning, as they would occur mainly on breeding areas because female cowbirds search for nests in the morning (Friedmann 1929, Hann 1941, Scott 1991). We could not definitively classify these as breeding areas because we rarely observed birds laying, which usually takes place at dawn (Burhans, Chapter 18, this volume) or nest searching. All observations after 1100 CST and before local sunset, irrespective of the observed behavior, were classified as being on afternoon ranges. Birds located during non-daylight hours were classified as being at roosts.

We plotted locations of several cowbirds from each study area on habitat maps to display general movement and habitat use patterns. We selected for display cowbirds that had average sample sizes and typical movement patterns for each area.

We determined the proportion of cowbirds on each study area that had distinct morning nonfeeding ranges, afternoon ranges, and evening ranges. We used multiple-response permutation procedures (MRPP; Mielke et al. 1981, Biondini et al. 1988, Mielke 1991; Blossom Software, National Ecology Research Center, U.S. Fish and Wildlife Service, Fort Collins, Colorado) to determine if observations from morning, afternoon, and evening were spatially distinct (see Thompson 1994 for details of this analysis). We treated each class of location (morning, afternoon, evening) as a group and compared all groups in two-way comparisons for each cowbird. A small P-value ($P < .05$) suggests locations were from distinct bivariate distributions, and we refer to these as spatially distinct. A difference can arise from some combination of differences in mean location and dispersion of locations in each group; it does not necessarily mean there is no overlap among groups (see Thompson 1994 for examples). For example, a significant difference could be detected if feeding and breeding locations were clustered and had minimal overlap, or if feeding locations were dispersed throughout a home range and breeding locations were clustered in a portion of the home range. All P-values are for individual tests; we did not control for experiment-wide error.

We calculated the distance moved between consecutive locations of cowbirds on morning and then afternoon ranges, afternoon and then evening ranges, and evening and then morning ranges within the same 24-hr period. The mean distance for each type of movement was calculated for individual cowbirds, and the mean and standard error of these for cowbirds from each landscape. We used one-way analysis of variance to test the hypothesis that mean home range size was equal among all three landscapes.

We calculated convex-polygon estimates (Mohr 1947, White and Garrott 1990) of the area of morning nonfeeding ranges, afternoon ranges, and home ranges (all locations). Home range, as used here, is consistent with the usual definition as the area used by an individual for its normal activities including feeding and breeding (Burt 1943). Other investigators studying cowbirds have also estimated the area of morning nonfeeding ranges because these likely correspond to breeding areas within which females locate and parasitize nests (Dufty 1982, Darley 1983, Rothstein et al. 1984, Teather and Robertson 1985). We compared the area of all three ranges among study areas. We tested for differences in home range size among study areas with Kruskal-Wallis tests (analysis of variance on ranked data) because distributions of home range sizes were non-normal.

We determined habitat preferences in each landscape by comparing the ranks of the differences between habitat use and availability (Alldredge and Ratti 1986, 1992). If cowbirds had no habitat preferences, we expected individuals to use habitats in proportion to their availability. Deviation in use from availability constitutes evidence of habitat selection and preference (Johnson 1980). We calculated habitat availability as the percentage of each landscape represented by each habitat. We mapped habitat patches larger than 0.5 ha on an aerial photographic image and calculated the total area of each habitat with a geographic information system.

Habitats were classified as feedlot, rowcrop, grass, shrubland, forest, and developed. The forest was 40- to 80-year-old oak-hickory dominated by *Quercus* spp. and *Carya* spp. Grassland habitat was primarily cool-season pasture (*Festuca* spp.) that was grazed or cut for hay. Cropland was recently tilled ground or corn, soybeans, or wheat. Shrubland was composed of sapling-sized trees and shrubs resulting from old-field succession or timber harvest. Developed habitats included urban areas, residences, farmyards, and roads. All types of animal pens in which livestock fed were classified as feedlots even if they were less than 0.5 ha.

We calculated habitat use by each cowbird as the percentage of all locations for that bird in each habitat. We calculated the difference between habitat use and availability for each cowbird, and ranked the differences to create a ranking of habitat preferences for each cowbird. We used a Friedman analysis of variance on ranked scores for each landscape to test if differences in use and availability were the same for all habitats. If the ranks differed, we used Fisher's least significant difference procedure to determine which habitats were preferred over others ($P < .05$).

Results

We radiotagged 132 female cowbirds but analyzed data only from 84 individuals for which we had 15 or more locations. This included 16 cowbirds at the Ashland site and 20 at the Carr Creek site in 1991; and 10, 14, and 27 cowbirds at the Ashland, Carr Creek, and Jonesboro sites, respectively, in 1992. We obtained 15 to 82 locations (mean = 42 ± 14 SD) of each cowbird in this sample. Of the 48 individuals not included in the analysis, 29 were not relocated regularly enough to obtain more than 15 locations, 1 removed its radio transmitter, and 18 died. Eight or more of the mortalities may have been caused by *Salmonella* poisoning possibly contracted from a cattle feeder, 2 individuals were killed or scavenged by cats, 3 individuals were killed by raptors, and 2 died from improperly fitted radio transmitters. The direct cause of the remaining mortalities was unknown. Our sample of radiotagged birds appeared to be a very small percentage of the local population. We rarely retrapped a radiotagged female or found more than one or two radiotagged individuals in afternoon flocks.

General Patterns

Cowbirds in all three landscapes demonstrated some temporal and spatial segregation of breeding, feeding, and roosting. Most morning observations were of a single female, and on average 1.5 males engaged in nonfeeding activities in the forest (Figure 10.1).

In the afternoon most females were located with flocks of feeding cowbirds in grassland habitats. Nearly all females, however, returned on occasion to forest habitats in midafternoon (before 1500 CST), well before they would roost for the evening. For instance, shaded symbols in the forest in the southeast quadrant of the Ashland study area and northeast quadrant of the Jonesboro study area in Figure 10.1 represent afternoon observations of females in the vicinity of their breeding areas.

At night cowbirds roosted singly or in small groups, with the exception of the Jonesboro site, where 15 of 28 radiotagged females were located throughout June and July 1992 at a communal roost of at least 200 cowbirds and more than 1,000 Common Grackles (*Quiscalus quiscula*). Thompson (1994) reports details on temporal patterns in behavior, habitat use, and group size.

Spatial Relationships of Breeding, Feeding, and Roosting

The distribution of locations for breeding, feeding, and roosting ranges was different for most cowbirds (MRPP, $P < .05$). The proportions of individuals on a study area with differing spatial distributions were smallest for the comparison of morning nonfeeding ranges and afternoon ranges (69–87%). Differences among study areas were greater for comparisons of evening and afternoon ranges and evening and morning ranges (54–100%, Table 10.1). Even though distributions of morning, afternoon, and evening activities were spatially distinct, there was still overlap in the areas used during these periods (Figure 10.1).

Cowbird Movements

Cowbird movements between morning and afternoon locations were similar for all study areas ($F = 1.47$, $P = .23$). Movements between afternoon and evening ranges differed among study areas ($F = 39.73$, $P < .001$); cowbirds on the Jonesboro site moved farther between feeding and roosting areas than those on the Carr Creek or Ashland sites (LSD, $P < .05$, Figure 10.2). Movements between evening and morning areas also differed among study areas ($F = 9.53$, $P < .001$, Figure 10.2); cowbirds on the Carr Creek site moved less than those on the Ashland or Jonesboro sites (LSD, $P < .05$). All other pairwise comparisons of movements by study area were not significantly different (Figure 10.2). Cowbirds also occasionally made afternoon (1100 to 1500 CST) movements from feeding ranges to forest habitats. These movements were less frequent than movements from morning ranges to feeding areas but were similar in length (0.1–4.4 km).

Home Ranges

Home range size was extremely variable, ranging from 15 to 4,241 ha. As a result of a small number of very large home ranges, the distribution of home range sizes on each study area was extremely non-normal. Therefore we used medians to compare home range sizes across sites. Median home range size ranged from 261 ha on the Ashland study area to 845 ha on the Jonesboro study area; breeding and feeding ranges were smaller (Table 10.2). While median home ranges varied among study areas, differences were significant only for entire home ranges (Table 10.2).

Habitat Preferences

Habitat availability varied greatly among study areas. Major components of the Ashland study area were forest (50%), grassland (32%), and rowcrop (13%). Major components of the Jonesboro study area were forest (55%) and rowcrop (36%). The Carr Creek area was predominately forest (93%) and limited grassland (4%) (Table 10.3).

Cowbirds demonstrated strong patterns in habitat use. In all landscapes approximately 60% or more of morning observations were in forest or shrubland habitats, and more than 50% of afternoon observations were in grass, rowcrop, or feedlot habitats (Table 10.3).

Female cowbirds had significant habitat preferences on morning and afternoon ranges (Table 10.3). Forest and shrubland habitats were generally the most preferred on morning ranges, and grass, feedlot, and developed habitats were most preferred on afternoon ranges. The exception to this was the Carr Creek study area, where grass and feedlot habitats had the highest preference rank in the morning, even though more than 50% of observations in the morning were in the forest (Table 10.3). We discuss possible reasons for this later.

Discussion

We do not believe that trapping and radiotracking activities greatly altered cowbird behavior. We did not detect any

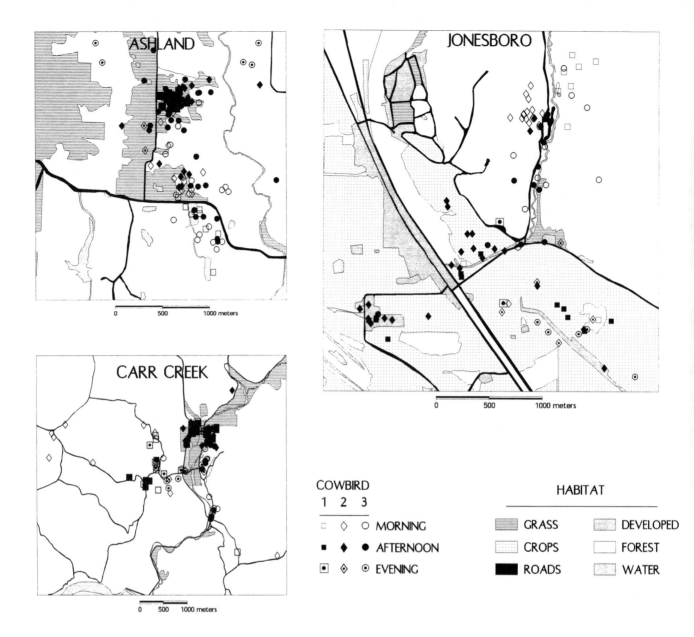

Figure 10.1. Locations of three female cowbirds on the Ashland, Missouri; Carr Creek, Missouri; and Jonesboro, Illinois, study areas. Individual cowbirds within each study area are represented by symbol shape, and symbol shading indicates type of observation: morning (before 1100 CST) nonfeeding observations, afternoon observations, and evening roosting observations.

Table 10.1. Percentage of Brown-headed Cowbirds in Three Midwestern Landscapes with Spatially Distinct Breeding, Feeding, and Roosting Areas

Locations Compared	Study Site		
	Ashland, MO	Carr Creek, MO	Jonesboro, IL
Morning, afternoon, evening	100	85	100
Morning, afternoon	80	69	87
Morning, evening	100	54	100
Afternoon, evening	100	58	96

Note: Determined by MRPP. Null hypothesis is that locations from each area have the same utilization distribution ($P < .05$).

shifts in areas used by cowbirds as a result of trapping. Behavior of radiotracked cowbirds did not appear different from the other birds that they were observed with. The only exception to this was the two individuals that died due to radio-transmitter harnesses that were not properly fitted. Other than these individuals, mortalities appeared to be unrelated to our methods.

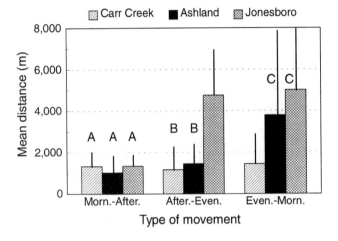

Figure 10.2. Mean distance moved by female Brown-headed Cowbirds between morning nonfeeding observations (Morn.), afternoon observations (After.), and evening roosting observations (Even.) in three midwestern landscapes. Vertical lines on bars represent one SD; bars with the same letter within a type of movement were not significantly different ($P > .05$).

The large number of individuals that we could not relocate regularly (29 of 132 individuals with fewer than 15 observations) is indicative of the difficulty of radiotracking a small bird with a very large home range. Possible explanations for our difficulty in relocating these individuals are that some cowbirds used inaccessible portions of the study areas in which we could not track them, or that they were nonlocal birds and we could not radiotrack over a large enough area to relocate them. Most of these individuals were primarily observed on feeding areas, and we were unable to locate their breeding ranges. Dufty (1982) similarly observed females at feeding sites believed to be nonresidents because they did not have nonfeeding ranges on his study area.

Distinct temporal and spatial patterns in behavior and habitat use occurred in all three study areas because cowbirds bred, fed, and roosted at different times and in different habitats. Female behavior and habitat use shifted from primarily nonfeeding behavior (and presumably breeding) in forest and shrub-sapling habitats in the morning to feeding in short grass, cropland, and feedlots in the afternoon. These diurnal patterns are similar to those reported in other parts of the cowbird's range (Dufty 1982; Darley 1983; Rothstein et al. 1980, 1984, 1986; Raim, Chapter 9, this volume).

Females also, on occasion, returned to forest habitats in the afternoon. The most likely explanation for this behavior is that females were returning to scout for host nests or to monitor the chronology of known nests. Females are also known to lay in the afternoon (S. K. Robinson pers. comm.), though this is probably uncommon. Females arrived on feeding areas as early as 0700 CST on some morn-

Table 10.2. Convex Polygon Estimates of Ranges of Female Brown-headed Cowbirds on Three Midwestern Study Areas, 1991, 1992

Study Area	N	Mean	SD	Median	Maximum	Minimum
Morning nonfeeding range						
Ashland, MO	24	298	706.4	43	2986	5
Carr Creek, MO	32	408	479.0	253	2412	10
Jonesboro, IL	21	366	396.1	182	1244	19
Afternoon range						
Ashland, MO	26	396	841.6	131	4302	4
Carr Creek, MO	30	302	665.7	89	3499	29
Jonesboro, IL	26	309	279.0	242	1076	22
Home range (all locations)						
Ashland, MO	25	564	889.5	261	4241	47
Carr Creek, MO	34	661	864.2	372	3895	15
Jonesboro, IL	27	1224	965.0	845[a]	3281	114

[a]Median home range size at Jonesboro was larger than at Ashland and Carr Creek (Kruskal-Wallis test, $P < .05$); all other comparisons were not significantly different.

Table 10.3. Habitat Availability and Habitat Use and Preference of Breeding Female Brown-headed Cowbirds in Three Midwestern Landscapes

Site	Habitat	% Avail.	Breeding Range[a]			Feeding Range[b]		
			% Use	Rank Pref.[c]		% Use	Rank Pref.[c]	
Ashland, MO	Feedlot	0.0	0.0	2.8	A	0.0	3.5	A
	Rowcrop	13.3	0.0	4.9		3.3	4.9	B
	Grass	31.6	8.2	5.8		47.3	2.5	C
	Developed	1.8	10.1	2.7	A	8.8	2.3	C
	Shrubland	2.1	16.6	2.4	A	4.8	3.3	A
	Forest	50.0	65.1	2.3	A	35.7	4.5	B
Carr Creek, MO	Feedlot	0.01	2.0	2.8	BC	2.3	2.6	
	Rowcrop	0.18	0.2	3.9	A	0.0	3.8	
	Grass	3.54	24.1	2.3	C	75.6	1.1	
	Developed	1.10	14.1	3.1	B	7.5	3.3	
	Shrubland	0.49	8.5	3.3	AB	1.3	4.2	
	Forest	93.09	51.4	5.5		13.3	6.0	
Jonesboro, IL	Feedlot	0.01	0.0	2.7	A	17.3	2.1	A
	Rowcrop	35.61	1.6	6.0		22.9	4.6	
	Grass	3.00	1.5	4.3		28.4	2.0	A
	Developed	3.01	5.0	3.6		8.3	2.9	
	Shrubland	2.00	10.2	2.7	A	2.1	3.6	
	Forest	55.04	81.2	1.7		20.9	5.8	

[a]Morning (<1100 CST), nonfeeding observations.

[b]Daytime feeding observations.

[c]Mean rank of the difference between habitat use and availability; lower ranks indicate greater preference. Ranks in the same column within a study site that are followed by the same letter are not significantly different.

ings, and these individuals may have had time to feed and return to the forest in the afternoon.

Spatial Relationships of Breeding, Feeding, and Roosting

Most cowbirds used spatially distinct areas for morning nonfeeding ranges, afternoon ranges, and evening (roosting). The ability of cowbirds to uncouple breeding and feeding and to commute between disjunct areas used for each activity has been reported by other investigators studying radiotagged cowbirds in the East (Dufty 1982), Midwest (Raim, Chapter 9, this volume), and West (Verner and Ritter 1983; Rothstein et al. 1984, 1986).

Our discovery that morning, afternoon, and evening ranges were distinct did not necessarily mean they did not overlap. For instance, evening roosts of some cowbirds on the Jonesboro study area (dotted circles, diamonds, and squares in Figure 10.1) were clustered with their home ranges. These locations overlapped the home range but had a different spatial distribution as determined by the MRPP test. Other roosts, however, were completely disjunct from the rest of the home ranges, as in the communal roost on the Jonesboro study area and locations on the Ashland study area (Figure 10.1, dotted circles).

A smaller proportion of cowbirds had spatially distinct morning, afternoon, and evening ranges on the Carr Creek study area than on the Ashland or Jonesboro study areas. Most of this difference was because females at Carr Creek tended to roost within their breeding and feeding areas, while individuals on the other study areas more often roosted in distinct locations (Figure 10.1). The range in the percentage of birds with distinct morning and afternoon ranges (69 to 87%, Table 10.1) was not large; however, it can be related to differences in resource distribution. The greatest amount of overlap in morning and afternoon ranges occurred on the Carr Creek study area, where cowbird numbers were lowest, cowbirds were most likely limited by availability of feeding habitat, and there were abundant forest and hosts in the landscape.

Cowbird Movements

It is noteworthy that cowbirds moved similar distances between breeding and feeding areas in all landscapes. Longer movements might be expected in landscapes where cowbirds are abundant and host competition is high. However, as demonstrated by the Jonesboro study area, these landscapes also tend to be highly fragmented, with feeding areas near most of the forest. Larger movements and home ranges on the Jonesboro study area and movements at the Ashland study area were largely the result of travel to or from roost sites. Additionally, 15 of 28 radiotracked cowbirds at the Jonesboro site roosted communally with more than 200 cowbirds and more than 1,000 grackles (Thompson 1994). It is not clear why female cowbirds should roost at sites more distant from their breeding ranges than their feeding ranges, or why this behavior was more prevalent in one landscape. Large movements to roost sites distant from breeding areas suggest they serve an important function (Thompson 1994), particularly given the benefits of arriving early on breeding areas (Burhans, Chapter 18, this volume). The longest movements we observed were to and from the large communal roost on the Jonesboro study area. The Jonesboro study area was also the only study area that had large flocks of other blackbirds that cowbirds could roost with. The communal roost was located in semiflooded forest in the Mississippi floodplain, habitat that was not available on the other study areas.

Home Ranges

Cowbirds had very large home ranges, particularly when compared to those predicted for similar-sized passerines (Schoener 1968). Their large home range reflected the distribution of three key resources: host-rich forest habitats, agricultural feeding areas, and secure roosting habitats.

Cowbird home ranges in our study were larger than those previously reported. Previous studies have reported a mean area of morning home ranges of 4–68 ha (Dufty 1982, Darley 1983, Rothstein et al. 1984, Teather and Robertson 1985). Rothstein et al. (1984) reported a mean total home range of 405 ha in the Sierra Nevada. The large home ranges in our study areas may partly be due to our methods; we used radiotelemetry and searched large areas to relocate birds (as did Rothstein et al. 1984). Home ranges based on visual resightings of birds may be underestimates.

Home ranges were very variable, making it difficult to detect significant differences among study areas (Table 10.2). The only significant difference among study areas was that home range size was larger on the Jonesboro study area than on the Carr Creek or Ashland study areas. This difference can be explained by the long flight most cowbirds on the Jonesboro area made to a communal roost 5–7 km south of their breeding and feeding areas.

The variability in home ranges, particularly morning non-feeding ranges, could indicate differences among individuals in territoriality, home range quality, and breeding status. Our methods did not allow us to determine if home ranges were actually defended territories. Seventeen of 77 morning ranges were ≤ 30 ha, which is within the range of sizes reported by Dufty (1982) and Darley (1983) as defended nonfeeding ranges. However, two to four morning ranges on each area were ≥ 1,000 ha and were likely not defendable.

The wide range in morning home range size could also indicate that some individuals had well-defined home ranges (i.e., open diamonds and squares on Jonesboro study area, Figure 10.1), while others floated or ranged over a wide area (open diamonds on Carr Creek study area, Figure 10.1). However, even individuals with large morning ranges often had areas of high use where many morning locations were clustered. These areas of high use may correspond to what other studies have mapped as morning ranges, and there may be less discrepancy in the size of morning ranges than there appears.

Variation in cowbird morning ranges may also have been the result of variation in the distribution of resources such as hosts. Host densities varied greatly in all study areas, most notably from ridgetops to ravines (S. K. Robinson unpubl. data, F. Thompson pers. obs.). However, we have no direct evidence that the home range size varied in response to host density.

Afternoon ranges were also not significantly different among study areas, and they too varied greatly among individuals. Some cowbirds restricted feeding activities to a few closely spaced locations, while others ranged over several disjunct areas. Afternoon ranges of many individuals approached the size of entire home ranges because some cowbirds returned to morning ranges at some point in the afternoon. For instance, 13–36% of afternoon locations were in mature forest (Table 10.3). These locations usually represented individuals that had fed in grassland, cropland, or feedlot habitats and then returned to the forest (Figure 10.1). Many afternoon ranges were also large because cowbirds changed their feeding sites in response to changes in habitat during the study period. For instance, cowbirds often moved to newly mowed fields to feed, or followed cattle when they were moved to a new pasture.

Habitat Preferences

Cowbird habitat use has been reported previously, but no studies have demonstrated habitat selection or preference. We believe that cowbirds preferred forest and shrubland habitats in the morning because of the comparatively high number of potential hosts in these habitats. While we did not survey host densities, our general observations were that there were few potential hosts in grass and cropland habitats. Cowbirds preferred grass, developed land, and feedlot habitats in the afternoon, when they primarily fed (Thompson 1994). This result is consistent with previously reported descriptions of feeding habitat (Friedmann 1929, Mayfield 1965, Rothstein et al. 1984). We believe that short grassland is a critical habitat for feeding cowbirds. It was among the most preferred afternoon habitats in all landscapes, even where availability was low. Use of feedlot and rowcrop habitats was greatest on the Jonesboro area where grass habitat was most limited (but still preferred). Cowbirds also had a strong afternoon preference for feedlots in the two landscapes they occurred in. These do not appear to be a critical habitat, because cowbirds were abundant on the Ashland study area where there were no feedlots.

Cowbirds on the Carr Creek study area did not show a morning preference for forest habitat but preferred grass habitats in the morning and afternoon. We believe that the low preference for forest habitats in the morning was due to real differences in cowbird behavior as well as to idiosyncrasies in the habitat preference analysis. We observed females on feeding areas at Carr Creek earlier in the day than on the other study areas, so cowbirds were detected less often in the forest. Cowbirds at Carr Creek, however, still used forest habitats more than any other habitat in the morning (51% of morning observations were in forest), but forest had a low preference rank because it was extremely abundant on the study area (93% of the area). Cowbirds may have been able to return earlier to feeding areas at Carr Creek because of low levels of competition for hosts resulting from low cowbird densities. Cowbird numbers and the cowbird/host ratio were much lower on the Carr Creek study area because cowbirds were likely limited by the availability of suitable feeding habitat (Thompson et al., Chapter 32, this volume).

Conclusions

There were many similarities in cowbird behavior and movements among study areas. Most cowbirds in all landscapes demonstrated temporal and spatial segregation of breeding and feeding activities. Mean movements between breeding and feeding ranges were similar, and home ranges were large and variable in all landscapes.

Habitat preferences, with one exception, were also similar. Morning habitat preferences for forest and shrubland habitats are likely the result of high host densities in these habitats (Rothstein et al. 1984). Strong afternoon habitat preferences of cowbirds for grass, feedlot, and developed habitats, even in heavily forested landscapes, are indicative of the species' dependence on agricultural and developed habitats. We believe that grass habitat is critical for feeding cowbirds. However, large areas of grass habitat may not be necessary to support a large cowbird population if other agricultural habitats are available, and if feeding habitat is dispersed in the landscape.

There were only a few significant differences in spatial patterns, home ranges, movements, and habitat preferences among the three study areas. These differences could all be related to the use of a distant, large communal roost on the Jonesboro study area. Given the large variation in the composition of these landscapes, the lack of more significant differences in cowbird behavior is noteworthy. While landscapes varied greatly, they all consisted of central hardwood forest fragmented by agricultural and developed habitats. The life-history strategy of female cowbirds on all study areas was similar; females preferred host-rich forest habitats during the morning but appeared dependent on open, agricultural, and developed habitats for feeding. We believe that similarities in spatial distribution, movements, home range, and habitat selection among study areas are a result of this common life-history strategy in fragmented midwestern forests.

While it was not directly measured in our study, one parameter that did vary greatly among study areas is cowbird density (Thompson et al., Chapter 32, this volume). We suggest that the cowbird breeding strategy in midwestern forests also acts as a constraint; local population size during the breeding season depends on the availability of suitable feeding habitat.

The dependence of cowbird populations on availability of feeding habitat in forested landscapes has important conservation implications for cowbird hosts. We believe the most effective method to reduce community-wide parasitism levels in forested habitats is to reduce the dispersion of feeding habitats. Dispersion of feeding sites in a landscape may be more important than the total area of feeding habitat because cowbirds are highly social in the afternoon, feed in flocks, and move limited distances between breeding and feeding areas.

Acknowledgments

The USDA Forest Service North Central Forest Experiment Station, the Mark Twain National Forest, and the Shawnee National Forest funded this research. The Missouri Department of Conservation provided field housing. K. Austin, M. Spanel, M. Mumford, R. Smith, G. Houf, and W. Alden facilitated work on Shawnee and Mark Twain national forests. We thank B. Edmond, L. Fray, T. Fredrickson, J. Gardner, B. Hartsell, C. Newbold, A. Taylor, and R. Weidel, who worked all hours radiotracking cowbirds. S. Robinson, J. Faaborg, S. Shifley, and T. Donovan provided assistance and advice.

References Cited

Alldredge, J. R., and J. T. Ratti. 1986. Comparison of some statistical techniques for analysis of resource selection. J. Wildl. Manage. 50:157–165.

———. 1992. Further comparison of some statistical techniques for analysis of resource selection. J. Wildl. Manage. 56:1–9.

Biondini, M. E., P. W. Mielke, and E. F. Redente. 1988. Permutation techniques based on Euclidean analysis spaces: A new and powerful statistical method for ecological research. Coenoses 3:155–174.

Burt, W. H. 1943. Territoriality and home range concepts as applied to mammals. J. Mammal. 24:346–352.

Darley, J. A. 1983. Territorial behavior of the female Brown-headed Cowbird (*Molothrus ater*). Can. J. Zool. 61:65–69.

Dufty, A. M., Jr. 1982. Movements and activities of radio-tracked Brown-headed Cowbirds. Auk 99:316–327.

Friedmann, H. 1929. The cowbirds: A study in the biology of social parasitism. C. C. Thomas, Springfield, IL.

Fuemmeler, W. J. 1992. Evaluation of techniques for estimating Mourning Dove population parameters in central Missouri. M.S. Thesis, University of Missouri, Columbia.

Hann, H. W. 1941. The cowbird at the nest. Wilson Bull. 53:211–221.

Johnson, D. H. 1980. The comparison of usage and availability measurements for evaluations of resource preference. Ecology 61:65–71.

Lowther, P. E. 1993. Brown-headed Cowbird (*Molothrus ater*). *In* The birds of North America 47 (A. Poole and F. Gill, eds). Academy of Natural Sciences, Philadelphia; American Ornithologists' Union, Washington, DC.

Mayfield, H. F. 1965. The Brown-headed Cowbird with old and new hosts. Living Bird 4:13–28.

Mielke, P. W., Jr. 1991. The application of multivariate permutation methods based on distance functions in the earth sciences. Earth-Sci. Rev. 31:55–71.

Mielke, P. W., K. J. Berry, P. J. Brockwell, and J. S. Williams. 1981. A class of nonparametric tests based on multiresponse permutation procedures. Biometrika 68:720–724.

Mohr, C. O. 1947. Table of equivalent populations of North American small mammals. Amer. Midl. Natur. 37:223–249.

Rothstein, S. I., J. Verner, and E. Stevens. 1980. Range expansion and diurnal changes in dispersion of the Brown-headed Cowbird in the Sierra Nevada. Auk 97:253–267.

———. 1984. Radio-tracking confirms a unique diurnal pattern of spatial occurrence in the parasitic Brown-headed Cowbird. Ecology 65:77–88.

Rothstein, S. I., D. A. Yokel, and R. C. Fleischer. 1986. Social dominance, mating and spacing systems, female fecundity, and vocal dialects in captive and free-ranging Brown-headed Cowbirds. Curr. Ornithol. 3:127–185.

Schoener, T. W. 1968. Sizes of feeding territories among birds. Ecology 49:123–141.

Scott, D. M. 1991. The time of day of egg-laying by the Brown-headed Cowbird and other icterines. Can. J. Zool. 69:2093–2099.

Stoddard, H. L. 1931. The Bobwhite quail: Its habits, preservation, and increase. Charles Scribner's Sons, New York.

Teather, K. L., and R. J. Robertson. 1985. Female spacing patterns in Brown-headed Cowbirds. Can. J. Zool. 63:218–222.

Thompson, F. R. III. 1994. Temporal and spatial patterns of breeding Brown-headed Cowbirds in the midwestern United States. Auk 111:979–990.

Verner, J., and L. V. Ritter. 1983. Current status of the Brown-headed Cowbird in the Sierra National Forest. Auk 100:355–368.

White, G. C., and R. A. Garrott. 1990. Analysis of wildlife radio-tracking data. Academic Press, San Diego, CA.

11. The Effects of Host Numbers on Cowbird Parasitism of Red-winged Blackbirds

CHRISTY A. CARELLO AND GREGORY K. SNYDER

Abstract

We studied local factors affecting parasitism of Red-winged Blackbird nests parasitized by Brown-headed Cowbirds. We compared frequencies of parasitism between Walden Ponds and Twin Lakes in Boulder County, Colorado, during two breeding seasons. Both sites had similar nest sites and large trees that could have been used as vantage points by cowbirds, yet the percentage of parasitism was much higher at Walden Ponds (40–100%), where small numbers of redwings bred, than at the larger colony at Twin Lakes (0–3%). Our results suggest that large colony size acts as a nest defense against cowbird parasitism in redwings. Nest arrangement and the presence of other aggressive species in the nest area may also contribute to defenses against cowbird parasitism.

Introduction

Brown-headed Cowbirds are a serious threat to the reproductive success of several migratory songbirds in North America (Brittingham and Temple 1983, Friedmann and Kiff 1985). Environmental conditions (e.g., vegetation composition, availability of cowbird perches, and proximity to cowbird feeding areas) and host behaviors (e.g., aggression, egg removal) affect parasitism frequencies (Wiley 1988, Freeman et al. 1990, Post et al. 1990, Wiley et al. 1990, Ortega and Cruz 1991, Neudorf and Sealy 1992, O'Conner and Faaborg 1992, Neudorf and Sealy 1994). The long-term success of efforts to protect songbirds from cowbird parasitism may depend on distinguishing which environmental and behavioral factors limit or facilitate nest parasitism (Kelly and DeCapita 1982, Beezley and Rieger 1987, Wiley et al. 1990). Therefore, the relative importance of these factors to the frequency of parasitism requires further investigation.

Red-winged Blackbirds (*Agelaius phoeniceus*) are suitable for studying the influence of environmental and behavioral factors on parasitism because they are frequent cowbird hosts (Ortega and Cruz 1991). Redwings are also aggressive toward female cowbirds (Ortega and Cruz 1991, Neudorf and Sealy 1992), and aggressive interactions between redwings and cowbirds may decrease the probability that a cowbird enters a potential host nest. In addition, redwings often nest in colonies (Linz and Bolin 1982, Ortega and Cruz 1991), which may further decrease the probability that a cowbird can enter a redwing nest (Linz and Bolin 1982, Freeman et al. 1990). Thus nest dispersion and colony size may facilitate nest defense.

Our purpose here is to investigate if colony size affects parasitism of Red-winged Blackbirds by cowbirds. We hypothesized that a greater number of nests at a given site would increase the chance of encounters between redwings and cowbirds, thereby decreasing parasitism frequency. We studied nest parasitism at two locations in Boulder County, Colorado: Twin Lakes, with a single large clump of redwing nests, and Walden Ponds, with four separate nest sites. At Walden Ponds the nests were spread along the fringes of different ponds, each pond containing a different number of nests. Twin Lakes and Walden Ponds have similar numbers of perches for cowbirds, and both locations are the same distance from cowbird feeding grounds. In addition, the dominant vegetation used for nesting by the redwings, cattails (*Typha* spp.), is similar at the two locations. Hence, differences in cowbird parasitism should reflect characteristics of the colonies themselves, such as nest number and nest arrangement.

Study Areas

Walden Ponds and Twin Lakes are both man-made. Twin Lakes, established in 1929, has two bodies of water, one of which is used for nesting by redwings. Walden Ponds consists of five reclaimed gravel pits. Marsh and Cottonwood ponds were excavated and filled in 1984. Bass, Island, and Duck ponds were excavated and filled in 1987.

Twin Lakes and Walden Ponds, which are approximately 2 km apart, are both surrounded by agricultural land. Both sites have Russian olive (*Elaeagnus angustifolia*) and plains cottonwood (*Populus sargentii*) trees at the edges of the cattails. The study site at Twin Lakes was in a clump of cattails 3 m from the lake shore (Figure 11.1A). The cattail clump had densely packed trees on two sides and sparsely distributed trees on the other side. At Walden Ponds, cattails were arranged in fringes along the edges of the ponds with trees distributed on the landward side of the cattail fringes (Figure 11.1B).

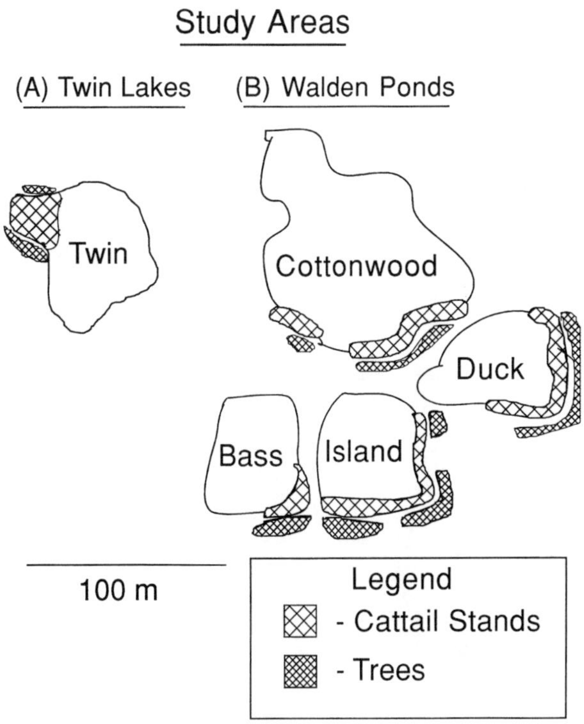

Methods

In 1993 and 1994, we checked Red-winged Blackbird nests for eggs at Twin Lakes and Walden Ponds. We also counted Common Grackle (*Quiscalus quiscula*) nests in the redwing nesting grounds. From 14 May until 29 July 1993, redwing nests were checked daily at Walden Ponds and every two days at Twin Lakes. At both sites, we tagged nests with red tape attached to a cattail 5 m from the nest or to a nearby tree. From 14 May until 20 July 1994, nests at both locations were checked every third day. In 1994, we measured distances to the nearest neighboring nests with a 50-m tape, cattail densities using the nearest neighbor method (Gysel and Lyon 1980), and tree height with a clinometer. Sample sizes for cattail density and nearest nest measurements were based on the number of cattail clumps in which there were nests except at Twin Lakes, where a subsample was used. Data are presented throughout as means ± SE.

Results

Similar nesting patterns were observed in 1993 and in 1994 (Figure 11.2). The first redwing nests with eggs were found during the second week in May, and the number of nests increased rapidly until 30 May. After 30 May there was a sharp decline in the number of nests at Walden Ponds and a more gradual decline at Twin Lakes. Nest number continued to decline through the third week in June at both sites (Figure 11.2). Nesting at Walden Ponds was completed by the end of June. However, at Twin Lakes there was a second period of nesting (data not shown). The pattern of nest development during this second nesting was similar to the first, and nesting ended by 19 July.

At Walden Ponds in 1993 and in 1994 parasitism started at the beginning of the nesting period and continued as long as there were active nests (Figure 11.2). Overall, between 40% and 100% of nests were parasitized (Table 11.1). At Twin Lakes, four cowbird eggs were found out of 106 redwing nests in 1993 (Figure 11.2, Table 11.1). The first of these cowbird eggs was not detected until 11 June, 25 days after the first cowbird egg was discovered at Walden Ponds and after the number of nests at Twin Lakes had declined (Figure 11.2). All of these cowbird eggs were found in nests on the edge of the colony. In addition, two of the eggs were

Study Areas

Figure 11.1. Maps of the two study sites in Boulder County, Colorado: (A) Twin Lakes and (B) Walden Ponds.

found in one nest, and a third egg was found within 3 m of that nest. No cowbird eggs were found during the second nesting period. In 1994, the peak number of nests available for parasitism at Twin Lakes was double that in 1993 (Figure 11.2, Table 11.1), yet no cowbird eggs were found. The number of redwing nests was the only environmental variable that was significantly correlated with the percentage of parasitism (Table 11.2).

In addition to the differences in percentage of parasitism between Twin Lakes and Walden Ponds, the frequency of parasitism within Walden Ponds also varied with the number of nests per pond (Figure 11.3). The pond with the fewest nests (Bass) had 100% parasitism in both years, while the ponds with 10 or more nests across the two years averaged only 59% parasitism.

Common Grackles nested in the cattails along with the redwings and in the surrounding trees at Twin Lakes, but not at Walden Ponds. In 1993, 8 grackle nests were found, and in 1994 18 nests were found. None of these were parasitized.

Discussion

We found much less frequent parasitism at Twin Lakes, the site with more breeding redwings, than at Walden Ponds. We also noted a trend for greater parasitism at ponds with fewer breeding redwings within the Walden Ponds area.

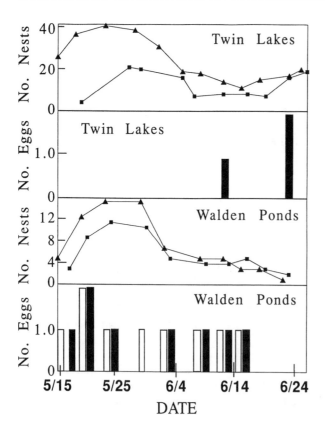

Figure 11.2. Numbers of Red-winged Blackbird nests available for cowbird parasitism (nests with at least one redwing egg) in 1993 (squares) and 1994 (triangles), and the numbers of Brown-headed Cowbird eggs laid in 1993 (filled bars) and 1994 (open bars) at Twin Lakes and Walden Ponds.

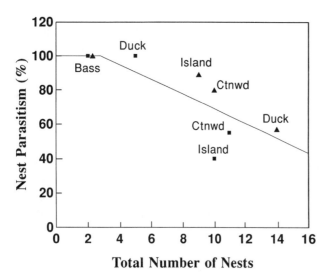

Figure 11.3. The percentage of Brown-headed Cowbird parasitism of Red-winged Blackbird nests at Walden Ponds in 1993 (squares) and 1994 (triangles). % parasitism = 112 - 4.3 (nests). F = 10.62; P = .01; r^2 = 0.64.

These patterns of parasitism were consistent over the two study years. We suggest that these results arose as a consequence of one or both of two factors. First, only a few individual female cowbirds may have visited each site to lay, and as a result, the percentage of parasitism was inversely related to host nest abundance at a site. Other studies, in contrast, have found positive effects of host abundance on the frequency of parasitism (Smith and Arcese 1994). Our explanation does not, therefore, apply generally across all species of cowbird hosts, even if it was the principal mechanism operating here. Second, aggression by redwings nesting in clumps at Twin Lakes may have made it difficult for female cowbirds to reach nests (Rothstein et al. 1980, Linz and Bolin 1982). This pattern may apply generally to the redwing as a host species, and may be the key mechanism driving percentage of parasitism in our study.

However, additional factors, such as nest dispersion and the presence of other species in the nesting grounds, may have contributed to the very low incidence of nest parasitism at Twin Lakes. One feature distinguishing Twin Lakes from Walden Ponds was the spatial distribution of nests. Nests at Twin Lakes were in a single clump, whereas nests at Walden Ponds were distributed along the pond fringes. Clumping may have disrupted the attempts of cowbirds to enter nests. We found, as have previous studies (Rothstein et al. 1980, Linz and Bolin 1982), that occasional parasitism that did occur at Twin Lakes was restricted to the periphery, where it may be easier for the cowbird to enter a nest.

Clumped nesting as a defense against nest parasitism warrants further study. Preliminary observations at Walden Ponds on Yellow-headed Blackbirds (*Xanthocephalus xanthocephalus*) suggested that clumped nests might also deter cowbirds in this species. Yellowheads nested in cattails bordering the redwing nest sites. However, unlike the redwing nests, the yellowhead nests were arranged in clumps, similar to the arrangement of redwing nests at Twin Lakes. At Walden Ponds, as in most other areas, yellowheads rarely suffer nest parasitism (Ortega and Cruz 1991), although they can be parasitized more frequently (Dufty, Chapter 12, this volume).

Finally, the presence of grackles at Twin Lakes may have contributed to the very low rate of nest parasitism. Clark and Robertson (1979) found that Yellow Warblers (*Dendroica petechia*) suffered from less cowbird parasitism when nesting near the more aggressive Red-winged Blackbird. Grackle nests at Twin Lakes were interspersed with those of the redwings. Since grackles are aggressive toward many passerine species (Davidson 1994), their presence may have interfered with cowbird access to redwing nests. Cowbirds often use observation posts such as trees to monitor host activity (Norman and Robertson 1975, Wiley 1988), and the availability of such posts has been shown to increase the frequency of nest parasitism (Freeman et al. 1990, Payne 1973). Some grackles at Twin Lakes nested in the trees sur-

Table 11.1. Nest Parasitism and Vegetation Analysis at Twin Lakes and Walden Ponds

		Twin Lakes	Duck Pond	Island Pond	Cottonwood Pond	Bass Pond
No. Nests	1993	106	5	10	11	2
	1994	126	14	9	10	2
% Parasitism	1993	4	100	40	55	100
	1994	0	57	89	80	100
No. multiple	1993	1	2	1	1	1
	1994	0	3	3	1	0
Cattail density in stems/m²		108.3 ± 15.4	79.8 ± 14.9	60.4 ± 12.5	102.3 ± 36.3	28.6 ± 6.3
(N)		(20)	(14)	(11)	(12)	(1)
Tree height in m		8.9 ± 1.6	8.7 ± 1.3	3.3 ± 0.1	6.0 ± 0.7	4.1
(N)		(12)	(12)	(26)	(12)	(1)
Nearest nest in m		1.1 ± 0.1	7.0 ± 1.7	7.1 ± 2.2	6.2 ± 2.0	1.4
(N)		(11)	(8)	(5)	(6)	(1)

Note: The values for nest parasitism include the number of Red-winged Blackbird nests with more than one cowbird egg (No. multiple). Vegetation values are given as means ± SE, with sample sizes in parentheses.

Table 11.2. Correlation Analysis of Percentage Parasitism and Nest Environment for Red-winged Blackbirds

	% Parasitism	Number of Nests per Site	Mean Tree Height	Cattail Density
Number of nests	−0.944			
	(0.01)			
Tree height	−0.833	0.630		
	(0.07)	(0.25)		
Cattail density	0.431	−0.622	−0.064	
	(0.46)	(0.26)	(0.91)	
Nearest redwing nest	−0.649	−0.579	−0.141	0.836
	(0.53)	(0.30)	(0.82)	(0.08)

Note: Values are Pearson correlation coefficients; values in parentheses are *P* values. *N* values for each variable are given in Table 11.1.

rounding the cattail beds and may have prevented female cowbirds from gaining access to the perch sites necessary to identify potential host nests. While further work is needed to verify a relationship between the presence of grackles and the access of cowbirds to redwing nests, it is noteworthy that cowbird eggs were found in redwing nests at Twin Lakes only after the grackles had completed nesting and left the area.

In conclusion, we found marked differences in parasitism levels at two nearby redwing breeding colonies in Colorado. Although we could not explain these differences fully because of our sampling design, we suspect that larger colony size and/or clumping of redwing nests were the principal determinants of reduced parasitism. These differences are of management interest for two reasons. First, they suggest that manipulating redwing nesting distribution may confer protection against brood parasitism of redwings by cowbirds. While the redwing is not a species of conservation interest in Colorado, it is of interest in Florida (Cruz et al., Chapter 4, this volume). Second, enhancing the size and altering the shape of colonies of the aggressive redwing might reduce the level of parasitism in nests of other marsh-nesting species of greater management concern.

Acknowledgments

We thank Kim Walker, Vikki Davis, Brian Cole and Keith Biggs for their invaluable assistance in the field. Sally Susnowitz, Jorge Moreno, Spencer Sealy, and Jamie Smith made helpful comments on the manuscript. This study was supported in part by National Science Foundation Grant IBN9120144 and NHLBI Grant HL32894 to GKS.

References Cited

Beezley, J. A., and J. P. Rieger. 1987. Least Bell's Vireo management by cowbird trapping. Western Birds 18:55–61.

Brittingham, M. C., and S. A. Temple. 1983. Have cowbirds caused forest songbirds to decline? BioScience 33:31–35.

Clark, K. L., and R. J. Robertson. 1979. Spatial and temporal multispecies nesting aggregations in birds as antiparasite and antipredator defenses. Behav. Ecol. Sociobiol. 5:359–371.

Davidson, A. H. 1994. Common Grackle predation on adult passerines. Wilson Bull. 106:174–175.

Freeman, S., D. F. Gori, and S. Rohwer. 1990. Red-winged Blackbirds and Brown-headed Cowbirds: Some aspects of a host-parasite relationship. Condor 92:336–340.

Friedmann, H., and L. F. Kiff. 1985. The parasitic cowbirds and their hosts. Proc. Western Found. Vert. Zool. 2:225–302.

Gysel, L. W., and L. J. Lyon. 1980. Habitat analysis and evaluation. In Wildlife management techniques manual. Wildlife Society, Washington, DC.

Kelly, S. T., and M. E. DeCapita. 1982. Cowbird control and its effect on Kirtland's Warbler reproductive success. Wilson Bull. 94:363–365.

Linz, G. M., and S. B. Bolin. 1982. Incidence of Brown-headed Cowbird parasitism on Red-winged Blackbirds. Wilson Bull. 94:93–95.

Neudorf, D. L., and S. G. Sealy. 1992. Reactions of four passerine species to threats of predation and cowbird parasitism: Enemy recognition or generalized response? Behaviour 123:84–105.

———. 1994. Sunrise nest attentiveness in cowbird hosts. Condor 96:162–169.

Norman, R. F., and R. J. Robertson. 1975. Nest-searching behavior in the Brown-headed Cowbird. Auk 92:610–611.

O'Conner, R. J., and J. Faaborg. 1992. The relative abundance of the Brown-headed Cowbird (Molothrus ater) in relation to exterior and interior edges in forests of Missouri. Trans. Missouri Acad. Sci. 26:1–9.

Ortega, C. P., and A. Cruz. 1991. A comparative study of cowbird parasitism in Yellow-headed Blackbirds and Red-winged Blackbirds. Auk 108:16–24.

Payne, R. 1973. The breeding season of a parasitic bird, the Brown-headed Cowbird, in central California. Condor 75:80–99.

Post, W., T. K. Nakamura, and A. Cruz. 1990. Patterns of Shiny Cowbird parasitism in St. Lucia and southwestern Puerto Rico. Condor 92:461–469.

Rothstein, S. I., J. Verner, and E. Stevens. 1980. Range expansion and diurnal changes in dispersion of the Brown-headed Cowbird in the Sierra Nevada. Auk 97:253–267.

Smith, J. N. M., and P. Arcese. 1994. Brown-headed Cowbirds and an island population of Song Sparrows: A 16-year study. Condor 96:916–934.

Wiley, J. W. 1988. Host selection by the Shiny Cowbird. Condor 90:289–303.

Wiley, J. W., W. Post, and A. Cruz. 1990. Conservation of the Yellow-shouldered Blackbird Agelaius xanthomus, an endangered West Indian species. Biol. Conserv. 55(191):119–138.

12. Cowbird Brood Parasitism on a Little-Used Host: The Yellow-headed Blackbird

ALFRED M. DUFTY, JR.

Abstract

Brood parasitism by Brown-headed Cowbirds has rarely been reported on Yellow-headed Blackbirds. However, during the 1991 and 1992 breeding seasons at a site near Boise, Idaho, a minimum of 21.2% and 17.2% of Yellow-headed Blackbird nests were parasitized, respectively. Most cowbird eggs laid in yellowhead nests were deposited after the peak in yellowhead clutch starts, whereas parasitism on Red-winged Blackbirds, which nested in the same area, occurred in proportion to the clutch starts of redwings. This suggests that yellowheads are not preferred hosts. A possible explanation for brood parasitism on this population of Yellow-headed Blackbirds is suggested, based on the possible effects of cattle grazing and suburban development on cowbird and host populations.

Introduction

The Yellow-headed Blackbird (*Xanthocephalus xanthocephalus*) is a locally abundant, marsh-nesting, colonial species of western North America (e.g., Burleigh 1972). Previous research has indicated that Yellow-headed Blackbird (henceforth yellowhead) colonies rarely suffer brood parasitism by Brown-headed Cowbirds (henceforth cowbirds). Willson (1966) found that 2/371 (0.5%) yellowhead nests were parasitized in eastern Washington, and Ortega and Cruz (1988) noted that none of 351 yellowhead nests in Colorado were naturally parasitized, despite their demonstration that yellowheads can raise cowbirds successfully. L. Beletsky and G. Orians (pers. comm.) found cowbird eggs in only 6/1,175 (0.5%) eastern Washington yellowhead nests from 1988 to 1992. Additional scattered instances have been collected by Friedmann (1963, Friedmann and Kiff 1985).

Hence, there is no evidence that cowbirds use yellowheads as hosts on a regular basis. This is particularly puzzling because Ortega and Cruz (1988) demonstrated that yellowheads accept cowbird eggs and successfully raise cowbird nestlings. Indeed, these authors conducted a study designed "to determine possible explanations for the *lack* of Brown-headed Cowbird . . . parasitism on Yellow-heads" (Ortega and Cruz 1991, p. 16, emphasis mine). I document the occurrence, during two breeding seasons, of much higher frequencies of cowbird brood parasitism on yellowheads than reported for any previous study of this species. I compare the parasitism on yellowheads with parasitism on a more commonly utilized host found in the same area, the Red-winged Blackbird (*Agelaius phoeniceus*, hereafter redwing) (Friedmann and Kiff 1985, Ortega and Cruz 1988, Orians et al. 1989), and I provide a possible explanation for the use of yellowheads as hosts in this instance.

Study Area and Methods

I performed this investigation during 1991 and 1992 in and around two abandoned lumber holding ponds separated from each other by an earthen berm approximately 10 m wide. These ponds are located adjacent to the Boise River, approximately 8 km east of Boise, Idaho. A thin riparian corridor runs along the river and is dominated by cottonwoods (*Populus* spp.), willows (*Salix* spp.) and native roses (*Rosa* spp.). A suburban housing development borders the corridor on the opposite side of the river from the study ponds and is set back approximately 60 m from the riverbank. This is a low-density residential area (fewer than 7 dwelling units/acre), begun in the mid-1980s and located less than a kilometer from the study ponds. It contains many lawns upon which cowbirds forage (pers. obs.). A small urban park and wildlife refuge lies directly opposite the study ponds. The vegetation understory is noticeably denser in the refuge than it is elsewhere along the river, but the riparian habitat does not extend further from the river.

An active lumber mill borders the study area, with freshly cut trees delivered and stacked throughout the avian breeding season. In addition, approximately 175 domestic cattle graze in the immediate vicinity of the study ponds (with access to the riparian corridor) during the late spring and summer months. At other times of the year these cattle graze in other pastures that surround the lumber mill. The pastures are closely cropped, with little habitat suitable for avian breeding (pers. obs.). They are dominated by grasses (e.g., *Bromus tectorum, Poa* spp.) and forbs (e.g., mullein, *Verbascum thapsus;* Russian thistle, *Salsola kali;* clover, *Trifolium* spp.; tumble mustard, *Sisymbrium* spp.; kochia, *Kochia scoparia;* puncture vine, *Tribulus terrestris*). These species colonize and dominate locations marked by high levels of disturbance (S. Novak pers. comm.). The nearby foothills area is sagebrush-grass habitat (*Artemisia tridentata–Pseudoregneria spicata* [= *Agropyron spicatum*]).

The total area of the study ponds is approximately 2 ha, and all yellowhead nests were within the pond boundaries. Nests were placed primarily in cattails (*Typha* spp.) and sedges (*Scirpus* spp.). Most redwing nests were associated with cattails in a canal less than 8 m wide that ran adjacent to the ponds. Additional redwing nests were interspersed among the yellowhead nests. All nest locations were marked with flagging tape.

Cowbird eggs were collected when discovered, as part of unrelated investigations that involved cowbird eggs (1991) and redwing, yellowhead, and cowbird eggs (1992). Cowbird eggs collected from nests of both species were replaced with redwing eggs in order to reduce the possibility of nest desertion. No experimenter-induced desertion was noted, although yellowheads apparently ejected many of the redwing eggs placed in their nests (Dufty 1994).

In 1991, redwing nests were examined beginning on 18 May, and nests were checked every 1–5 days. The most advanced nests were being incubated by 18 May, but most females were still laying or building nests. Yellowhead nests were not examined until 3 June 1991 because of the expected absence of brood parasitism. However, when the first yellowhead nest examined contained a cowbird egg, these nests were checked every 1–4 days. Four active yellowhead nests contained chicks when first located; the remaining nests were at earlier stages of development. Nests were considered to be active if they contained at least one yellowhead, redwing, or cowbird egg or nestling. Nest checks ended on 10 July.

In 1992, nests were monitored from 7 May to 1 July. Yellowhead nests usually were checked every day, although sometimes every 2 days. Redwing nests were also usually checked daily, although sometimes every 2–3 days.

Clutch initiation in both years was defined as the date the first egg was laid (= Day 1). I estimated the onset of laying for advanced nests (discovered late in the nesting cycle) by backdating from the date of hatch or fledging, using a 12-day incubation period (Fautin 1941, Payne 1969) and an 11-day nestling period (Beletsky and Orians 1987, Gori 1988) for both blackbird species. Advanced nests that were depredated or abandoned before these estimates could be made

were not considered when calculating nest-initiation dates. Because this dating procedure is not exact, the breeding season was partitioned into 1-week intervals for analysis.

The laying date for cowbird eggs discovered in advanced nests was estimated in one of two ways. First, by estimating the nest-initiation date (as described above) and assigning the date the cowbird egg was laid as the middle of the egg-laying period for that nest. Second, some cowbird eggs were artificially incubated, and the laying date was calculated by backdating from date of hatch, assuming a 12-day incubation period. Artificially incubated eggs in my laboratory took longer to hatch than the 10–11 days (Briskie and Sealy 1990) required for naturally incubated eggs.

Results

1992 Breeding Season

Data for 1992 are more complete than those for 1991 and are presented first (Figure 12.1). Nesting in both blackbird species peaked in the first two weeks of May and declined thereafter. A small secondary increase in yellowhead nesting activity occurred in late June. The difference between the breeding seasons in the two species is not significant (maximum $D = 0.190$, Kolmogorov-Smirnov test).

Five of 29 (17.2%) active yellowhead nests were parasitized by cowbirds in 1992, each with 1 parasitic egg. Fourteen of 38 (36.8%) redwing nests were parasitized in this area with 16 cowbird eggs (2 nests had 2 cowbird eggs each). Although this is more than double the parasitism frequency found in yellowheads, the difference is not significant ($G_{adj} = 3.13$, 1 df, $P < .1$).

Laying dates were estimated for 13/16 cowbird eggs found in redwing nests. The pattern of cowbird parasitism of redwing nests did not differ significantly from the pattern of redwing nest initiations (maximum $D = 0.350$, Figure 12.2A). In contrast, the pattern of cowbird parasitism of yellowhead nests was significantly different from the pattern of yellowhead nest initiations (maximum $D = 0.793$, $P < .01$, Figure 12.2B), with brood parasitism concentrated at the end of the yellowhead breeding season, after their peak in nest initiations. Similarly, the pattern of parasitism of redwing nests differed from the pattern for yellowhead nests (maximum $D = 0.723$, $P < .05$).

1991 Breeding Season

Thirty-three yellowhead nests were located in 1991, of which 7 (21.2%) were parasitized with 9 cowbird eggs (2 nests were parasitized twice). The difference in parasitism frequencies between the two years is not significant for yellowheads ($G_{adj} = 0.150$, 1 df, $P < .7$). Combining data for 1991 and 1992, 12/62 (19.4%) yellowhead nests were parasitized with 14 cowbird eggs.

Five redwing nests were located in the yellowhead breeding area (the canal adjacent to the ponds was not monitored

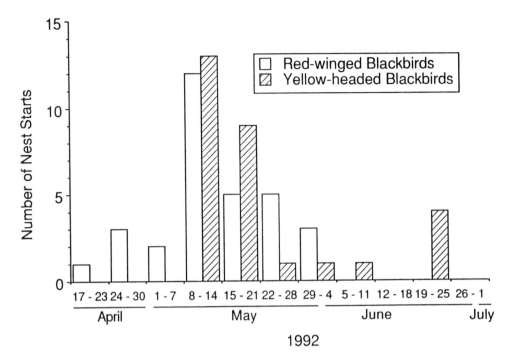

Figure 12.1. Seasonal pattern of nest starts for Red-winged Blackbirds and Yellow-headed Blackbirds. See text for an explanation of the method used to estimate dates.

in 1991). Three (60.0%) of these were parasitized with four cowbird eggs.

Discussion

Brood Parasitism on Red-winged Blackbirds and Yellow-headed Blackbirds

Brood parasitism on redwings was higher than that on yellowheads in both years, although the number of nests examined in 1991 was small. However, Red-winged Blackbirds are frequent cowbird hosts (Friedmann and Kiff 1985, Ortega and Cruz 1988, Orians et al. 1989), and the observed levels of parasitism on redwings in this study are not unusual.

Conversely, Yellow-headed Blackbirds have rarely been reported as cowbird hosts (e.g., Willson 1966, Ortega and Cruz 1988). Therefore, the elevated levels of cowbird parasitism I recorded are noteworthy. Furthermore, they indicate that brood parasitism on Yellow-headed Blackbirds is not always "accidental," as has been suggested previously (Friedmann et al. 1977, Ortega and Cruz 1988). Nonetheless, despite the observed frequency of brood parasitism on yellowheads, the results suggest that yellowheads are not preferred hosts for cowbirds, even in this study area. For example, in 1992 cowbirds did not begin to parasitize yellowheads until late in the breeding season, after the peak in clutch initiation dates for yellowheads and at a time when few redwing nests were available. This result contrasts with the situ-

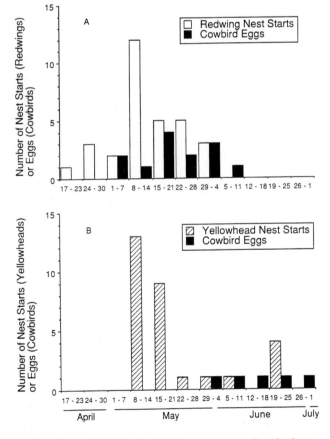

Figure 12.2. Seasonal patterns of host nest starts and cowbird parasitism in (A) Red-winged Blackbirds and (B) Yellow-headed Blackbirds.

ation for redwings, where parasitism coincided with red-wing clutch initiation, and suggests that cowbirds use yellowheads only if preferred hosts, such as redwings, are unavailable. A similar phenomenon was noted by Briskie et al. (1990), where most Least Flycatcher (*Empidonax minimus*) nests were parasitized after the peak in flycatcher nest initiation, when few nests of a preferred host, the Yellow Warbler (*Dendroica petechia*), were available.

The observed amount of brood parasitism probably represents the minimum level because some cowbird eggs may have been ejected prior to discovery of the yellowhead nest (Dufty 1994), and some parasitized yellowhead nests may have been lost prior to their discovery (Mayfield 1965).

The amount of parasitism on yellowheads in this study is substantially higher than that reported in other yellowhead populations (reviewed in Friedmann 1963, Friedmann et al. 1977, Friedmann and Kiff 1985; also Ortega and Cruz 1991, Beletsky and Orians pers. comm.). However, Lincoln (1920), writing of the Clear Creek District of Colorado, gives this account of the cowbird (and its hosts), quoted in its entirety (p. 69): "Cowbird—Summer resident, common. Red wings, Yellow-heads and Yellow-throats seem to be the species most generally imposed upon in this region." This brief but intriguing report has been dismissed by current researchers (Friedmann et al. 1977) because of its lack of supporting data. However, in view of my results, Lincoln's description should be given more credence. Further, it suggests that my results do not represent an isolated phenomenon.

Why Are Yellowheads Parasitized?

I believe that the relatively high rate of parasitism of yellowheads in this population is due to local habitat effects that produce a depauperate avifauna generally, but which may support a relatively large cowbird population.

The thin riparian corridor running along the Boise River constitutes the major breeding area for cowbirds and their hosts at my study site (in addition to the ponds and the canal). In the northwestern United States, most land birds breed in riparian habitat, and over one-third of all avian species breed exclusively in riparian areas (Mosconi and Hutto 1982). Smith and Schaefer (1993, see also Temple 1986) showed that riparian stream corridors bordered by residential housing developments, such as the one in my study, exhibit reduced bird species diversity and evenness compared to rural riparian communities. Although Smith and Schaefer (1993) found bird density to be high in urban corridors, the species that contributed to this effect are not regular hosts of the cowbird (Friedmann et al. 1977, Friedmann and Kiff 1985) and/or are not present in my study area. Further, several neotropical species, which are potential cowbird hosts (Friedmann et al. 1977, Friedmann and Kiff 1985), are either at lower density or absent from urban riparian corridors (Smith and Schaefer 1993). If the situation described by Smith and Schaefer (1993) also occurs in

my study area, then it would put increased parasitic pressure on the remaining set of potential hosts because the intensity of brood parasitism is inversely related to the density of available nests (Zimmerman 1983, Hoover and Brittingham 1993).

In addition to the possible effects of urbanization, avian populations in the study area probably also suffered from the effects of livestock grazing. Riparian areas are very sensitive to the effects of grazing, particularly year-long and spring-summer grazing (see Kauffman and Krueger 1984 for a review). Extensive grazing can reduce avian species diversity and alter overall species composition by reducing the heterogeneity of the habitat (e.g., Wiens and Dyer 1975, Mosconi and Hutto 1982, Taylor 1986, Bock et al. 1993). Livestock have grazed in the study area for approximately 20 years, and prior to that the fields were cut for hay (D. Harris pers. comm.). The resulting fields provide a very homogeneous habitat (pers. obs.). The combined negative effects of suburban development and livestock grazing on the avian community may have severely reduced the availability of potential host species to cowbirds.

Interestingly, factors that may have reduced general species diversity may also have contributed to conditions that support a relatively large cowbird population. Cowbirds are tolerant of a wide range of habitat conditions (Douglas et al. 1992, Yahner and DeLong 1992, Zimmerman 1992) and are especially attracted to pastures, livestock, and human habitations that provide foraging habitat (e.g., Friedmann 1929, Robinson et al. 1993). The intensity of brood parasitism is positively correlated with such human- and livestock-induced disturbance (Airola 1986; see also Bock et al. 1993).

The cowbird population on my study site may have been enhanced by the urbanization and grazing activities that surround it. This urbanization, if combined with a reduced number of suitable host species (due to the same factors, as described above), could result in intense parasitic pressure on preferred hosts. Indeed, anecdotal evidence suggests that brood parasitism rates on preferred hosts were high. Three nests of the Song Sparrow (*Melospiza melodia*), a frequently parasitized host (Friedmann 1929, 1963; Friedmann et al. 1977), were discovered while searching for yellowhead and redwing nests. All three nests were parasitized: one contained a single cowbird egg and four host eggs, the second contained four cowbird eggs and two Song Sparrow eggs, and the third contained two nestling cowbirds ready to fledge and no host young. In conjunction with elevated parasitism rates on preferred hosts, the relative scarcity of such hosts could force cowbirds to parasitize less-preferred species such as the Yellow-headed Blackbird (cf. Briskie et al. 1990) and may account for the unusually high use of yellowheads by cowbirds in this study. These suggestions await confirmation by future research.

Acknowledgments

Brian Bizik, Susan Loper, and Jack Small helped to locate and monitor nests. I thank L. D. Beletsky, T. L. Cook, C. P. Ortega, and S. G. Sealy for their helpful comments on an earlier version of this paper, and S. Novak for vegetation identification. This study was supported by a Faculty Research Grant from Boise State University and by funds from the Idaho EPSCOR Minigrant Program.

References Cited

Airola, D. A. 1986. Brown-headed Cowbird parasitism and habitat disturbance in the Sierra Nevada. J. Wildl. Manage. 50:571–575.

Beletsky, L. D., and G. H. Orians. 1987. Territoriality among male Red-winged Blackbirds, 1: Site fidelity and movement patterns. Behav. Ecol. Sociobiol. 20:21–34.

Bock, C. E., V. A. Saab, T. D. Rich, and D. S. Dobkin. 1993. Effects of livestock grazing on neotropical migratory landbirds in western North America. Pp. 296–309 in Status and management of neotropical migratory birds (D. M. Finch and P. W. Stangel, eds). USDA For. Serv. Gen. Tech. Rep. RM-229.

Briskie, J. V., and S. G. Sealy. 1990. Evolution of short incubation periods in the parasitic cowbirds, Molothrus spp. Auk 107:789–794.

Briskie, J. V., S. G. Sealy, and K. A. Hobson. 1990. Differential parasitism of Least Flycatcher and Yellow Warblers by the Brown-headed Cowbird. Behav. Ecol. Sociobiol. 27:403–410.

Burleigh, T. D. 1972. Birds of Idaho. Caxton Printers, Caldwell, ID.

Douglas, D. C., J. T. Ratti, R. A. Black, and J. R. Aldredge. 1992. Avian habitat associations in riparian zones of Idaho's Centennial Mountains. Wilson Bull. 104:485–500.

Dufty, A. M., Jr. 1994. Rejection of foreign eggs by Yellow-headed Blackbirds. Condor 96:799–801.

Fautin, R. W. 1941. Incubation studies of the Yellow-headed Blackbird. Wilson Bull. 53:107–122.

Friedmann, H. 1929. The cowbirds: A study in the biology of social parasitism. C. C. Thomas, Springfield, IL.

———. 1963. Host relations of the parasitic cowbirds. U.S. Natl. Mus. Bull. 233.

Friedmann, H., and L. F. Kiff. 1985. The parasitic cowbirds and their hosts. Proc. Western Found. Vert. Zool. 2:225–302.

Friedmann, H., L. F. Kiff, and S. I. Rothstein. 1977. A further contribution to knowledge of the host relations of the parasitic cowbirds. Smithson. Contrib. Zool. 235.

Gori, D. F. 1988. Adjustment of parental investment with mate quality by male Yellow-headed Blackbirds (Xanthocephalus xanthocephalus). Auk 105:672–680.

Hoover, J. P., and M. C. Brittingham. 1993. Regional variation in cowbird parasitism of Wood Thrushes. Wilson Bull. 105:228–238.

Kauffman, J. B., and W. C. Krueger. 1984. Livestock impacts on riparian ecosystems and streamside management implications: A review. J. Range Manage. 37:430–438.

Lincoln, F. C. 1920. Birds of Clear Creek district, Colorado. Auk 37:60–77.

Mayfield, H. 1965. Chance distribution of cowbird eggs. Condor 67:257–263.

Mosconi, S. L., and R. L. Hutto. 1982. The effects of grazing on land birds of a western Montana riparian habitat. Pp. 221–233 in Wildlife-livestock relationships symposium: Proc. 10. University of Idaho Forest, Wildlife, and Range Experiment Station, Moscow, ID.

Orians, G. H., E. Røskaft, and L. D. Beletsky. 1989. Do Brown-headed Cowbirds lay their eggs at random in the nests of Red-winged Blackbirds? Wilson Bull. 101:599–605.

Ortega, C. P., and A. Cruz. 1988. Mechanisms of egg acceptance by marsh-dwelling blackbirds. Condor 90:349–358.

———. 1991. A comparative study of cowbird parasitism in Yellow-headed Blackbirds and Red-winged Blackbirds. Auk 108:16–24.

Payne, R. B. 1969. Breeding seasons and reproductive physiology of Tricolored Blackbirds and Red-winged Blackbirds. Univ. Calif. Publ. Zool. 90:1–137.

Robinson, S. K., J. A. Grzybowski, S. I. Rothstein, M. C. Brittingham, L. J. Petit, and F. R. Thompson III. 1993. Management implications of cowbird parasitism on neotropical migrant songbirds. Pp. 93–102 in Status and management of neotropical migratory birds (D. M. Finch and P. W. Stangel, eds). Gen. Tech. Rep. RM-229. USDA Forest Service, Rocky Mountain Forest and Range Experiment Station, Fort Collins, CO.

Smith, R. J., and J. M. Schaefer. 1993. Avian characteristics of an urban riparian strip corridor. Wilson Bull. 104:732–738.

Taylor, D. M. 1986. Effects of cattle grazing on passerine birds nesting in riparian habitat. J. Range Manage. 39:254–258.

Temple, S. A. 1986. Predicting impacts of habitat fragmentation on forest birds: A comparison of two models. Pp. 301–304 in Wildlife 2000: Modeling habitat relationships of terrestrial vertebrates (J. Verner, M. L. Morrison, and C. J. Ralph, eds). University of Wisconsin Press, Madison.

Wiens, J. A., and M. I. Dyer. 1975. Rangeland avifaunas: Their composition, energetics, and role in the ecosystem. Pp. 146–182 in Symposium on management of forest and range habitats for nongame birds. USDA Forest Serv. Gen. Tech. Rep. WO-1.

Willson, M. F. 1966. Breeding ecology of the Yellow-headed Blackbird. Ecol. Monogr. 36:51–77.

Yahner, R. H., and C. A. DeLong. 1992. Avian predation and parasitism on artificial nests and eggs in two fragmented landscapes. Wilson Bull. 104:162–168.

Zimmerman, J. L. 1983. Cowbird parasitism of Dickcissels in different habitats and at different nest densities. Wilson Bull. 95:7–22.

———. 1992. Density-independent factors affecting the avian diversity of the tallgrass prairie. Wilson Bull. 104:85–94.

13. Host Selection in the Forest Interior: Cowbirds Target Ground-nesting Species

D. CALDWELL HAHN AND JEFF S. HATFIELD

Abstract

We investigated patterns of cowbird host selection in a large (1,300 ha) unfragmented forest in eastern New York in 1992–1993. In particular, we examined whether cowbird parasitism rates are due to host species-specific traits or to other features associated with nest sites. Nest height was significantly associated with parasitism (P = .003) in this community of 23 host species (N = 430 nests, 23% parasitized). However, the difference in mean nest heights between parasitized and unparasitized nests was due to species identity. Within each species, there was no difference in mean nest heights between parasitized and unparasitized nests. These results imply that during 1992–1993 cowbirds in this forest specialized on species that had low nests and did not select low nests regardless of species. This interpretation was further supported by a negative association across all 23 species between mean nest height and parasitism rate (P = .03).

Thus, although most of the forest-nesting species in this community experienced cowbird parasitism, parasitism rates were higher on low-nesting species such as the Ovenbird, Black-and-white Warbler, Louisiana Waterthrush, Veery, and Hermit Thrush. The Wood Thrush, a mid-height nester that is heavily parasitized in southern Illinois, experienced only 10% parasitism at our site and ranked ninth in parasitism rate, although its nests were found most often by us. A long-term study is necessary to determine whether cowbirds at this forested site consistently parasitize ground-nesting host species more often than species nesting at higher levels, or whether they periodically shift among hosts nesting at different heights and in different habitats.

Introduction

Brown-headed Cowbirds were once thought to be a field- and edge-nesting species (Friedmann 1929, 1963; Mayfield 1965). Now that cowbirds are frequently found in frag-mented woodlands (Brittingham and Temple 1983, Rothstein et al. 1984, Wilcove 1985, Lynch 1987, Robinson 1992), investigators are trying to determine which forest-nesting hosts are most vulnerable (Robinson et al. 1993). Knowing the mechanisms of host selection may help managers anticipate which forest-nesting hosts will be most frequently parasitized. Such knowledge might then be used to reduce or avoid host reproductive failure due to cowbird parasitism.

Forest-nesting neotropical migrants may be parasitized more frequently than traditional field and edge hosts, because they have less experience with cowbirds and fewer defenses (Robbins et al. 1989a,b; Terborgh 1989; Robinson 1992; Robinson et al., Chapter 33, this volume). Some studies have concluded that ground-nesting forest species are less vulnerable to cowbird parasitism (Robinson 1992, Petit and Petit unpubl. manuscript) due to nest crypsis or to cowbirds' inability to search effectively on the forest floor. Host species vulnerability is partly determined by cowbirds' host selection patterns and partly by host defenses.

We tested two alternate hypotheses: cowbird parasitism rates depend on either species-specific host identity or on particular features associated with nest sites. We investigated patterns of cowbird host selection in a large unfragmented forest where we had already found cowbird parasitism throughout the habitat (Hahn and Hatfield 1995) by examining the relative importance of nest height and species identity to parasitism rate.

Methods

The study was done during 1992 and 1993 at Millbrook, New York (51° 50' N, 73° 45' W), on a 400-ha study site located within a 1,300-ha oak-maple-hemlock forest. Stands ranged in age from 70 to 150 years and have experienced little recent use or disturbance by humans (Figure 13.1). The study area lies within a 30,000-ha township, of which 55% is forested. The remaining area is a mosaic of equal parts pasture, livestock, and suburban development

(Glitzenstein et al. 1990). Both the study area and nest searching methods are described in detail elsewhere (Hahn and Hatfield 1995).

Avian Community

The principal songbirds nesting in the forest community were Wood Thrush (*Hylocichla mustelina*), American Redstart (*Setophaga ruticilla*), Veery (*Catharus fuscescens*), Ovenbird (*Seiurus aurocapillus*), and Red-eyed Vireo (*Vireo olivaceus*), but we included all nests of known cowbird hosts

(Friedmann 1963) in our sample. We did not include nests of species known to reject cowbird eggs such as the American Robin (*Turdus migratorius*) and Gray Catbird (*Dumetella carolinensis*) (Rothstein 1975a).

Fisher's exact test was used to test for differences in parasitism rate among species, and the Jonckheere-Terpstra test was used to test for an association between parasitism rate and mean nest height across species. Both tests are nonparametric and take into account the small numbers of nests detected for some species (Lehmann 1975).

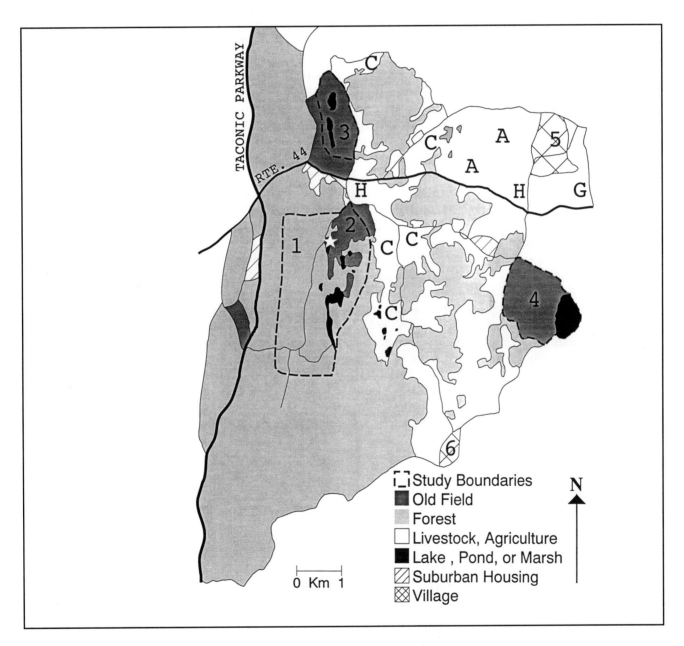

Figure 13.1. The study area, with the forested section of the area enclosed within the dashed lines. A, agricultural fields; C, cattle and dairy farms; G, golf course; H, horse farms; star, field station. From Hahn and Hatfield 1995, courtesy Blackwell Scientific Publishers.

Nest Height and Species Identity

We analyzed the relationship between parasitism rate, nest height, and species identity. In both 1992 and 1993 nest height was estimated by eye for each nest. We recorded nest height as zero only if the nest was located directly on the ground, as in leaf litter (e.g., Ovenbird), top of rocky outcroppings (e.g., Hermit Thrush), or among roots on a stream bank (e.g., some Louisiana Waterthrushes, *Seiurus motacilla*). Nests located among the roots of overturned trees (e.g., some Louisiana Waterthrushes) or on rock ledges on the face of a large rock outcropping (e.g., Eastern Phoebe, *Sayornis phoebe,* or some Hermit Thrushes, *Catharus guttatus*) were recorded as having height greater than zero, measured as the nearest distance to soil.

We used a two-way analysis of variance (ANOVA) to compare mean nest heights of parasitized nests and unparasitized nests in 1992 and 1993. Subsequently we used a three-way ANOVA to compare mean nest heights versus the effects of species identity ($N = 23$ species), parasitism, year, and the interactions of these factors. Type IV sums of squares were used to control for an unbalanced design due to some species being absent in one of the years.

In addition we estimated the association between percentage of parasitism and average nest height of each species using a weighted correlation coefficient. Weighting was by the number of nests of each species, and we used the Jonckheere-Terpstra test to test the significance of this correlation.

We also used the Wilcoxon signed-rank test (Lehmann 1975) to perform a simple, nonparametric comparison of average nest heights among species between parasitized and unparasitized nests. Although this test does not control for differences among years or allow for interactions among year, species, and parasitism, it has the advantage of assuming only that differences in average nest heights between parasitized and unparasitized nests within species are independent. (The above tests assume that nests within species, year, and parasitism categories are independent.) We pooled data over years, calculated the differences in average nest height for each of the 16 species with both parasitized and unparasitized nests, and then calculated the signed-rank statistic for these differences.

Results

Parasitism rate varied from 0 to 61.5% among species (Fisher's exact test, $P = .000$, Table 13.1). The Red-eyed Vireo (*Vireo olivaceus*), Ovenbird, and Black-and-white Warbler (*Mniotilta varia*) all experienced $\geq 50\%$ parasitism.

Parasitized nests were significantly lower ($P = .003$) than unparasitized nests (two-way ANOVA, Table 13.2). There was no significant difference between years and no significant interaction between year and parasitism. The effect of nest height corresponded to the high rates of parasitism

we observed on several species of low- and ground-nesting birds. Ovenbird, Black-and-white Warbler, Louisiana Waterthrush, and Hermit Thrush all had parasitism rates over 40% (Table 13.1), while only one species in the mid-height and high-nesting species, Red-eyed Vireo, had such a high parasitism rate (61.5%).

We further examined the relationship of nest height to parasitism rate by using three-way ANOVA to determine the relative contributions of species identity, parasitism, year, and their interactions. These analyses revealed that the significant difference in the mean nest heights of parasitized and unparasitized nests was due to a species effect ($F = 24.45$, df $= 22$, $P = .000$), not to nest height ($P = .60$). Also, there were no differences in mean nest height between parasitized and unparasitized nests, after controlling for species effects and pooling data over years ($P = .10$, $N = 16$ species, Wilcoxon signed-rank test).

The interaction between species and parasitism in the three-way ANOVA was also not significant ($P = .81$), implying that mean nest heights of parasitized and unparasitized nests were similar within each species. These results, together with the other results of the two-way and three-way ANOVAs, imply that cowbirds specialized on species that had low nests and did not select low nests regardless of species. This implies that ground-nesting species like Ovenbird and Worm-eating Warbler were parasitized by cowbirds more frequently than mid-height nesting species like the Wood Thrush, even when the latter species used lower nest sites.

To confirm these results, we tested for an association between percent parasitism and average nest height of each species (Table 13.1). The weighted correlation coefficient between these two variables is $r = -0.33$ (Jonckheere-Terpstra test, $P = .03$).

Mid-height nesting species that serve as common cowbird hosts in other sites experienced both low parasitism rates (e.g., Wood Thrush 10.0%, Eastern Phoebe 12.2%) and moderate to high parasitism rates (American Redstart 31.8%, Red-eyed Vireo 61.5%) at Millbrook.

Discussion

Most of the forest-nesting species at Millbrook experienced cowbird parasitism. Every acceptor species (Rothstein 1975a) for which we found 5 or more nests was parasitized ($N = 14$ species, Table 13.1). The substantial parasitism rates on most of these forest-nesting birds corresponded to our repeated observations that cowbirds were found throughout our 400-ha study area. Nest searchers often saw or heard both male and female cowbirds during the morning as well as the afternoon in all parts of the forest. Radiomarked female cowbirds consistently spent their morning (egg-laying) hours in the forest (Hahn unpubl. data).

Selective parasitism of ground-nesting forest bird species has not been reported before. In fact, other investigators

Table 13.1. Summary Statistics by Host Species for Nest Height and Cowbird Parasitism

Species[a]	Number of Nests	Mean Nest Height ± SD (m)	Percentage Parasitized[b]
Ovenbird	28	0.00 ± 0.00	53.57
Worm-eating Warbler	11	0.00 ± 0.00	18.18
Black-and-white Warbler	8	0.06 ± 0.18	50.00
Rufous-sided Towhee	13	0.20 ± 0.38	15.38
Veery	66	0.36 ± 0.54	21.21
Louisiana Waterthrush	14	0.45 ± 0.41	42.86
Carolina Wren	4	0.50 ± 1.00	25.00
Hermit Thrush	20	0.55 ± 0.84	40.00
Chestnut-sided Warbler	2	0.75 ± 0.35	0.00
Great Crested Flycatcher	2	1.51 ± 0.02	0.00
Yellow-rumped Warbler	1	2.13	0.00
Eastern Phoebe	49	2.36 ± 1.47	12.24
Wood Thrush	90	2.45 ± 1.24	10.00
Northern Cardinal	9	2.48 ± 1.04	11.11
Solitary Vireo	3	2.64 ± 1.57	33.33
Rose-breasted Grosbeak	16	3.01 ± 1.60	6.25
Red-eyed Vireo	13	3.05 ± 2.32	61.54
American Redstart	63	3.78 ± 1.92	31.75
Scarlet Tanager	5	7.51 ± 2.63	20.00
Least Flycatcher	3	10.67 ± 3.21	0.00
Eastern Wood Peewee	4	14.30 ± 11.55	0.00
Blue-gray Gnatcatcher	5	14.40 ± 8.70	0.00
Warbling Vireo	1	15.00	0.00

[a]The species are ordered from the lowest to highest mean nest height.
[b]Fisher's exact test yielded $P = .000$ for a test of whether the percentage of parasitism differs among host species. The weighted correlation coefficient across species between the percentage of parasitism and mean nest height is $r = -0.33$; Jonckheere-Terpstra test, $P = .03$, one-tailed test.

Table 13.2. Heights of Parasitized and Unparasitized Nests in 1992 and 1993

	1992		1993	
	N	Mean Nest Height ± SD (m)	N	Mean Nest Height ± SD (m)
Parasitized nests	55	1.49 ± 1.98	44	1.35 ± 1.48
Unparasitized nests	183	2.46 ± 3.54	148	2.48 ± 3.04

Note: A two-way ANOVA yielded $F = 9.08$, df = 1, $P = .003$ for equality in mean nest heights of parasitized and unparasitized hosts, and $F = 0.03$, df = 1, $P = .86$ for differences between years in the mean nest heights. The interaction between year and parasitism was not significant, $F = 0.05$, df = 1, $P = .82$.

have suggested that ground-nesting species are often less parasitized (Robinson 1992), or are parasitized at the same frequency as birds nesting at other levels (Martin 1993). The frequent parasitism of ground-nesting species, particularly Ovenbird, Black-and-white Warbler, Hermit Thrush, and Carolina Wren, was also consistent with field observations by nest searchers and radiotrackers that cowbirds often walked on the forest litter layer or perched on low (1 m or lower) perches in the woods. These data suggest that cowbirds in eastern New York readily penetrate all levels of the forest, even in unfragmented forests of more than 1,000 ha. Askins et al. (1990) pointed out that such forests were understudied. This result occurred at a site where cowbirds are exploiting forest hosts significantly more often than old-field hosts available in adjacent sites (Hahn and Hatfield 1995).

Why are Wood Thrushes not parasitized more frequently in our site? Wood Thrush was the most abundant forest-nesting bird at this site (N = 90 nests monitored), and nest searchers found their bulky, mid-height nests more conspicuous than those of many other species. Yet they were parasitized at a rate of only 10% and ranked ninth in parasitism rate among 23 forest species we studied. Researchers in southern Illinois (Robinson 1992; Robinson et al. 1995; Trine, Chapter 15, this volume), and at some sites in Maryland (Dowell et al., Chapter 29, this volume), and Ohio (Petit and Petit, Chapter 31, this volume) have found frequent and intense parasitism of Wood Thrushes. It may be useful for managers to know that the Wood Thrush is not particularly vulnerable to parasitism, nor is it always selected by cowbirds throughout its range. This finding serves as a warning that patterns observed for cowbirds in one region cannot be directly applied to other regions without field confirmation.

Is our finding of higher parasitism on ground-nesting species an anomaly in forest habitats where cowbirds are active? This is difficult to evaluate, because few quantitative studies have been done of parasitism rates on ground-nesting forest birds, and even fewer comparative studies have been done of relative parasitism rates at different forest levels. Here we review six studies of parasitism on ground-nesting forest birds, three of single species and three comparative studies.

During a 3-year study in a southern Michigan forest, Hann (1937) found 52% parasitism on Ovenbirds (N = 48 nests), a very similar parasitism rate to the 53.6% we observed in Millbrook, New York. Hann's behavioral observations established, as ours did, that cowbirds perched at low heights to watch Ovenbirds nest build, and fed on the forest floor while walking and watching Ovenbirds. Because Hann studied only the Ovenbird, his study does not indicate how this parasitism rate compares to the parasitism rates on other host species in that forest community.

Martin (1993), in a comparative study of predation and parasitism at different nesting levels, cites only Hann (1937) and two other studies as the three data points in the category of ground-nesting forest birds. The other two studies, one of Wilson's Warbler and one of Dark-eyed Junco, are both cited as examples of zero cowbird parasitism on ground-nesting species, and these data points, together with Hann's report of 52% parasitism, influenced Martin's finding of no relationship between nest height and parasitism rate. However, close examination of these studies suggests that they should not be used as examples of zero parasitism on ground-nesting species. The study of Wilson's Warbler in the Sierra Nevada in California mentions that cowbird parasitism on Wilson's Warbler did occur (Stewart et al. 1978:90–91) but was not quantified, perhaps because it was not a focus of their study; nevertheless, parasitism on Wilson's Warbler at that site was not zero. The study of Dark-eyed Junco (Smith and Anderson 1982) in northern Utah does not mention that cowbirds were observed in the study area. Perhaps the lack of parasitism here was due to the rarity or absence of cowbirds, rather than to parasitism of other hosts in the community. Consequently, this study too should perhaps not be used as a data point demonstrating that cowbirds failed to parasitize a ground-nesting forest species.

Peck and James (1987) provided extensive data on parasitism of ground-nesting birds in Ontario. They used nest records to calculate overall levels of cowbird parasitism on all host species reported during a 125-year period. We extracted the data on 16 of the 17 species found in both Ontario and our study in Millbrook, New York, to see whether ground-nesting species in Ontario were exempt from parasitism or frequently used (Table 13.3). (There were no data from Ontario for the seventeenth species, the Worm-eating Warbler.) Although 6 ground-nesting species were shared across both study sites, only three of them had adequate numbers of nests (N ≥ 20) in both samples. The Veery had a similar rate of parasitism (19.0% in Ontario versus 21.2% in Millbrook), and the Ovenbird had a much higher rate at Millbrook (53.6%) than in Ontario (11.9%).

Three comparative studies, like the present study, provide information on relative parasitism rates on forest birds nesting at different levels. Robinson (1992:414) categorized species in his study site as ground nesters or mid- and high-level nesters. He reported a 41% parasitism rate on 17 nests of 4 ground-nesting species and a 75% parasitism rate on 56 nests of 7 mid-height and high-nesting species (our calculations from data presented in his Table 5). He concluded that ground-nesting species were less frequently parasitized than those at higher nest levels, but since his conclusion was not based on comparisons of nest heights, we tested the strength of his conclusion by reanalyzing his data. When we used the order of the nest heights that we observed for species in Millbrook and tested only the 7 species found in both his study and ours, we found no significant correlation between

Table 13.3. Percentage of Nests Parasitized in Ontario for Sixteen Forest-nesting Species Found in Both Ontario and Millbrook, New York

	Nests (*N*)	Parasitized Nests (*N*)	Parasitized Nests (%)	Mean Group Parasitism Rate ± SD
Low- and ground-nesting species				
Ovenbird	260	31	11.9	
Chestnut-sided Warbler	211	45	21.3	
Hermit Thrush	154	11	7.1	
Carolina Wren	17	0	0	
Veery	368	70	19.0	
Black-and-white Warbler	43	9	20.9	
Louisiana Waterthrush	8	2	25.0	
				15.03 ± 8.35
Mid- and high-nesting species				
Scarlet Tanager	36	7	19.4	
Least Flycatcher	99	5	5.1	
Red-eyed Vireo	354	136	38.4	
Eastern Phoebe	1349	162	12.0	
American Redstart	285	57	20.0	
Wood Thrush	195	53	27.2	
Rose-breasted Grosbeak	275	18	7.5	
Eastern Wood-Pewee	117	6	5.1	
Solitary Vireo	44	2	4.5	
				15.46 ± 11.12

Source: Data are from Peck and James's (1987) summary of nest record cards at the Royal Ontario Museum, covering a 125-year span.

parasitism rate and nest height ($P > .05$, Jonckheere-Terpstra test).

Martin (1993:906–907) compared the relative parasitism rates on 56 species nesting at different levels by drawing on the published results from 70 studies conducted at different sites across the United States over a 55-year period (1937–1992). Many of the studies included in the analysis were ones in which the authors had measured parasitism rates on single species without reporting parasitism rates on other species in the community, so the analysis of relative parasitism rates is across *studies*, not across a natural community. Based on this analysis, Martin (1993) concluded that parasitism was not differentially associated with any forest level. However, this conclusion deserves further consideration for two reasons. First, it is based on a sample size of 3 ground-nesting forest species, 2 of which appear to be wrongly listed as cases of zero parasitism. Second, a comparison of parasitism rates across species introduces noise associated with many uncontrolled variables such as differences in parasitism patterns between years and between sites. Variability in

parasitism rates from year to year, host to host, and among neighboring avian communities has been reported in prairie and grassland communities (Wiens 1963, Zimmerman 1983). Consequently, it seems likely that an analysis using multiple studies done in different sites and different years, comprised primarily of single-species studies at each site, could fail to detect a pattern that might be measurable in individual cross-community studies at single sites.

Peck and James's (1987) data set, like Martin's (1993), represents a set of parasitism rates on species collected at widely varying times (125 years versus Martin's 55 years) and places (across Ontario versus across the United States). Our calculations, using Peck and James's (1987) data, compare the parasitism levels on the 16 species found both in Ontario and at Millbrook. We found that all ground-nesting species had notable levels of parasitism and that ground- and aboveground-nesting species had similar average rates of parasitism (15.0 ± 8.35% for ground-nesting species versus 15.5 ± 11.12% for mid- and high-level nesting species, Table 13.3).

Our review of these 6 studies suggests that we do not yet have enough quantitative information to evaluate (a) how frequently cowbird parasitism affects ground-nesting forest species and (b) the relative impact of parasitism on ground nesters versus species nesting at other levels in forests. Parasitism of ground-nesting species may occur relatively infrequently in the literature because human observers have more difficulty finding ground nests than cowbirds do. Alternatively, parasitism of ground-nesting forest birds could reflect the length of time over which cowbirds have exploited host species in forests. Frequent parasitism could indicate either that parasitism is so recent that hosts have not yet developed defenses (Rothstein 1975b, Rohwer and Spaw 1988) or that cowbird populations have adapted to the host defenses over a long time. As studies of cowbird parasitism patterns in forest communities accumulate, comparative data should clarify whether parasitism rates at different levels in the vegetative structure can be predicted. Only a long-term study would indicate whether the population of cowbirds studied here consistently parasitizes the forest-nesting hosts as we observed, or whether they periodically shift among hosts in different habitats across the local landscape. Our radiotracking data (Hahn unpubl.) demonstrates that several individual females occupied the same forest site for 2–3 years. This result suggests that the pattern of parasitism throughout the forest might be stable across years.

Management Implications

Our data suggest that managers cannot use studies from one region or site to predict which species will be more or less heavily parasitized in another without local field confirmation. To gain a national perspective on the host selection patterns of cowbirds, managers need results from a network of long-term field studies, strategically sited in different regions and habitats. These studies should be focused at the community level on all potential cowbird hosts, and at a landscape scale that crosses forest-field boundaries. The results would show whether there are consistent host selection patterns within each site and how similar they are in different sites and regions. Now that managers are alerted that cowbird parasitism may be a problem for forest-nesting neotropical migrants, they must collaborate with scientists in commissioning studies that assess local cowbird parasitism to determine whether a cowbird problem exists. These studies can then determine whether management actions are needed and what the nature of these actions should be.

Acknowledgments

This study benefited from the excellent field skills of many nest searchers. We thank J. Boone, J. Cherry, K. Corey, M. DiDomenico, J. Goodell, J. Koloszar, P. Osenton, G. Oines, S. Plentovich, and J. Sedgwick. P. Osenton, L. Williams, and J. Boone assisted with data analysis. We thank Rockefeller University Field Research Center (RUFRC), the Institute of Ecosystem Studies, and Mr. Bruce Kovner for permission to use their land, and RUFRC for use of their facilities and the help of their staff.

References Cited

Askins, R. A., J. F. Lynch, R. Greenberg. 1990. Population declines in migratory birds in eastern North America. Curr. Ornithol. 7:1–57.

Brittingham, M. C., and S. A. Temple. 1983. Have cowbirds caused forest songbirds to decline? BioScience 33:31–35.

Friedmann, H. 1929. The cowbirds: A study in the biology of social parasitism. C. C. Thomas, Springfield, IL.

———. 1963. Host relations of the parasitic cowbirds. Smithsonian Institution, U.S. Natl. Mus. Bull. 233:1–276.

Glitzenstein, J. S., C. D. Canham, M. J. McDonnell, and D. R. Streng. 1990. Effects of environment and land-use history on upland forests of the Cary Arboretum, Hudson Valley, New York. Bull. Torrey Botanical Club 117:106–122.

Hahn, D. C., and J. S. Hatfield. 1995. Parasitism at the landscape scale: Do cowbirds prefer forest? Conserv. Biol. 9:1415–1424.

Hann, H. W. 1937. Life history of the Oven-bird in southern Michigan. Wilson Bull. 49(3):145–237.

Lehmann, E. L. 1975. Nonparametrics: Statistical methods based on ranks. McGraw-Hill, New York.

Lynch, J. F. 1987. Responses of breeding bird communities to forest fragmentation. Pp. 123–140 in Nature conservation: The role of remnants of native vegetation (D. A. Saunders, G. W. Arnold, A. A. Burbridge, and A. J. M. Hopkins, eds). Surrey Beatty and Sons, Norton, New South Wales.

Martin, T. E. 1993. Nest predation among vegetation layers and habitat types: Revising the dogmas. Amer. Natur. 141:897–913.

Mayfield, H. F. 1965. The Brown-headed Cowbird with old and new hosts. Living Bird 4:13–28.

Peck, G. K., and R. D. James. 1987. Breeding birds of Ontario: Nidiology and distribution, vol. 2: Passerines. Royal Ontario Museum, Toronto.

Robbins, C. S., D. K. Dawson, and B. A. Dowell. 1989a. Habitat area requirements of breeding forest birds of the middle Atlantic states. Wildl. Monogr. 103:1–34.

Robbins, C. S., J. R. Sauer, R. S. Greenberg, and S. Droege. 1989b. Population declines in North American birds that migrate to the neotropics. Proc. Natl. Acad. Sci. USA 86:7658–7662.

Robinson, S. K. 1992. Population dynamics of breeding neotropical migrants in a fragmented Illinois landscape. Pp. 455–471 in Ecology and conservation of neotropical migrant landbirds. (J. M. Hagan and D. W. Johnston, eds). Smithsonian Institution Press, Washington, DC.

Robinson, S. K., J. Grzybowski, S. I. Rothstein, L. J. Petit, and F. Thompson. 1993. Management implications of cowbird parasitism for neotropical migrants. Pp. 93–102 in Status and management of neotropical migratory birds. (D. M. Finch and P. W. Stangel, eds). USDA Gen. Tech. Rept. RM-229.

Rohwer, S., and C. D. Spaw. 1988. Evolutionary lag versus bill-size constraints: A comparative study of the acceptance of cowbird eggs by old hosts. Evol. Ecol. 2:27–36.

Rothstein, S. I. 1975a. An experimental and teleonomic investigation of avian brood parasitism. Condor 77:250–271.

———. 1975b. Evolutionary rates and host defenses against avian brood parasitism. Amer. Natur. 109:151–176.

Rothstein, S. I., J. A. Verner, and E. Stevens. 1984. Radio-tracking confirms a unique diurnal pattern of spatial occurrence in the Brown-headed Cowbird. Ecology 65:77–88.

Smith, K. G., and D. C. Anderson. 1982. Food, predation, and reproductive ecology of the Dark-eyed Junco in northern Utah. Auk 99:650–661.

Stewart, R. M., R. P. Henderson, and K. Darling. 1978. Breeding ecology of the Wilson's Warbler in the high Sierra Nevada, California. Living Bird 16:83–102.

Terborgh, J. 1989. Where have all the birds gone? Princeton University Press, Princeton, NJ.

Wiens, J. A. 1963. Aspects of cowbird parasitism in southern Oklahoma. Wilson Bull. 75:130–138.

Wilcove, D. S. 1985. Nest predation in forest tracts and the decline of migratory songbirds. Ecology 66:1211–1214.

Zimmerman, J. L. 1983. Cowbird parasitism of Dickcissels in different habitats and at different nest densities. Wilson Bull. 95:7–22.

14. Reproductive Interactions between Brown-headed Cowbirds and Plumbeous Vireos in Colorado

JAMESON F. CHACE, ALEXANDER CRUZ,

AND REBECCA E. MARVIL

Abstract

We studied the impact of Brown-headed Cowbird parasitism on the reproductive success of Plumbeous Vireos nesting in the foothills of the Rocky Mountains west of Boulder, Colorado. Initial work took place from 1984 to 1986 (Marvil and Cruz 1989) and further work in 1992 and 1993. This chapter summarizes results for all five years of study and discusses landscape patterns affecting parasitism frequency. During all years of the study 47.0% of 132 nests were parasitized, and parasitism was consistent across years. Parasitized nests had smaller clutches (3.22 vs. 3.73), lower hatching success (49.3% vs. 74.0%), and lower fledging success (16.2% vs. 57.0%) than unparasitized nests. The number of vireo fledglings per nest was 76% lower in parasitized nests (0.52 vs. 2.14). Growth of vireo nestlings in parasitized nests was slower than for vireo nestlings in unparasitized nests. Landscape measurements at nest sites in 1992–1993 showed that nests near roads and residential areas had more frequent parasitism than nests in the foothills. Based on this evidence, we recommend that land managers in Boulder County strive to minimize fragmentation of the foothills by roads and residential development. Reduced fragmentation may reduce parasitism frequencies on Plumbeous Vireos and possibly on other forest dwelling, open-cup nesting birds in Boulder County.

Introduction

The Plumbeous Vireo (*Vireo plumbeus,* formerly *V. solitarius*) is sympatric with the Brown-headed Cowbird in eastern and western North America (AOU 1983). Although Plumbeous Vireos were known cowbird hosts (Friedmann 1971, Friedmann et al. 1977, Friedmann and Kiff 1985), little information existed on the timing, frequency, or effects of cowbird parasitism until the work of Marvil and Cruz (1989). Expansion of the elevations occupied by cowbirds in Colorado brought them into contact with several montane host species including the Plumbeous Vireo by 1980 (Hanka 1985). Even though Plumbeous Vireos are fairly common in Colorado, the Colorado Nongame Advisory Council listed the species as being of special concern in 1985 (Winternitz and Crumpacker 1985) because of the frequent parasitism (48.7% of nests) reported by Marvil and Cruz (1989). Further monitoring of frequency and effect of brood parasitism on Plumbeous Vireos in Boulder County began in 1992. Our objectives here are to compile all data on the effects of cowbird parasitism on Plumbeous Vireos in Boulder County and to recommend ways to reduce frequencies of brood parasitism.

Methods

Data on Plumbeous Vireo nests were collected during the summers of 1984–1986 and 1992–1993 in the foothills of the Rocky Mountain Front Range west of the city of Boulder, Colorado (40° 00' N, 105° 20' W) (Figure 14.1). Study sites ranged in elevation from 1,800 m to 2,400 m and had a parklike appearance dominated by open-canopy ponderosa pine (*Pinus ponderosa*). Sites also contained scattered Douglas firs (*Pseudotsuga menziesii*) and an understory dominated by chokecherry (*Prunus virginiana*), wax currant (*Ribes cereum*), skunkbrush (*Rhus aromatica*), small ninebark (*Physocarpus monogynus*), Oregon grape (*Mahonia repens*), kinnikinnick (*Arctostaphylos uva-ursi*), and various grasses (*Bromus* spp., *Stipa* spp.) and herbs (*Achillea* spp., *Artemisia* spp.).

Local cowbird hosts (Friedmann et al. 1977) breeding in the study area are listed in the Appendix. Plumbeous Vireos were chosen for this study because their low nests are relatively easy to locate and monitor. Additionally, Plumbeous Vireos are fairly abundant on the study sites and are known acceptors of cowbird eggs (Friedmann et al. 1977).

An index of Plumbeous Vireo densities was estimated in 1992 from a variable distance circular-plot census (Reynolds et al. 1980). Seven points were censused 14 times between 10 June and 25 July 1992. Points were more than

250 m apart to ensure independence, and all censuses occurred between 0.5 hour before and 3 hours after sunrise (between 0500 and 0900 hr). All individual birds detected during 8-minute periods were recorded. This estimate was compared to a 1984 estimate using the Emlen transect method reported by Thompson and Strauch (1985). Both censuses occurred in the same habitat and at approximately the same elevation within the City of Boulder Open Space.

Vireo nests were found during all stages of the nesting cycle and subsequently visited once every 2–4 days. Care was taken to minimize disturbance and attraction of nest predators to the nest site (Major 1990, Ralph et al. 1993). Outcome of each clutch (i.e., parasitism, predation, abandonment, and fledging) was scored. Nest success was estimated using the Mayfield method (Mayfield 1975). Nest appearance and mode of disturbance were used to determine whether predators disturbed nests. Nests that were found empty before the oldest vireo nestling was 12 days old or

cowbird was 9 days old were scored as preyed upon. Nests found empty on or after that point were determined to have been preyed upon if adults gave no alarm calls and no juveniles could be found in the nest area after a careful search.

Measurements of mass and ulna-radius length were analyzed for 20 nestlings spanning all years of study. Nestlings from parasitized vireo nests included 4 individuals from 4 nests in which the cowbird egg hatched within two days of the vireos and fledged. Nestlings from unparasitized vireo nests included 9 vireos that fledged from 4 unparasitized nests. Seven cowbird nestlings measured all fledged from separate vireo nests. Masses were recorded to the nearest 0.1 g with 10-g and 50-g Pesola spring scales. Ulna-radius lengths were measured to the nearest 0.1 mm with a dial caliper and ruler. We used Ricklefs's (1967, also see Weatherhead 1989) method to fit logistic growth curves for each vireo nestling based on more than 3 measurements per nestling and to calculate the rate constant (K) for each nes-

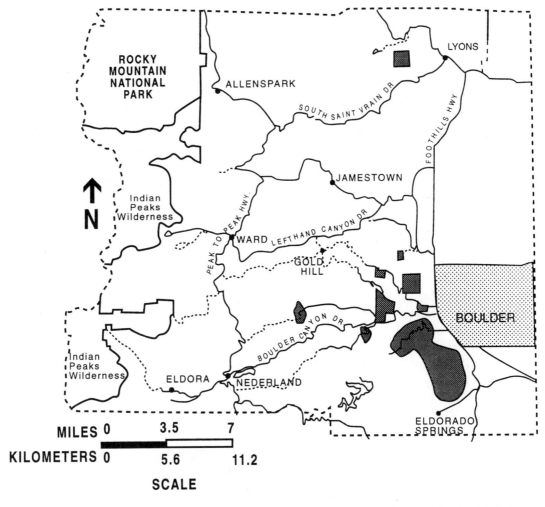

Figure 14.1. Western Boulder County, Colorado. Stippled areas represent study sites for all years.

tling's growth. The rate constants were used for statistical analysis by t-tests, with alpha set at 0.05.

For vireo nest sites studied in 1992–1993, we measured distances to the nearest road, trail, and residence. Frequencies of parasitism were measured for nests within 50 m of trails, 100 m of roads and residences (nearest permanently occupied house), and remaining nests, in both the lower-elevation (below 1,860 m) and higher-elevation (above 1,860 m) foothills of Boulder County. A G-test was used to test for differences between sites, with alpha set at 0.05. Only nests in which the final outcome was known were included in the analysis.

Results

Plumbeous Vireo densities were similar during all years of the study. Breeding Bird Censuses conducted on the 1992 census site have shown a consistent number of Plumbeous Vireo pairs since 1984 (see Herring in Van Velzen and Van Velzen 1984). In contrast, cowbird numbers may have increased. Thompson and Strauch (1985) found 0.28 ± 0.23 (mean $\pm$ SE) cowbirds/ha, and we found 0.67 ± 0.15 cowbirds/ha in 1992. Brown-headed Cowbirds were not detected during local BBS censuses, but Thompson and Strauch (1985) detected 0.24 ± 0.32 cowbirds/ha in a larger area.

Brown-headed Cowbirds parasitized 47.0% of the 132 Plumbeous Vireo nests monitored during all years, and parasitism was independent of year (Table 14.1). Most cowbird eggs (87.5%, $N = 72$) were laid during vireo egg laying. Cowbird eggs hatched after about 11 days of incubation and fledged about 11 days later. Plumbeous Vireo eggs hatched after about 16 days of incubation and fledged about 14 days later.

Most parasitized nests ($N = 62$) contained 1 cowbird egg (71.4%, $N = 40$), 15 nests (26.9%) contained 2 cowbird eggs, and 1 nest (1.8%) contained 3 cowbird eggs.

The mean clutch size of parasitized vireo nests was significantly smaller than in unparasitized nests (Table 14.2). Parasitized clutches averaged 0.51 fewer vireo eggs. This result suggests that cowbirds removed host eggs from about 50% of parasitized nests (see Sealy 1992). Egg removal was suspected in 8 cases: 4 cases where eggshell fragments were found near the nest, and 4 other cases where eggs were removed following parasitism of an initiated clutch. Parasitized nests had significantly lower hatching success, fledging success per egg, and fewer eggs hatched than unparasitized nests (Table 14.3). Unparasitized nests fledged over four times as many vireos (2.14) as parasitized nests (0.52, Table 14.3).

A successful nest was one that fledged at least one vireo. Using the Mayfield method, nests in 1984–1986 had a significantly higher probability of nest success than nests in 1992–1993 (G-test, $G = 5.306$, df = 1, $P < .05$). For both

1984–1986 and 1992–1993, unparasitized nests had a significantly higher probability of nest success than parasitized nests (1984–1986, 0.86 vs. 0.30, $G = 15.03$, df = 1, $P < .001$; 1992–1993, 0.45 vs. 0.18, $G = 4.37$, df = 1, $P < .05$).

Only 4 of 130 nests were deserted, 2 cases each at parasitized and unparasitized nests. Nest predation was higher during 1992–1993 (32.7%) than in 1984–1986 (18.5%) ($G = 3.127$, df = 1, $P > .05$). Predation did not differ significantly between parasitized (21 of 62) and unparasitized nests (18 of 70) ($G = 1.05$, df = 1, $P > .05$).

Growth was analyzed for 13 vireo nestlings spanning all years of study. Development of vireo chicks in parasitized nests ($N = 4$; mass: mean $K \pm$ SE, 0.164 ± 0.07; length: 0.150 ± 0.04) was significantly slower than for vireo young in unparasitized nests ($N = 9$; mass: $K = 0.364 \pm 0.015$, $t = 3.888$, df = 11, $P < .005$; length: 0.353 ± 0.03, $t = 3.616$, df = 11, $P < .005$; Figures 14.2 and 14.3). Growth differences between vireos and cowbirds probably account for the lower fledging success of vireos in parasitized nests (Figure 14.2). Cowbirds ($N = 7$) grew faster (mass: $K = 0.408 \pm 0.02$, $t = 7.278$, df = 9, $P < .001$; length: $K = 0.426 \pm 0.04$, $t = 4.772$, df = 9, $P < .002$) and, as expected from differences in adult mass, were much larger than their vireo nest mates. In effect, cowbirds out-competed the host juveniles for food (Figures 14.2 and 14.3). Cowbird young, however, did not grow significantly faster than vireos in unparasitized nests (mass: $t = 1.679$, df = 14, $P > .05$; length: $t = 1.454$, df = 14, $P > .05$. Fifteen vireo nestlings starved in parasitized nests and none in unparasitized nests.

Parasitized nests in which cowbirds hatched two or more days ahead of the host young ($N = 38$) resulted in 4 vireos and 17 cowbirds fledging. In four nests where the cowbird and vireos were the same age, 7 vireos and 4 cowbirds fledged. In the latter case, both vireos and cowbirds successfully fledged young from each nest (1.8 and 1.0 young, respectively). Cowbirds did not fledge from the remaining 13 nests because the eggs were laid too late (3 cases), they did not hatch (2 cases), or they were depredated (8 cases).

The frequency of cowbird parasitism on Plumbeous Vireo nests in Boulder County was higher in the settled sites at lower elevations (Table 14.4). There was a significantly higher frequency of cowbird parasitism in residential areas ($G = 7.01$, df = 1, $P < .01$) and along roads ($G = 4.42$, df = 1, $P < .05$) than in the low-elevation foothills. Additionally, there was significantly higher nest parasitism in residential areas ($G = 6.66$, df = 1, $P < .01$) and along roads ($G = 4.0$, df = 1, $P < .05$) than in the high-elevation foothills.

Discussion

Plumbeous Vireos nesting along the Colorado Front Range are suffering from Brown-headed Cowbird parasitism. Plumbeous Vireos accept cowbird eggs, and nests that were parasitized had significantly smaller clutches than unparasi-

Table 14.1. Frequency of Parasitism on Plumbeous Vireo Nests, Boulder County, Colorado

	1984	1985	1986	1992	1993	Total
Unparasitized nests	11	12	17	5	25	70
Parasitized nests	9	12	17	3	21	62
% parasitism[a]	45.0	50.0	50.0	37.5	45.6	47.0

[a]Frequency of parasitism is independent of year ($G = 2.69$, df = 1, $P < .75$).

Table 14.2. Clutch Size of Plumbeous Vireos in Parasitized and Unparasitized Nests, Boulder County, Colorado (1984–1986 and 1992–1993)

	Clutch Size					
	1	2	3	4	5	Mean ($\pm$ SE)
Unparasitized nests	0	0	19	48	1	3.73 (0.06)
Parasitized nests	3	7	21	21	2	3.22 (0.12)
Total	3	7	40	69	3	3.50 (0.07)

Note: Differences in clutch size between unparasitized and parasitized nests are significant (Wilcoxon two-sample test, $Z = -3.57$, $P = .000$).

Table 14.3. Nesting Success in Unparasitized and Parasitized Nests of Plumbeous Vireos, Boulder County, Colorado, 1984–1986, 1992–1993

	Unparasitized	Parasitized	All	Cowbird
No. active nests	66	51	117	
Total eggs	235	154	389	74
Total hatched	174	76	250	46
Total fledged	134	25	159	26
Hatching success (%)	74.0	49.3[a]	64.3	62.2
Fledging success (%)	57.0	16.2[a]	40.9	35.1
Fledglings/egg hatched (%)	77.0	32.9[a]	63.6	56.5
Fledglings/active nest	2.14	0.52[b]	1.35	0.51

Note: Includes only nests found at egg-laying stage and followed to fledging or failure.
[a]Differences between parasitized and unparasitized nests are significant for hatching success ($G = 24.56$, $P < .001$), fledging success ($G = 68.59$, $P < .001$), and fledgling/egg hatched success ($G = 43.94$, $P < .001$).
[b]Differences between parasitized and unparasitized nests are significant (Wilcoxon two-sample test, $Z = -5.45$, $P < .001$).

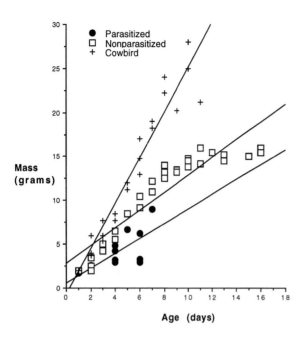

Figure 14.2. Regression of growth rate of nestling mass for cowbirds ($N = 7$) and for vireos from unparasitized ($N = 9$) and parasitized ($N = 4$) nests. All regressions were significant. Masses of cowbirds ($y = 2.597x - 0.805$, $r^2 = 0.93$, $P = .000$), unparasitized vireos ($y = 1.011x + 2.759$, $r^2 = 0.85$, $P = .000$), and parasitized vireos ($y = 0.840x + 0.541$, $r^2 = 0.56$, $P = .002$).

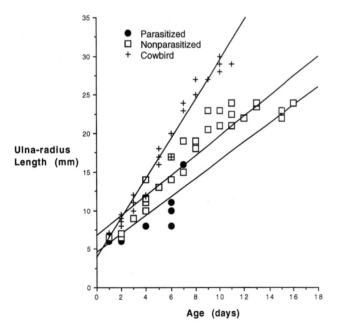

Figure 14.3. Regression of growth rate of ulna-radius length for cowbirds ($N = 7$), and for vireos from unparasitized ($N = 9$) and parasitized ($N = 4$) nests. Regressions were significant for ulna-radius lengths of cowbirds ($y = 2.564x + 3.872$, $r^2 = 0.97$, $P = .000$), unparasitized vireos ($y = 1.290x + 6.713$, $r^2 = 0.87$, $P = .000$), and parasitized vireos ($y = 1.198x + 4.508$, $r^2 = 0.65$, $P = .001$).

tized nests, probably due to cowbird host-egg removal (see Sealy 1992). Hatching success in parasitized nests was further reduced by the shorter incubation period of the cowbird. Vireo nestlings that hatched in parasitized nests grew slower than vireos in unparasitized nests and rarely fledged. Vireos that fledged from parasitized nests typically were from nests where cowbird eggs did not hatch, or where cowbird eggs were laid 2 or more days before the vireo eggs. Parasitized nests had significantly lower nesting success than unparasitized nests. Brood parasitism accounted for nearly half of the reproductive failures of Plumbeous Vireos nesting in Boulder County over the past 10 years.

Curiously, despite the nearly 50% parasitism frequency near Boulder, the Solitary Vireo complex is one of the few neotropical migrant taxa that has shown significant population increases over the past 20 years along continental Breeding Bird Survey routes (Sauer and Droege 1992). The Plumbeous Vireo was listed as a Colorado species of special concern in 1985 based on the report of the high frequency of parasitism in this population (Winternitz and Crumpacker 1985). Given the high frequencies of nest parasitism and low vireo fledgling success (2.1 per unparasitized nest) at our sites, the Colorado Front Range may be a population sink for Plumbeous Vireos. If so, numbers are maintained only by emigrants from areas with less intense parasitism (May and Robinson 1985, Robinson et al. 1993). Local population estimates of Plumbeous Vireos in Boulder County suggest a population size that is stable due to immigration. Furthermore, Plumbeous Vireos have higher densities in disturbed sites than undisturbed sites (Szaro and Balda 1979, Medin 1985). Immigration to areas disturbed by roadways and residential development could increase vireo numbers locally; i.e., urban development may act as an ecological trap. Because of high levels of nest parasitism, vireos in such areas may produce fewer young than needed to compensate for adult mortality. Other songbirds attracted to edges in fragmented landscapes may have decreased reproductive success without increases in cowbird numbers (Gates and Gysel 1978, Robinson et al. 1993, Thompson 1991).

Brown-headed Cowbirds were ubiquitous throughout our ponderosa pine forest study sites. However, we found that cowbird parasitism on Plumbeous Vireos is variable in space, being heaviest in sites fragmented by residential and road disturbance. Fragmentation of the ponderosa pine forest by urbanization (e.g., roads and residences) may increase the ratio of forest edge to interior area. This could result in vireos nesting in small forest fragments nearer to disturbed habitat, which might then increase the likelihood of vireo nests being parasitized by cowbirds (Brittingham and Temple 1983, Temple and Carey 1988, Terborgh 1989, Finch 1991, Robinson 1992). Vireos nesting in disturbed areas had significantly lower nesting success for a given frequency of parasitism. We therefore recommend that efforts be made to maintain large undisturbed areas of forest to act as net producers of young.

Fragmentation and disturbance of habitat are associated with high parasitism frequencies elsewhere (Rothstein et al. 1980, Brittingham and Temple 1983). Landscape-level management represents the best long-term strategy for cowbird control (Harris 1991, Robinson et al. 1993). Managers of ponderosa pine forests should strive to reduce fragmentation in conjunction with disturbances such as roadways and residential areas. By doing so, they may reduce parasitism frequencies and consequently enhance the reproductive success of Plumbeous Vireos and other breeding migrant birds. A landscape-level approach should strive to maintain large, intact stands of ponderosa pine (20,000–50,000 ha) as prescribed for bird populations in the Midwest (Robinson et al. 1993). Such large stands are probably necessary due to the open savannah-like structure of ponderosa pine forests in which grassy areas create natural edge habitat and increase the ratio of edge to interior area (Temple and Carey 1988, Robinson et al. 1993). Cowbird feeding areas, such as livestock grazing areas and feedlots, should be identified and, if possible, removed from protected forest stands (Rothstein et al. 1980, Robinson et al. 1993). Maintaining large areas of intact ponderosa pine habitat isolated from human disturbance will probably increase the reproductive success of other migrant songbirds through reduced parasitism.

Acknowledgments

We thank the City of Boulder Mountain Parks and Open Space, Boulder County Open Space, and Roosevelt National Forest for allowing us research access. We also thank D. Bennett and S. Severs for their assistance in the field. R. A. Bernstein, T. Cook, D. M. Cruz, T. Prince, and S. G. Sealy offered helpful suggestions for improving the manuscript. This research was funded by grants from the University of Colorado Graduate School and Department of EPO Biology, and the Boulder County Nature Association.

Table 14.4. Frequencies of Cowbird Parasitism on Plumbeous Vireos in Relation to Nest Location, 1992–1993

Location	No. Nests	% Parasitism
Road	7	71.4
Residential	3	100
Trails	14	42.9
Low-elevation foothills	16	25.0
High-elevation foothills	15	26.7

APPENDIX

Actual and Potential Cowbird Hosts Found on the Study Area in Boulder County

Plumbeous Vireo, *Vireo plumbeus**
Warbling Vireo, *V. gilvus**
Yellow-rumped Warbler, *Dendroica coronata*
MacGillivray's Warbler, *Oporornis tolmiei*
Virginia's Warbler, *Vermivora virginiae**
Black-headed Grosbeak, *Pheucticus melanocephalus*
Lazuli Bunting, *Passerina amoena**
Green-tailed Towhee, *Pipilo chlorurus*
Spotted Towhee, *P. maculatus*
Chipping Sparrow, *Spizella passerina**
Dark-eyed Junco, *Junco hyemalis*
Western Tanager, *Piranga olivacea**
Pine Siskin, *Carduelis pinus*
Lesser Goldfinch, *C. psaltria**

*Species known to be parasitized by cowbirds.

References Cited

American Ornithologists' Union. 1983. Check-list of North American birds, 6th edition. American Ornithologists' Union, Washington, DC.

Brittingham, M. C., and S. A. Temple. 1983. Have cowbirds caused forest songbirds to decline? BioScience 33:31–35.

Finch, D. M. 1991. Population ecology, habitat requirements, and conservation of neotropical migratory birds. USDA Gen. Tech. Rep. RM-205. 26 pp.

Friedmann, H. 1971. Further information on the host relations of the parasitic cowbirds. Auk 88:239–255.

Friedmann, H., and L. F. Kiff. 1985. The parasitic cowbirds and their hosts. Proc. Western Found. Vert. Zool. 2:225–302.

Friedmann, H., L. F. Kiff, and S. I. Rothstein. 1977. A further contribution to knowledge of the host relations of the parasitic cowbirds. Smithsonian Contr. Zool. 235:1–75.

Gates, J. E., and L. W. Gysel. 1978. Avian nest dispersion and fledging success in field forest ecotones. Ecology 59:871–883.

Hanka, L. R. 1985. Recent range expansion by the Brown-headed Cowbird in Colorado. Western Birds 16:183–184.

Harris, J. H. 1991. Effects of brood parasitism by Brown-headed Cowbirds on Willow Flycatcher nesting success along the Kern River, California. Western Birds 22:13–26.

Major, R. E. 1990. The effects of human observers on the intensity of nest predation. Ibis 132:608–612.

Marvil, R. E., and A. Cruz. 1989. Impact of Brown-headed Cowbird parasitism on the reproductive success of the Solitary Vireo. Auk 106:476–480.

May, R. M., and S. K. Robinson. 1985. Population dynamics of avian brood parasitism. Amer. Natur. 126:475–494.

Mayfield, H. F. 1975. Suggestions for calculating nest success. Wilson Bull. 87:456–466.

Medin, D. E. 1985. Breeding bird responses to diameter-cut logging in west-central Idaho. USDA Res. Paper INT-355. 13 pp.

Ralph, J. C., G. R. Guepel, P. Pyle, T. E. Martin, and D. F. DeSante. 1993. Handbook of field methods for monitoring landbirds. USDA Gen. Tech. Rep. PSW-GTR-144. 41 pp.

Reynolds, R. T., J. M. Scott, and R. A. Nussbaum. 1980. A variable circular-plot method for estimating bird numbers. Condor 82: 309–313.

Ricklefs, R. E. 1967. A graphical method of fitting equations to growth curves. Ecology 48:978–983.

Robinson, S. K. 1992. Population dynamics of breeding neotropical migrants in a fragmented Illinois landscape. Pp. 408–418 *in* Ecology and conservation of neotropical migrant landbirds (J. M. Hagan III and D. W. Johnston, eds). Smithsonian Institution Press, Washington, DC.

Robinson, S. K., J. A. Grzybowski, S. I. Rothstein, M. C. Brittingham, L. J. Petit, and F. R. Thompson. 1993. Management implications of cowbird parasitism on neotropical migrant songbirds. Pp. 93–102 *in* Status and management of neotropical migratory birds (D. M. Finch and P. W. Stangel, eds). USDA Gen. Tech. Rep. RM-229.

Rothstein, S. I., J. Verner, and E. Stevens. 1980. Range expansion and diurnal changes in dispersion of the Brown-headed Cowbird in the Sierra Nevada. Auk 97:253–267.

Sauer, J. R., and S. Droege. 1992. Geographic patterns in population trends of neotropical migrants in North America. Pp. 26–42 *in*

Ecology and conservation of neotropical migrant landbirds (J. M. Hagan III and D. W. Johnston, eds). Smithsonian Institution Press, Washington, DC.

Sealy, S. G. 1992. Removal of Yellow Warbler eggs in association with cowbird parasitism. Condor 94:40–54.

Szaro, R. C., and R. P. Balda. 1979. Bird community dynamics in a ponderosa pine forest. Studies in Avian Biol. 3:1–66.

Temple, S. A., and J. R. Cary. 1988. Modeling dynamics of habitat-interior bird populations in fragmented landscapes. Conserv. Biol. 2:340–347.

Terborgh, J. W. 1989. Where have all the birds gone? Princeton University Press, Princeton, NJ.

Thompson, F. R. III. 1991. Simulated responses of a forest-interior bird population to forest management options in central hardwood forests of the United States. Conserv. Biol. 7:325–333.

Thompson, R. W., and J. G. Strauch, Jr. 1985. Habitat use by breeding birds on city of Boulder open space, 1984. Western Ecosystems, Lafayette, CO.

Van Velzen, W. T., and A. C. Van Velzen (eds). 1984. Forty-seventh breeding bird census. Amer. Birds 38:64.

Weatherhead, P. J. 1989. Sex ratios, host-specific reproductive success, and impact of Brown-headed Cowbirds. Auk 106:358–366.

Winternitz, B. L., and D. W. Crumpacker (eds). 1985. Colorado wildlife workshop, species of special concern. Colorado Nongame Advisory Council, Colorado Division of Wildlife, Denver. 92 pp.

15. Effects of Multiple Parasitism on Cowbird and Wood Thrush Nesting Success

CHERYL L. TRINE

Abstract

Large hosts of the Brown-headed Cowbird such as the Wood Thrush are less affected by parasitism than are smaller hosts. Larger hosts in some areas, however, are more likely to be multiply parasitized. I studied the impact of multiple parasitism on the reproductive success of Wood Thrushes and cowbirds in southern Illinois. Brood losses due to cowbird parasitism occurred throughout the nesting cycle. Nests lost an estimated 0.3 or 0.4 Wood Thrush eggs per cowbird egg received due to egg removal by female cowbirds. Six percent of the host eggs in multiply parasitized nests disappeared during incubation, whereas less than 1% of eggs disappeared from unparasitized or singly parasitized nests. Thrush hatching success declined by 4% per cowbird egg present. The probability of a Wood Thrush nestling surviving to fledge in a nondepredated nest decreased about 5% per cowbird nestmate. Using the most conservative estimates of losses at each stage, overall host fledging success was reduced about 18% for each cowbird egg received. Because these nests typically received 2–4 cowbird eggs, these losses seriously affected host productivity. Each addition of a cowbird egg reduced cowbird hatching success by 8–10%. In nests with more than 5 cowbird eggs, or more than 2 cowbird nestlings, cowbird egg and nestling survival rates were considerably lower than in nests with fewer eggs or nestlings. The number of cowbird fledglings produced per successful nest increased with the number of cowbird eggs incubated, up to 4 cowbird eggs/nest. Nests with 4 or more cowbird eggs, however, all produced about the same number of cowbird fledglings (1.8 fledglings/successful nest). I conclude that multiple parasitism entails serious costs for hosts, including large species such as the Wood Thrush, and that extreme multiple parasitism may also be maladaptive for cowbirds.

Introduction

A nest parasitized by Brown-headed Cowbirds often receives only one cowbird egg. Multiple parasitism, however, is by no means rare (Friedmann 1963), but few detailed accounts of the extra costs of multiple parasitism on host or cowbird fledging success have been published. Friedmann believed that nests with more than three cowbird eggs generally fledged no host young, and Temple and Cary (1988) thought more than two cowbird eggs usually resulted in host failure, but neither considered the consequences for the cowbirds. Although many studies provide information about cowbird and host success in parasitized and unparasitized nests (e.g., see May and Robinson 1985), few distinguish between parasitized nests with different numbers of cowbird eggs (but see Smith and Arcese 1994).

The Wood Thrush (*Hylocichla mustelina*) is a widespread species with considerable geographic variation in the intensity of cowbird parasitism (Hoover and Brittingham 1993). Although its reproductive behavior and ecology have been well studied (Brackbill 1943, 1948, 1958; Longcore and Jones 1969; Roth and Johnson 1993), little information is available on the effects of cowbird parasitism on its breeding success. Wood Thrushes regularly fledge mixed broods from parasitized nests (Friedmann 1963, Rothstein 1975) and are typically double-brooded (Roth and Johnson 1993). Whereas single parasitism may have little effect on Wood Thrush fledging success except for losses resulting from egg removal by female cowbirds, multiple parasitism may pose more severe problems and affect population viability (Trine et al. 1998). In this chapter I examine the intensity and consequences of multiple parasitism for each stage of the nest cycle in a southern Illinois Wood Thrush population where multiple parasitism is the rule rather than the exception.

Study Area and Methods

The study area consisted of three ravine study sites (Dutch Creek, DC; Pine Hills, PH; and South Ripple Hollow, SRH) located on three of the largest forested tracts (1,500–3,000 ha) in Shawnee National Forest. These tracts (described in more detail in Robinson et al., Chapter 33, this volume) are near Jonesboro in southern Illinois. The topography consists of long, narrow ravines and low but steep-sided ridges. Oak-hickory is the predominant forest type, but mixed mesic hardwoods are found in the ravines along the streambeds.

The Wood Thrush populations in PH and SRH ravines were stable during the study, whereas the DC population was in decline from the beginning of the study and almost disappeared.

Fieldwork was conducted from late April through July in 1989–1992. This period encompassed over 99% of annual breeding for Wood Thrushes (known egg dates ranged from April 26 to August 1) and extended beyond the laying period of cowbirds, which ended in mid- to late July (C. L. Trine and S. K. Robinson, unpubl. data). Sample sizes differed among ravines because thrush numbers did. Daily nest searches were conducted, and nest contents were checked every 2–4 days, using a mirror on an extension pole, until each nest's fate was determined. I considered a nest successful if it produced at least one fledgling of either species. Cowbird nestlings were considered to have fledged if they remained in the nest until 8 days of age, whereas Wood Thrush nestlings were considered to have fledged if they were present until 10 days of age. I saw nestlings of both species leave disturbed nests at these ages and survive at least several days.

Egg removal by female cowbirds usually occurs within a day or two of when they lay (Hann 1937, Nolan 1978, Smith 1981, Sealy 1992). Therefore, nest contents were characterized as the number of eggs found in the nest three days after the last parasitism event or three days into incubation, whichever occurred later. Because sample sizes from DC and SRH were too small, only the PH sample was used for detailed analyses of egg removal and nest predation. Very few unparasitized nests occurred in this study, so an estimate of the frequency of egg removal could not be based on the difference in number of host eggs in unparasitized and parasitized nests. I therefore used two alternative methods of estimating egg removal: (1) comparison of average Wood Thrush clutch size after the start of incubation in this study with unparasitized clutch sizes from other studies, and (2) linear regression of the mean number of thrush eggs present relative to the number of cowbird eggs received.

Data were pooled from all sites and years and then analyzed for effects of multiple parasitism at different nesting stages. Both hatching success and overall success from egg to fledging were analyzed using data from nests that were not depredated during the stages being studied. Data for egg survival rates during incubation were obtained from nests found before the start of incubation. Nests observed from the time of hatching provided data for nestling survival rates. Daily survival rates and their standard errors were calculated using Mayfield's (1975) method as modified by Hensler and Nichols (1981). Significance tests of the differences among survival rates used the methods of Sauer and Williams (1989) and were conducted using the computer program CONTRAST (J. E. Hines and J. R. Sauer software, and unpubl. software documentation).

All other statistical tests were done by SYSTAT (Wilkinson 1990). Pairwise comparisons of the percentage-parasitism levels were made using Fisher's exact tests. A Kruskal-Wallis one-way analysis of variance tested for differences in cowbird egg distributions among years. The data were then pooled and a chi-square goodness-of-fit test was used to test for departures from a Poisson distribution (Mayfield 1965).

Proportionate changes in dependent variables were measured by conventional regression coefficients. Elements of nonnormality in the frequency data make the usual variance test of significance of regression unreliable. The regression-like trends in the number of Wood Thrush eggs incubated, hatching success, and overall fledging success were therefore tested for significance by using ordered contingency tables. For these tables, I calculated Goodman and Kruskal's gamma coefficient (Goodman and Kruskal 1954), a variant of Kendall's (1949) tau. The significance of gamma was determined by testing its numerator, which is identical to the numerator of tau (Kendall 1955).

Results

Cowbirds parasitized about 91% of all nests. Thus, the percentage of nests parasitized did not vary significantly among sites within years and within sites between years (Table 15.1; Fisher's exact test, $P > .05$ in all comparisons). Nests averaged more cowbird than host eggs at all sites in most years. The number of Wood Thrush eggs in parasitized nests was similar among study sites, but the intensity of parasitism varied (Table 15.1; for most years, DC << PH < SRH).

Cowbird eggs were distributed randomly among Wood Thrush nests at PH. The distribution of cowbird eggs among nests at PH did not differ among years (Figure 15.1A; Kruskal-Wallis one-way ANOVA, df = 2, $P = .76$), and the pooled sample showed no significant departure from the Poisson distribution ($\chi^2 = 3.592$, df = 5, $P > .5$).

Because there were no significant differences in nest predation rates among years for the PH study site (CONTRAST, $\chi^2 = 2.93$, df = 3, $P = .40$), years were pooled for testing the significance of the relationship between nest predation rates and multiple parasitism (Table 15.2). Predation rates did not differ among nests receiving different numbers of cowbird eggs (CONTRAST, $\chi^2 = 3.04$, df = 5, $P = .69$), or among nests with different numbers of cowbird nestlings (CONTRAST, $\chi^2 = 1.64$, df = 3, $P = .65$).

Costs of Multiple Parasitism for Wood Thrushes

In successful nests observed from the start of incubation, the mean number of Wood Thrush fledglings produced per nest was inversely related to the number of cowbird eggs present (Figure 15.2A; linear regression, $P = .003$). Success per thrush egg was also inversely related to the intensity of cowbird parasitism at the start of incubation (Goodman and Kruskal's gamma = 0.364, $P < .002$). The percentage of Wood Thrush eggs that fledged from successful nests declined by about 9% for each additional cowbird egg present (Figure 15.2B; $y = 81.1 - 9.2x$, $P = .002$), in addition to any losses from egg removal in the laying stage.

The same average numbers of Wood Thrush eggs (3.3 per unparasitized nest) were found in each of three season-long studies of Wood Thrushes conducted where cowbird parasitism was low (Brackbill 1958, Maryland, $N = 16$ nests; Longcore and Jones 1969, Delaware, $N = 74$; J. Hoover unpubl. data, Pennsylvania, $N = 111$). Thus, assuming stable clutch sizes across the middle latitudes of the Wood Thrush's range, cowbirds removed an average of 1.0–1.2 Wood Thrush eggs per nest in this study (Table 15.1, years combined). In the parasitized PH nests found at the start of incubation ($N = 89$), there were 183 Wood Thrush eggs (mean = 2.1/nest) and 278 cowbird eggs (mean = 3.1/nest). Assuming an unparasitized clutch of 3.3, there should have been around 294 Wood Thrush eggs present. Each cowbird egg was associated with the removal of 0.4 Wood Thrush egg [(294 - 183)/278], a 12% reduction in Wood Thrush clutch

Table 15.1. Frequency of Brown-headed Cowbird Parasitism and Contents of Wood Thrush Nests on Three Study Sites in the Shawnee National Forest, 1989–1991

	Study Sites					
	DC		PH		SRH	
% parasitized nests (N)[a]						
1989	83.3	(12)	93.5	(31)	94.4	(18)
1990	75.0	(20)	92.6	(54)	92.9	(28)
1991	100.0	(13)	91.8	(49)	95.2	(21)
1992	100.0	(2)	87.8	(49)	85.7	(14)
All years	85.0	(47)	91.3	(183)	92.6	(81)
Mean no. Wood Thrush eggs/parasitized nest (N)[b]						
1989	2.3	(3)	2.1	(12)	1.7	(6)
1990	2.5	(6)	2.3	(22)	2.2	(10)
1991	1.7	(3)	1.9	(31)	2.2	(10)
1992	1.0	(1)	2.0	(24)	1.0	(2)
All years	2.2	(13)	2.1	(89)	2.0	(28)
Mean no. cowbird eggs/parasitized nest[b,c]						
1989		2.0		3.9		2.8
1990		1.5		3.1		3.6
1991		2.0		3.1		4.4
1992		3.0		2.8		3.0
All years		2.1		3.1		3.7

[a]Nests found before hatching in which the clutch was known to have been completed.
[b]Nests found before incubation started which survived long enough for the clutch to be completed.
[c]Same nest sample sizes as for Wood Thrush eggs.

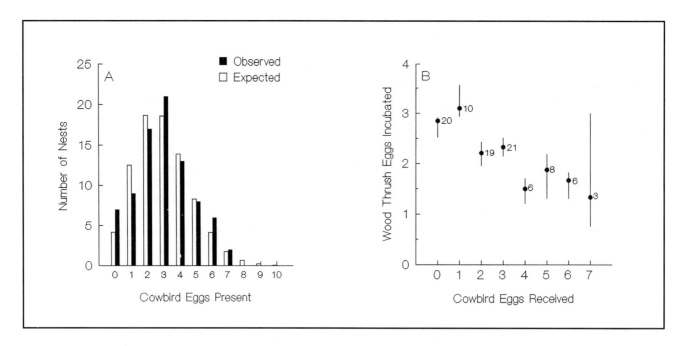

Figure 15.1. Wood Thrush nest contents at the start of incubation (after parasitism and egg removal). (A) Observed and expected (Poisson) distribution of cowbird eggs in nests at the start of incubation. Data from PH for 1990–1992. Classes with more than 5 cowbird eggs per nest were pooled for a goodness-of-fit test.

(B) Mean number of Wood Thrush eggs/nest ± SE relative to the number of cowbird eggs received. Asymmetric error bars were determined by calculating variance estimates separately for each tail. Nest sample sizes are adjacent to means. Data from PH 1989–1992.

Table 15.2. Daily Survival Rates Relative to the Number of Cowbird Eggs Received and the Number of Cowbird Nestlings Present[a]

	Total Nests (N)	Nests Depredated (N)	Exposure Days	Daily Survival Rate	SE
Cowbird eggs received (N)					
0	15	6	217.0	0.972	0.011
1	21	11	321.5	0.966	0.010
2	40	23	573.0	0.960	0.008
3	42	19	711.5	0.973	0.006
4	32	16	528.5	0.970	0.007
≥5	28	11	478.5	0.977	0.007
Cowbird nestlings present (N)					
0	21	6	199.0	0.970	0.012
1	44	16	385.0	0.959	0.010
2	40	13	363.0	0.964	0.010
≥3	23	10	180.0	0.944	0.171

[a]Calculated by Mayfield's (1975) method.

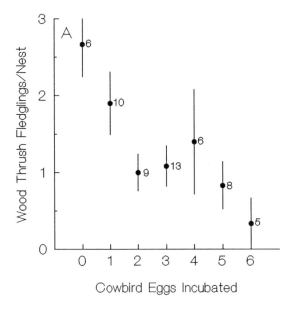

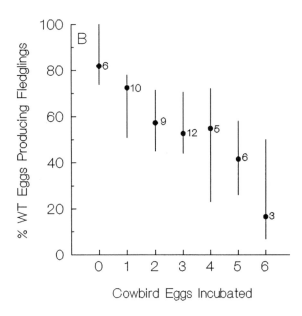

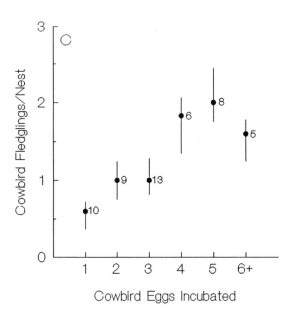

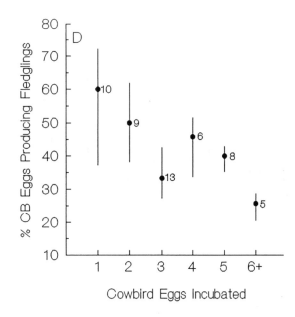

Figure 15.2. Mean fledging success ± SE relative to the number of cowbird eggs incubated from nests observed from the start of incubation that fledged at least one nestling. Asymmetric error bars were determined by calculating variance estimates separately for each tail. Nest sample sizes are adjacent to means. (A) Mean number of Wood Thrush fledglings per nest. (B) Mean percentage of Wood Thrush eggs in successful nests that actually produced fledglings. (C) Mean number of cowbird fledglings per nest. (D) Mean percentage of cowbird eggs in successful nests that actually produced fledglings.

size. Alternatively, linear regression suggested a 9% reduction in the number of Wood Thrush eggs incubated for each additional cowbird egg in a nest ($y = 3.0 - 0.3x$, $P < .001$; Figure 15.1B).

During incubation, 7.2% of 305 Wood Thrush eggs disappeared from 151 active nests. This frequency of egg loss was not significantly different from that for cowbird eggs (5.2%, $N = 423$ eggs; Fisher's exact test, $P = .273$). Egg loss during incubation was not related to total clutch size. Daily survival rates of Wood Thrush eggs in unparasitized nests and singly parasitized nests did not differ significantly (Figure 15.3A). These rates were different, however, from the daily survival rates of eggs in multiply parasitized nests (CONTRAST, $P = .03$). Ninety-nine percent of the Wood Thrush eggs in unparasitized and singly parasitized nests remained in the nest throughout incubation, whereas 93% remained in multiply parasitized nests. The apparent trend for decreasing survival with an increase in the number of cowbird eggs in multiply parasitized nests was not significant (Goodman and Kruskal's gamma, $P > .05$).

Hatching success of Wood Thrush eggs present in the nest at hatching time was inversely related to the number of cowbird eggs present throughout incubation (Figure 15.4A; gamma = 0.238, $P < .01$). Using each nest as a sampling unit, linear regression ($y = 87.4 - 4.3x$, $P = .008$) suggested that 87% of eggs in unparasitized nests should hatch, and that each addition of a cowbird egg reduced this frequency

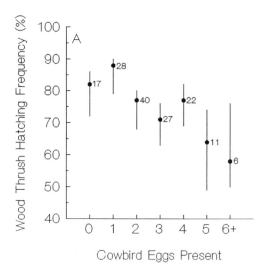

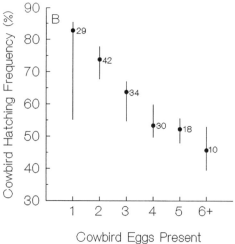

Figure 15.4. Mean hatching frequency ± SE for eggs present in the nest at hatching time relative to the number of cowbird eggs present. Asymmetric error bars were determined by calculating variance estimates separately for each tail. Sample sizes in terms of numbers of eggs are adjacent to means. Data are from nests in which at least one egg hatched. (A) Wood Thrush. (B) Cowbird.

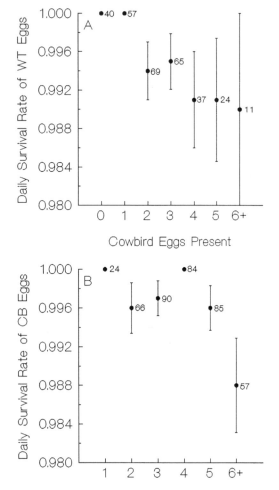

Figure 15.3. Mean daily survival rates ± SE for eggs during incubation. Sample sizes in terms of numbers of eggs are adjacent to means. Rates were calculated using Mayfield's (1975) method. (A) Wood Thrush eggs. (B) Cowbird eggs.

by 4%. Neither the number of Wood Thrush eggs present nor total clutch size were significant factors for hatching success (Goodman and Kruskal's gamma, $P > .05$).

The daily survival rate of Wood Thrush nestlings in nondepredated nests was inversely related to the number of cowbird nestmates (Figure 15.5A; CONTRAST, $X^2 = 12.76$, df = 3, $P = .005$). The probability of a newly hatched Wood Thrush nestling surviving to fledge was about 97% in unparasitized nests and decreased by about 5% per cowbird nestmate.

Costs of Multiple Parasitism for Cowbirds

Because I did not mark eggs, I could not quantify cowbird egg loss resulting from cowbirds removing conspecific eggs.

Both broken and intact cowbird eggs, however, were occasionally found beneath active Wood Thrush nests, often accompanied by Wood Thrush eggs and fragments, which suggests that cowbirds did remove each other's eggs. These losses occurred primarily during the laying and early incubation (first or second day) stages in heavily parasitized nests, or in nests containing only cowbird eggs.

The daily survival rate of cowbird eggs during incubation in nests with six or more cowbird eggs was significantly lower than the survival rates in nests with fewer cowbird eggs (Figure 15.3B). Among those with five or fewer eggs, however, the rates were not correlated with the number of cowbird eggs in the nest.

Hatching success varied inversely with the number of cowbird eggs present (gamma = 0.30, P < .001), declining about 10% for each additional cowbird egg up to four, after which the costs per additional egg decreased. Mean hatching success was 59.5%. Using each nest as a sampling unit, linear regression ($y = 88.9 - 7.9x$, P < .001) suggested that hatching success should be about 81% in singly parasitized nests, and that each additional cowbird egg present reduced hatching success by 8% (Figure 15.4B).

Daily survival rates of cowbird nestlings in active nests did not differ among nests with different numbers of cowbird nestlings (Figure 15.5B; CONTRAST, $\chi^2 = 5.34$, df = 2, P = .07). Daily survival rates also did not differ with total brood size (CONTRAST, $\chi^2 = 4.33$, df = 3, P = .23).

Not surprisingly, the number of cowbird fledglings produced increased with the number of cowbird eggs incubated in successful nests followed from the start of incubation (Figure 15.2C). Linear regression suggested an increase of 0.27 cowbird fledgling per additional cowbird egg ($y = 0.38 + 0.27x$, P < .001). Cowbird eggs and/or nestlings in multiply parasitized nests did not fare as well as those in singly parasitized nests, where 60% of the eggs produced fledglings (Figure 15.2D). The percentage of cowbird eggs that produced fledglings showed a declining trend with increasing numbers of cowbird eggs in the nest (gamma = 0.180, P > .05).

Discussion

The high parasitism and predation rates in my study areas precluded experimental manipulations of clutches. Nonetheless, my results show that multiple parasitism decreases Wood Thrush nesting success and may also decrease cowbird fledging success, although the evidence for the latter effect is not as strong. Reduced productivity resulted from egg removal by female cowbirds, damage to eggs (expressed as egg loss and reduced hatching success), and reduced nestling survival. First, I discuss some of these topics with particular reference to host species. Next, I consider the potential costs to cowbirds. Lastly, I discuss the relationship between nest predation and brood parasitism.

Extent of Egg Removal

Removal of host eggs by female cowbirds produced the greatest cost of parasitism to Wood Thrushes in this study. Both estimates of removal (30% and 40%) may be conservative for three reasons. (1) Parasitism of Wood Thrush nests was most intense during May, when Wood Thrushes produce larger clutches (Longcore and Jones 1969; 3.7 eggs in May, 2.8 eggs in both June and July). Because almost all early-season nests were parasitized, early (May) nests were virtually absent from the sample of unparasitized nests used in the regression analysis. (2) The probability of thrush-egg removal may be greatest with the addition of the first few cowbird eggs and may decrease for later cowbird eggs. If removal of Wood Thrush eggs is less likely as the number of cowbird

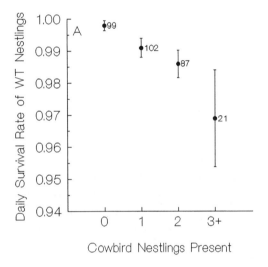

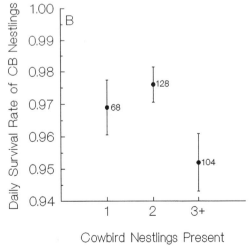

Figure 15.5. Daily survival rates ± SE for nestlings in nests not terminated by nest predation. Sample sizes in terms of numbers of nestlings are adjacent to means. Rates were calculated using Mayfield's (1975) method. (A) Wood Thrush nestlings. (B) Cowbird nestlings.

eggs increases, it would account for my finding of similar Wood Thrush clutch sizes among ravines even though parasitism intensities differed (Table 15.1). (3) An unknown number of cowbird eggs were also removed. The first two reasons would cause an underestimate of removal of Wood Thrush eggs, whereas the third would cause an underestimate of the frequency with which cowbirds removed eggs.

The two estimated frequencies of egg removal (30% and 40% of the time) were low compared to frequencies (from 13% to almost 100%) reported from several host species (Hann 1937, Hofslund 1957, Nolan 1978, Burgham and Picman 1989, Røskaft et al. 1990, Sealy 1992). Sealy (1992) found that cowbirds removed eggs from host nests about 30% of the time in a Yellow Warbler (*Dendroica petechia*) population in Manitoba, but the extent of egg removal varied significantly from year to year. Sealy explored various hypotheses for why female cowbirds remove eggs. Two hypotheses offer possible explanations for annual variation in the extent of egg removal. Perhaps the cowbird population is polymorphic with respect to egg removal: some females always remove eggs while others never remove eggs. Studies of egg removal by individual cowbirds are needed to shed light on this question. Alternatively, removed eggs may serve as food for the parasite. Perhaps the nutritional status and availability of calcium affect the extent to which cowbirds remove eggs. When resources are poor, female cowbirds might be more inclined to remove eggs and eat them as a nutritional supplement, whereas eggs are not needed nutritionally when resources are good. Scott et al. (1992) estimated, however, that cowbirds ate only about 50% of the eggs they removed, and concluded that nourishment is not the primary cause of egg removal by Brown-headed Cowbirds.

Egg Loss during Incubation

After the beginning of incubation, hosts only occasionally receive additional cowbird eggs (Trine pers. obs.; see also Hann 1941, Sealy 1992); most eggs lost during incubation, therefore, are probably not removed by female cowbirds. Multiple parasitism (but not single parasitism) appears to be associated with decreased daily survival rates of Wood Thrush eggs, resulting in about a 6% loss of eggs during incubation. Eggs may be removed by the host because of damage caused by three mechanisms associated with parasitism. (1) The actions of female cowbirds at the time of parasitism may damage eggs (Røskaft et al. 1990). (2) The unusually thick shells of cowbird eggs (Spaw and Rohwer 1987, Picman 1989) may damage eggs in the nest while the cowbird egg is being laid or during the incubation period (Blankespoor et al. 1982, Weatherhead 1991). (3) Thicker shells may make puncture-ejection of cowbird eggs by the host more difficult; thus, the host may damage its own eggs while attempting to remove the cowbird egg (Rothstein 1977, Spaw and Rohwer 1987, Rohwer et al. 1989). One might expect the effects of

such mechanisms to be cumulative, but in my study egg survival rates did not differ among nests containing two, three, or more cowbird eggs.

Reduced Hatching Success

The hatching success for Wood Thrush eggs in unparasitized nests estimated by linear regression (87%) is similar to the 89–96% found in other Wood Thrush studies (Brackbill 1958, Longcore and Jones 1969, Hoover 1992). The inverse relationship between the number of cowbird eggs present and hatching success of both Wood Thrush and cowbird eggs is probably a result of parasitism. In addition to the causes of egg damage discussed above, hatching success may decrease because the host female reduces incubation after the first egg, often a cowbird's, hatches (Petit 1991). Alternatively, incubation efficiency is reduced in enlarged clutches, either through insufficient heat (Hofslund 1957) or inability of the host to turn the eggs properly. Experimental clutch manipulations (e.g., Peer and Bollinger, Chapter 21, this volume) would help distinguish among these possibilities.

Little is known about the factors associated with egg mortality in the nest. Nolan (1978) did not find any difference in Prairie Warbler (*Dendroica discolor*) hatching success between eggs in unparasitized (95% hatched) and parasitized nests (92% hatched). Others, however, have found such differences (e.g., Red-winged Blackbird *Agelaius phoeniceus*, Røskaft et al. 1990; Yellow Warbler, but not Red-winged Blackbird, Weatherhead 1991). More frequent embryonic deaths occurred in parasitized Prothonotary Warbler (*Protonotaria citrea*) nests (Petit 1991). These deaths were at least partially responsible for the reduced hatching success in parasitized nests (73%, vs. 89% for unparasitized nests). Both Petit's and Røskaft et al.'s studies suggest that the presence of the cowbird egg during incubation, rather than actions of the female cowbird, was responsible for the reduced hatching success.

Cost of Multiple Parasitism to the Host

My results indicate that some "large" birds that are regularly multiply parasitized (e.g., Wood Thrush, Scarlet Tanager *Piranga olivacea*, Summer Tanager *P. rubra*) can suffer significant reproductive losses due to cowbird parasitism. For the Wood Thrush, the cumulative estimate of the cost per cowbird egg is 18–24%. In this study, the average nest receives about three cowbird eggs, and thus, fledgling production may be reduced 50–60%. The cost may be overestimated, however, because all the cowbird eggs in the nest may not hatch; thus, the host's nestling survival rate may be higher. The results show that even large host species can be only partially resistant to the effects of cowbird parasitism. Losses may be tolerable in singly or even doubly parasitized nests. As parasitism on the same nest increases, however, the cumulative impact may become great.

Management Implications

For hosts that tolerate low levels of parasitism, a high frequency of multiple parasitism may not cause a population decline if other sources of mortality are low and competition for breeding territories is strongly regulated. If a population is declining, however, other sources of mortality may be extremely high and swamping the effects of parasitism (e.g., the DC Wood Thrush population, where nest predation has been up to 80%; C. L. Trine and S. K. Robinson unpubl. data). In situations where partially resistant host populations are declining and experiencing severe parasitism, cowbird control efforts might not need to be thorough to provide considerable benefits.

Costs of Multiple Parasitism to the Parasite

I noted two costs of multiple parasitism for cowbird nesting success: reduced hatching success in nests with more cowbird eggs present and decreased egg and nestling survival in heavily parasitized nests. Egg removal by conspecifics would also increase costs of multiple parasitism (see also Hann 1937, Klaas 1975, Elliott 1977, Scott 1977).

Fledgling production appears to level off when four or more cowbird eggs are present (see Figure 15.2C). If cowbird fledgling production per nest reaches an asymptote in nests receiving four or more cowbird eggs, female cowbirds should select nests with fewer than four cowbird eggs to maximize productivity per cowbird egg. The blue Wood Thrush eggs are distinctly different from the cowbird egg, so a cowbird should have no difficulty distinguishing between nests with few and many cowbird eggs.

Departures from the Poisson would suggest that cowbirds are being selective in the nests they use (Preston 1948, Mayfield 1965, Elliott 1977). My sample shows a good fit to the Poisson, suggesting that cowbirds distribute their eggs randomly in Wood Thrush nests, although their success may be lower in nests with many cowbird eggs. Where cowbird densities are high, unparasitized nests are rarely available. Selecting against already parasitized nests of large hosts may not be advantageous if large hosts are better able to raise multiple cowbird fledglings. Perhaps their nests serve as "dump nests," where cowbirds can put excess eggs that have at least some chance of producing fledglings, rather than multiply parasitizing nests of smaller hosts that usually do not raise more than one cowbird.

Nest Predation and Parasitism

Parasitism and nest predation were not correlated. Perhaps nests conspicuous to cowbirds are not necessarily obvious to a predator, and vice versa. Alternatively, because of the presence of a rich guild of nest predators, predictably safe nest sites may not exist (see Filliater et al. 1994). Consequently, all Wood Thrush nests may be equally susceptible to predation. Although cowbird nestlings are more active and much noisier than Wood Thrush nestlings (pers. obs.), this did not significantly affect nest predation rates. Several studies of other species have also found no relationship between parasitism and predation (Southern 1958, Mayfield 1960, Klaas 1975, Elliott 1978, Petit 1991). Smith (1981), in his study of Song Sparrows (*Melospiza melodia*), suggested that cowbird females may have selected nests more likely to be successful. Newer data from the same ongoing study (Smith and Arcese 1994) do not support this, but do suggest that cowbirds may be acting as both parasites and nest predators, in agreement with Smith's 1981 suggestion. Other studies have found higher predation rates on parasitized nests (Nice 1937, Finch 1983). Dissimilar habitats and assemblages of nest predators (see Miller and Knight 1993) may cause these different results. In some study areas, cowbirds and predators may overlap considerably in search methods resulting, in predation and parasitism being correlated. In other study areas, however, a different suite of predators might have little overlap with cowbirds in search methods.

Acknowledgments

I greatly appreciate the advice, comments, and encouragement of S. K. Robinson and A. W. Ghent. S. G. Sealy and J. N. M. Smith made many helpful suggestions on the manuscript. S. Amundsen, S. Bailey, T. Berto, K. Bruner, R. Jack, L. Lee, and S. Morse were especially helpful in collecting the data. I am grateful to M. Mumford for permission to use the Shawnee National Forest sites and the facilities of the Dutch Creek conservation camp. This research was supported by the American Museum of Natural History (Chapman Memorial Fund); American Ornithologists Union (Alexander Wetmore Memorial Award); Sigma Xi (Grant-in-aid of Research); Champaign County Audubon Society (Kendeigh Memorial Fund); University of Illinois, Graduate College and Department of Ecology, Ethology, and Evolution (all to C. L. Trine); National Science Foundation BSR 9101211 DIS (to S. K. Robinson and C. L. Trine); Illinois Department of Energy and Natural Resources; and the U.S. Forest Service (to S. K. Robinson).

References Cited

Blankespoor, G. W., J. Oolman, and C. Uthe. 1982. Eggshell strength and cowbird parasitism of Red-winged Blackbirds. Auk 99:363–365.

Brackbill, H. 1943. A nesting study of the Wood Thrush. Wilson Bull. 55:72–87.

———. 1948. A singing female Wood Thrush. Wilson Bull. 60:98–102.

———. 1958. Nesting behavior of the Wood Thrush. Wilson Bull. 70:70–89.

Burgham, M. C. J., and J. Picman. 1989. Effect of Brown-headed Cowbirds on the evolution of Yellow Warbler anti-parasite strategies. Anim. Behav. 38:298–308.

Elliott, P. F. 1977. Adaptive significance of cowbird egg distribution. Auk 94:590–593.

———. 1978. Cowbird parasitism in the Kansas tallgrass prairie. Auk 95:161–167.

Filliater, T. S., R. Breitwisch, and P. M. Nealen. 1994. Predation on Northern Cardinal nests: Does choice of nest site matter? Condor 96:761–768.

Finch, D. M. 1983. Brood parasitism of the Abert's Towhee: Timing, frequency, and effects. Condor 85:355–359.

Friedmann, H. 1963. Host relations of the parasitic cowbirds. U.S. Natl. Mus. Bull. 233:1–276.

Goodman, L. A., and W. H. Kruskal. 1954. Measures of association for cross classifications. J. Amer. Statisticians Assoc. 49:732–764.

Hann, H. W. 1937. Life history of the Oven-bird in southern Michigan. Wilson Bull. 49:146–237.

———. 1941. The cowbird at the nest. Wilson Bull. 53:211–221.

Hensler, G. L., and J. D. Nichols. 1981. The Mayfield method of estimating nesting success: A model, estimators, and simulation results. Wilson Bull. 93:42–53.

Hofslund, P. B. 1957. Cowbird parasitism of the Northern Yellowthroat. Auk 74:42–48.

Hoover, J. P. 1992. Nesting success of Wood Thrush in a fragmented forest. M.S. Thesis. Pennsylvania State University, State College.

Hoover, J. P., and M. C. Brittingham. 1993. Regional variation in cowbird parasitism of Wood Thrushes. Wilson Bull. 105:228–238.

Kendall, M. G. 1949. Rank and product-moment correlation. Biometrika 36:177–193.

———. 1955. Rank correlation methods. Charles Griffin, London.

Klaas, E. E. 1975. Cowbird parasitism and nesting success in the Eastern Phoebe. Occas. Papers, Museum of Natural History, University of Kansas 41:1–18.

Longcore, J. R., and R. E. Jones. 1969. Reproductive success of the Wood Thrush in a Delaware woodlot. Wilson Bull. 81:39–46.

May, R. M., and S. K. Robinson. 1985. Population dynamics of avian brood parasitism. Amer. Natur. 126:475–494.

Mayfield, H. F. 1960. The Kirtland's Warbler. Cranbrook Institute of Science, Bloomfield Hills, MI.

———. 1965. Chance distribution of cowbird eggs. Condor 67:257–263.

———. 1975. Suggestions for calculating nest success. Wilson Bull. 87:456–466.

Miller, C. K., and R. L. Knight. 1993. Does predator assemblage affect reproductive success in songbirds? Condor 95:712–715.

Nice, M. M. 1937. Studies in the life history of the Song Sparrow. Part 1. Trans. Linn. Soc. N.Y. 4:1–247.

Nolan, V., Jr. 1978. The ecology and behavior of the Prairie Warbler *Dendroica discolor*. Ornithol. Monogr. 26:1–595.

Petit, L. J. 1991. Adaptive tolerance of cowbird parasitism by Prothonotary Warblers: A consequence of nest-site limitations? Anim. Behav. 41:424–432.

Picman, J. 1989. Mechanism of increased puncture resistance of eggs of Brown-headed Cowbirds. Auk 106:577–583.

Preston, F. W. 1948. The cowbird (*M. ater*) and the cuckoo (*C. canorus*). Ecology 29:115–116.

Rohwer, S., C. D. Spaw, and E. Røskaft. 1989. Costs to Northern Orioles of puncture-ejecting parasitic cowbird eggs from their nests. Auk 106:734–738.

Røskaft, E., G. H. Orians, and L. D. Beletsky. 1990. Why do Red-winged Blackbirds accept eggs of Brown-headed Cowbirds? Evol. Ecol. 4:35–42.

Roth, R. R., and R. K. Johnson. 1993. Long-term dynamics of a Wood Thrush population breeding in a forest fragment. Auk 110:37–48.

Rothstein, S. I. 1975. An experimental and teleonomic investigation of avian brood parasitism. Condor 77:250–271.

———. 1977. Cowbird parasitism and egg recognition of the Northern Oriole. Wilson Bull. 89:21–32.

Sauer, J. R., and B. K. Williams. 1989. Generalized procedures for testing hypotheses about survival or recovery rates. J. Wildl. Manage. 53:137–142.

Scott, D. M. 1977. Cowbird parasitism on the Gray Catbird at London, Ontario. Auk 94:18–27.

Scott, D. M., P. J. Weatherhead, and C. D. Ankney. 1992. Egg-eating by female Brown-headed Cowbirds. Condor 94:579–584.

Sealy, S. G. 1992. Removal of Yellow Warbler eggs in association with cowbird parasitism. Condor 94:40–54.

Smith, J. N. M. 1981. Cowbird parasitism, host fitness, and age of the host female in an island Song Sparrow population. Condor 83:152–161.

Smith, J. N. M., and P. Arcese. 1994. Brown-headed Cowbirds and an island population of Song Sparrows: A 16-year study. Condor 96:916–934.

Southern, W. E. 1958. Nesting of the Red-eyed Vireo in the Douglas Lake region, Michigan. Jack-Pine Warbler 36:105–130.

Spaw, C. D., and S. Rohwer. 1987. A comparative study of eggshell thickness in cowbirds and other passerines. Condor 89:307–318.

Temple, S. A., and J. R. Cary. 1988. Modeling dynamics of habitat-interior bird populations in fragmented landscapes. Conserv. Biol. 2:340–347.

Trine, C. L., W. D. Robinson, and S. K. Robinson. 1998. Consequences of Brown-headed Cowbird brood parasitism for host population dynamics. Pp. 273–295 *in* Parasitic birds and their hosts (S. I. Rothstein and S. K. Robinson, eds). Oxford University Press, New York.

Weatherhead, P. J. 1991. The adaptive value of thick-shelled eggs for Brown-headed Cowbirds. Auk 108:196–198.

Wilkinson, L. 1990. SYSTAT: The system for statistics. SYSTAT, Evanston, IL.

16. Comparing the Relative Effects of Brood Parasitism and Nest Predation on Seasonal Fecundity in Passerine Birds

JOSEPH A. GRZYBOWSKI AND CRAIG M. PEASE

Abstract

The conservation and management of passerines requires an understanding of the relative impacts of brood parasitism and nest predation on seasonal fecundity (young raised per female per year). Yet few empirical studies successfully separate these effects, and most measure nest success rather than seasonal fecundity. Using a model of Pease and Grzybowski (1995), we quantify the relative impacts of brood parasitism and nest predation on seasonal fecundity. We then evaluate how these effects change as a function of breeding season length, the severity of host brood reduction from parasitism, and the probability of abandonment in response to parasitism. In general, mortality that causes greater (or total) brood reduction and/or consumes more days of the breeding season will have the greater relative effect in reducing seasonal fecundity. A single nest predation event generally reduces seasonal fecundity more than a single brood parasitism event because predation typically destroys the entire brood or clutch, while parasitized nests often produce some host young. Similarly, parasitism increasingly reduces seasonal fecundity as the magnitude of brood reduction due to parasitism increases. Nest predation can occur at any time in the nesting cycle, whereas parasitized nests that are abandoned are lost early in the nesting cycle; consequently, the average depredated nest consumes more days and has a greater impact on seasonal fecundity than a parasitized nest that is abandoned. However, because parasitized nests that are not abandoned and fledge young use the entire nesting cycle, they use more breeding days than are lost for the average nest that is depredated. For this reason, when brood reduction is severe and few or no parasitized nests are abandoned, parasitism can be relatively more important than predation in depressing seasonal fecundity. When brood reduction is severe, abandoning parasitized nests improves seasonal fecundity compared to not abandoning; with low brood reduction, the converse is true. Surprisingly, at moderate to high levels of brood reduction, abandoning or not abandoning may have similar effects on seasonal fecundity. Seasonal fecundity is most sensitive to changes in predation and parasitism when the intensities of each are high and when the breeding season is short. Empirical measures of nest success implicitly overlook how predation and parasitism influence the number of days available for renesting. Nest success is thus an imperfect surrogate of seasonal fecundity.

Introduction

Brood parasitism and nest predation have been widely implicated in the decline of neotropical songbirds in North America (Terborgh 1989). To conserve and manage neotropical migrants, it is thus important to understand the relative effects of brood parasitism by cowbirds and nest predation on passerine breeding productivity.

Nearly all studies that compare the effects of these two mortality sources use nest or egg success as an index of reproductive success (summaries in Ricklefs 1969, 1973; Martin 1992; Robinson et al. 1995). However, there is a major problem with this. Songbirds often renest after a nest failure, and this is not accounted for when one simply measures nest or egg success rather than seasonal fecundity. Also, many migrant passerines have short breeding seasons that allow only one brood to be raised a year. In these species, a female with one or more nest failures may incur little or no decrease in reproductive success, as a single successful nesting effort will overcome all previous losses. Focusing on egg or nest mortality assumes that the sole effect of nest failure caused by predation and parasitism is brood reduction, which is untrue. Nest success (or mortality) overlooks how nest losses cause loss of part of the breeding season.

Because nest predation generally involves loss of the entire clutch or brood, a female whose nest is depredated must renest to produce any offspring. By contrast, a female whose nest is parasitized has a choice; she may abandon the nest and immediately renest, or she can accept the parasitism event and attempt to complete the nesting cycle. A third option of ejecting the parasite egg is available mostly to larger host species.

When brood parasitism reduces host brood size substantially, females that do not immediately abandon parasitized nests and instead fledge parasite young lose much of the breeding season with little or no gain in their own reproductive success. Abandoning the parasitized nest short-circuits the loss in breeding days, but at the cost of losing any host young that would have been fledged from the parasitized nest. Thus, the benefit of abandonment in enhancing seasonal fecundity is uncertain (Rothstein 1975, 1990).

Our mathematical model (Pease and Grzybowski 1995) quantifies the reduction in passerine seasonal fecundity caused by nest predation and brood parasitism, and explicitly accounts for renesting after both nest failure and fledging a brood. Because various parameters describing passerine breeding biology can be readily manipulated in the model, it allows one to compare directly the impacts of parasitism and predation on seasonal fecundity. Such comparisons are almost impossible in empirical studies because of the difficulty of empirically manipulating the rates of parasitism and predation, and then determining the consequences of these manipulations on seasonal fecundity. Moreover, critical parameters such as the length of the breeding season and nesting cycle are impossible to manipulate empirically. Here, we use this model to examine the relative impact on seasonal fecundity of different levels of predation and parasitism and to explore how this relation is moderated by breeding-season length, the level of brood reduction from parasitism, and the probability of abandonment in response to parasitism.

Methods

As specified by Pease and Grzybowski (1995), parameters describing passerine breeding biology include those specifying (1) timing of events in the nesting cycle, (2) the length and other characteristics of the breeding season, and (3) the average brood sizes at fledging of parasitized and unparasitized nests. Table 16.1 shows the model's parameters and numerical values used here; symbols are consistent between this chapter and Pease and Grzybowski (1995). The values used here are typical of emberizids capable of double brooding (Ehrlich et al. 1988) but do not represent any particular species.

The breeding season (s_s) is the number of days each year that females have available for breeding. To a good approximation, s_s is the number of days between the average dates that females initiate their first nests in the spring and their last nests in the summer. For our comparisons, we set the breeding season at 50, 100, or 150 days (see Table 16.1).

The nesting cycle starts at the initiation of nest-building and ends when the young become independent of their parents. It includes inactivity before nest-building, nest-building, egg-laying, incubation, care of young in the nest, and care of the young after they leave the nest (Pease and Grzybowski 1995). For a typical emberizid, the time from the start of the nesting cycle through the end of the nest-building period is six days (e.g., Graber 1961, Nolan 1978). Because cowbirds sometimes lay their eggs on the day before host laying begins, the window of susceptibility to brood parasitism begins on the start of the sixth day of the nesting cycle (t_e) and extends to one day after egg-laying is completed (t_i; see Nolan 1978, Pease and Grzybowski 1995). As-

Table 16.1. Model Parameters and Values Used in Comparisons

Parameter	Values Used	
ρ	Brood parasitism rate	Allowed to vary[a]
d	Nest predation rate	Allowed to vary[a]
a	Probability of abandonment	0.45 or 0.0
t_e	Beginning of susceptibility window	Fixed at start of day 6
t_i	End of susceptibility window for parasitism	Fixed at end of day 11
t_f	End of susceptibility window for predation	Fixed at end of day 30
t_r	Time when successful females renest	Fixed at start of day 41
s_s	Breeding season (measured as length of period for nest initiations)	50, 100, or 150 days
Numbers of host young fledged from successful:		
f_u	Unparasitized nest	Fixed at 3.5
f_p	Parasitized nest	3.0, 1.0, or 0.0

Source: Pease and Grzybowski 1995.

[a]Values used corresponded to 10, 20, 30 . . . or 90% of nests affected; see Table 16.2.

suming a typical emberizid clutch contains four eggs, we set t_i at the end of the eleventh day (one day for each egg laid, plus one day before and one day after egg-laying). The window of susceptibility to nest predation starts at the same day of the nesting cycle as does the parasitism window but extends to fledging (t_f). Nest predation in this period normally causes loss of the entire brood or clutch, while partial brood loss is more typical after fledging. After fledging a brood, females may initiate additional nesting attempts; our analysis assumes a female can initiate another nesting cycle 10 days after fledging young (e.g., Scott et al. 1987); we specify this as day t_r of the nesting cycle.

Pease and Grzybowski (1995) express the intensity of nest predation and brood parasitism as instantaneous rates. For example, a parasitism rate of 0.02 per day corresponds to about 2% of the available nests being parasitized in a day. However, it is inappropriate to use these instantaneous rates directly when comparing the impacts of brood parasitism and nest predation on seasonal fecundity, since nests are typically susceptible to brood parasitism for many fewer days than they are susceptible to nest predation.

To account for this, we define the nest predation and brood parasitism rates as being of equal intensity when the fraction of nests parasitized in a 6-day window equals that depredated in a 25-day window. In doing these calculations, we assume that either brood parasitism or nest predation occurs by itself. Table 16.2 gives the equivalent rates (i.e., affecting the same percentage of nests) of brood parasitism (ρ) and nest predation (d), assuming these two windows of susceptibility. We express levels of parasitism or predation as percentages of nests affected within the relevant window of susceptibility.

We first examined the effects of predation and parasitism separately by setting one to zero and seeing how seasonal fecundity is reduced as the other increases. Because predation and parasitism usually occur together, we also calculated seasonal fecundity with either predation or parasitism varying while holding the other at a moderate level of 50% (Nice 1957, Martin 1992). We made additional calculations at a predation level of 80% to investigate how abandonment influences seasonal fecundity.

Results and Explanations

For breeding season and nesting cycle parameters typical of many emberizids, nest predation is generally more influential than an equivalent level of brood parasitism in reducing seasonal fecundity (Figure 16.1). Most obviously, nest predation generally causes loss of the entire clutch or brood, while in many circumstances, some host young are produced from a parasitized nest. However, as brood reduction becomes severe, parasitism can become more important (Figures 16.1 and 16.2).

In addition, the relative impacts of predation and parasi-

Table 16.2. Instantaneous Parasitism Rates (ρ) and Predation Rates (d) That Correspond to Various Percentage Losses over an Entire Window of Susceptibility

% Nests Depredated or Parasitized during Appropriate Window of Susceptibility	Instantaneous Rate (per Day)	
	ρ	d
10	0.0175	0.0042
20	0.0365	0.0089
30	0.0577	0.0142
40	0.0815	0.0202
50	0.1090	0.0273
60	0.1415	0.0360
70	0.1817	0.0470
80	0.2350	0.0623
90	0.3190	0.0880

Note: The calculations assume a window of susceptibility to parasitism of 6 days and a window of susceptibility to predation of 25 days.

tism are determined by the number of days of the breeding season each utilizes. Brood parasitism usually occurs during or near host egg-laying at the beginning of the nesting cycle. When parasitized nests are abandoned, the host can renest immediately. By comparison, nest predation can occur anytime in the nesting cycle until fledging. Consequently, on average, more days of the breeding season are lost to a nest predation event than to a brood parasitism event that is abandoned, causing nest predation to have a relatively greater impact on seasonal fecundity.

However, when few parasitized nests are abandoned, parasitized nests can actually have a longer average lifespan than depredated nests, increasing the relative importance of parasitism. This is because successful parasitized nests consume the entire nesting cycle. (We define a successful nest as one that fledges at least one young, be it host or parasite.) Because of this, provided that the abandonment probability is not high, parasitism that causes complete host brood loss will generally depress seasonal fecundity to a greater extent than will the complete loss of eggs or young caused by predation (Figure 16.1). In summary, parasitism is more important when brood reduction is severe and parasitized nests are abandoned less frequently (Figures 16.1 and 16.2).

Predation becomes less important relative to parasitism in longer breeding seasons (Figures 16.1 and 16.2). This occurs because nesting attempts that consume more nesting cycle days are more important than those that consume fewer days. Because unsuccessful nests are replaced at a faster rate than successful nests, proportionately more renesting attempts occur after nest failures than after young

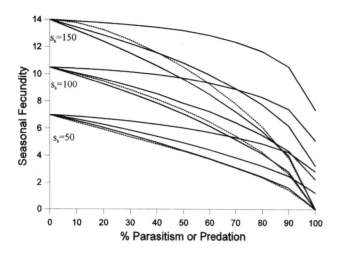

Figure 16.1. Influence of parasitism or predation on seasonal fecundity when $a = 0.45$ and $f_u = 3.5$ and the level of the other is held at 0%. Solid lines show influence of parasitism when $f_p = 3.0$, 1.0, or 0.0 (upper to lower lines) for each of three values of breeding season length (s_s). Dashed lines show influence of predation. See Table 16.1 for symbol designations.

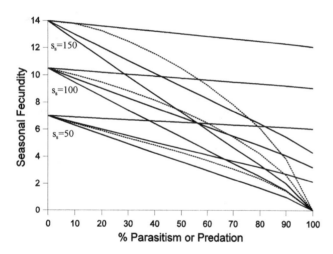

Figure 16.2. Influence of parasitism or predation on seasonal fecundity when $a = 0.0$ and the level of the other is held at 0%. Other parameter values as in Figure 16.1.

Additionally, the impact of parasitism in depressing seasonal fecundity increases slightly as breeding season length increases. Successful nests always use more days of the nesting cycle than nest failures. Consequently, because longer breeding seasons afford increased opportunities for renesting attempts, attended parasitized nests become more important than abandoned parasitized nests or depredated nests. When parasitism greatly reduces brood size, the relative impact of parasitism in depressing seasonal fecundity increases.

The results cited above were obtained by setting either predation or parasitism to zero and varying the other. We also held parasitism or predation at a constant moderate level (50%) while changing the other. The patterns in this case (Figure 16.3, Tables 16.3 and 16.4) are similar to those above (Figures 16.1 and 16.2); predation causes a larger relative decrease in seasonal fecundity than does parasitism, except when brood reduction from parasitism is extreme.

Our model can separate the effects of parasitism and predation on seasonal fecundity. A case study in Tables 16.3 and 16.4 considers a breeding season of 50 days, with high brood reduction ($f_p = 1$ versus $f_u = 3.5$), and assumes that parasitism and predation each affect 50% of the nests. The seasonal fecundity in this case is 3.02 young/female. Since, for these parameter values, a maximum production of 7.00 young/female is possible with no predation or parasitism (Figure 16.1), the overall reduction in seasonal fecundity caused by the effect of both parasitism and predation is 3.98 young/female, or 56.9%.

We can use Tables 16.3 and 16.4 to separate the effects of parasitism and predation on seasonal fecundity; however,

of successful nests become independent. Renesting attempts replace some nest failures with nesting attempts that eventually will be successful, ameliorating the effect of the nest failures on seasonal fecundity. In long breeding seasons, there is increased opportunity for renesting after nest failures due to predation, diminishing the relative importance of nest predation. A corollary here is that a larger percentage of the nests active towards the end of the season will eventually fledge young, in comparison to nests active early in the breeding season.

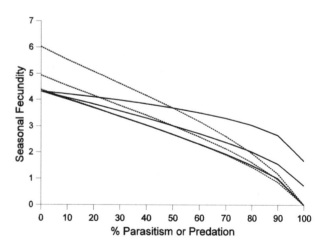

Figure 16.3. Influence of parasitism or predation on seasonal fecundity when $a = 0.45$, $f_u = 3.5$, $s_s = 50$, and the level of the other is held at 50%. Solid lines depict parasitism; dashed lines depict predation. The three pairs of lines crossing at parasitism or predation levels of 50% are for f_p (from top to bottom) of 3.0, 1.0, or 0.0. See Table 16.1 for symbol designations.

Table 16.3. Seasonal Fecundity as a Function of the Level of Brood Parasitism[a]

% Nests Parasitized	s_s = 50 Days			s_s = 100 Days			s_s = 150 Days		
	$f_p = 3$	$f_p = 1$	$f_p = 0$	$f_p = 3$	$f_p = 1$	$f_p = 0$	$f_p = 3$	$f_p = 1$	$f_p = 0$
0	4.32	4.32	4.32	7.37	7.37	7.37	10.51	10.51	10.51
10	4.22	4.08	4.02	7.21	6.98	6.86	10.28	9.95	9.78
20	4.11	3.84	3.70	7.03	6.57	6.34	10.03	9.37	9.04
30	3.99	3.58	3.38	6.84	6.15	5.80	9.76	8.77	8.27
40	3.85	3.31	3.04	6.63	5.70	5.24	9.46	8.14	7.48
50	3.69	3.02	2.69	6.39	5.23	4.66	9.13	7.48	6.65
60	3.51	2.72	2.32	6.12	4.73	4.04	8.75	6.77	5.78
70	3.30	2.39	1.93	5.79	4.18	3.38	8.29	6.00	4.85
80	3.04	2.02	1.51	5.37	3.56	2.66	7.71	5.12	3.82
90	2.66	1.56	1.01	4.75	2.79	1.81	6.84	4.01	2.60
% reduction at 90% parasitism	38.4	63.9	76.6	35.5	62.1	75.4	34.9	61.8	75.3

[a]Predation is held at 50%, and the probability of abandonment in response to parasitism is 0.45.

Table 16.4. Seasonal Fecundity as a Function of the Level of Nest Predation[a]

% Nests Depredated	s_s = 50 Days			s_s = 100 Days			s_s = 150 Days		
	$f_p = 3$	$f_p = 1$	$f_p = 0$	$f_p = 3$	$f_p = 1$	$f_p = 0$	$f_p = 3$	$f_p = 1$	$f_p = 0$
0	6.03	4.94	4.39	9.67	7.94	7.06	13.20	10.81	9.61
10	5.54	4.54	4.04	9.06	7.42	6.60	12.56	10.29	9.15
20	5.09	4.17	3.71	8.45	6.92	6.15	11.85	9.71	8.63
30	4.63	3.79	3.37	7.80	4.39	5.68	11.03	9.04	8.04
40	4.17	3.42	3.04	7.13	5.84	5.19	10.14	8.30	7.38
50	3.69	3.02	2.69	6.39	5.23	4.66	9.13	7.48	6.65
60	3.18	2.60	2.31	5.51	4.56	4.06	7.98	6.54	5.82
70	2.62	2.14	1.91	4.64	3.80	3.38	6.68	5.47	4.86
80	1.97	1.62	1.44	3.55	2.91	2.58	5.13	4.20	3.73
90	1.19	0.98	0.87	2.18	1.79	1.59	3.17	2.59	2.31
% reduction at 90% predation	80.3	80.2	80.2	77.5	77.5	77.5	76.0	76.0	76.0

[a]Parasitism is held at 50%, and the probability of abandonment in response to parasitism is 0.45.

we find different answers. Table 16.3 assumes that the effect of predation is accounted for first; using it, we calculate that the reduction due to predation is 7.00 - 4.32 or 2.68 young/female (38.3%), and that due to parasitism is 4.32 - 3.02 or 1.30 young/female (18.6%). Table 16.4 makes the converse assumption; using it, the reduction due to predation is 4.94 - 3.02 or 1.92 young/female (27.4%), and that due to parasitism is 7.00 - 4.94 or 2.06 young/female (29.4%).

This discrepancy occurs because parasitism and predation interact to determine seasonal fecundity. Considering a single nesting attempt, there are three sources of interaction: (1) Some nests that are parasitized but not abandoned are eventually depredated. (2) Some nests are depredated before they would have been parasitized. (3) Some nests are parasitized and abandoned before they would have been depredated. Because predation generally causes loss of the entire brood, we need not consider depredated nests that are later parasitized. However, as we have stressed, one cannot fully understand the effect of parasitism and predation by looking only at their effects on a single nesting attempt. The interaction also affects seasonal fecundity by altering the average number of breeding season days lost due to parasitism and predation, and thus altering the number of renesting attempts during the breeding season. The overall effect of this interaction is that the effects of parasitism and predation acting together are not as severe as expected from summing the effects of each acting alone.

For our example, the interaction can be estimated as the difference between the reduction in seasonal fecundity due to parasitism with no predation, and the reduction due to parasitism in the presence of 50% predation; that is, (7.0 - 4.94) - (4.32 - 3.02) = 0.76 young/female (see Tables 16.3 and 16.4), or 10.9%. This is equivalent to the statistical estimate for interaction: (7.00 - 4.94 - 4.32 + 3.02) = 0.76 young/female (Stuart and Ord 1991, p. 1113). The overall reduction of 56.9% can thus be partitioned into an effect of predation alone (2.68 - 0.76 = 1.92, or 27.4%), an effect of parasitism alone (2.06 - 0.76 = 1.30, or 18.6%), and an interaction of 10.9%.

Calculating similarly for breeding season length of 100 days where maximum production is 10.50 young/female/ season, the overall reduction in seasonal fecundity is 50.2%, including a 25.8% reduction due to predation alone, a 20.4% reduction due to parasitism alone, and an interaction of 4.0%. For season length of 150 days where maximum production is 14 young/female/season, the overall reduction is 46.6%, including a 23.8% reduction due to predation alone, a 21.6% reduction due to parasitism alone, and an interaction of just 1.1%.

Thus, the magnitude of the interaction between parasitism and predation declines in longer breeding seasons. The interaction arises from the effects on seasonal fecundity of nests that are, or would have been, subject to both parasi-

tism and predation. When renesting is possible, a parasitized nest that is depredated will more likely be replaced by a nest that itself will either be successful, or will only be depredated or parasitized. Because a renesting can effectively cancel out the effect of the previous nest failure, and extended breeding seasons allow for more renesting attempts, the interaction effect on seasonal fecundity declines markedly.

Abandonment in response to parasitism is generally considered an adaptation for reducing the impact of parasitism on seasonal fecundity (Rothstein 1975, 1990; Payne 1977). Our results show that level of brood reduction, length of breeding season, and levels of parasitism and predation affect the costs and benefits of abandonment in ways that may counter each other and are, thus, not entirely obvious. At low to moderate levels of brood reduction in parasitized nests, seasonal fecundity is generally higher with no abandonment than with a moderate level of abandonment (Figure 16.4, Table 16.5). The cost of abandoning a nest is loss of part of the breeding season, and in these circumstances, this cost is not overcome by the potentially (but not certainly) larger brood that might result from renesting after abandonment. The benefits of abandonment decrease as host productivity in parasitized nests increases. With moderate to relatively high brood reduction (50–70% reduction), the results of a moderate level of abandonment (0.45 probability) are roughly equal to those for no abandonment (Table 16.5, Figure 16.4B). It is only at very high levels of brood reduction that the host always benefits from abandoning parasitized nests.

As breeding season length decreases, abandoning parasitized nests becomes less effective in enhancing seasonal fecundity (Table 16.5) because in short breeding seasons there is less time to renest. This is another example of how long breeding seasons accentuate the relative effects of nesting outcomes that take more time. Similarly, as the intensity of predation or parasitism increases, the benefits of abandonment become less pronounced (Table 16.5), because subsequent nesting attempts are increasingly likely to be depredated or parasitized themselves.

Some of these effects counter each other. Looking at these factors together, abandonment after parasitism is a poorer reproductive strategy during short breeding seasons when brood reduction from parasitism is minimal and levels of predation and parasitism are high. Conversely, abandonment is most effective when brood reduction is severe, and during long seasons when the level of predation is low and the level of parasitism is moderate (Table 16.5).

As levels of nest predation increase, an increased percentage of nests will be lost. However, this effect is tempered when predation intensity is high, because nest losses to predation occur, on average, earlier in the nesting cycle, and less of the potential nesting cycle is used by each failed nesting attempt. Although fewer nesting attempts are successful,

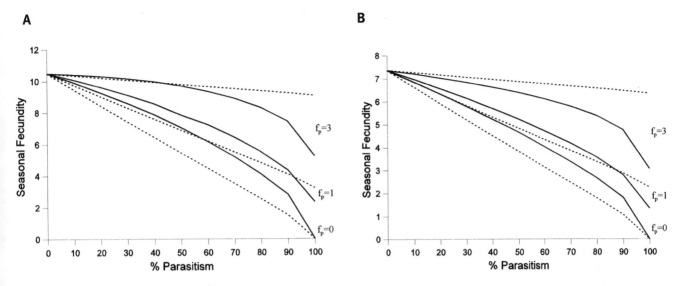

Figure 16.4. Influence of parasitism on seasonal fecundity when $a = 0.45$ (solid lines) or 0.0 (dashed lines) and $s_s = 100$. Predation held at (a) 0% or (b) 50%.

Table 16.5. Percentage Differences in Seasonal Fecundity between Abandonment Probabilities of 0.45 and 0.0

f_p	Brood Reduction	Predation = 0% S_s 50	100	150	Predation = 50% S_s 50	100	150	Predation = 80% S_s 50	100	150
Level of parasitism = 90%										
3	Low	−27	−17	−12	−26	−23	−23	−31	−29	−29
1	High	−3	+2	+5	−3	−1	−1	−6	−5	−4
0	Severe	+8	+12	+14	+9	+10	+10	+7	+8	+8
Level of parasitism = 50%										
3	Low	−7	−1	+1	−8	−7	−6	−10	−9	−9
1	High	+5	+10	+12	+4	+5	+6	+3	+3	+3
0	Severe	+11	+15	+17	+10	+11	+11	+9	+9	+9
Level of parasitism = 10%										
3	Low	0	+1	+1	−1	−1	−1	−1	−1	−1
1	High	+2	+3	+3	+2	+2	+2	+1	+1	+1
0	Severe	+4	+4	+4	+3	+3	+3	+2	+3	+3

Note: Positive values indicate that seasonal fecundity (s_f) was higher when abandonment probability (a) = 0.45. Negative values indicate that s_f was lower when a = 0.45. Symbols as in Table 16.1.

Percentage difference in seasonal fecundity is measured as: $[(\text{condition } s_f / \max s_f)(a = 0.45) \times 100] - [(\text{condition } s_f / \max s_f)(a = 0.0) \times 100]$.

more attempts can be made. Thus, increasing levels of predation do not produce proportional decreases in seasonal
fecundity, so that the relation between nest success and seasonal fecundity becomes curvilinear (Figures 16.1 and 16.2,
Table 16.3). Being single-brooded intensifies this curvilinear
effect, since in such species, the part of the breeding season
that would have been used to raise second broods cannot be
lost.

Discussion

The relative effects of predation and parasitism on seasonal
fecundity of songbirds obviously depend on the levels of
brood parasitism and nest predation actually present in nature—when the parasitism level is much higher than the
level of nest predation, brood parasitism will likely be most
important. Our central question is subtler. We compare the
relative effects of predation and parasitism, assuming reasonable and equivalent magnitudes of each. We define
"equivalent" above to account for the fact that nests are susceptible to parasitism for many fewer days than they are susceptible to predation. A full understanding of the relative effects of parasitism and predation on seasonal fecundity
requires both an understanding of the levels of parasitism
and predation typically found in nature, and an understanding of why they do not impact seasonal fecundity equally.

Parasitism levels are sometimes high enough to reduce
significantly host seasonal fecundity, population size, and/
or geographic range. Examples here include the Kirtland's
Warbler (*Dendroica kirtlandii*), Least Bell's Vireo (*Vireo bellii
pusillus*), Southwestern Willow Flycatcher (*Empidonax traillii extimus*), and Black-capped Vireo (*V. atricapillus;* reviewed
by Robinson et al. 1995; see also Part V, this volume). These
four species are particularly susceptible to high brood
parasitism rates because few host young are fledged from
nests that fledge cowbird young. Cowbird removal has generally been a successful management strategy for them, although the Kirtland's Warbler population size did not initially increase after cowbird removal was started, probably
because insufficient habitat was available.

In contrast, Stutchbury (1997) found that although cowbird removal dramatically decreased parasitism rates in
Hooded Warbler (*Wilsonia citrina*) nests, it did not have
much impact on the average number of young fledged per
nest because many nests were still lost to predation. By separating the effects of brood parasitism and nest predation
on seasonal fecundity (rather than nest success), our model
helps to explain the population consequences of findings
such as Stutchbury's.

A Single Nesting Attempt versus the Entire Breeding Season

When comparing the relative importance of brood parasitism and nest predation in reducing passerine fecundity,
many authors have focused on how these nest mortality

sources impact individual nesting attempts. By contrast, our
results show that nest mortality (or success) is only indirectly related to overall reproductive success. The patterns
relating nest mortality and reproductive success are best explained by considering the effects of both predation and
parasitism on the number of offspring fledged from the affected nests, and comparing how the possible outcomes
consume days in the breeding season.

The proportion of nests in a sample that are depredated or
parasitized has been frequently used as an index of seasonal
fecundity (e.g., McGeen 1972, Zimmerman 1983, Wilcove
1985, Robinson 1992, Martin 1992). However, to estimate
population growth rate accurately, one needs to know how
seasonal fecundity varies with different levels of parasitism
and predation. Thus, even if the proportions of all nest
starts depredated or parasitized are accurately estimated, the
issue remains of how to translate such estimates of nest or
egg mortality into estimates of the reduction in seasonal
fecundity.

It is obvious that seasonal fecundity decreases as nest or
egg success decreases. Perhaps less obviously, the results presented here show that the relationship between nest success
and seasonal fecundity is not proportional or linear. A key
here is that the theoretical relationship between nest success
and seasonal fecundity varies greatly depending on nesting
cycle and breeding season parameters, as well as on the interaction of nest predation and brood parasitism. While we
do not generally expect a close linear relationship between
egg or nest success and seasonal fecundity, we know of only
one useful empirical example. In this case, there was a fairly
linear relationship between nesting success and yearling recruitment ($r^2 = 0.57$) in American Redstarts (*Setophaga ruticilla*). The study plot here was broadly representative of the
region, and the breeding season was relatively short (Sherry
and Holmes 1992).

Sampling Biases in Measuring Nest or Egg Success

To assess the relative importance of parasitism and predation, many field workers measure how these sources of mortality decrease nest success. To interpret such data correctly,
however, one must account for several inherent and widespread biases, which greatly complicate any attempt to draw
an inference about seasonal fecundity from empirical measurements on egg or nest success (Pease and Grzybowski
1995).

If a parasitized nest is abandoned in response to a parasitism event, the abandonment will usually occur early in the
nesting cycle during the egg-laying period. These abandonment events occur in the period sampled least effectively by
protocols that involve obtaining a sample of active nests at
all stages of the nesting cycle. This causes an underestimate
in the true frequency at which nests are parasitized, in turn
causing the impact of parasitism to be underestimated. This
bias can be large if the probability of nest abandonment is

moderate or large (Pease and Grzybowski 1995). In the relatively few studies where the parasitism rate has been accurately reported, the estimates were obtained from a sample of nests followed from nest-building (e.g., Nolan 1978, Smith 1981).

Many sampling protocols provide inaccurate estimates of predation for similar reasons (i.e., the samples include nests found through the nesting cycle). The proportion of nests in a sample that are depredated underestimates the actual level of predation because successful nests are active over a longer period than unsuccessful nests and are thus much more likely to be found.

Even when one is following nests longitudinally, biased estimates of daily parasitism and predation rates can arise for two reasons. (1) If all nests in the sample are not found very near nest initiation, the sample will miss some early nest failures (which are more likely to be due to parasitism). Consequently, predation rate will be overestimated in comparison to parasitism rate. (2) Some nests that are parasitized and abandoned would have otherwise been depredated, so that predation will be underestimated relative to parasitism. The bias created by (1) will not necessarily compensate for that created by (2), even though they act in opposite directions.

Interspecific Comparisons

We have outlined problems in estimating the effects of parasitism and predation and in using them to predict population phenomena. These relations are not straightforward, but the difficulties of comparing parasitism and predation across communities are even more acute. Because different species are subject to different sampling biases, incur different levels of brood reduction, and have nesting cycles and breeding seasons that differ in length, interspecific generalizations on the relative effects of parasitism and predation on nest success (or mortality) can be inaccurate. Where any factors change the relative impacts of predation and parasitism, the conclusions will be suspect. Some studies pre-date the Mayfield (1975) approach and thus provide biased estimates of predation. Most studies of parasitized species underestimate parasitism (Pease and Grzybowski 1995). Arguments that assume that all depredated (or parasitized) nests would have produced young in the absence of predation (or parasitism) essentially ignore the influence of predation (or parasitism) on the number of renesting attempts possible and ignore the interaction between parasitism and predation. All of these difficulties seriously confound interspecific comparisons.

Complexities of Abandonment

The reduction in seasonal fecundity caused by parasitism is determined in a complex way by the extent of brood reduction and the percentage of parasitized nests that are abandoned. When brood reduction is severe (i.e., no host young are produced), a parasitism event causes complete brood loss, as does a nest predation event. However, unlike a nest predation event, the nesting cycle is usually not terminated by a parasitism event. When a parasitized nest is not abandoned, renesting cannot occur until any parasite young have fledged or the nest has been depredated. Consequently, when the abandonment probability is low, the average amount of the breeding season consumed by a brood parasitism event will generally be larger than that lost to the average predation event, causing brood parasitism to have a relatively greater impact on seasonal fecundity.

Rothstein (1975, 1990) considered the strategy of immediately abandoning parasitized nests as a potential adaptation for reducing the impact of brood parasitism on seasonal fecundity, albeit a strategy that bears a cost to the host. Because less of the breeding season remains for successful breeding, Clark and Robertson (1981) argued that abandonment late in the season would be more costly than in the beginning. Graham (1988) postulated that the proportion of nests abandoned after a parasitism event would be different in early versus late season nesting attempts, but found no difference.

Our model shows here: (1) When a host abandons a parasitized nest, it loses any young that may have fledged, as well as losing the number of days of the nesting cycle up to the time of abandonment. Consequently, not abandoning a parasitized nest sometimes results in higher seasonal fecundity than with moderate levels of abandonment. (2) For a broad array of conditions, including relatively high levels of brood reduction, short to long breeding season lengths, and moderate to high levels of predation, the differences in seasonal fecundity between moderate levels of abandonment and nonabandonment are small. (3) Consequently, nest abandonment is highly advantageous to the host only when brood reduction is extreme.

Our model allows some empirical tests of theoretical predictions concerning abandonment in response to parasitism. The Dickcissel (*Spiza americana*) is a heavily parasitized grassland species of the Great Plains with a long history of contact with cowbirds. It maintains low probabilities of nest abandonment in response to brood parasitism—0.06 for the parasitized nests observed by Elliott (1978), and 0.10 and 0.15 in nests studied by Zimmerman (1983). Because cowbirds cause only a modest reduction in the number of Dickcissel young fledged from a nest, our model predicts that the highest seasonal fecundity will occur with no abandonment. The Northern Cardinal (*Cardinalis cardinalis*) also shows modest levels of brood reduction in parasitized nests. However, it exhibits modest levels of abandonment (*a* = 0.32–0.52; Graham 1988, Scott and Lemon 1996) and has a relatively long breeding season. The predictions of our model in this case are ambiguous; the long breeding season favors abandonment, while low brood reduction favors no abandonment. The Prairie Warbler (*Dendroica discolor*) also has a moderate level of abandonment ($a = 0.46$); brood reduction from parasitism is high but not complete (73%; Nolan 1978, Pease and Grzybowski 1995). Our model predicts that seasonal fecundity would be nearly unchanged for a broad array of abandonment probabilities and levels of parasitism and predation. One case clearly contradicts the predictions of our model. The Black-capped Vireo has modest levels of abandonment ($a = 0.43$; Graber 1961, Pease and Grzybowski 1995), but Black-capped Vireo nests suffer complete brood reduction. In this case, our model predicts that a higher level of abandonment would increase seasonal fecundity.

Management Considerations

Management of neotropical migrants often involves determining what type of intervention is appropriate in light of a population decline. Brood parasitism and nest predation on the breeding grounds are thought to play a critical role in many of the observed population declines of neotropical migrants (Terborgh 1989).

The approach we use for understanding the relative impacts of nest predation and brood parasitism offers several advantages (in addition to those above) over the use of other estimators. (1) While it is logistically difficult to follow color-banded females through the nesting season and to measure seasonal fecundity directly, our model can estimate seasonal fecundity from nest history data that are much more readily gathered (Pease and Grzybowski 1995). The model first estimates predation and parasitism rates from the available data, and then estimates seasonal fecundity using these and estimates of parameters describing the nesting cycle and breeding season. (2) Accurate estimates of seasonal fecundity can be combined with data on survivorship, dispersal, and immigration to estimate intrinsic population growth rates. These can then predict the change in population size that will be observed next year, thereby reducing time lags in making management decisions. (3) Seasonal fecundity directly measures reproductive performance. The approach we advocate allows one to estimate, prior to initiating actual management procedure, how much seasonal fecundity will increase when parasitism and/or predation are reduced by a given amount. In this regard, our approach shifts the emphasis from examining whether predation or parasitism has more influence on seasonal fecundity to implementing management actions to elevate seasonal fecundity to make intrinsic population growth rates stable or positive.

To date, the currencies for comparing the effects of nest predation and brood parasitism have been primarily nest success or nest mortality (summary in Martin 1992), less frequently seasonal fecundity (Ricklefs 1969, Nolan 1978, Smith 1981). Our results suggest that partitioning the effects of predation and parasitism by looking at their effects on egg or nest success will generally be inaccurate. We recommend, instead, that managers and researchers jointly consider the effects of predation or parasitism on seasonal fecundity in their decision-making, as seasonal fecundity is a direct measure of reproductive success and fitness.

Acknowledgments

Our research was supported in part by funding from the National Fish and Wildlife Foundation, Texas Nature Conservancy, and a Reeder Fellowship from the University of Texas (the latter awarded to CMP). We thank D. M. Scott for use of his unpublished data on cardinals. William J. Radke and Jamie Smith provided comments useful in improving this manuscript.

References Cited

Clark, K. L., and R. J. Robertson. 1981. Cowbird parasitism and evolution of anti-parasite strategies in the Yellow Warbler. Wilson Bull. 93:249–258.

Ehrlich, P. R., D. S. Dobkin, and D. Wheye. 1988. The birder's handbook: A field guide to the natural history of North American birds. Simon and Schuster, New York.

Elliott, P. F. 1978. Cowbird parasitism in the Kansas tallgrass prairie. Auk 95:161–167.

Graber, J. W. 1961. Distribution, habitat requirements, and life history of the Black-capped Vireo (Vireo atricapillus). Ecol. Monogr. 31:313–336.

Graham, D. S. 1988. Responses of five host species to cowbird parasitism. Condor 90:588–591.

Martin, T. E. 1992. Breeding productivity considerations: What are the appropriate habitat features for management. Pp. 455–473 in Ecology and conservation of neotropical migrant landbirds (J. M. Hagan III and D. W. Johnston, eds). Smithsonian Institution Press, Washington, DC.

Mayfield, H. 1975. Suggestions for calculating nest success. Wilson Bull. 73:255–261.

McGeen, D. S. 1972. Cowbird-host relationships. Auk 89:360–380.

Nice, M. M. 1957. Nesting success in altricial birds. Auk 74:305–321.

Nolan, V., Jr. 1978. The ecology and behavior of the Prairie Warbler, Dendroica discolor. Ornithol. Monogr. 26:1–595.

Payne, R. B. 1977. The ecology of brood parasitism. Annu. Rev. Ecol. Syst. 8:1–28.

Pease, C. M., and J. A. Grzybowski. 1995. Assessing the consequences of brood parasitism and nest predation on seasonal fecundity in passerine birds. Auk 112:343–363.

Ricklefs, R. E. 1969. An analysis of nesting mortality in birds. Smithsonian Contr. Zool. 9:1–48.

———. 1973. Fecundity, mortality, and avian demography. Pp. 366–435 in Breeding biology of birds (D. S. Farner, ed). National Academy of Science, Washington, DC.

Robinson, S. K. 1992. Population dynamics of breeding neotropical migrants in a fragmented Illinois landscape. Pp. 408–418 in Ecology and conservation of neotropical migrant landbirds (J. M. Hagan III and D. W. Johnston, eds). Smithsonian Institution Press, Washington, DC.

Robinson, S. K., S. I. Rothstein, M. C. Brittingham, L. J. Petit, and J.

A. Grzybowski. 1995. Ecology and behavior of cowbirds and their impacts on host populations. Pp. 428–460 *in* Ecology and management of neotropical migratory birds (T. E. Martin and D. M. Finch, eds). Oxford University Press, New York.

Rothstein, S. I. 1975. An experimental and teleonomic investigation of avian brood parasitism. Condor 77:250–271.

———. 1990. A model system for coevolution: Avian brood parasitism. Annu. Rev. Ecol. Syst. 21:481–508.

Scott, D. M., and R. E. Lemon. 1996. Differential reproductive success of Brown-headed Cowbirds with Northern Cardinals and three other hosts. Condor 98:259–271.

Scott, D. M., R. E. Lemon, and J. A. Darley. 1987. Relaying interval after nest failure in Gray Catbirds and Northern Cardinals. Wilson Bull. 99:708–712.

Sherry, T. W., and R. T. Holmes. 1992. Population fluctuations in a long-distance migrant: Demographic evidence for the importance of breeding season events in the American Redstart. Pp. 431–442 *in* Ecology and conservation of neotropical migrant landbirds (J. M. Hagan III and D. W. Johnston, eds). Smithsonian Institution Press, Washington, DC.

Smith, J. N. M. 1981. Cowbird parasitism, host fitness and age of the host female in an island Song Sparrow population. Condor 83:153–161.

Stuart, A., and J. K. Ord. 1991. Kendall's advanced theory of statistics, vol. 2. Oxford University Press, Oxford.

Stutchbury, B. J. M. 1997. Effects of female cowbird removal on reproductive success of Hooded Warblers. Wilson Bull. 109:74–81.

Terborgh, J. 1989. Where have all the birds gone? Princeton University Press, Princeton, NJ.

Wilcove, D. S. 1985. Nest predation in forest tracts and the decline of migratory songbirds. Ecology 66:1211–1214.

Zimmerman, J. L. 1983. Cowbird parasitism of Dickcissels in different habitats and at different nest densities. Wilson Bull. 95:7–22.

PART III

Host-Cowbird
Behavioral
Interactions

17. Introduction

JAMES N. M. SMITH AND SPENCER G. SEALY

Brood parasites and their hosts interact behaviorally while the parasites search for nests or engage in parasitic activities at or near nests. They also interact through the ecological effects of parasite on host, and vice versa (see Part IV). Finally, they interact on an evolutionary time scale as hosts and parasites evolve together, e.g., when hosts evolve egg ejection, and parasites counter by evolving mimetic eggs (Rothstein 1990, General Introduction to this volume). The chapters in this short section deal primarily with behavioral interactions, while touching on some ecological effects and evolutionary consequences of brood parasitism.

When a female cowbird interacts with a host individual, she is either directly attempting to lay or seeking information that may lead to successful parasitic laying. Hosts, on the other hand, often respond aggressively to cowbirds (e.g., Neudorf and Sealy 1992), and aggression and other protective behaviors are often interpreted as defenses against the cowbird. Because of the recent range expansion of the Brown-headed Cowbird (Rothstein 1994), many new host species and races have been exposed to cowbird parasitism. Interactions between cowbirds and these new hosts may not involve finely tuned or adaptive behaviors by either cowbird or host (Rothstein 1982, 1990).

Host-Parasite Interactions at the Time of Egg Laying

Two of the four chapters in this part deal with behavioral interactions during the egg-laying period of the host. It has long been known that cowbirds lay most of their eggs during the laying period of hosts (Friedmann 1929). However, it has only recently been learned that cowbirds lay their eggs only in the few minutes of twilight just before sunrise (Scott 1991, see below). In addition, laying is rapid, typically lasting less than a minute (Sealy et al. 1995). Whether appropriate host actions during this short time window can thwart parasitism is the topic of Chapter 18 and Chapter 19. The results in the two chapters support the idea that dawn laying by cowbirds is a cryptic adaptation to which some small hosts have no effective response.

In the first chapter, Burhans explores whether differential host aggression explains the frequent parasitism on the Indigo Bunting and infrequent use of the co-occurring Field Sparrow. Both species recognized cowbird models as enemies, but neither behaved very aggressively toward them in Burhans's tests. Dawn observations at the nests of the two species revealed that female Field Sparrows arrived at the nest area before the cowbirds' brief laying period but did not defend nests actively, while female Indigo Buntings arrived after the cowbird laying period and thus rarely met laying cowbirds.

Most Field Sparrow females whose nests were parasitized immediately abandoned their clutches. This observation is relevant to the measurement of the frequency of parasitism. Nests of species that are abandoned immediately when a cowbird lays in them are likely to be underrepresented in results of monitoring programs, and thus the effects of parasitism on such species will be underestimated (see also Grzybowski and Pease Chapter 16).

In a related chapter, Sealy et al. examine whether nesting female Yellow Warblers begin to sit tightly on their nests before sunrise at the onset of laying and thus prevent laying by cowbirds. Yellow Warbler females, however, did not modify their nest attentiveness markedly at the onset of laying. Instead, nest attendance increased steadily through the egg-laying period. The authors interpret this increased nest attendance as the normal hormonally mediated transition from laying to incubation seen in all birds, rather than as a specific defense against parasitism. Yellow Warblers do respond adaptively to cowbird parasitism by burying clutches containing cowbird eggs, and they attack cowbirds near their nests vigorously (see the Appendix to Chapter 19 for graphic descriptions of these encounters), but they are seldom physically able to prevent the female cowbird from laying.

Larger cowbird hosts, which can perhaps fend off a laying cowbird, might benefit more from dawn vigilance. Several such large host species, however, do not show greater vigilance at dawn than Yellow Warblers, and a large host female sitting in her nest cannot always deter a female cowbird from laying (Neudorf and Sealy 1994).

Egg Puncture and Egg Removal

The other two chapters in this part deal with two cowbird behaviors that are associated with parasitism, removal and puncture of host eggs. Egg removal by Brown-headed Cowbirds occurs to a varying extent among host species (Sealy 1992). Eggs are removed either before or after the cowbird lays her egg but not at the actual time of laying (Sealy 1992). In contrast, Shiny Cowbirds rarely remove host eggs, but they puncture host eggs and leave them in nests (Post and Wiley 1977). Occasional egg puncture is also associated with parasitism in Brown-headed Cowbirds (e.g., Smith and Arcese 1994).

Chapter 20 by Nakamura and Cruz quantifies how egg puncture by Shiny Cowbirds in southwestern Puerto Rico has changed over time as cowbird abundance has increased, and how it varies among host species. Egg puncture is closely associated with nest parasitism, as it is not seen in nests of unparasitized species. Unlike egg removal by Brown-headed Cowbirds, egg puncture by Shiny Cowbirds occurs during the host incubation period, rather than during laying. It is particularly common in multiply parasitized host nests of the more commonly parasitized host species and is often associated with complete nesting failure.

Total nest failure clearly benefits neither host nor parasite, so egg puncture, unlike egg removal, is not an immediately adaptive behavior performed by the laying parasite. Nakamura and Cruz feel that egg puncture is either an incidental effect of parasitic activity at host nests in dense parasite populations, or a behavior by which Shiny Cowbirds interfere with the egg-laying activities of conspecifics and perhaps enhance their future laying opportunities. Arcese et al. (1996) have argued that clutch destruction and killing of host nestlings serves the latter function in Brown-headed Cowbirds.

Peer and Bollinger in Chapter 21 test an idea developed from earlier work by Davies and Brooke (1988). They propose that removal of a host egg increases cowbird hatching efficiency in nests of larger hosts. When the cowbird egg is smaller than the host eggs, it makes less contact with the host female's brood patch and is thus slow to develop. Peer and Bollinger show that model cowbird eggs (mostly the similar-sized eggs of the House Sparrow) placed in nests of the Common Grackle, a large and rarely used host of the cowbird, hatch earlier in smaller clutches. In nests of larger hosts, early hatching gives the small cowbird chick a critical head start over its larger brood mates. The significance of their experimental work is that egg removal is viewed as a trait of past adaptive value in large hosts that were once used commonly but are now rarely used. Egg removal thus need not be adaptive in the smaller hosts currently used by most cowbird females.

References Cited

Arcese, P., J. N. M. Smith, and M. I. Hatch. 1996. Nest predation by cowbirds, and its consequences for passerine demography. Proc. Nat. Acad. Sci. USA. 93:4608–4611.

Davies, N. B., and M. de L. Brooke. 1988. Cuckoos vs. Reed Warblers. Adaptations and counter-adaptations. Anim. Behav. 36:262–284.

Friedmann, H. 1929. The cowbirds: A study in the biology of social parasitism. C. C. Thomas, Springfield, IL.

Neudorf, D. L., and S. G. Sealy. 1992. Reactions of four passerine species to threats of predation and cowbird parasitism: Enemy recognition or generalized response? Behaviour 123:84–105.

———. 1994. Sunrise nest attentiveness in cowbird hosts. Condor 96:162–169.

Post, W., and J. W. Wiley. 1977. Reproductive interactions of the Shiny Cowbird and the Yellow-shouldered Blackbird. Condor 79:176–184.

Rothstein, S. I. 1982. Successes and failures in avian egg recognition with comments on the utility of optimality reasoning. Amer. Zool. 22:547–560.

———. 1990. A model system for coevolution: Avian brood parasitism. Annu. Rev. Ecol. Syst. 21:481–508.

———. 1994. The cowbird's invasion of the far West: History, causes, and consequences experienced by host species. Studies in Avian Biol. 15:301–315.

Scott, D. M. 1991. The time of day of egg-laying by the Brown-headed Cowbird and other icterines. Can. J. Zool. 69:2093–2099.

Sealy, S. G. 1992. Removal of Yellow Warbler eggs in association with cowbird parasitism. Condor 94:40–54.

Sealy, S. G., D. L. Neudorf, and D. P. Hill. 1995. Rapid laying by Brown-headed Cowbirds *Molothrus ater* and other parasitic birds. Ibis 137:76–84.

Smith, J. N. M., and P. Arcese. 1994. Brown-headed Cowbirds and an island population of Song Sparrows: A 16-year study. Condor 94:916–934.

18. Morning Nest Arrivals in Cowbird Hosts: Their Role in Aggression, Cowbird Recognition, and Host Response to Parasitism

DIRK E. BURHANS

Abstract

Indigo Buntings (*Passerina cyanea*) nesting in old-field habitats in central Missouri are parasitized at least four times as often as Field Sparrows. I used model cowbirds placed near nests to test if host aggression explained this difference. Although both Field Sparrows and Indigo Buntings responded to Brown-headed Cowbird models with significantly more chips than to a Fox Sparrow (*Passerella iliaca*) control, only one female Field Sparrow and one Indigo Bunting pair attacked the cowbird model.

Because the utility of aggression and cowbird recognition may depend on a host's actually encountering laying cowbirds, I also observed arrival times of hosts and cowbirds at nests near dawn during egg laying. Laying female Indigo Buntings rarely encountered laying cowbirds; the mean arrival time for Brown-headed Cowbirds at host nests was 11.4 min before sunrise ($N = 8$), and female Indigo Buntings arrived 16.7 min after sunrise ($N = 6$). Laying Field Sparrows arrived at nests before or at about the same time as cowbirds (mean arrival 17.4 min before sunrise, $N = 16$). Most parasitized Field Sparrow nests were abandoned in 1992 and 1993, suggesting that detection of female cowbirds at the nest causes Field Sparrows to desert their nests. These results suggest that aggression rarely prevents cowbirds from laying in nests of these hosts, but that recognition of brood parasites during laying elicits host nest desertion by Field Sparrows. Additionally, nest desertion by hosts may lead to underestimates of parasitism frequencies, because deserted nests are less likely to be found. Nest arrival times may limit the efficacy of host defenses against cowbirds, especially if cowbirds must commute to breeding areas from distant communal roosts.

Introduction

Many host species recognize Brown-headed Cowbirds as a threat (Robertson and Norman 1976, 1977; Smith et al.

1984; Briskie and Sealy 1989; Burgham and Picman 1989; Hobson and Sealy 1989; Neudorf and Sealy 1992). Although how aggression affects parasitism rates has not been measured, it may contribute to varying parasitism frequencies among species. For example, Briskie et al. (1990) found that Yellow Warblers (*Dendroica petechia*) were parasitized six times more frequently than Least Flycatchers (*Empidonax minimus*) nesting in the same habitat, although the Yellow Warblers were poorer hosts because they often bury cowbird eggs (Clark and Robertson 1981). The authors suggested that more aggressive nest defense by Least Flycatchers was one reason why they were infrequently parasitized.

Robertson and Norman (1976) hypothesized that aggression is selected for in frequently parasitized species and that it is proportional to the frequency of parasitism on that host, especially for birds that do not regularly eject cowbird eggs (see also Neudorf and Sealy 1992). Robertson and Norman's hypothesis was supported by a direct relationship between aggression and frequencies of parasitism within taxonomic groups. Thus, although Brown-headed Cowbirds are generalist brood parasites, parasitism frequently differs among suitable hosts nesting in the same habitat (Friedmann 1929, Elliott 1977, Southern and Southern 1980, Briskie et al. 1990, Ortega and Cruz 1991), and these differences may be related to levels of aggression in the host.

However, the way in which aggression functions, its value to the host, and the period when it is most effective are not clear. Hosts acting aggressively toward cowbirds during the nest-building stage may warn cowbirds that they are vigilant nest defenders or, perhaps more likely, advertise their susceptibility to parasitism, as stated by the nesting-cue hypothesis (Robertson and Norman 1977). Host aggression can thus tip off cowbirds to the location of a host's nest, possibly resulting in higher frequencies of parasitism for that host species (Robertson and Norman 1976, 1977; Smith et al. 1984). Alternatively, no aggression may be the best response for susceptible hosts (Neudorf and Sealy 1992). McLean (1987) found that communally breeding

Whitehead (*Mohoua albicilla*) females responded to models of parasitic Long-tailed Cuckoos (*Eudynamys taitensis*) by remaining inconspicuous and hiding during the incubation period. Secretive behavior may be the best way to avoid parasitism for hosts that recognize brood parasites (McLean 1987, McLean and Rhodes 1991).

Because cowbirds usually lay during the host's laying period (Friedmann 1929), aggression after the laying period usually occurs too late to benefit hosts, even though many hosts still respond aggressively to cowbird models at this stage (Hobson and Sealy 1989, Neudorf and Sealy 1992, Bazin and Sealy 1993). If aggressive defense is to be effective, it may have to function best during the laying period, so that it repels female cowbirds that are trying to lay. Because female cowbirds lay before dawn (Scott 1991), Neudorf and Sealy (1994) recorded dawn attentiveness of 10 cowbird hosts, as well as cowbird laying times. They predicted that those species that accept cowbird eggs (acceptors) would be more vigilant than species that reject cowbird eggs (rejectors) and that rarely parasitized acceptors would be more vigilant than frequently parasitized acceptors. While they did not find the above relationships, they found that hosts that roost on nests overnight were in the best position to thwart laying cowbirds (Neudorf and Sealy 1994).

Field Sparrows and Indigo Buntings nest in the same old-field habitats in central Missouri, and their nests are often placed in the same species of plant at similar heights. Although parasitism frequencies on the two species vary depending on the study or region concerned (see Friedmann 1963 and references therein), Indigo Buntings at Missouri sites are parasitized more than four times as frequently as Field Sparrows. To determine how aggression might influence this difference, I placed model cowbirds at nests of both species and recorded host responses. I also observed nests of both species during their egg-laying periods, because this is the time during which cowbirds are most likely to lay in host nests (Friedmann 1929). If hosts do not respond to cowbird eggs and other defenses are employed, the host must be at the nest early enough to encounter laying cowbirds (Neudorf and Sealy 1994). Thus, if the aggressive defense hypothesis (Robertson and Norman 1977) applies, Indigo Buntings would either not be at their nests when cowbirds arrived to lay or would mount a less effective defense against cowbirds than Field Sparrows, based on their response to a cowbird model.

Methods

Aggression Experiments

Nests were found on old-field sites on and adjoining the Thomas S. Baskett Wildlife Research and Education Center in Boone County, Missouri (38° 45′ N, 92° 12′ W), from April through June in 1992 and 1993. Five old fields ranging in size from 2.8 ha to 16.3 ha were searched daily for nests. Once found, each nest was individually numbered and flagged from at least 5 m distance. Nests initiated prior to and including the day the last cowbird egg was laid were used for estimating parasitism frequencies.

Although testing aggression toward cowbirds during the host's laying period is the best way to simulate the natural phenology of cowbird egg laying, Field Sparrows and Indigo Buntings rarely visit their nests during this period except while actually laying eggs (Payne 1990, Carey et al. 1994, Burhans pers. obs.). Thus, aggression experiments were performed on nests of both species during the incubation period. Although host aggression may vary through the nesting cycle (Hobson and Sealy 1989, Bazin and Sealy 1993), I had no reason to think that there would be a period-dependent difference in host aggression between the host species. While most hosts were not color banded prior to testing, experiments were carried out during a short period (Field Sparrows, 3 May to 15 May; Indigo Buntings, 23 May to 7 June). During these periods, nesting activity overlapped temporally, so no birds were tested twice.

Camouflaged blinds were set 10 to 20 m from each nest at least half an hour before presentations. An upright freeze-dried female cowbird model was placed within 0.5 m of each nest so that it faced into the nest cup. Because nests were often hard to see from the blind, models were mounted on a camouflaged telescoping brass rod so that height of the model was kept at 1 m above each nest rim. To determine if host response was directed to cowbirds or any bird at the nest, a Fox Sparrow (*Passerella iliaca*) model was presented at each nest as a control. Fox Sparrows are about the same size as cowbirds, but are not known to prey on nests of these hosts (Hobson and Sealy 1989, Bazin and Sealy 1993). Both models were presented at each nest for 5 min. Order of presentation was random, and the second model was not presented until 15 min after the returning hosts had stopped chipping in response to the first model. Only females incubate in both species (Carey et al. 1994, Payne 1990), and they were sometimes flushed off nests during model placement, but all birds directed their attention to models during experiments and not to the blind. In all cases hosts stopped chipping soon after the first model was removed, and in almost every instance females returned to sit on the nest; a few left the immediate nest area to feed nearby.

The following responses to models were spoken into a tape recorder for each 5-min period: (1) number of chip calls (Payne 1990, Carey et al. 1994), (2) number of "eeee" calls (Payne 1990, Carey et al. 1994), (3) number of swoops or close passes at the model, and (4) number of hits (i.e., the model was contacted by the flying host). Because the effectiveness of aggression in deterring parasitism may be related to whether both parents respond to the cowbird, the actions and vocalizations of both parents were recorded. I could not always record movement data for individual Field Sparrows, as most were not color banded. Indigo Buntings are sexually dimorphic, so I could distinguish responses by male and female parents.

Data were analyzed with nonparametric statistical tests (Siegel and Castellan 1988). Within-species comparisons between responses to Fox Sparrow and Brown-headed Cowbird models were analyzed with Wilcoxon signed-rank tests. Wilcoxon rank sum tests were used to test for species differences in response to Brown-headed Cowbird models; the number of parents responding (one or both) was tested with Fisher's exact test. I also compared chipping responses between parasitized and unparasitized Indigo Buntings with Wilcoxon rank sum tests. None of the Field Sparrow nests tested were parasitized.

Morning Nest Arrivals

Blinds were placed for observations 10 to 20 m from nests on the afternoon of the day the host laid its first egg, at which time the nest contents were examined for presence of cowbird eggs. On the following day, observers entered the blind at least 30 min before scheduled sunrise (SR - 30 min). Because of the possibility of disturbing roosting females, nests were not inspected at this time. As other studies have found that cowbirds lay before dawn (Scott 1991, Neudorf and Sealy 1994), I assume that cowbird eggs found after the observation period were laid by cowbirds that we observed and were not present prior to the observation period.

Observers spent at least one hour in the blind, recording any arrival times of hosts and female cowbirds. Upon arriving, female hosts and cowbirds sometimes perched on vegetation near the nest before flying directly to it. Where possible, arrival time was based on the time that birds were seen to fly directly to the nest, rather than arrival at the nest area. In most instances, cowbirds flew to vegetation several meters from the nest and perched for several minutes before flying to nests. If nests were obscured by vegetation, the time at which the female cowbird was seen landing on the nest plant was recorded as the arrival time.

After the observation period, nests were inspected for the presence of host and cowbird eggs. Local sunrise times were obtained from the National Weather Service office at Columbia, Missouri, and are accurate to the nearest minute; host and cowbird arrival times were also rounded to the nearest minute. As with aggression experiments, each host was tested only once, based on either color-banding information or overlap of nest phenology. Cowbirds were not color marked, so it is not known if all cowbird eggs were laid by different individuals.

Times of arrival of hosts and cowbirds in relation to sunrise were analyzed using a Kruskal-Wallis test, and differences between each species were compared using multiple comparisons analysis (Neter et al. 1990).

Results

Parasitism Frequencies

Backdating from hatching date using 11 days as the estimated incubation period for cowbirds (Lowther 1993, Burhans pers. obs.) revealed that in 1992 the last nest was parasitized by a cowbird on July 7. The sample of all nests for 1992 includes nests of Field Sparrows and Indigo Buntings that were initiated prior to this date. The last cowbird egg found in 1993 was laid on the morning of July 11, so the sample of all nests includes nests initiated prior to this date for 1993. Samples for the "all nests" category from both years include nests of Field Sparrow pairs whose nests were found soon after abandonment.

Based on nests found during building as the most conservative measure of parasitism frequency (see Discussion), Indigo Buntings were parasitized 4.4 times more frequently than Field Sparrows for the two years combined ($X^2 = 29.0$, df = 1, $P < .001$; Table 18.1).

Aggression Experiments

Experiments were performed on 14 Field Sparrow nests and 13 Indigo Bunting nests during the incubation period. Chipping frequencies for both hosts (Table 18.2) toward the cowbird model were significantly higher than toward the Fox Sparrow (Wilcoxon signed rank test: Field Sparrows, $Z = -3.23$, $P = .001$; Indigo Buntings, $Z = -2.90$, $P = .004$). Only 1 of 14 Field Sparrow responses involved an attack resulting in "eeee" calls, swoops, or hits on the cowbird model; this attack was by a female that had been incubating. No Field Sparrows attacked the Fox Sparrow model. Only 1 of 13 Indigo Bunting tests elicited an attack on the cowbird model (Table 18.2); in this case both the male and female re-

Table 18.1. Frequency (%) of Parasitized Nests by Stage (Parasitized/Total Nests Found in Stage)

Year	Stage Found	Field Sparrow	Indigo Bunting
1992	Building	28 (5/18)	83 (5/6)
	Before last egg	22 (5/23)	67 (6/9)
	All nests	11 (7/62)	65 (24/37)
1993	Building	13 (3/24)	92 (11/13)
	Before last egg	12 (5/42)	80 (20/25)
	All nests	15 (13/86)	54 (36/67)

Note: Includes only nests found during seasonal period of cowbird activity. "All nests" includes five Field Sparrow nests of known active pairs whose nests were found after abandonment.

sponded with "eeee" calls, swoops, and hits. No Indigo Buntings attacked the Fox Sparrow control.

Field Sparrow and Indigo Bunting responses to cowbird models did not differ statistically. In 11 of 14 Field Sparrow nests both parents responded to the cowbird model, compared to 9 of 13 bunting nests with both parents responding (Fisher's exact test, $P = .29$). Field Sparrows and Indigo Buntings responded to the cowbird model with an equivalent number of chips per 5-min period (Table 18.2; Wilcoxon rank sum test, $Z = -0.58$, $P = .56$). As stated above, only 1 of 14 Field Sparrow nests involved attacks on the model, compared to 1 of 13 Indigo Bunting nests (Fisher's exact test, $P = .52$). Parasitized and unparasitized Indigo Buntings did not differ in their chipping response to the cowbird model, although sample sizes were small (parasitized Indigo Buntings 244.20 ± 68.45 chips/5 min, $N = 5$; unparasitized buntings 273.38 ± 51.36 chips/5 min, $N = 8$; $Z = -0.29$, $P = .77$).

Morning Nest Arrivals

Field Sparrows arrived at nests to lay their second eggs on average 17.4 ± 2.3 (SE) min before sunrise ($N = 16$). The earliest arrival during the observation period was 25 min before sunrise (18 May, 0529 CDT), and the latest was 10 min after sunrise (15 May, 0606 CDT). Seven birds flew directly to nests, and 8 of 16 landed on vegetation near nests before flying to them. The latter birds perched an average of 35.25 ± 7.4 sec before landing at nests ($N = 8$, range 12–62 sec). A Field Sparrow that was on the nest before the observation may have roosted there overnight and was removed from the sample, as was a Field Sparrow that arrived to lay after the observation period was over. Field Sparrow laying bouts, defined as the interval from the time a female lands on the nest, deposits an egg, and departs (Sealy et al. 1995), averaged 42.1 ± 3.2 min ($N = 14$, range 20–60 min). The latter sample does not include one bird that was still on the nest at the end of the observation period and one that was chased off by a cowbird (see below).

Of 10 Indigo Bunting nests observed, 4 were abandoned, apparently before the observation took place; 6 other females arrived during the observation period. Average arrival time for Indigo Buntings was 16.7 ± 5.2 min after sunrise ($N = 6$, range SR - 1 min to SR + 30 min). Two birds flew directly to their nests, whereas 4 of 6 perched in the nest vicinity for an average of 223.50 ± 87.7 sec before flying to the nest ($N = 4$, range 15–431 sec). Abandonment of the other bunting nests may have been due to our setting blinds near them early in the nesting period, as Indigo Buntings rarely abandon nests due to cowbird parasitism unless the nest is parasitized before the first egg is laid (Payne 1990, Burhans pers. obs.). As far as I know, our blinds did not affect arrival times of buntings that did not desert nests, with the possible exception of a bird that appeared to chip at the blind. This female remained in the nest area for 431 sec before flying to the nest. As Indigo Buntings arrived comparatively late, we did not always record the lengths of laying bouts; most birds were still on the nest by the end of the observation period.

Table 18.2. Summary of mean ($\pm$ SE) aggressive responses of Field Sparrows ($N = 14$ nests) and Indigo Buntings ($N = 13$ nests) to Brown-headed Cowbird and Fox Sparrow Models over 5-min Periods

Response	Model	Field Sparrow	Indigo Bunting
Nests with both parents responding	Cowbird	11/14 nests	9/13 nests
	Fox Sparrow	10/13 nests	4/13 nests
Chips	Cowbird	220.4 ± 32.7	262.2 ± 39.6
	Fox Sparrow	56.0 ± 15.0	151.0 ± 20.4
Eeee call	Cowbird	2.2 ± 2.2	0.5 ± 0.5
	Fox Sparrow	0	0
Swoops	Cowbird	0.6 ± 0.6	0.9 ± 0.9
	Fox Sparrow	0	0
Hits	Cowbird	0.3 ± 0.3	0.7 ± 0.7
	Fox Sparrow	0	0

One female bunting observed leaving the nest had been on the nest for 33 min; other female buntings had been on nests for an average of 23.40 ± 5.8 min ($N = 5$, range 3–38 min) when observers left the blinds.

Female cowbirds laid during eight nest observations; seven times at Indigo Bunting nests and once at a Field Sparrow nest. The average nest arrival time of female cowbirds was 11.4 ± 1.8 min before sunrise ($N = 8$, range SR - 20 min to SR - 5 min). Two abandoned Indigo Bunting nests were parasitized, and one active Indigo Bunting nest was parasitized twice in one morning before the female bunting arrived. Buntings that were parasitized never encountered cowbirds. Mean difference between departure time of cowbirds and arrival time of female buntings at the same nest was 34 ± 5.1 min ($N = 5$, range 20–48 min). Seven of eight female cowbirds perched in trees or shrubs near nests for some time before flying to them. The average time cowbirds spent perching near nests within view of the blinds was 190.50 ± 57.6 sec ($N = 6$, range 48–426 sec), not including one cowbird that encountered a Field Sparrow at its nest (see below). Two cowbirds flew to the host plant or near to it after perching nearby, but then moved through the immediate vegetation for some time until they apparently located nests. We could not tell exactly when the latter birds landed on host nests. Thus a conservative estimate for the average cowbird laying bout is 97.5 ± 32.6 sec ($N = 8$, range 16–275 sec).

Differences in nest arrivals of Field Sparrows, Brownheaded Cowbirds and Indigo Buntings were significant (Kruskal-Wallis test, $\chi^2 = 16.86$, df = 2, $P = .000$). Field Sparrows tended to arrive at nests earlier than cowbirds (.15 < P < .20, multiple comparison procedure for Kruskal-Wallis test), and Indigo Buntings tended to arrive at nests later than cowbirds (.10 < P < .15).

The Field Sparrow whose nest was parasitized during the observation (D. Martasian pers. obs.) arrived to lay 21 min before sunrise (arrival time 05:22:12 CDT). At 05:28:30, a female cowbird arrived about 10 m from her nest and was approached by a chipping Field Sparrow, presumably the mate of the laying female. The cowbird was briefly out of sight, then flew to the ground within 1 m of the nest while the male Field Sparrow continued to chip. At 05:34:55, the female cowbird chased the female Field Sparrow off the nest and flew after her for 10 m. The female Field Sparrow was on the nest for 12 min and 7 sec before being chased off. The female cowbird landed on the nest at 05:35:50 and flew off the nest 16 sec later. The Field Sparrows flew about the nest vicinity for the next 2 min but then left the nest area and were not seen again during the observation. The nest contained one cowbird egg and one Field Sparrow egg after the observation, so the Field Sparrow apparently did not get a chance to lay her second egg in the nest. The nest was abandoned, and no Field Sparrows were seen at or near it during the next several days.

Discussion

Both Field Sparrows and Indigo Buntings clearly recognized cowbirds as threats, because of differences in chipping frequencies toward the cowbird and Fox Sparrow models (Table 18.2). If these hosts respond to live cowbirds as they do to models, neither host response could be expected to deter laying cowbirds. Actual attacks by both host species were rare. Other studies using cowbird models have documented low aggression levels for some cowbird hosts (Robertson and Norman 1976, Smith et al. 1984, Hobson and Sealy 1989, Neudorf and Sealy 1992). Other hosts, however, may attack cowbird models vigorously (Robertson and Norman 1976, Folkers and Lowther 1985, Briskie and Sealy 1989, Neudorf and Sealy 1992, Bazin and Sealy 1993).

Observations on nests of larger hosts such as Gray Catbirds (*Dumetella carolinensis*) and American Robins (*Turdus migratorius*) indicate that these species use aggression to repel live cowbirds from nests (Friedmann 1929, Scott 1977). In one case, a robin injured the cowbird (Leathers 1956). Smaller hosts may have more difficulty; Clay-colored Sparrows (*Spizella pallida*) drove away a cowbird, but the cowbird immediately returned to parasitize the nest successfully (Neudorf and Sealy 1994; see also Hill and Sealy 1994). During three observations at Yellow Warbler nests, host females that were on the nests attacked cowbirds when the cowbird arrived. Parasitism occurred at two nests, followed by acceptance of cowbird eggs by the hosts. One attempt may have been thwarted (Sealy et al., Chapter 19, this volume). S. I. Rothstein and A. O'Loghlen (unpubl. data) found that cowbirds attracted by playbacks to sites defended by Western Wood-Pewees (*Contopus sordidulus*) did not retreat despite repeated attacks. When attracted to areas defended by larger species such as American Robins, Redwinged Blackbirds, or Brewer's Blackbirds (*Euphagus cyanocephalus*), cowbirds always retreated. Thus, larger hosts can mount more effective nest defenses than smaller hosts.

Differences in aggression between Indigo Buntings and Field Sparrows do not seem to account for differences in parasitism frequency, because their chipping rates and frequency of attacks were so similar. Field Sparrows at these sites probably encounter laying cowbirds most of the time. If aggression directed toward models is similar to that directed toward live cowbirds, they are probably not able to repel laying cowbirds. Additional nest observations of cowbird interactions with these hosts, collected since the present study, support this interpretation (Burhans unpubl. data). Indigo Buntings probably rarely encounter laying cowbirds at their nests, and the rarity of attacks in my experiments suggests that they would do little to deter them if they did. Neither host at the Field Sparrow nest that was visited by a laying cowbird attempted to drive her away, although the male did approach her when she initially flew to the nest area.

Field Sparrows have been observed chasing female cowbirds during nest building (Crooks 1948, D. Dearborn pers. obs.). This is thought to be the time when cowbirds cue on host nests (Friedmann 1929, Hann 1941, Norman and Robertson 1975). Aggression in this situation can be counterproductive, because cowbirds may use such behavior to locate host nests, as predicted by the nesting-cue hypothesis (Robertson and Norman 1977, S. A. Gill et al. unpubl. data). Walkinshaw (1968) noted that Field Sparrows dropped nest material and began to feed in the presence of cowbirds. Although we never observed such behavior during many observations of Field Sparrow nest building (Carey et al. 1994), inconspicuous behavior during nest building should be advantageous to hosts, if it reduces the likelihood of detection by cowbirds.

Neudorf and Sealy (1994) defined a critical period of 30 min around sunrise during which parasitism was most likely to occur, based on acts of cowbird parasitism observed in Manitoba. They found no relationship between presence of hosts near their nests and frequency of parasitism for the ten acceptor species they observed, nor did acceptors spend more time guarding their nests during this period than rejectors. However, females that roosted overnight on nests were more likely to be on their nests during the critical period for cowbird parasitism, and females that did not roost overnight were not likely to be near the nest area at all (Neudorf and Sealy 1994). Field Sparrows rarely roost overnight on nests and may be unusual in arriving to lay before cowbirds. In observations at nests of Clay-colored Sparrows, congeners of Field Sparrows, five out of seven different females arrived at the nest during the critical period for parasitism. This includes a female that roosted overnight, was accidentally flushed from the nest, and returned to encounter a cowbird (S. G. Sealy pers. comm.).

Although cowbirds perched briefly in trees, shrubs, or fence lines before flying to nests, their manner of flight from the perch suggests that cowbirds had located nests in advance, as also suggested by Neudorf and Sealy (1994). Perching nearby in shrubs may allow female cowbirds to see if the host is on the nest, giving the cowbird the option of parasitizing the nest later if the host is present. This provides another possible explanation for differences in parasitism between Field Sparrows and Indigo Buntings. Female cowbirds may observe Field Sparrows flying toward or sitting on nests and decide not to parasitize these nests at the risk of eliciting desertion by the host (J. N. M. Smith pers. comm.). Placing Indigo Bunting models on nests before female buntings arrive and comparing parasitism frequencies on these nests versus unmanipulated nests could test this hypothesis.

As in Neudorf and Sealy's (1994) observations, cowbirds did not vocalize during flight near nests. Although mean time for cowbird laying bouts in my study was greater than in Neudorf and Sealy's study (63 ± 12.3 sec vs. 97.5 ± 32.6 sec in this study), our difficulty in detecting exactly when cowbirds landed on some nests may account for this difference. Even the two female cowbirds that moved about the nest plant before laying flew directly to the nest plant or its immediate vicinity. Silent, direct flight and rapid laying all suggest that cowbirds try to remain undetected when they parasitize nests, either because hosts sometimes thwart parasitism attempts (Neudorf and Sealy 1994, Sealy et al. 1995) or to avoid nest abandonment by hosts (see below).

The time of nest arrival around sunrise has other consequences besides aggression for host responses to parasitism. Field Sparrows regularly abandon parasitized nests, at least throughout the midwestern portions of their range. At Missouri sites, Field Sparrows deserted 6 of 8 (75%) parasitized nests found during nest building in 1992–1993 and deserted 15 out of all 20 (75%) parasitized nests found. Similar desertion frequencies have been found in Michigan and Illinois (Walkinshaw 1968, Best 1978).

Nest desertion has been widely reported among cowbird hosts, particularly in small birds (Friedmann 1963, Clark and Robertson 1981, Graham 1988, Sedgwick and Knopf 1988), which presumably have difficulty in grasp-ejecting cowbird eggs. Puncture ejection, however, has recently been documented in a small (15 g) host (Sealy 1996). However, with few exceptions (Rothstein 1976, Burgham and Picman 1989), nest desertion is not induced in small hosts by experimental addition of real or artificial cowbird eggs (Rothstein 1975, Hill and Sealy 1994, Sealy 1995), while grasp-ejection is easily documented experimentally in grasp-ejecting hosts (Rothstein 1975).

Further experiments with Field Sparrows have indicated that they do not desert simply in response to addition of artificial or real cowbird eggs (Burhans unpubl. data). It may be that detection of laying parasites at the nest is the cue required by Field Sparrows to induce nest desertion. Additional observations of Field Sparrow–cowbird interactions at the nest also support this possibility (Burhans unpubl. data).

Methodological and Conservation Implications

Nest desertion may serve as an index of the vulnerability of host species to parasitism. Many studies present parasitism frequencies that include nests found during all stages of the nesting cycle. However, if hosts are prone to deserting parasitized nests, such a sample likely underestimates the true parasitism frequency, because deserted nests that lack host activity are usually hard for field workers to find. In my study, this held true for Field Sparrow nests (Table 18.1). Of nests found during nest building, 19.0% (8/42) were later parasitized. The number decreases to 15.4% (10/65) for all nests found prior to incubation and decreases further to 13.5% (20/148) for the total sample of nests, even including 5 parasitized nests in the last group that were abandoned before they were found. Such a trend may prevail for reasons other than nest abandonment, including egg rejection, partial clutch loss due to predation, or brood reduction (Briskie et al. 1990, Hill and Sealy 1994, Pease and Grzybowski 1995). While a tendency for parasitism frequencies to decline over the sampling period was also observed for Indigo Buntings, which do not usually abandon parasitized nests (Table 18.1), nest-deserting species are likely to exaggerate the trend (Sedgwick and Knopf 1988). For these reasons, researchers may prefer to use the cohort parasitism fraction, or percentage of a cohort of nests started, to measure parasitism frequencies more conservatively (Pease and Grzybowski 1995; Grzybowski and Pease, Chapter 16, this volume).

In addition, vulnerability to parasitism may be affected by the interaction of nest arrival time with either nest defenses or nest desertion behavior. Defense of the nest is probably most useful if hosts are at or near the nest when cowbirds arrive to lay (Neudorf and Sealy 1994). Similarly, if nest desertion is a viable defense and detection of the cowbird stimulates nest desertion, species that arrive later to lay may be more vulnerable to parasitism. Although data are sparse, there may be considerable regional variability in cowbird nest arrival times. Scott's (1991) review documented 20 direct observations of cowbird laying, with mean time of laying at 11.00 ± 4.05 min before sunrise, not including one record of a cowbird that arrived 3 hr after sunrise; Scott's nest arrival times agree closely with those of this study (11.4 ± 1.8 min). However, mean arrival time for cowbirds at Delta Marsh, Manitoba, was 35.6 ± 2.4 min before sunrise (range 44–25 min before sunrise, $N = 7$, Neudorf and Sealy 1994). Earlier arrival times could be partly due to earlier onset of civil twilight in relation to sunrise in northern latitudes. Onset of civil twilight in June occurs between 43 and 45 min before sunrise in Manitoba, compared to 31 and 32 min in Missouri (United States Naval Observatory 1946). However, this difference does not account for all of the disparity in cowbird arrival times between Manitoba and Missouri. Earlier arrival times may also be the result of cowbirds roosting closer to host nest sites at Delta Marsh and not having to spend as much time traveling to nests (S. G. Sealy pers. comm.).

Recent radiotelemetry studies of cowbirds in the Midwest indicate that most cowbirds do not roost where they breed, but either have distinct roosting areas or roost near feeding areas (Thompson 1994; Thompson and Dijak, Chapter 10, this volume). At one site, a communal roost of more than 200 Brown-headed Cowbirds was discovered. Hosts near such large roosts may be parasitized earlier in the day and may arrive to lay their own eggs after cowbirds have already visited the nest. For hosts that do not regularly roost overnight on the nest during laying, proximity of the nest to cowbird roosts may thus be crucial in limiting the effectiveness of their defenses, regardless of whether the defense is aggression, protection behavior, or desertion.

Further study of the relationship between cowbird roost distance and cowbird arrival times at nests may help to understand host defenses. If nest defense and nest desertion by hosts do sometimes deter cowbirds, managers may want to consider cowbird and host arrival times, as well as proximity of cowbird roosts, when assessing vulnerability of local host populations to parasitism. Local host populations that are close to cowbird roosts may benefit from roost disruption and cowbird control to a greater degree than populations far from roosts, because hosts nesting farther from cowbird roosts may have a better chance of detecting cowbirds at dawn and using their own defenses.

Acknowledgments

The USDA Forest Service North Central Forest Experiment Station through the assistance of F. Thompson III funded this research. J. Faaborg, F. Thompson, T. Donovan, C. Galen, M. Ryan, C. Gerhardt, and D. Dearborn gave helpful comments on early drafts, and S. Sealy gave invaluable advice on many aspects of the study. D. Dearborn, C. Freeman, and D. Martasian worked very early in the morning gathering field observations. T. Hayden, J. Cornelius, and D. Willard loaned specimens for the models, and N. Unklesbay generously loaned the use of freeze-drying equipment. E. Wiggers made us welcome at the refuge, and C. Freiling allowed us to work on his property. I thank S. Sealy and S. Rothstein for helpful reviews of an earlier draft.

References Cited

Bazin, R. C., and S. G. Sealy. 1993. Experiments on the responses of a rejector species to threats of predation and cowbird parasitism. Ethology 94:326–338.

Best, L. B. 1978. Field Sparrow reproductive success and nesting ecology. Auk 95:9–22.

Briskie, J. V., and S. G. Sealy. 1989. Changes in nest defense against a brood parasite over the breeding cycle. Ethology 82:61–67.

Briskie, J. V., S. G. Sealy, and K. A. Hobson. 1990. Differential parasitism of Least Flycatchers and Yellow Warblers by the Brown-headed Cowbird. Behav. Ecol. Sociobiol. 27:403–410.

Burgham, C. J., and J. Picman. 1989. Effect of Brown-headed Cowbirds on the evolution of Yellow Warbler anti-parasite strategies. Anim. Behav. 38:298–308.

Carey, M., D. E. Burhans, and D. A. Nelson. 1994. Field Sparrow (Spizella pusilla). In The birds of North America 103 (A. Poole and F. Gill, eds). Academy of Natural Sciences, Philadelphia; American Ornithologists' Union, Washington, DC.

Clark, K. L., and R. J. Robertson. 1981. Cowbird parasitism and evolution of anti-parasite strategies in the Yellow Warbler. Wilson Bull. 93:249–258.

Crooks, M. P. 1948. Life history of the Field Sparrow Spizella pusilla pusilla (Wilson). M.S. Thesis, Iowa State College, Cedar Falls.

Elliott, P. F. 1977. Adaptive significance of cowbird egg distribution. Auk 94:590–593.

Folkers, K. L., and P. E. Lowther. 1985. Responses of nesting Red-winged Blackbirds and Yellow Warblers to Brown-headed Cowbirds. J. Field Ornithol. 56:175–177.

Friedmann, H. 1929. The cowbirds: A study in the biology of social parasitism. C. C. Thomas, Springfield, IL.

———. 1963. Host relations of the parasitic cowbirds. US Natl. Mus. Bull. 233:1–276.

Graham, D. S. 1988. Responses of five host species to cowbird parasitism. Condor 90:588–591.

Hann, H. W. 1941. The cowbird at the nest. Wilson Bull. 53: 211–221.

Hill, D. P., and S. G. Sealy. 1994. Desertion of nests parasitized by cowbirds: Have Clay-colored Sparrows evolved an anti-parasite defense? Anim Behav. 48:1063–1070.

Hobson, K. A., and S. G. Sealy. 1989. Responses of Yellow Warblers to the threat of cowbird parasitism. Anim. Behav. 38:510–519.

Leathers, C. L. 1956. Incubating American Robin repels female Brown-headed Cowbird. Wilson Bull. 68:68.

Lowther, P. 1993. Brown-headed Cowbird (*Molothrus ater*). *In* The birds of North America 47 (A. Poole and F. Gill, eds). Academy of Natural Sciences, Philadelphia; American Ornithologists' Union, Washington, DC.

McLean, I. 1987. Response to a dangerous enemy: Should a brood parasite be mobbed? Ethology 75:235–245.

McLean, I. G., and G. Rhodes. 1991. Enemy recognition and response in birds. Curr. Ornithol. 8:173–211.

Neter, J., W. Wasserman, and M. H. Kutner. 1990. Applied linear statistical models. Richard D. Irwin, Homewood, IL.

Neudorf, D. L., and S. G. Sealy. 1992. Reactions of four passerine species to threats of predation and cowbird parasitism: Enemy recognition or generalized response? Behaviour 123:84–105.

———. 1994. Sunrise nest attentiveness in cowbird hosts. Condor 96:162–169.

Norman, R. F., and R. J. Robertson. 1975. Nest-searching behavior in the Brown-headed Cowbird. Auk 92:610–611.

Ortega, C. P., and A. Cruz. 1991. A comparative study of cowbird parasitism in Yellow-headed Blackbirds and Red-winged Blackbirds. Auk 108:16–24.

Payne, R. B. 1990. Indigo Bunting (*Passerina cyanea*). *In* The birds of North America 4 (A. Poole, P. Stettenheim, and F. Gill, eds). Academy of Natural Sciences, Philadelphia; American Ornithologists' Union, Washington, DC.

Pease, C. M., and J. A. Grzybowski. 1995. Assessing the consequences of brood parasitism on seasonal fecundity in passerine birds. Auk 112:343–363.

Robertson, R. J., and R. F. Norman. 1976. Behavioral defenses to brood parasitism by potential hosts of the Brown-headed Cowbird. Condor 78:166–173.

———. 1977. The function and evolution of aggressive host behavior towards the Brown-headed Cowbird (*Molothrus ater*). Can. J. Zool. 55:508–515.

Rothstein, S. I. 1975. An experimental and teleonomic investigation of avian brood parasitism. Condor 77:250–271.

———. 1976. Experiments on defenses Cedar Waxwings use against cowbird parasitism. Auk 93:675–691.

Scott, D. M. 1977. Cowbird parasitism on the Gray Catbird at London, Ontario. Auk 94:18–27.

———. 1991. The time of day of egg laying by the Brown-headed Cowbird and other icterines. Can. J. Zool. 69:2093–2099.

Sealy, S. G. 1995. Burial of eggs by parasitized Yellow Warblers: An empirical and experimental study. Anim. Behav. 49:877–889.

———. 1996. Evolution of host defenses against brood parasitism: Implications of puncture-ejection by a small passerine. Auk 113:346–355.

Sealy, S. G., D. L. Neudorf, and D. P. Hill. 1995. Rapid laying by Brown-headed Cowbirds *Molothrus ater* and other parasitic birds. Ibis 137:76–84.

Sedgwick, J. A., and F. L. Knopf. 1988. A high incidence of Brown-headed Cowbird parasitism of Willow Flycatchers. Condor 90:253–256.

Siegel, S., and N. J. Castellan, Jr. 1988. Nonparametric statistics for the behavioral sciences. McGraw-Hill, New York.

Smith, J. N. M., P. Arcese, and I. G. McLean. 1984. Age, experience, and enemy recognition by wild Song Sparrows. Behav. Ecol. Sociobiol. 14:101–106.

Southern, W. E., and L. K. Southern. 1980. A summary of the incidence of cowbird parasitism in northern Michigan from 1911–1978. Jack-Pine Warbler 58:77–84.

Thompson, F. R. III. 1994. Temporal and spatial patterns of breeding Brown-headed Cowbirds in the midwestern United States. Auk 111:979–990.

United States Naval Observatory. 1946. Tables of sunrise, sunset, and twilight: Supplement to the American ephemeris. U.S. Government Printing Office, Washington, DC.

Walkinshaw, L. H. 1968. Eastern Field Sparrow. *In* Life histories of North American cardinals, grosbeaks, buntings, towhees, finches, sparrows, and allies (O. L. Austin, ed). U.S. Natl. Mus. Bull. 237:1217–1235.

19. Yellow Warbler Nest Attentiveness before Sunrise: Antiparasite Strategy or Onset of Incubation?

SPENCER G. SEALY, D. GLEN MCMASTER,

SHARON A. GILL, AND DIANE L. NEUDORF

Abstract

We watched Yellow Warbler nests before sunrise from nest building to the first day of incubation. Our aim was to test if owners guarded their nests on the days of the nesting cycle and at the time of day that Brown-headed Cowbirds are most likely to parasitize nests. Previous studies had shown that Yellow Warblers respond to Brown-headed Cowbirds with unique behaviors that may thwart cowbird parasitism. To deter cowbirds, Yellow Warblers should attend their nests prior to sunrise, the time of day cowbirds lay their eggs, before clutch initiation and during early egg laying. We tested two competing hypotheses: (1) nest attentiveness prior to sunrise is an antiparasite adaptation that places a Yellow Warbler in a position to prevent cowbird parasitism, and (2) increasing nest attentiveness is a by-product of the physiological mechanisms that bring about incubation. Our results support the second hypothesis. Yellow Warblers increased their nest attentiveness daily over the laying period to a maximum as incubation began. This pattern of attentiveness is probably a manifestation of the onset of incubation and not an antiparasite strategy, although the behavior may occasionally allow warblers to thwart parasitism.

Introduction

To many host species, avian brood parasitism is costly because the hosts raise parasitic young at the expense of some or all of their own young. Not surprisingly, many hosts have sophisticated behaviors that terminate or prevent parasitism. The most obvious defense that hosts have against brood parasitism is ejection of the parasitic egg (Rothstein 1990). Most hosts of the generalist brood parasite the Brown-headed Cowbird (henceforth cowbird), however, accept parasitism, which is puzzling because cowbird eggs are distinguishable from the eggs of most host species (Rothstein 1975).

Yellow Warblers (*Dendroica petechia*) reject cowbird para-

sitism in a unique way, by burying the cowbird egg (and sometimes also eggs of their own) under a new nest lining. They then initiate a new clutch in the superimposed nest (Friedmann 1929, Clark and Robertson 1981, Burgham and Picman 1989, Sealy 1995). Although the buried eggs do not hatch, burial eliminates parasitism less consistently than ejection of the cowbird egg. When a species shows ejection, this behavior usually occurs at all or most parasitized nests (e.g., Rothstein 1975, Rohwer and Spaw 1988, Sealy and Bazin 1995). By contrast, Yellow Warblers tend to bury only those cowbird eggs laid before the midpoint of their laying cycle. At nests where burial occurs, there are additional time and energy costs due to the need to construct a new nest cup (Clark and Robertson 1981; Sealy 1992, 1995).

Nest defense directed specifically toward a brood parasite would minimize the likelihood of being parasitized in the first place and reduce the need for egg recognition (Neudorf and Sealy 1992, 1994). Indeed, nesting Yellow Warblers are defensive toward Brown-headed Cowbirds when they encounter them near their nests. Most individuals utter a unique alarm call, and the females generally rush to their nests, sit tightly in them, and stop vocalizing (Hobson and Sealy 1989, Gill and Sealy 1996). These behaviors apparently evolved in response to cowbird parasitism, because Yellow Warblers that nest where there are no cowbirds rarely perform these behaviors (Briskie et al. 1992). Implicitly, this behavior allows Yellow Warblers to thwart parasitism attempts, but its effectiveness in this regard is not known. Cowbirds have been recorded parasitizing between 20% and 80% of nests in some populations of Yellow Warblers (e.g., Schrantz 1943, Cannings et al. 1987, Clark and Robertson 1981, Sealy 1992, J. N. M. Smith pers. comm.).

Regardless of the sophistication of their nest-defense system, Yellow Warblers must be at or near their nests at the proper time to intercept a laying cowbird. Cowbirds lay their eggs within seconds (Sealy et al. 1995), a few minutes before sunrise (Scott 1991, Neudorf and Sealy 1994). Neudorf and Sealy (1994) found that most female Yellow Warblers that attended their nests before sunrise prior to laying their second eggs had roosted in the nest the previous night. Roosting females usually remained in the nest until just before sunrise, a few minutes after the cowbird's normal laying time.

We hypothesized that nest attentiveness prior to sunrise is an antiparasite adaptation that puts a potential host in a position to prevent cowbird parasitism. Based on ultimate causation, we predicted that females would be in or near their nests each morning throughout the expected time of arrival of laying cowbirds. At Delta Marsh, Manitoba, about 75% of the Yellow Warbler nests that are parasitized receive the cowbird egg before the hosts have laid their second eggs (Sealy 1992, 1995). If nest attentiveness prior to sunrise is an antiparasite strategy, we predicted that Yellow Warblers would attend their nests from the time they were lined and continue this behavior throughout the laying period. Females could safely leave their nests when the males are nearby, if a coordinated system of nest guarding exists, as described by Slack (1976) for Gray Catbirds (*Dumetella carolinensis*).

Alternatively, based solely on proximate considerations, increasing predawn nest attentiveness may be a by-product of the physiological mechanisms that bring about incubation. If they are influenced by hormones and the stimulation of the increasing clutch (Mead and Morton 1985, Book et al. 1991, Hébert and Sealy 1992), female Yellow Warblers should gradually increase their nest attentiveness, possibly by extending nighttime roosting, over the egg-laying period.

We tested these competing hypotheses by watching Yellow Warbler nests around dawn, when nests were most susceptible to parasitism (prelaying through the first day of incubation). We recorded the attentiveness of female Yellow Warblers in and near their nests before sunrise. When a female roosted overnight, we recorded when she first left the nest in the morning to determine if her departure was after the time cowbirds parasitize nests on the study area (see Neudorf and Sealy 1994). For a female that did not roost in the nest, we recorded whether she was near the nest when parasitism might occur, and when she first entered the nest in the morning. We also recorded the presence of male Yellow Warblers near their nests. If females roosted in their nests in response to varying temperatures, we assumed they made the decision to roost around sunset. We therefore recorded sunset temperatures to determine whether females roosted in response to lower temperatures.

Methods

We worked in 1992 and 1993 at Delta Marsh, Manitoba (50° 11' N, 98° 19' W), in the forested dune ridge that separates the marsh and Lake Manitoba (descriptions of study area and nesting habitat in MacKenzie 1982, MacKenzie et al. 1982). Beginning about 25 May each year, we located and

numbered Yellow Warbler nests placed less than about 2 m high in the shrub layer of the ridge forest. We are confident that nearly all nests available were found during construction and were watched regardless of their degree of concealment.

We inspected each nest daily until most nests were unlined cups, 1–2 days from receiving their first warbler eggs. On the evening before each nest watch, we set up a blind 5–10 m from the nest. We entered the blind the next morning and watched the nest around sunrise, from 0330 to 0430 (Central Standard Time), when we expected cowbirds to parasitize nests (see Scott 1991, Neudorf and Sealy 1994). We watched a sample of nests from 4 days before clutch initiation (e.g., 4 days before the first egg was laid = Day -4, etc.), on the morning the first Yellow Warbler egg was laid (Day 0), and until 1 day after clutches of 4 eggs (Day 4) or 5 eggs (Day 5) were laid, or until the nest failed. Upon our arrival at the nest we determined, sometimes with a flashlight, whether the female was in the nest. We did not flush any females from their nests. We assumed that females in their nests at this time had roosted there overnight (see Nolan 1978) and that none had left the nest before 0330. We considered warblers to be in a position to guard the nest if either the male or female was within 5 m of the nest (including time spent in the nest).

Using Scott's (1991) method, we related to sunrise (SR) the times cowbirds arrived at Yellow Warbler nests to parasitize them and the times roosting females left their nests or first arrived at them in the morning if they had not roosted in them. We determined the exact time of sunrise at Delta Marsh for the days we observed Yellow Warbler nests using the 1991 *Observer's Handbook of the Royal Astronomical Society of Canada*. Sunrise times for a given date exhibit little year-to-year variation in the Temperate Zone for a given latitude. Based on 11 acts of parasitism involving 4 host species at Delta Marsh (Neudorf and Sealy 1994; see Appendix), we defined a 30-min critical period for parasitism (hereafter "critical period") during which Yellow Warblers should be in or near their nests if they are to thwart cowbird parasitism. This 30-min period spanned 0343 to 0413 (SR - 42 min to SR - 12 min) and ended on average 12 min before sunrise (range SR - 19 min to SR - 9 min).

During each nest watch, we recorded seven variables: (1) whether the female had roosted in the nest the previous night, (2) the amount of time the female spent in the nest during the critical period for parasitism (0343 to 0413), (3) the amount of time the female spent within 5 m of the nest during the critical period, (4) the time of day relative to sunrise when the female first left the nest, if she had roosted, (5) the time of day relative to sunrise when the female first arrived at the nest, if she had not roosted, (6) the amount of time the male spent within 5 m of the nest during the critical period, and (7) the time of day relative to sunrise when the male first arrived at the nest.

The time a female Yellow Warbler spends in and near the nest (variables 2 and 3 above) is additive; that is, for every minute she is in the nest there is 1 minute less available to her to spend within 5 m of it. For example, if a female spends 20 min of the 30-min critical period in the nest, then 10 min are available to be spent within 5 m. Therefore, recording female attentiveness within 5 m of the nest relative to the amount of time available to spend near the nest (i.e., the amount of time *not* in the nest) is more appropriate than an absolute measure (variable 3). We calculated a relative measure of female attentiveness within 5 m of the nest using the equation:

$$NA = \frac{A}{30 - B} \times 100$$

where *NA* is the percentage of the critical period that the female spent out of the nest that was nevertheless spent within 5 m of it, *A* is the time spent by the female within 5 m of the nest during the critical period, 30 is the duration of the critical period in minutes, *B* is the time the female spent in the nest during the critical period.

We examined whether the eight variables above, including the time a female spends in and near her nest, differed through the laying and incubation stages. We analyzed females that roosted in their nest and those that roosted elsewhere separately to ascertain if nest-roosting females were in better position to defend their nests against parasitism during the critical period. We sorted the eight activity variables by the day of the nesting cycle and then divided the variables on the basis of whether the female had roosted in the nest the previous night. Because few females roosted on Days -4 to -1 and many roosted late in the period (i.e., Days 2 to 5), we compared the behavior of roosting and nonroosting warblers on Day 0 (*N* = 20 and *N* = 10, respectively) and Day 1 (*N* = 18 and *N* = 31, respectively).

We observed four acts of successful laying by cowbirds at four Yellow Warbler nests, and one visit to a nest by a female cowbird that may have been a thwarted parasitism attempt. At each of these five nests, we recorded the cowbird's behavior as she approached and left the nest and the warblers' responses to the cowbird. Our analyses do not include data for these parasitized nests, because parasitism may have influenced the warblers' behavior.

From the meteorological station at the University of Manitoba Field Station (Delta Marsh), we obtained ambient temperatures at sunset for the nest-watch periods of both years. Temperatures were not available for three dates, so we omitted those data points from the analyses.

We used Friedman's two-way ANOVA to examine whether the activity variables differed over the egg-laying cycle (Conover and Iman 1981). When significant differences (*P* < .05) occurred, we used Fisher's protected least signifi-

cant difference test to identify the days on which warbler behavior varied significantly. We used Wilcoxon two-sample tests to examine differences between roosting and nonroosting females on Days 0 and 1 (Conover and Iman 1981).

Results

Female Attentiveness during the Critical Period

We conducted 166 watches at 67 nests between 4 June and 8 July 1992 (23 nests) and 28 May and 30 June 1993 (44 nests). Female warblers initially spent little time in their nests during the critical period, but they spent significantly more time there as the nesting cycle progressed (Figure 19.1; Friedman's two-way ANOVA, *F* = 17.39, *P* = .000), especially once egg laying had begun. The amount of time females spent within 5 m of their nests was not influenced by the day of nesting cycle (*F* = 1.41, *P* = .20). If females did not sit in their nests during the critical period, they spent little time near them (Figure 19.2). However, the percentage of time the female did not spend in the nest but did spend within 5 m of it increased over the nesting cycle (Table 19.1). The increase approached significance (*F* = 1.75, *P* = .08).

More female Yellow Warblers roosted overnight as successive eggs in their clutches were laid (Figure 19.3; *F* = 12.21, *P* = .000). Few females roosted before Day 0, but most did so after Day 1 (Figure 19.3). The time of departure of roosting females (Table 19.1) did not vary significantly over the lay-

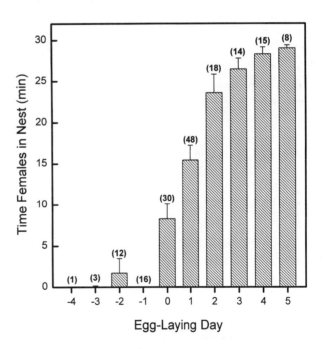

Figure 19.1. Mean (± SE) number of minutes spent by female Yellow Warblers in the nest during the 30-min critical period for parasitism. Day 0 is the day the first warbler egg was laid. Sample sizes (number of nests) in parentheses.

ing period ($F = 1.69$, $P = .15$). However, the time that non-roosting females arrived at the nest did vary significantly (Figure 19.4; $F = 2.68$, $P = .03$). Females that did not roost on Day -2 arrived at the nest significantly earlier than females that did not roost on Day 1, but significantly later than females that did not roost on Days -3 and 0.

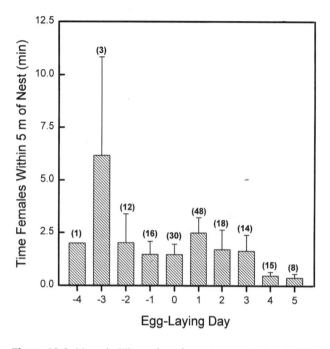

Figure 19.2. Mean ($\pm$ SE) number of minutes spent by female Yellow Warblers within 5 m of the nest during the critical period for parasitism. Sample sizes in parentheses.

Roosting versus Nonroosting Females

Females that roosted overnight in the nest spent more time in their nests during the critical period on Day 0 (Wilcoxon two-sample test, $Z = 4.321$, $P = .000$) and Day 1 ($Z = 5.42$, $P = .000$) than females that did not roost (Figure 19.5). However, the mean times that roosting and nonroosting females spent within 5 m of the nest during the critical period did not differ on Day 0 (1.0 ± 0.6 min for roosting versus 1.7 ± 0.7 min for nonroosting females, $Z = 0.754$, $P = .45$) or on Day 1 (2.2 ± 0.9 min versus 3.0 ± 1.4 min, $Z = 0.207$, $P = .84$).

A comparison of the time relative to sunrise that roosting females left the nest with the time that nonroosting females first arrived at the nest showed that on Days 0 and 1, roosting females left their nests on average 5–13 min before the end of the critical period. By contrast, nonroosting females arrived at the nest on average 0–6 min before the end of that period. Therefore, roosting females spent more time in the nest during the critical period than females that did not roost.

Male Attentiveness during the Critical Period

Males spent about 5–10% of their time near the nest during the critical period (Figure 19.6). The amount of time male warblers spent within 5 m of their nest did not vary significantly over the egg-laying period ($F = 1.8$, $P = .08$). Male attentiveness within 5 m of the nest did not vary significantly in relation to whether their female had roosted in the nest or not. (Day 0: 0.1 ± 0.1 min versus 0.2 ± 0.1 min, respec-

Table 19.1. Timing of Female and Male Yellow Warbler Behaviors at Nests over the Egg-laying Cycle

Day of Laying Cycle[a]	% Female's Time off Nest but within 5 m during Critical Period (N)	Mean Departure Time of Roosting Females Relative to Sunrise $\pm$ SE in Min (N)	Mean Arrival Time of Male Relative to Sunrise $\pm$ SE in Min (N)
−4	6.6 (1)	—	+23 (1)
−3	20.7 (3)	—	−34.0 $\pm$ 10 (3)
−2	6.7 (12)	−39.5 $\pm$ 11.5 (2)	−20.3 $\pm$ 8 (12)
−1	4.9 (16)	—	−7.8 $\pm$ 6.7 (16)
0	12.2 (30)	−25.4 $\pm$ 4.5 (10)	+16.8 $\pm$ 5.2 (30)
1	29.3 (48)	−17.7 $\pm$ 2.9 (31)	+14.8 $\pm$ 5.3 (48)
2	27.3 (18)	−9.9 $\pm$ 5.1 (16)	+2.9 $\pm$ 9.6 (18)
3	68.7 (14)	−11.5 $\pm$ 5.4 (13)	−1.9 $\pm$ 7.8 (14)
4	51.5 (15)	−15.7 $\pm$ 4.2 (14)	−1.6 $\pm$ 9.8 (15)
5	60.0 (8)	−9.8 $\pm$ 5.6 (8)	+10.1 $\pm$ 11.5 (8)

[a]Day 0 is the day the first warbler egg was laid.

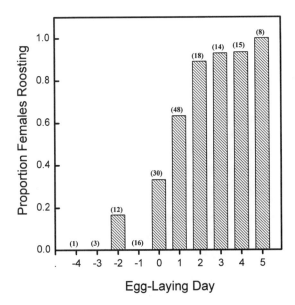

Figure 19.3. The proportion of female Yellow Warblers roosting overnight in the nest relative to their egg-laying period. Sample sizes in parentheses.

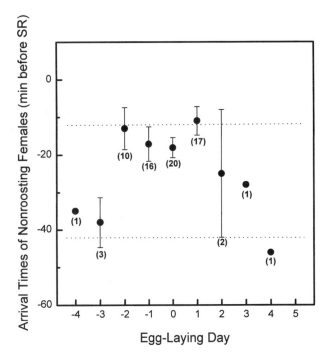

Figure 19.4. Mean (± SE) number of minutes before sunrise that nonroosting female Yellow Warblers arrived at the nest over the egg-laying period. Sample sizes in parentheses. Dotted lines demark the beginning (lower line) and end (upper line) of the critical period for parasitism.

tively; $Z = 0.136$, $P = .89$. Day 1: 0.3 ± 0.2 min versus 2.2 ± 1.7 min; $Z = 0.209$, $P = .83$.) The time at which males arrived at the nest was not significantly influenced by whether their female had roosted in the nest or not. (Day 0: SR + 15.0 ± 9.7 min versus SR + 17.9 ± 6.2 min, respectively; $Z = 0.101$, $P = .92$. Day 1: SR + 19.1 ± 6.9 min versus SR + 6.6 ± 7.9 min; $Z = 1.016$, $P = .31$.) Males arrived at the nests on average after the end of the critical period on most days; see Table 19.1 for exceptions on Days -3 ($N = 3$) and -2 ($N = 12$) of the egg-laying period (SR - 12 min = end of critical period).

Observed Acts of Cowbird Parasitism

From 1990 to 1992, Neudorf and Sealy (1994) observed seven acts of cowbird parasitism on three host species other than the Yellow Warbler. Cowbirds arrived at these nests between 0344 (SR - 44 min) and 0400 (SR - 25 min), with a mean arrival time of SR - 35.6 ± 2.4 min. In 1993, we observed four successful cases of parasitism on Yellow Warblers and one possibly unsuccessful attempt (details in Appendix). At the four parasitized nests, the cowbirds arrived on average SR - 24.8 ± 2.6 min (range: SR - 33 min to SR - 19 min). Parasitism occurred at two of these nests even though the female warblers had roosted and were in the nests when the cowbirds arrived, and physical encounters ensued between host and parasite. The warblers were presumably out of the area at the two other nests and parasitism occurred

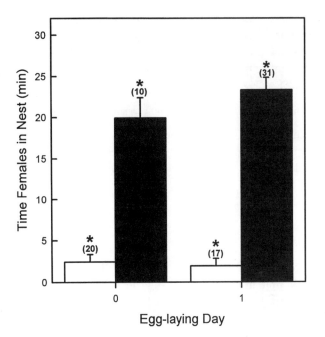

Figure 19.5. Mean (± SE) number of minutes spent by nonroosting (open bar) and roosting (solid bar) female Yellow Warblers in the nest during the critical period for parasitism on Days 0 and 1. Asterisks indicate significant differences ($P < .05$) in the behavior of roosting and nonroosting females on each day.

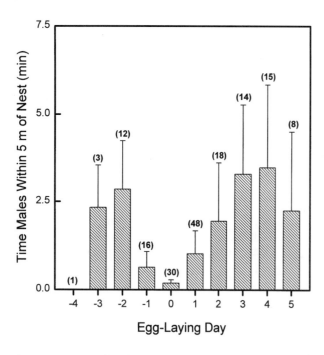

Figure 19.6. Mean (± SE) number of minutes spent by male Yellow Warblers within 5 m of the nest during the critical period for parasitism. Sample sizes (observation sessions) in parentheses.

unchallenged. The cowbird whose parasitism attempt may have been thwarted arrived at the nest at 0343 (SR - 42 min). Laying cowbirds spent an average of 21.8 ± 3.9 sec in the nest, after which they flew silently and directly away. The overall parasitism frequency on Yellow Warblers at Delta Marsh from 1974 to 1987 was 21% (Sealy 1992). The parasitism frequencies in 1992 and 1993 were 28.6% of 126 and 20.5% of 249 nests, respectively.

Effects of Sunset Temperature on Overnight Roosting

Female warblers that roosted overnight were not exposed to significantly lower sunset temperatures than females that did not roost. (Day 0: 15.4 ± 0.8°C versus 16.2 ± 0.7°C, respectively; $Z = 0.454$, $P = .65$. Day 1: 16.7 ± 0.7°C versus 17.1 ± 1.1°C; $Z = 0.175$, $P = .86$.)

Discussion

Before clutches were initiated, female Yellow Warblers spent little time in their nests during the 30-min critical period for parasitism (Figure 19.1). After clutch initiation, however, females spent more time in their nests each day, until a plateau was reached by Day 3 or 4, by which time egg laying was nearly complete and the threat of parasitism was almost nil (Sealy 1992, 1995). When females were not on their nests, they spent little time within 5 m of them. Thus, females off their nests before clutch initiation usually were out of sight and may not have been able to defend nests

against cowbird parasitism. Most males first arrived at their nests after the critical period for parasitism (Table 19.1), and those that did show up during this period spent little time near their nests (Figure 19.6). Therefore, males too were seldom in a position to defend their nests during the critical period. These results suggest that nest attentiveness prior to sunrise by Yellow Warblers is not an antiparasite strategy but rather is part of the general incubation behavior in the female. Males may visit nest areas at dawn to begin guarding females against mating attempts by rival males. Nevertheless, while increasing hormone levels and other factors may cause the increased attentiveness, it presumably also contributes to the eventual success of nesting attempts.

Cowbirds parasitized most Yellow Warbler nests during the prelaying or early egg-laying periods (Sealy 1992, 1995), when females generally did not attend their nests (Figures 19.1–19.3). If this relationship were causal, cowbirds would be expected to parasitize nests before or at the beginning of the laying period to avoid encountering warblers at the nest. Attentiveness late in the egg-laying period may allow warblers to thwart cowbird parasitism. However, we believe that the timing of cowbird parasitism and female attentiveness are not coadaptations (see also Scott 1991). Parasitism frequency declines midway through the warbler laying period probably because of the risk of parasitizing clutches too late (see Briskie and Sealy 1990), even though Yellow Warblers are more likely to accept cowbird eggs at this time (Sealy 1992, 1995).

Although it varied among females, nest attentiveness increased as egg laying progressed, probably in response to the needs of incubation. Hébert and Sealy (1992) manipulated the attentiveness of female Yellow Warblers during their egg-laying period by changing the number of eggs in nests. This is indirect evidence that attentiveness by females is controlled by tactile stimuli from within the nest (see also Book et al. 1991), rather than by the threat of cowbird parasitism. Female Yellow Warblers similarly increased their nest attentiveness during the afternoon over the laying period (Hébert and Sealy 1992, McMaster unpubl. data), at a time of the day when cowbirds do not parasitize nests (Scott 1991). However, cowbirds do sometimes visit nests in the afternoon to inspect them or to remove host eggs (Mayfield 1961, Sealy 1992, Scott et al. 1992).

Roosting females generally left their nests about 10 min before the end of the critical period for parasitism. Still, their presence on nests during most of the critical period may afford nests some protection from cowbird parasitism. Two observed acts of parasitism (see Appendix) suggest that when a female warbler is in the nest when a cowbird arrives to lay, the warblers cannot ward it off. Cowbirds parasitized two Yellow Warbler nests even though female warblers were in them when the cowbird arrived and fought with the cowbird (see also Neudorf and Sealy 1994). At a third nest, the female cowbird landed on the nest edge but after a brief pause flew away without attempting to lay. The female Yellow Warbler was in her nest when the cowbird arrived, and stayed there.

These observations show that female Yellow Warblers consistently performed nest-protection behavior, whereas the cowbirds' behavior varied while they attempted to gain access to the nest. At other times of the day, incubating female Yellow Warblers remained silent and crouched lower in their nests when approached by cowbirds (Gill unpubl. data), perhaps to be inconspicuous. The behavior of female cowbirds may reflect their urgency to parasitize a nest. Cowbirds whose first parasitism attempt at another nest was thwarted earlier in the morning, or that found a nest depredated, may be more willing to risk an encounter with hosts during an attempt to parasitize another nest. Our observation of a possible thwarted parasitism attempt involved a cowbird that arrived very early at the host nest (SR - 43 min, see Appendix); perhaps the cowbird chose to visit an unoccupied nest to parasitize rather than risk an encounter at the occupied nest. (This nest was never parasitized.) The two observations of encounters at occupied nests occurred later (SR - 19 min and SR - 33 min).

Female Yellow Warblers that roosted in their nests early in the egg-laying period were more attentive at the nests the following morning than females that did not roost in their nests (see also Neudorf and Sealy 1994). While roosting and pre-sunrise nest attentiveness together may signal the onset of incubation, observations on roosting behavior by open-cup nesting passerines (e.g., Brackbill 1985) give little indication as to why some females begin to roost earlier in the nesting cycle than others. Roosting in the nest during early laying could play a role in thermoregulation (Mueller et al. 1982). On cold nights warblers might conserve energy by roosting in their nests because nest insulation reduces heat loss (Skowron and Kern 1980, Walsberg and King 1978). Species nesting in cavities or covered nests often roost in their nests during the nest-building and early egg-laying stages (Skutch 1989). Roosting female Yellow Warblers did not experience significantly colder temperatures than non-roosting females on the first and second days of laying. Therefore, while females likely do conserve energy by roosting in their nests, some other variable apparently controls the onset of roosting.

Moksnes and Røskaft (1989) found that Meadow Pipits (*Anthus pratensis*) responded aggressively to Common Cuckoo (*Cuculus canorus*) models at the nest, but responded more intensely when both parent birds were present. This suggests that nest guarding coordinated between members of a pair increases the effectiveness of defense against brood parasitism (see also Montgomerie and Weatherhead 1988). Most male Yellow Warblers, however, were not near their nests during the critical period for parasitism (see also Neudorf and Sealy 1994). In studies of nest defense, male Yellow Warblers responded with fewer risky behaviors and in fewer trials (Hobson and Sealy 1989, Gill et al. unpubl. data). Because some responses may be learned through experience (Smith et al. 1984, Hobson and Sealy 1989) or cultural transmission (Curio et al. 1978, McLean and Rhodes 1991), the low responsiveness by males may be due to inexperience caused by the lack of nest attentiveness at the time cowbirds parasitize nests.

In summary, predawn nest attentiveness in Yellow Warblers apparently did not evolve in response to cowbird parasitism. Female Yellow Warblers spent little time at their nests before sunrise prior to clutch initiation, when the threat of cowbird parasitism was greatest. If nest attentiveness had been shaped by natural selection to serve as a host defense, it should have reached a high level on the days the first and second warbler eggs are laid, as these are the days when most parasitism occurred. Instead, nest attentiveness increased to a plateau about one day before clutches were complete, when incubation had begun (Hébert and Sealy 1992). This pattern is probably linked to the gradual onset of incubation and is unlikely to be an antiparasite strategy. Male Yellow Warblers spent little time near the nest before sunrise during egg laying, and at one nest, the male did not defend the nest during a parasitism event. When on the nest, female Yellow Warblers may occasionally thwart cowbird parasitism, although they did not do so in two of three observed acts of parasitism. Even if aspects of nest attentiveness are adaptive responses to cowbird parasitism, the resulting defense is ineffective. Four of five cowbirds that visited Yellow Warbler nests were able to lay their eggs. Thus, when analyzed critically, some putative host defenses may not have evolved in response to brood parasitism (see also Smith et al. 1984, Rothstein 1986).

Acknowledgments

We thank the staff of the University of Manitoba Field Station (Delta Marsh) for providing logistical and material support during the fieldwork. The officers of the Portage Country Club permitted us to conduct some work on their property. Diane Beattie, Kim Caldwell, Doug Froese, Gloria Goulet, Paula Grieef, and Lisa Zdrill got up early in the morning and helped us watch nests. Llwellyn M. Armstrong (University of Manitoba Statistical Services) provided important statistical advice. Stephen I. Rothstein and James N. M. Smith critically read drafts of the manuscript and offered many suggestions that improved it. Financial support was provided by the Natural Sciences and Engineering Research Council of Canada (research grant A9556 to SGS, postgraduate scholarship to SAG) and a University of Manitoba Graduate Fellowship to DGM. This paper is contribution number 241 of the University of Manitoba Field Station (Delta Marsh).

APPENDIX

Observed Acts of Parasitism

The act of oviposition by Brown-headed Cowbirds has been witnessed directly and described or mentioned in the literature 21 times from around the Great Lakes region (Scott 1991) and 7 times (in 1991) at Delta Marsh (Neudorf and Sealy 1994). Below we describe four acts of oviposition by cowbirds at Yellow Warbler nests and one instance in which a possible attempt to parasitize a nest may have been thwarted (all acts took place in 1993).

Nest 93-G2

The first warbler egg was laid on 31 May. When Doug Froese arrived at the nest at 0330 on 30 May, the female warbler was in it but left at 0403. At 0410 a cowbird flew to a perch 4 m from the nest, remained there several seconds and then moved to a perch about 1 m from the nest before flying to it. The cowbird reached the nest edge about 0411 (SR - 19 min). Moments before the cowbird arrived at the nest, the female warbler rushed to the nest from a perch about 2 m from the cowbird, vigorously uttering "seet" calls (Hobson and Sealy 1989). By the time the cowbird was beside the nest the warbler was sitting deeply in it, now not vocalizing. While still outside the nest, the cowbird pecked at the warbler's head for a few seconds until the warbler left the nest. The warbler immediately pecked the cowbird's head, but the cowbird eventually drove the warbler away, entered the nest, and laid its egg over the next 15 sec. The female warbler returned, however, and while perched on the cowbird's back pecked the cowbird on the head as she laid her egg. At this time, the male warbler landed less than 1 m from the nest but did not attack. The cowbird then left without vocalizing. We removed the cowbird egg. The next day the warbler initiated her clutch, which was depredated during incubation.

Nest 93-E76

This nest was parasitized before the warbler had laid her first egg. The female warbler was not in the nest when SAG arrived to begin the watch, just before 0330 on 9 June. In fact, the female did not approach the nest that day until 0429, 29 min after the nest had been parasitized. At 0400 (SR - 24 min), a cowbird flew to the nest's edge, paused there for 2 sec before entering, and laid her egg over the next 33 sec. The cowbird did not vocalize when she entered or left the nest. The warbler buried the cowbird egg and on 11 June initiated a 4-egg clutch from which 4 young fledged.

Nest 93-G130

The nest was parasitized on 14 June, four days before the warbler initiated her clutch. Diane Beattie watched the empty nest from 0330 to 0359, when a cowbird flew directly to it without vocalizing (SR - 23 min). The cowbird stood above it for about 6 sec, looked into it once, and then sat

deeply in the nest for 14 sec and laid her egg. The cowbird then flew directly away from the nest, again without vocalizing. The female warbler did not visit the nest until 0425 (SR + 3 min). The cowbird egg was accepted, but the nest failed during incubation.

Nest 93-G165

The first warbler egg was laid on 18 June. At 0330 on 20 June, SGS found the female warbler roosting in the nest. She was still there at 0346, when a female cowbird landed about 10 cm above the ground, about 2 m from the nest. The cowbird perched motionless and silently for 45 sec before it began to walk on the ground toward the nest. After moving about 0.5 m, the cowbird stopped for a few seconds and then walked again. It repeated this scenario several times, sometimes rustling the vegetation, until it hopped to a branch 30 cm from the nest and paused. At 0349 (SR - 33 min), the cowbird hopped to the edge of the nest and paused there for 21 sec before it entered the nest. The female warbler stayed in the nest and pecked the cowbird vigorously, but the cowbird pushed the warbler onto the edge of the nest. Over the next 19 sec, the cowbird laid her egg. When the cowbird flew from the nest, without vocalizing, the warbler slipped over the side of the nest, but got back into it immediately. The female warbler left the nest at 0403 but returned at 0436 (SR + 14 min), and during the next 21 min she laid her third egg. The male warbler was not seen within 5 m of the nest until 0417. The warbler accepted the cowbird egg (later switched with an artificial cowbird egg) and added two more eggs to the clutch. The nest was depredated during incubation.

Nest 93-E1

At 0330 on 30 May, Kim Caldwell found the female warbler roosting in the empty nest. At 0347 (SR - 43 min), a female cowbird flew directly to a branch about 30 cm from the nest, called once, and then moved to the edge of the nest. The warbler immediately crouched lower and did not vocalize. Seconds later the cowbird flew away, without vocalizing. Still in the nest, the warbler rose at 0355, and at 0402 she left the nest and foraged within 5 m. Over the next 51 min the female warbler twice visited the nest and molded its lining. She entered the nest at 0453 (SR + 23 min) and over the next 25 min laid her first of 5 eggs. The nest was not visited again by a cowbird during watches on each of the next five mornings. Warbler young fledged from the nest.

References Cited

Book, C. M., J. R. Millam, M. J. Guinan, and R. L. Kitchell. 1991. Brood patch innervation and its role in the onset of incubation in the turkey hen. Physiology and Behavior 50:281–285.

Brackbill, H. 1985. Initiation of nest-roosting by passerines with open nests. J. Field Ornithol. 56:71.

Briskie, J. V., and S. G. Sealy. 1990. Evolution of short incubation periods in the parasitic cowbirds, *Molothrus* spp. Auk 107:789–794.

Briskie, J. V., S. G. Sealy, and K. A. Hobson. 1992. Behavioral defenses against avian brood parasitism in sympatric and allopatric host populations. Evolution 46:334–340.

Burgham, M. C. J., and J. Picman. 1989. Effect of Brown-headed Cowbirds on the evolution of Yellow Warbler anti-parasite strategies. Anim. Behav. 38:298–308.

Cannings, R. A., R. J. Cannings, and S. G. Cannings. 1987. Birds of the Okanagan Valley, British Columbia. Royal British Columbia Museum, Victoria, BC, Canada.

Clark, K. L., and R. J. Robertson. 1981. Cowbird parasitism and evolution of anti-parasite strategies in the Yellow Warbler. Wilson Bull. 93:249–258.

Conover, W. J., and R. L. Iman. 1981. Rank transformations as a bridge between parametric and nonparametric statistics. Amer. Statistician 35:124–132.

Curio, E., U. Ernst, and W. Vieth. 1978. Cultural transmission of enemy recognition: One function of mobbing. Science 202:899–901.

Friedmann, H. 1929. The cowbirds: A study in the biology of social parasitism. C. C. Thomas, Springfield, IL.

Gill, S. A., and S. G. Sealy. 1996. Nest defense by Yellow Warblers: Recognition of a brood parasite and an avian nest predator. Behaviour 133:263–282.

Hébert, P. N., and S. G. Sealy. 1992. Onset of incubation in Yellow Warblers: A test of the hormonal hypothesis. Auk 109:249–255.

Hobson, K. A., and S. G. Sealy. 1989. Responses of Yellow Warblers to the threat of cowbird parasitism. Anim. Behav. 38:510–519.

MacKenzie, D. I. 1982. The dune-ridge forest, Delta Marsh: Overstory vegetation and soil patterns. Can. Field-Natur. 96:61–68.

MacKenzie, D. I., S. G. Sealy, and G. D. Sutherland. 1982. Nest-site characteristics of the avian community in the dune-ridge forest, Delta Marsh, Manitoba: A multivariate analysis. Can. J. Zool. 60:2212–2223.

Mayfield, H. F. 1961. Vestiges of proprietary interest in nests by the Brown-headed Cowbird parasitizing the Kirtland's Warbler. Auk 78:162–166.

McLean, I. G., and G. Rhodes. 1991. Enemy recognition and response in birds. Curr. Ornithol. 8:173–211.

Mead, P. S., and M. L. Morton. 1985. Hatching asynchrony in the Mountain White-crowned Sparrow (*Zonotrichia leucophrys oriantha*): A selected or incidental trait? Auk 102:25–37.

Moksnes, A., and E. Røskaft. 1989. Adaptations of Meadow Pipits to parasitism by the Common Cuckoo. Behav. Ecol. Sociobiol. 24:25–30.

Montgomerie, R. D., and P. J. Weatherhead. 1988. Risks and rewards of nest defence by parent birds. Q. Rev. Biol. 63:167–187.

Mueller, H. C., N. S. Mueller, and K. D. Meyer. 1982. Unusual nest attentiveness of an Eastern Phoebe. J. Field Ornithol. 53:421–422.

Neudorf, D. L., and S. G. Sealy. 1992. Reactions of four passerine species to threats of predation and cowbird parasitism: Enemy recognition or generalized response? Behaviour 123:84–105.

———. 1994. Sunrise nest attentiveness in hosts of the Brown-headed Cowbird. Condor 96:162–169.

Nolan, V. 1978. The ecology and behavior of the Prairie Warbler *Dendroica discolor*. Ornithol. Monogr. 26:1–595.

Rohwer, S., and C. D. Spaw. 1988. Evolutionary lag versus bill-size constraints: A comparative study of the acceptance of cowbird eggs by old hosts. Evolutionary Ecol. 2:27–36.

Rothstein, S. I. 1975. An experimental and teleonomic investigation of avian brood parasitism. Condor 77:250–271.

———. 1986. A test of optimality: Egg recognition in the Eastern Phoebe. Anim. Behav. 34:1109–1119.

———. 1990. A model system for coevolution: Avian brood parasitism. Annu. Rev. Ecol. Syst. 21:481–508.

Schrantz, F. G. 1943. Nest life of the Eastern Yellow Warbler. Auk 60:367–387.

Scott, D. M. 1991. The time of day of egg laying by the Brown-headed Cowbird and other icterines. Can. J. Zool. 69:2093–2099.

Scott, D. M., P. J. Weatherhead, and C. D. Ankney. 1992. Egg-eating by female Brown-headed Cowbirds. Condor 94:579–584.

Sealy, S. G. 1992. Removal of Yellow Warbler eggs in association with cowbird parasitism. Condor 94:40–54.

———. 1995. Burial of cowbird eggs by parasitized Yellow Warblers: An empirical and experimental study. Anim. Behav. 49:877–889.

Sealy, S. G., and R. C. Bazin. 1995. Low frequency of observed cowbird parasitism on Eastern Kingbirds: Host rejection, effective nest defense, or parasite avoidance? Behavioral Ecology 6:140–145.

Sealy, S. G., D. L. Neudorf, and D. P. Hill. 1995. Rapid laying by Brown-headed Cowbird *Molothrus ater* and other brood parasitic birds. Ibis 137:76–84.

Skowron, C., and M. Kern. 1980. The insulation in nests of selected North American songbirds. Auk 97:816–824.

Skutch, A. F. 1989. Birds asleep. University of Texas Press, Austin.

Slack, R. D. 1976. Nest guarding behavior by male Gray Catbirds. Auk 93:292–300.

Smith, J. N. M., P. Arcese, and I. G. McLean. 1984. Age, experience, and enemy recognition by wild Song Sparrows. Behav. Ecol. Sociobiol. 14:101–106.

Walsberg, G. E., and J. R. King. 1978. The energetic consequences of incubation for two passerine species. Auk 95:644–655.

20. The Ecology of Egg-puncture Behavior by the Shiny Cowbird in Southwestern Puerto Rico

TAMMIE K. NAKAMURA AND ALEXANDER CRUZ

Abstract

We examined the distribution and biology of egg puncture by brood parasitic Shiny Cowbirds in southwestern Puerto Rico. The frequency of egg puncture increased from 2.8% before 1975 to 29% in 1982–1987. Egg puncture was limited to the nests of parasitized species and occurred in seven of the eight species parasitized. Host and cowbird eggs were punctured at similar frequencies, 34.4% for host eggs and 26.7% for cowbird eggs. Egg puncture was independent of host clutch sizes, but it was strongly related to the number of cowbird eggs. Parasitized nests were more likely to contain punctured eggs than unparasitized nests. Egg puncture reduced viability of host and cowbird eggs in nests of the Yellow Warbler, Yellow-shouldered Blackbird, and Black-whiskered Vireo. Clutches with punctured eggs were abandoned within three days following egg puncture by all parasitized species except Black-whiskered Vireos. Desertion may be a response to egg and nest soiling, to the inability of the adult to turn eggs in the nest, and to infestation of nests by formicid ants. Hatching failure in punctured clutches was independent of nest abandonment and predation and is probably related to reduced viability of soiled eggs. Cowbirds punctured eggs throughout incubation and may have returned to parasitize subsequent renesting attempts by the host. The greater incidence of egg puncture in the nests of frequently used hosts and the association of egg puncture with multiple parasitism suggest that egg puncture reflects competitive interactions between laying cowbirds. Our results underscore the benefits of reducing cowbird numbers in nesting areas where multiple parasitism by Shiny Cowbirds is common.

Introduction

Avian brood parasitism is associated with unusual behaviors that adapt the species to the parasitic way of life. Egg puncture by Brown-headed, Shiny, and Bronzed Cowbirds is one such behavior (Friedmann 1929; Post and Wiley 1976, 1977a,b; Mason 1980; Fraga 1985; Carter 1986). Explanations for this behavior have ranged from insignificant "habits" to factors that influence the reproductive success and evolution of the species. For example, Hoy and Ottow (1964) found punctured eggs after the end of the cowbird egg-laying season and suggested that egg puncture is a random and general habit. On the other hand, egg puncture by brood parasites occurs in a variety of circumstances, some of which suggest that egg puncture is an attempt to manipulate the host's production of young (Payne 1977, Post and Wiley 1977b, Carey 1986, Davies and Brooke 1988, Lerklund et al. 1993).

What advantage does a cowbird gain by puncturing the egg of its host? Egg puncture may reduce egg hatchability (Blankespoor et al. 1982, Carey 1986), and nesting birds often remove damaged eggs from the nest (see Kemal and Rothstein 1988). Since host eggs are punctured more frequently than cowbird eggs (Mason 1980, Carter 1986), this behavior may reduce competition for the cowbird young in the host nest (Friedmann 1929, Hoy and Ottow 1964, Scott 1977, Blankespoor et al. 1982). On the other hand, egg puncture may not increase fledging success of cowbirds. Nesting birds often abandon clutches containing punctured eggs for two reasons. First, abandonment may be a general response to apparent nest predation (Hoy and Ottow 1964, King 1973, Post and Wiley 1977a, Mason 1980, Fraga 1985, Carter 1986, Røskaft et al. 1990). Second, abandonment may occur because the parent cannot turn eggs in the nest when the nest and eggs are soiled by leakage from broken eggs (Nice 1929, Post and Wiley 1977b, Kemal and Rothstein 1988). These observations suggest that egg puncture is a risky behavior with significant potential to cause complete failure of a nesting attempt.

Egg-puncture behavior may be related to cowbird densities. Post and Wiley (1977b) found high rates of egg puncture in nests of Yellow-shouldered Blackbirds (*Agelaius xanthomus*) in areas of Puerto Rico with high densities of Shiny

Cowbirds. Cowbird densities were higher in the east regions of the island because the species invaded that area in the early 1960s at least 10 years earlier than it did the west (Post and Wiley 1977a). Egg puncture was the main cause of blackbird nest failure in the northeast of Puerto Rico but not in the southwest. Even though both blackbird populations suffered 100% brood parasitism, 66.7% (N = 18) of the clutches in the northeast were punctured and subsequently abandoned, in contrast to 2.8% (N = 35) in the southwest. These authors attributed the difference in egg puncture to greater frequencies of multiple parasitism (the average number of cowbird eggs per nest) and lower host densities in the northeast. Thus, egg-puncture behavior may be related in some way to competition among cowbirds for host nests.

This study examines the distribution and biology of egg puncture in southwestern Puerto Rico, approximately 10 years after the studies by Post and Wiley (1977a,b). The southwest region experienced increasing cowbird densities combined with diminishing host numbers throughout the 1970s and early 1980s (Post and Wiley 1976, 1977a; Post 1981; U.S. Fish and Wildlife and Department of Natural Resources, Puerto Rico, unpubl. data). If the frequency of egg puncture increases with cowbird density, then puncture frequencies should be higher than found in studies before 1975. We also explored the ecology of egg puncture by examining the occurrence of the behavior throughout the passerine community nesting in southwestern Puerto Rico. Even though Shiny Cowbirds may consume eggs (Wiley 1982, Nakamura pers. obs.), it is not known if punctured eggs commonly result from the actions of other egg predators. If egg puncture reflects general egg predation, we expected to find punctured clutches distributed throughout the community. Finally, we examined the reproductive consequences of egg puncture to both cowbird and its host.

Study Area and Methods

The breeding biology of Shiny Cowbirds was studied in southwestern Puerto Rico during 1982–1988 (see also Cruz et al. 1989, Post et al. 1990). The 75-km² study area ranged 5 km south and west of the Sierra Bermeja to the Caribbean coast (including the Cabo Rojo National Wildlife Refuge), north and west to Boquerón and east to Parguera. Efforts to find nests were concentrated in areas where Post and Wiley (1977a,b) and Wiley (1982) conducted similar studies from 1972 to 1981. The area is mainly subtropical dry forest and includes coastal mangrove (*Avicennia germinans* and *Rhizophora mangle*) and dry evergreen scrub dominated by mesquite (*Prosopis juliflora*), acacia (*Acacia farnesiana*) and ucar (*Bucida buceras*). The climate is moderately arid with seasonal rains from March to May and September to October. Total annual precipitation from 1982 through 1988 was 928.5 ± 61.1 mm/year (Cabo Rojo National Wildlife Refuge unpubl. records).

The nesting season for most passerine species is during spring and early summer (March–July), and some species continue to lay eggs as late as November. Cowbird parasitism occurred from May through early June and continued through September in years with fall rains (Wiley 1982, Post et al. 1990). Study areas were searched daily for nests, and active nests were checked every 2–4 days. To determine if egg puncture resulted in particular types of nesting failure, we classified nest failures into two categories. Nests were scored as abandoned when the eggs were cold and the clutch was unattended by the adult on at least two consecutive inspections of the nest. Nests were scored as depredated when eggshell fragments were found in or around the nest, or when the nest or portions of it were physically destroyed. We were not able to determine the cause of most nest failures, so the above numbers of abandoned and depredated nests are minimum estimates.

We assumed that all observations of egg puncture were associated with cowbird activities at the nest and were not due to puncture ejection of cowbird eggs by the nesting host. Using Rothstein's (1971) method of parasitizing nests experimentally, Post et al. (1990) showed that parasitized host species in Puerto Rico do not eject cowbird eggs or damage their own eggs in response to brood parasitism. In addition, to reduce the possibility of including egg damage by other generalist predators (such as *Rattus* spp.), we did not include cracked and bruised eggs in our analysis. Such damage is characteristic of sheer or crushing trauma and is unlikely to be related to damage from puncture trauma (tensile failure, Voisey and Hunt 1969, 1974; Voisey et al. 1979; Hamilton 1982). We used egg-puncture criteria similar to those described by Hunt et al. (1977). Puncture trauma is a clean hole with no bruising or concentric cracks radiating from the point of impact (Tyler and Moore 1965). Puncture holes ranged from 1 to 14 mm in diameter and were triangular or circular in shape. In this chapter, eggs with damaged shell membranes were not treated separately in the analysis.

Observations of cowbirds removing or puncturing eggs are also included in this chapter. These data were obtained during direct observation of host nests in other studies. Direct observations of egg-laying, incubation, and nestling care behaviors in Yellow Warblers (*Dendroica petechia*), Black-whiskered Vireos (*Vireo altiloquus*), and Yellow-shouldered Blackbirds were taken from blinds placed 3–6 m from the nest (Nakamura unpubl. data).

Results

The Incidence of Egg Puncture

We estimated the frequency and occurrence of egg puncture throughout the avian community by surveying 2,965 nests of 30 species (Table 20.1). Egg puncture was clearly restricted to species parasitized by the Shiny Cowbird and thus was not due to general egg depredation. Of eight parasitized species, seven had nests with punctured eggs. The percentage of clutches containing one or more punctured eggs ranged from 2.1% (Greater Antillean Grackle *Quiscalus niger*) to 19.6% (Yellow-shouldered Blackbird). The frequency of egg puncture in Yellow-shouldered Blackbird nests increased from 2.8% of 35 clutches in the southwest study area in 1975 (Post and Wiley 1977b) to 19.6% of 194 nests in our study (G-test with Yates correction, $G = 9.78$, 1 df, $P = .002$). Our puncture frequencies of eggs in blackbird nests, however, were much lower than those reported by Post and Wiley for northeastern Puerto Rico (66.7% of 18 nests, 1 df, $G = 26.71$, $P = .001$). Even though the local frequency of egg punctures within our study area reached 29.0% (18 of 44 nests), we found no evidence for the high frequencies reported in the northeast.

In general, more parasitized nests than unparasitized nests contained punctured host and/or cowbird eggs (parasitized, 18.0% of 423 nests, 81.7% of all 93 punctured clutches; unparasitized, 3.9% of 442 nests; $G = 47.94$, 1 df, $P = .000$).

However, the relationship between egg puncture and nest parasitism was less apparent for some host species. Rates of egg puncture differed between parasitized and unparasitized nests in the Puerto Rican Flycatcher (*Myiarchus antillarum*; 58.3% of 12 parasitized versus 1.6% of 64 unparasitized nests; $G = 24.54$, 1 df, $P = .000$), and in the Yellow Warbler (14.6% of 123 parasitized versus 2.9% of 139 unparasitized nests; $G = 12.42$, 1 df, $P = .001$). No significant differences, however, were found for three species: the Puerto Rican Vireo (*Vireo latimeri*; 17.6% of 34 parasitized versus 0% of 5 unparasitized nests), Black-whiskered Vireo (13.6% of 81 parasitized versus 13.3% of 30 unparasitized nests; $G = 0.001$, 1 df, $P = .97$), and Yellow-shouldered Blackbird (19.8% of 167 parasitized versus 18.5% of 27 unparasitized nests; $G = 0.02$, 1 df, $P = .88$). High frequencies of egg puncture in unparasitized nests of the last three species may indicate intense cowbird activity around all nests of heavily parasitized species (87.1% of 39 Puerto Rican Vireo nests, 73.0% of 111 Black-whiskered Vireo nests, and 86.3% of 194 Yellow-shouldered Blackbird nests). There may be less cowbird activity near nests of the two species that were parasitized less frequently (16.3% of 75 Puerto Rican Flycatcher nests and 46.9% of 262 Yellow Warbler nests).

Eggs were punctured indiscriminately, and cowbird eggs were just as likely to be punctured as host eggs. Of 47 punctured clutches containing at least one host and one cowbird egg, 38.3% of 18 nests contained at least 1 punctured cowbird egg, 46.8% of 22 nests contained at least one damaged host egg, and 14.9% of 7 nests contained punctured host and cowbird eggs. Host and cowbird eggs were punctured at similar frequencies (34.4% of 122 host eggs and 26.7% of 120 cowbird eggs, $G = 1.72$, 1 df, $P = .19$). Most often, only a single host or cowbird egg was punctured (in 54.5% of 55 nests), but the entire clutch was punctured in 5 of the nests.

The incidence of egg puncture was independent of host clutch size but was clearly associated with the number of cowbird eggs in the host nest. Average host clutch sizes were similar whether clutches did or did not contain punctured eggs ($F = 1.48$, 1 df, $P = .22$, two-factor Kruskal-Wallis ANOVA, Zar 1984). While clutch size varied strongly among species ($F = 19.08$, 2 df, $P = .000$) there was no significant interaction between effects of species and egg puncture ($F = 1.41$, 2 df, $P = .23$). On the other hand, the average number of cowbird eggs per nest differed between punctured and unpunctured clutches ($F = 50.16$, 1 df, $P = .000$). There were strong differences between species ($F = 34.39$, 2 df, $P = .000$), and there was a significant interaction between species identity and egg puncture ($F = 4.36$, 2 df, $P = .002$). Multiple parasitism, i.e., clutches containing 2 or more cowbird eggs, was most common in the nests of the large Yellow-shouldered Blackbird (78.1% of 114 parasitized nests had up to 15 cowbird eggs/nest). Lower levels occurred in the smaller Puerto Rican Flycatcher (33.3% of 12 nests had 2–3 eggs), Black-whiskered Vireo (24.7% of 81 nests had 2–3 eggs), Yellow Warbler (9.0% of 106 nests had 2 eggs), and Puerto Rican Vireo (5.9% of 34 nests had 2 eggs).

The proportion of clutches of all species containing punctured eggs was lower in unparasitized clutches (5.3% of 266 eggs) than in singly parasitized clutches (10.0% of 21 eggs, Figure 20.1). However, the difference only approached statistical significance ($G = 3.20$, 1 df, $P = .07$). Unparasitized and singly parasitized nests were less likely to contain punctured eggs than nests containing 2 or more cowbird eggs (6.8% of 476 versus 31.8% of 152 eggs, respectively; $G = 76.45$, $P = .000$). Thus, the addition of one or more eggs following the initial parasitism of the nest was associated with much more frequent egg puncture.

Reproductive Consequences of Egg Puncture

Egg puncture reduced viability of both host and cowbird eggs in nests of Yellow Warblers, Yellow-shouldered Blackbirds, and Black-whiskered Vireos. All 54 punctured host eggs and 93.3% of 30 cowbird eggs failed to hatch. Undamaged host and cowbird eggs in punctured clutches hatched less often (11.9% of 84 host eggs and 7.4% of 94 cowbird eggs) than eggs in unpunctured clutches, where 43.3% of 745 host eggs and 52.0% of 277 cowbird eggs hatched (both comparisons, $P = .000$, G-tests). The more frequent hatching failure of young from punctured clutches caused significantly higher rates of total nest failure (Table 20.2). Yellow Warbler and Yellow-shouldered Blackbird nests containing punctured eggs failed at least twice as often as did nests with unpunctured clutches. Nest failure was associated with greater incidence of nest abandonment for these two species.

The combined reproductive consequences of egg puncture and multiple parasitism significantly reduced hatching success for both host and cowbird. For unpunctured clutches, the mean number of cowbird young hatched per egg did not

Table 20.1. Incidence of Egg Puncture in a Community of Nesting Passerines in Southwestern Puerto Rico, 1982–1988

Species	No. Nests	% Nests Par[a]	% Nests Punct[b]	Total No. Eggs	% Eggs Punct[c]	Ave. No. Eggs Punct/ Nest	Range No. Eggs
Gray Kingbird (*Tyrannus dominicensis*)	125	0.0	0.0	411	0.0	0.0	—
Loggerhead Kingbird (*Tyrannus caudifasciatus*)	8	0.0	0.0	31	0.0	0.0	—
Puerto Rican Flycatcher (*Myiarchus antillarum*)	76	16.3	10.5	243	6.2	1.9	1–3
Lesser Antillean Pewee (*Contopus latirostris*)	21	0.0	0.0	63	0.0	0.0	—
Caribbean Elaenia (*Elaenia martinica*)	53	0.0	0.0	185	0.0	0.0	—
Northern Mockingbird (*Mimus polyglottos*)	135	0.0	0.0	406	0.0	0.0	—
Red-legged Thrush (*Mimocichla plumbea*)	5	0.0	0.0	17	0.0	0.0	—
Caribbean Martin (*Progne dominicensis*)	86	0.0	0.0	274	0.0	0.0	—
Bananaquit (*Coereba flaveola*)	110	0.0	0.0	274	0.0	0.0	—
Puerto Rican Vireo (*Vireo latimeri*)	39	87.1	15.4	131	8.4	1.8	1–2
Black-whiskered Vireo (*Vireo altiloquus*)	111	73.0	13.5	331	10.9	2.4	1–4
Yellow Warbler (*Dendroica petechia*)	262	46.9	8.4	697	3.9	1.2	1–2
Adelaide's Warbler (*Dendroica adelaidae*)	33	3.0	3.0	70	1.4	1.0	1
Greater Antillean Grackle (*Quiscalus niger*)	137	2.9	2.1	457	0.7	1.3	1
Black-cowled Oriole (*Icterus dominicensis*)	2	0.0	0.0	6	0.0	0.0	—
Troupial (*Icterus icterus*)	12	8.3	0.0	24	0.0	0.0	—
Yellow-shouldered Blackbird (*Agelaius xanthomus*)	194	86.3	19.6	609	7.7	1.2	1–2
Blue-hooded Euphonia (*Euphonia musica*)	6	0.0	0.0	16	0.0	0.0	—
Stripe-headed Tanager (*Spindalis zena*)	1	0.0	0.0	3	0.0	0.0	—
Puerto Rican Bullfinch (*Loxigilla portoricensis*)	19	0.0	0.0	61	0.0	0.0	—
Yellow-faced Grassquit (*Tiaris olivacea*)	92	0.0	0.0	253	0.0	0.0	—
Black-faced Grassquit (*Tiaris bicolor*)	147	0.0	0.0	490	0.0	0.0	—
Grasshopper Sparrow (*Ammodramus savannarum*)	3	0.0	0.0	8	0.0	0.0	—
Orange-cheeked Waxbill (*Estrilda melpoda*)	43	0.0	0.0	126	0.0	0.0	—
Black-rumped Waxbill (*Estrilda troglodytes*)	3	0.0	0.0	11	0.0	0.0	—
Red Avadavat (*Amandava amandava*)	89	0.0	0.0	289	0.0	0.0	—
Warbling Silverbill (*Lonchura malabarica*)	109	0.0	0.0	486	0.0	0.0	—
Bronze Mannikin (*Lonchura cucullata*)	79	0.0	0.0	339	0.0	0.0	—
Nutmeg Mannikin (*Lonchura punctulata*)	39	0.0	0.0	141	0.0	0.0	—
Chestnut Mannikin (*Lonchura malacca*)	926	0.0	0.0	3625	0.0	0.0	—

[a]Percentage of all nests parasitized by the Shiny Cowbird.
[b]Percentage of nests containing one or more punctured eggs (parasitized or unparasitized).
[c]Percentage of all eggs punctured.

differ significantly between species (warbler 0.63 ± 0.48, blackbird 0.52 ± 0.44, vireo 0.55 ± 0.48; $F = 0.52$, 2 df, $P = .59$). After correcting for differences between species, the average number of cowbird young hatched per egg was slightly less in nests containing more than 1 cowbird egg (0.46 ± 0.41) than in nests with single cowbird eggs (0.61 ± 0.48; $F = 3.60$, 1 df, $P = .06$). On the other hand, multiply

parasitized and punctured clutches hatched only 0.05 ± 0.21 cowbird eggs per nest. Differences between punctured and unpunctured clutches were significant ($F = 8.78$, 1 df, $P = .004$) after correcting for differences between species and number of cowbird eggs. Thus, egg puncture may represent one of two scenarios: (a) female cowbirds that both parasitize and puncture eggs reduce the hatching probability of

Table 20.2. Percentage of Nests Failing to Hatch at Least One Young with Punctured and Unpunctured Eggs

| | Nest Failure | | | | | | Proportion Abandoned[d] | | | |
| | Punctured | | Unpunctured[b] | | | | Punctured | | Unpunctured | |
	(%)	(N)[a]	(%)	(N)	G[c]	P	(%)	(N)	(%)	(N)
Yellow Warbler	95.5	(22)	43.2	(197)	16.23	<0.001	70.0	(10)	15.2	(33)
Shiny Cowbird	88.9	(18)	36.6	(82)	14.36	<0.001				
Yellow-shouldered Blackbird[e]	88.2	(17)	40.5	(42)	12.36	0.001	91.7	(12)	30.0	(10)
Shiny Cowbird	88.9	(18)	34.2	(41)	15.00	<0.001				
Black-whiskered Vireo	66.7	(15)	34.9	(83)	5.34	0.021	20.0	(5)	22.2	(9)
Shiny Cowbird	75.0	(12)	40.3	(62)	4.99	0.027				

[a]Sample sizes vary because some active nests did not have host or cowbird eggs.

[b]Parasitized and unpunctured nests only.

[c]G statistic with Yates correction for 2 × 2 analysis for differences between punctured and unpunctured nests.

[d]Nest failure associated with clutch abandonment relative to nest depredation (proportion expressed as a percentage).

[e]Yellow-shouldered Blackbird sample from 1982–1984 only; does not include nests in which cowbird and damaged blackbird eggs were removed as a part of a conservation program for the species (see Post and Wiley 1976, Wiley et al. 1991).

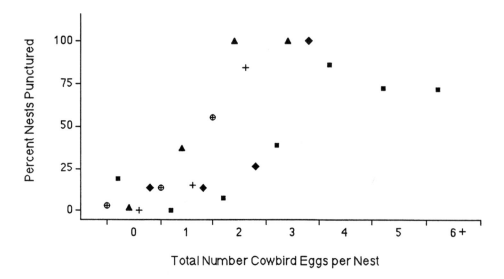

Figure 20.1. Relationship between number of cowbird eggs and proportion of nests containing punctured eggs ($r^2 = 0.46$, $0.02 < P < 0.05$). Square, Yellow-shouldered Blackbird. Circled cross, Yellow Warbler. Triangle, Puerto Rican Flycatcher. Cross, Puerto Rican Vireo. Diamond, Black-whiskered Vireo.

their own eggs by 89.1%, or as seems much more likely, (b) females that puncture eggs do not lay their eggs in the punctured clutch.

Why Do Damaged Clutches Fail?

Clutches with punctured eggs were abandoned within three days following egg puncture by all parasitized species except the Black-whiskered Vireo (see below). Desertion may be a response to egg and nest soiling and to the inability of the adult to turn eggs in the nest. Punctured clutches were abandoned more often (41.3% of 46 cases) than were unpunctured clutches (9.4% of 106 cases; $G = 19.5$, 1 df, $P = .000$). Abandonment of damaged clutches may also be related to infestation of the nest by formicid ants. Damaged eggs (punctured or cracked) were found in all nest cases of nest infestation ($N = 18$, 6 species). On the other hand, there is no evidence that nests were abandoned because they were at higher risk of predation. Predation of punctured and abandoned clutches (17.4% of 46 nests) occurred less often than it did for unpunctured and unabandoned clutches (39.6% of 106 clutches; $G = 12.0$, 1 df, $P = .001$).

Hatching failure was related to reduced viability of soiled eggs independently of nest abandonment. The Black-whiskered Vireo did not abandon nests in response to egg puncture. One of five abandoned clutches was punctured versus two of nine unpunctured ones ($G = 0.01$, 1 df, $P = .92$). We have also observed Black-whiskered Vireos incubating clutches containing more than one punctured egg for up to 22 days. The failure of Black-whiskered Vireos to abandon nests in response to egg damage did not increase the likelihood that a cowbird chick would hatch from a punctured clutch (Table 20.2). Punctured and unabandoned clutches failed to hatch cowbird eggs more often (75.0% of 12 clutches) than parasitized and unpunctured clutches (40.3% of 63 cases). Egg and nest soiling are probably responsible for the high hatching failure in punctured clutches because soiling reduces egg viability.

When Are Eggs Punctured?

Cowbirds did not puncture eggs at the time of egg laying, but did so later throughout the incubation period. In 18 observed acts of parasitism, females never punctured eggs before or up to 3 days following egg deposition (Nakamura unpubl. data). On the other hand, we observed 8 cases of nonlaying females puncturing eggs. One case was in a Yellow Warbler nest, 2 in Black-whiskered Vireo nests, and 5 in Yellow-shouldered Blackbird nests. All these acts of egg puncture occurred after host incubation had commenced (2 cases on Days 1–4, 1 case on Day 5–6, and 3 cases on Days 8–10; in the final 2 cases the observation day was unknown). Figure 20.2 summarizes damage and removal of cowbird or host eggs relative to the day of incubation of Yellow-shouldered Blackbird nests. The proportions of clutches punctured during early (35.3% of 17 clutches) and late incubation (64.7% of 17 clutches) did not differ from random ($G = 0.78$, 1 df, $P = .38$). However, there was a tendency for egg puncture combined with egg removal to occur more often late in incubation ($G = 2.72$, 1 df, $P = .10$).

We have anecdotal observations of two marked female cowbirds that suggest egg-puncturing females return to parasitize the renesting attempt of the host following nest abandonment. In the first observation, a single female cowbird entered the blackbird nest on incubation Day 3–4 and punctured 3 eggs (2 of 4 blackbird and 1 of 2 cowbird eggs). The nesting birds abandoned the punctured clutch and 4 days later began constructing a new nest 2 m from the abandoned nest (blackbirds have high fidelity to nest boxes, Department of Natural Resources unpubl. data). We assume that these unbanded blackbirds were the same individuals. The marked female cowbird was observed monitoring nest-

building activities and parasitized the new blackbird nest on the second day of egg laying. A second banded female punctured 1 of 3 cowbird eggs in a different nest during incubation Days 5–7. The hosts abandoned this nest in one day, and a second nest was completed on Day 12, 10 m from the abandoned nest. The banded cowbird was flushed from the new nest on the third and final day of egg laying by the hosts, and she had added a new egg.

Discussion and Management Implications

Egg puncture by cowbirds had significant reproductive consequences for hosts. We have no evidence to support the notion that egg puncture increases the chances of a cowbird fledging from the nest. Eggs were punctured indiscriminately, and cowbird eggs were just as likely to be damaged as host eggs. Egg puncture led to a high frequency of failure of cowbird eggs because most nests were abandoned and the viability of eggs was reduced. We have indirect evidence that eggs are punctured by nonlaying cowbirds. This was inferred from the observations of punctured eggs in unparasitized nests, the puncture of cowbird eggs only, and the destruction of all eggs (host and cowbird) in the nest. Female cowbirds have not been observed puncturing eggs simultaneously with the act of laying, and eggs are punctured throughout the incubation period. These observations are in contrast to those of Sealy (1992), who reported that 83.3% of the removal of Yellow Warbler eggs by Brown-headed Cowbirds occurred by the third day of incubation. This study suggests that in Puerto Rico, cowbirds either monitor nests throughout incubation and return later to puncture eggs, or that egg puncture reflects random and destructive encounters at the nest by nonlaying individuals.

Egg puncture may be linked with the number of cowbirds in the nesting area. The incidence of egg puncture increased during the twenty-year period following the invasion of southwestern Puerto Rico by cowbirds. The greater incidence of egg puncture in the nests of frequently used hosts and the association of the behavior with multiple parasitism suggest that egg puncture reflects competitive interactions between cowbirds for parasitism opportunities. This was demonstrated by patterns of egg puncture in Yellow-shouldered Blackbird nests in relation to changes in cowbird abundance. Throughout Puerto Rico, the blackbird experienced a rapid decline in population number due to brood parasitism by the Shiny Cowbird (Post and Wiley 1976). From 1982 to 1986, the island population was estimated to be fewer than 1,200 individuals, approximately half the number estimated in 1975 (Wiley et al. 1991). Winter roost counts in the southwestern part of the island from 1974 to 1975 estimated blackbird and cowbird numbers at 1,163 and 1,166 individuals, respectively. However, in similar counts from 1981 to 1982, cowbirds were five times more common than blackbirds (1,387 and 266, respectively) (Wiley et al.

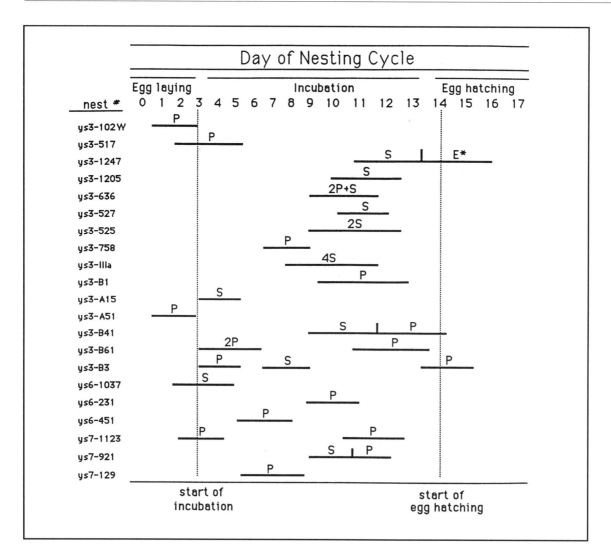

Figure 20.2. Day of egg puncture and egg removal by cowbirds from Yellow-shouldered Blackbird nests. E* indicates viable host egg disappeared following hatching of cowbird chick. Horizontal lines represent intervals between nest inspections, and letters indicate if an egg was punctured (P) or removed (S) from the nest. The midpoint of the observation interval was used to estimate time of egg puncture relative to early (Days 1–6) and late (Days 7–14) incubation, as described in text.

1991). Brood parasitism of blackbird nests was the most intense of all host species in the community. Average parasitism of blackbird nests was high in 1975–1977 (3.1 cowbird eggs/nest, 89% parasitism, *N* = 66) and in 1982–1988 (3.7 eggs/nest, 96.8% parasitism, *N* = 100). However, it declined for other host species, such as the Yellow Warbler (1975–1977, 2.9 eggs/nest, 80% parasitism, *N* = 20; 1982– 1988, 1.1 eggs/nest, 46.9% parasitism, *N* = 262; Wiley 1982, Post et al. 1990, Nakamura unpubl. data). Rates of egg puncture increased in southwestern Puerto Rico from 2.8% prior to 1977 (Post and Wiley 1977a) to 19.6% from 1982 to 1986. Subsequent to long-term cowbird removals in southwestern

Puerto Rico (1983 to present), egg-puncture rates have decreased dramatically (1991, 10.3%, 7/68 nests; 1992, 1.6%, 2/125 nests; 1993, 3.7%, 7/187 nests; Department of Natural Resources unpubl. reports). Thus egg puncture appears to depend on densities of cowbirds relative to those of hosts.

Alternatively, patterns in egg puncture described by this study may be unique to Puerto Rico and may represent a behavior that increases the colonization success of cowbirds in new areas. The Shiny Cowbird first colonized the island sometime in the 1960s (Post and Wiley 1977a, Cruz et al. 1989), and Post and Wiley (1977b) reported the highest known frequencies of egg puncture in blackbird nests

(66.7%) for the northeastern part of the island. Puncture rates were lower in the southwest (2.8%), presumably because the cowbird invaded the area at least 10 years later than it did in the northeast. However, approximately one decade after the initial studies by Post and Wiley, we failed to detect the high puncture rates reported for the northeastern part of the island. Egg-puncture behaviors may be more common following the initial invasion of cowbirds into a new habitat. We also assume that egg puncture is a self-limiting trait that is constrained by competition among parasites at high population densities, and that this explains why puncture rates were lower in our study.

If it is true that egg puncture is more common in recently colonized populations of cowbirds, it may also be assumed that egg puncture increases parasitism opportunities for cowbirds. The reproductive opportunities created by puncturing behavior depend on the potential of the host species to renest following the abandonment of punctured clutches, and we assume that renesting potential is highest in neotropical habitats. It is significant that egg puncture has been reported more commonly for species of cowbirds that occur in warmer regions (i.e., Mason 1980, Fraga 1985, Carter 1986), because these areas present different reproductive opportunities for brood parasites relative to northern temperate habitats. In neotropical communities, some host species have prolonged nesting seasons. In Puerto Rico, nesting by several potential host species begins as early as March and continues through November, and nesting birds may attempt 3–5 clutches each year (Wiley 1982, Nakamura unpubl. data). We observed two cases where egg-puncturing cowbirds returned to parasitize renesting attempts by the host. These observations may simply reflect patterns of host use within an individual cowbird's breeding range rather than deliberate manipulation of the host. However, if renesting by breeding hosts provides reproductive opportunities for brood parasites, and egg-puncturing individuals are more successful at parasitizing hosts than nonpuncturing individuals, the incidence of egg puncture will increase within a population. Furthermore, this behavior can allow the parasite to synchronize its reproduction with novel hosts. This, in turn, may explain why egg puncture was more common when the Shiny Cowbird first invaded Puerto Rico.

Egg puncture has serious implications for the breeding biology of some host species in Puerto Rico. For example, the Black-whiskered Vireo is a seasonal migrant in the Greater Antilles (Lack 1976, Bond 1983). In our study area, nesting begins in May and most breeding activity is completed by June (Raffaele 1989, Nakamura pers. obs.). Black-whiskered Vireos seldom abandon nests with damaged eggs and renest. We assume that this behavior reflects the shortness of the breeding season for this species. The negative effects of egg puncture on the reproductive success of the Black-whiskered Vireo may be severe in comparison to other nonmigratory residents (such as Yellow Warblers), which have a long nesting season that extends outside the main cowbird egg-laying period. Reducing the effects of brood parasitism and egg puncture may therefore be critical for the conservation of migratory breeding birds like the Black-whiskered Vireo.

These results have two implications for the design of management strategies of parasitized species. First, significant reductions in reproductive success of nesting birds due to multiple parasitism and egg puncture underscore the importance of reducing cowbird numbers in host nesting areas. Nesting Yellow-shouldered Blackbird populations in northeastern Puerto Rico have been virtually extirpated, and egg puncture and parasitism by cowbirds contributed to the decline of the species (Wiley et al. 1991). Cowbird removal programs in southwestern Puerto Rico (1983–present) have significantly reduced cowbird numbers, and puncture of eggs of the Yellow-shouldered Blackbird has declined (Department of Natural Resources unpubl. report).

Second, management programs must accommodate potential negative impacts of egg puncture on nesting birds under conditions of high cowbird densities. Ortega et al. (1994) demonstrated that addition of real and artificial Brown-headed Cowbird eggs to Red-winged Blackbird (*Agelaius phoeniceus*) nests in Colorado reduced subsequent nest parasitism, whereas cowbird egg removal may promote further parasitism of the nest. In Puerto Rico, the incidence of multiple parasitism indicates that a cowbird egg in a host nest did not preclude subsequent parasitism of the nest. The association of egg puncture with multiple parasitism suggests that the addition of artificial eggs may adversely impact the success of nests. Even though we have data that suggest that egg removal increases the risk of nest parasitism, we found no association of egg removal with increased risks of egg puncture (Post, Nakamura, and Cruz unpubl. data). Thus removing cowbird eggs from parasitized nests may reduce the risk of egg puncture.

In summary, egg-puncture behavior is closely associated with brood parasitism by the Shiny Cowbird in Puerto Rico. The behavior eliminates the production of both host and cowbird young in that nest and may be related to encounters of nonlaying cowbirds at the nest. Egg puncture may reflect interference competition among cowbirds for opportunities to parasitize nests under high cowbird densities. The destructive consequences of egg puncture and association of the behavior with multiple parasitism emphasize the important of reducing cowbird numbers in nesting areas to conserve species at risk.

Acknowledgments

This work was supported by National Science Foundation Grant PRM-812194 to the University of Colorado, U.S. Fish and Wildlife Service, Frank M. Chapman Memorial Fund of the American Museum of Natural History, Alexander Wetmore Memorial Fund of the American Ornithologists' Union, and the Caribbean Islands National Wildlife Refuge. We thank R. A. Andrews, D. Bennett, W. Childress, S. Furniss, P. Gertler, C. Harvey, P. Hoges, F. Lopez, C. P. Ortega, B. Pace, W. Post, G. Potter, C. Ricart, S. Silander, and A. Valido for logistical and field assistance. We also thank S. G. Sealy, J. N. M. Smith, S. Susnowitz, and R. Bernstein for helpful comments on earlier drafts of this manuscript.

References Cited

Blankespoor, G. W., J. Oolman, and C. Uthe. 1982. Eggshell strength and cowbird parasitism of Red-winged Blackbirds. Auk 99:363–365.

Bond, J. 1983. Birds of the West Indies. Collins, London.

Carey, C. 1986. Possible manipulation of eggshell conductance of host eggs by Brown-headed Cowbirds. Condor 88:388–390.

Carter, M. D. 1986. The parasitic behavior of the Bronzed Cowbird in south Texas. Condor. 88:11–25.

Cruz, A., J. Wiley, T. Nakamura, and W. Post. 1989. The Shiny Cowbird Molothrus bonariensis in the West Indian region: Biogeographical and ecological implications. In Biogeography of the West Indies: Past, present, and future (C. A. Woods, ed). E. J. Brill, Leiden.

Davies, N. B., and M. de L. Brooke. 1988. Cuckoos versus Reed Warblers: Adaptations and counter adaptations. Anim. Behav. 36: 777–796.

Fraga, R. M. 1985. Host-parasite interactions between Chalk-browed Mockingbirds and Shiny Cowbirds. In Neotropical ornithology (P. A. Buckley and M. S. Foster, eds). Ornithol. Monogr. 36:829–844.

Friedmann, H. 1929. The cowbirds: A study in the biology of social parasitism. C. C. Thomas, Springfield, IL.

Hamilton, R. M. G. 1982. Methods and factors that affect the measurement of egg shell quality. Poultry Sci. 61:2022–2039.

Hoy, G., and J. Ottow. 1964. Biological and oological studies of the molothrine cowbirds (Icteridae) of Argentina. Auk 81:186–203.

Hunt, J. R., P. W. Voisey, and B. K. Thompson. 1977. Physical properties of eggshells: A comparison of the puncture and compression tests for estimating shell strength. Can. J. Anim. Sci. 57:329–338.

Kemal, R. E., and S. I. Rothstein. 1988. Mechanisms of avian egg recognition: Adaptive responses to eggs with broken shells. Anim. Behav. 36:175–183.

King, J. R. 1973. Reproductive relationships of the Rufous-collared Sparrow and the Shiny Cowbird. Auk 90:19–34.

Lack, D. 1976. Island biology. University of California Press, Berkeley.

Lerkelund, H. E., A. Moksnes, E. Røskaft, and T. H. Ringsby. 1993. An experimental test of optimal clutch size of the Fieldfare, with a discussion on why brood parasites remove eggs when they parasitize a host species. Ornis Scand. 24:95–102.

Mason, P. 1980. Ecological and evolutionary aspects of host selection in cowbirds. Ph.D. Thesis., University of Texas, Austin.

Nice, M. M. 1929. Some cowbird experiences in Columbus, Ohio. Wilson Bull. 41:42.

Ortega, C. P., J. C. Ortega, and A. Cruz. 1994. Use of artificial Brown-headed Cowbird eggs as a potential management tool in deterring parasitism. J. Wildl. Manage. 58:488–492.

Payne, R. B. 1977. The ecology of brood parasitism in birds. Annu. Rev. Ecol. Syst. 8:1–28.

Post, W. 1981. Biology of the Yellow-shouldered Blackbird: Agelaius xanthomus on a tropical island. Bull. Fla. State. Mus. 26:1–202.

Post, W., T. K. Nakamura, and A. Cruz. 1990. Patterns of Shiny Cowbird parasitism in St. Lucia and southwestern Puerto Rico. Condor 92:461–469.

Post, W., and J. W. Wiley. 1976. The Yellow-shouldered Blackbird: Present and future. Amer. Birds 30:13–20.

———. 1977a. The Shiny Cowbird in the West Indies. Condor 79:119–121.

———. 1977b. Reproductive interactions of the Shiny Cowbird and the Yellow-shouldered Blackbird. Condor 79:176–184.

Raffaele, H. A. 1989. A guide to the birds of Puerto Rico and the Virgin Islands. Princeton University Press, Princeton, NJ.

Røskaft, E., G. H. Orians, and L. D. Beletsky. 1990. Why do Red-winged Blackbirds accept eggs of Brown-headed Cowbirds. Evolutionary Ecol. 4:35–42.

Rothstein, S. I. 1971. Observations and experiment in the analysis of interactions between brood parasites and their hosts. Amer. Natur. 105:71–74.

Scott, D. M. 1977. Cowbird parasitism on the Gray Catbird at London, Ontario. Auk 94:18–27.

Sealy, S. G. 1992. Removal of Yellow Warbler eggs in association with cowbird parasitism. Condor 94:40–54.

Tyler, C., and D. Moore. 1965. Types of damage caused by various cracking and crushing methods used for measuring eggshell strength. Br. Poultry Sci. 6:175–182.

Voisey, P. W., R. M. G. Hamilton, and B. K. Thompson. 1979. Laboratory measurements of eggshell strength, 2: The quasistatic compression, puncture, nondestructive deformation, and specific gravity methods applied to the same egg. Poultry Sci. 58:288–294.

Voisey, P. W., and J. R. Hunt. 1969. Effect of compression speed on the behaviour of eggshells. J. Agric. Eng. Res. 14:40–46.

———. 1974. Measurement of eggshell strength. J. Texture Studies 5:135–182.

Wiley, J. W. 1982. Ecology of avian brood parasitism at an early interfacing of host and parasite populations. Ph.D. Thesis, University of Miami, Coral Gables, FL.

Wiley, J. W., W. Post, and A. Cruz. 1991. Conservation of the Yellow-shouldered Blackbird Agelaius xanthomus, an endangered West Indian species. Biol. Conserv. 55:119–137.

Zar, J. H. 1984. Biostatistical analysis. Prentice Hall, NJ.

21. Why Do Female Brown-headed Cowbirds Remove Host Eggs?

A Test of the Incubation Efficiency Hypothesis

BRIAN D. PEER AND ERIC K. BOLLINGER

Abstract

Female Brown-headed Cowbirds often remove eggs from nests they parasitize, and several hypotheses have been proposed for the origin and adaptive significance of this behavior. The incubation efficiency hypothesis (Peer and Bollinger 1997a) states that the size and number of host eggs affects the incubation efficiency of the parasitic egg. We tested this hypothesis by transferring House Sparrow eggs and cowbird eggs, when available, into the nests of Common Grackles. There was a significant positive relationship between the incubation length for the "parasitic" House Sparrow eggs and grackle clutch size. In addition, the House Sparrow eggs hatched before grackle eggs more often and displayed the largest hatching differential in the smaller clutches. The percentage of unhatched House Sparrow eggs that were considered fertilized was also greater in the larger clutches. We obtained incubation lengths for five cowbird eggs. These also required less time to hatch in smaller grackle clutches. These data support the hypothesis that Brown-headed Cowbirds parasitized larger species more frequently earlier in their evolutionary history. To ensure the adequate incubation of their smaller eggs, female cowbirds may have been forced to remove eggs of their larger hosts.

Introduction

Many avian brood parasites possess unique behaviors with clear adaptive values. For example, nestling Common Cuckoos (*Cuculus canorus*) eject host eggs and nestlings (Jenner 1788, Friedmann 1968), and young honeyguides (*Indicator* spp.) kill their nestmates with specialized mandibular hooks (Friedmann 1955). Both behaviors eliminate competition and improve the parasites' chances of survival. Although parasitic Brown-headed Cowbirds often remove host eggs, the adaptive value of this trait is less apparent (Sealy 1992). The frequency of egg removal tends to vary with the host species. Cowbirds usually remove at least one egg from the nests of Red-winged Blackbirds (*Agelaius phoeniceus*; Blankespoor et al. 1982, Røskaft et al. 1990), whereas they remove an egg from only one third of parasitized Yellow Warbler (*Dendroica petechia*) nests (Clark and Robertson 1981, Sealy 1992).

Several hypotheses have been proposed to explain the origin of host-egg removal by cowbirds (see Sealy 1992). One that has received support is the host incubation limit hypothesis (Davies and Brooke 1988). This hypothesis states that the chance of the parasitic egg hatching increases if a host egg is removed, compared to nests in which no host eggs are removed. Hence, egg removal functions to prevent the clutch size from exceeding a host's capacity to incubate the entire clutch successfully. To date, support for the host incubation limit hypothesis has been based on anecdotal observations (Davies and Brooke 1988) or on clutch augmentations with same-sized eggs (Lerkelund et al. 1993).

Peer and Bollinger (1997a) artificially parasitized Common Grackle (*Quiscalus quiscula*) nests and found that cowbird eggs hatched only in clutches that contained three or fewer eggs. Peer and Bollinger suggested that both the large egg size of the grackle and the number of eggs in a clutch may affect the incubation efficiency of the parasitic cowbird egg. This incubation efficiency hypothesis differs from the host incubation limit hypothesis in that it considers the effects of both host egg size and the number of host eggs on the effectiveness of incubation of the parasitic egg. This hypothesis predicts that an enlarged clutch volume will have an adverse effect on the incubation efficiency of the parasitic egg when it is smaller than the host's eggs, rather than affecting the incubation of the entire clutch. It also proposes no discrete clutch size limit to effective incubation.

The incubation efficiency hypothesis conforms to speculations that Brown-headed Cowbirds parasitized larger species more frequently in the past (Rothstein 1975, Mason 1980, Peer and Bollinger 1997a). Most rejectors of Brown-headed Cowbird eggs are larger species (Rothstein 1975, Peer and Bollinger 1997a), and cowbirds may have subsequently been forced to parasitize smaller species after the frequently parasitized larger species began rejecting cowbird eggs. Currently, both Shiny (Post and Wiley 1977, Mason 1986a,b) and Bronzed Cowbirds (Carter 1986) parasitize larger hosts more often than Brown-headed Cowbirds. Most species that reject eggs of these cowbirds are also larger species (Carter 1986, Mason 1986b).

If Brown-headed Cowbirds parasitized larger hosts more frequently in the past, as both the Shiny and Bronzed Cowbirds do today, then the explanation for egg removal by female cowbirds is more obvious. A female cowbird that consistently parasitizes larger hosts would benefit by removing a host egg to increase contact of her own smaller egg with the host's brood patch. For host species with smaller eggs this advantage is reduced or absent, because the cowbird egg is larger. Presumably, the larger cowbird egg would receive sufficient heat from the host's brood patch without egg removal (e.g., Hofslund 1957). Thus, our objective here was to test the incubation efficiency hypothesis, by testing if the egg and clutch size of a large host (Common Grackle) affects the incubation length and incubation efficiency of Brown-headed Cowbird eggs.

Methods

This study was conducted in Coles County, Illinois, from 1 April to 1 July 1993. The primary study sites were a Christmas tree farm and a cemetery. The cemetery contained scattered rows of northern white cedar (*Thuja occidentalis*), 2–3 m in height. Deciduous trees were also found throughout the cemetery. The Christmas tree farm contained evenly spaced rows of Scotch pine (*Pinus sylvestris*), 2–2.5 m in height.

Due to the asynchrony of cowbird and grackle breeding seasons (Peer and Bollinger 1997a), we used House Sparrow eggs (*Passer domesticus*) instead of cowbird eggs in most experiments (see Rothstein 1977). Real cowbird eggs were used when they were available. House Sparrow and cowbird eggs have similar spotting patterns and dimensions. The House Sparrow eggs used in this study averaged 21.95 (SE = 0.12) × 15.52 mm (SE = 0.08; N = 49) and cowbird eggs averaged 20.38 (SE = 0.27) × 16.05 mm (SE = 0.15; N = 19). Both egg types were collected daily to ensure that their laying dates were known and that they had not been incubated. We transported the eggs in a lined egg carton to reduce the risk of damage, and transferred all eggs to new nests within 15 min of removal.

We added grackle eggs collected from nonexperimental nests to other grackle nests. We created clutch sizes of one to seven grackle eggs (in addition to the parasitic egg). This is the normal range of grackle clutch sizes. The upper and lower extremes are rare, and the modal clutch size is five eggs (Peer and Bollinger 1997b). We added the parasitic eggs only during the grackles' laying period (before incubation had

begun) and visited nests each day before 1200 (CDT) to record hatching dates. We considered eggs that did not hatch to have been inefficiently incubated only if they contained an embryo. If not, we considered the eggs to be infertile. It is possible that we did not detect all fertilized eggs because the eggs were only viewed macroscopically. Therefore, we analyzed the total number of eggs that did not hatch in addition to only those eggs considered to be fertile.

We defined the length of incubation as the time from when the female began incubating a clutch to the day that the first egg hatched. Grackles typically begin incubation after the penultimate egg has been laid in clutches of five or more eggs, whereas they begin incubation after the last egg is laid in smaller clutches (Howe 1978). We confirmed this pattern of incubation in our study by checking the eggs for warmth.

Results

We obtained incubation lengths for 62 House Sparrow eggs. There was a significant positive relationship between the incubation length for the "parasitic" House Sparrow eggs and grackle clutch size (Spearman's rank correlation, r_s = 0.376, df = 61, P < .005) (Figure 21.1). In contrast, the incubation lengths of grackle eggs did not vary with clutch size (r_s = 0.044, df = 92, P > .50, Figure 21.1). Thus, larger host clutches resulted in longer incubation lengths for the parasitic eggs. Furthermore, incubation lengths for House Sparrow eggs in clutches containing five or more grackle eggs (mean = 13.9 d, SE = 0.28, N = 16) or four grackle eggs (mean = 13.6 d, SE = 0.30, N = 21) were significantly greater than those for House Sparrow eggs in clutches of three or fewer grackle eggs (mean = 12.9 d, SE = 0.17, N = 25; Mann-Whitney tests, W = 438, P = .005; W = 590, P < .05). Cowbird eggs followed the same trend as the House Sparrow eggs, requiring more time to hatch in larger grackle clutches, though the difference was not statistically significant (r_s = 0.76, df = 4, P < .10, Figure 21.2).

House Sparrow incubation lengths ranged from 11 days, in clutches of two and four grackle eggs, to 16 days in clutches of four and five grackle eggs. The incubation lengths for cowbird eggs ranged from 10 days in a clutch of two grackle eggs, to 13.5 days in a clutch of four grackle eggs. The average incubation length for House Sparrow eggs from all clutch sizes (13.44 d, SE = 0.15) was essentially equal to that of the grackle eggs (13.46 d, SE = 0.07). The average incubation length of the cowbird eggs was 12.1 d (SE = 0.60).

The difference in incubation lengths between House Sparrow eggs and grackle eggs within a clutch was significantly related to clutch size (r_s = -0.424, df = 50, P < .005). This difference was greatest in grackle clutches of 1–2 eggs, in which the sparrow eggs hatched a full day earlier, followed by grackle clutches of 3 eggs, in which the sparrow eggs still hatched nearly three-quarters of a day sooner than the grackle eggs. However, for grackle clutches of 4 or more eggs, the situation was reversed, and grackle eggs hatched almost half a day earlier than the House Sparrow eggs. The proportion of House Sparrow eggs that hatched before grackle eggs was also greater in clutches of 3 or fewer eggs (60.9%, 14 of 23), compared to clutches of 4 (14.3%, 2 of 14) or clutches of 5 or more eggs (20.0%, 3 of 15) (χ^2 = 10.63, df = 2, $P < .005$).

The percentage of unhatched House Sparrow eggs (that we considered to have been fertilized) was higher in clutches of five or more eggs (20.0%, 4 of 20) compared to clutches of four or fewer eggs (5.9%, 3 of 51), (χ^2 = 3.22, df = 1, $P = .035$). However, when all of the unhatched House Sparrow eggs were included in the analysis (including eggs we considered as infertile), the trend was no longer evident (χ^2 = 0.08, df = 1, $P > .75$).

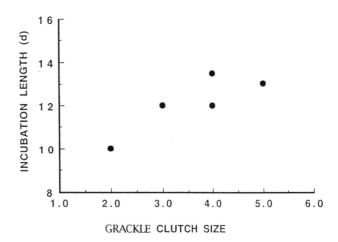

Figure 21.2. The relationship between incubation length of Brown-headed Cowbird eggs ($N = 5$) and clutch size of Common Grackles.

Discussion

Host Clutch Size and Incubation Length of Parasitic Eggs

After a host has accepted a cowbird egg, it must hatch to have any chance of success. However, the time at which the parasite egg hatches relative to the host's eggs is almost as significant as whether it hatches or not. Hatching earlier allows a brood parasite to gain an advantage over its nestmates and permits the cowbird to compensate for its smaller size in the case of larger hosts (Peer and Bollinger 1997a). In the case of smaller hosts, hatching earlier allows the cowbird to increase its size advantage over the smaller nestlings and thereby monopolize parental care (Wolf 1987).

We found that in experimentally reduced clutches the parasitic eggs had shorter incubation lengths, hatched before grackle eggs more often, and displayed the largest hatching differentials. The probability of hatching was also greater in smaller clutches for the sparrow eggs we considered to be fertile. However, when all unhatched House Sparrow eggs were analyzed, there was no trend. This fact suggests that contrary to expectation, there was a higher proportion of infertile House Sparrow eggs in the smaller clutches.

Thus, by removing an egg of a large host, the female cowbird increases the likelihood that her egg will hatch and decreases the length of time required for the egg to hatch. Host egg removal also helps to ensure that the parasitic egg will hatch first and increases the disparity between the hatching of the parasite and the host's nestlings. Therefore, these results support the incubation efficiency hypothesis as an explanation for egg removal in Brown-headed Cowbirds.

Results from Other Studies

Studies of House Wrens (*Troglodytes aedon*; Baltz and Thompson 1988) and Blue Tits (*Parus caeruleus*; Smith 1989) found that clutches enlarged with conspecific eggs (hence, same-sized eggs) were just as effectively incubated as control clutches. However, they noted that incubation length increased by half a day in the enlarged clutches. In contrast, experimentally enlarged clutches of Collared Flycatchers (*Ficedula albicollis*; Moreno et al. 1991) and Fieldfares (*Turdus pilaris*; Lerkelund et al. 1993) experienced a higher incidence of unhatched eggs. Lerkelund et al. (1993) also noted increased "wear and tear" on the eggs in the larger clutches and that this damage further reduced their hatching success.

The incubation efficiency of cowbird eggs has also been estimated for two larger hosts. Ortega and Cruz (1991) found that eggs of Brown-headed Cowbirds (males 49 g, females 39 g; adult body masses from Dunning 1993) were incubated effectively in the nests of Yellow-headed Blackbirds

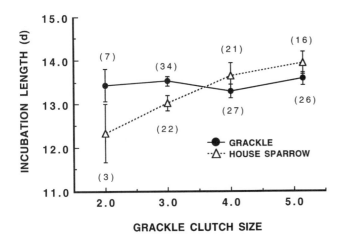

Figure 21.1. The mean incubation length (± SE) of House Sparrow and Common Grackle eggs in relation to clutch size of Common Grackles. Sample sizes are given in parentheses.

(*Xanthocephalus xanthocephalus*; males 80 g, females 49 g), although the blackbird eggs were 11% wider than the cowbird eggs. We report only the differences in egg widths because it is likely that egg width, rather than egg length, is the critical parameter in determining the amount of contact an egg has with the brood patch (S. I. Rothstein pers. comm.). Wiley (1985), on the other hand, recorded a higher incidence of unhatched Shiny Cowbird (males 39 g, females 32 g) eggs in the nests of Greater Antillean Grackles (*Q. niger*) (males 86 g, females 62 g) than in the nests of other host species. He suggested that the smaller cowbird eggs were not adequately heated by the brood patch of the grackle and thus did not hatch. Greater Antillean Grackle eggs are 20% wider than the eggs of the Shiny Cowbird. The difference in width is even greater (33%) between Common Grackle (males 127 g, females 100 g) and Brown-headed Cowbird eggs.

Great Spotted Cuckoos (*Clamator glandarius*) also parasitize larger hosts, yet it is unclear how often these cuckoos remove host eggs (see Soler 1990). Soler (1990) recorded approximately one less host egg in parasitized versus unparasitized nests of four host species. He suggested that cuckoo eggs striking and damaging host eggs during laying caused this. Presumably, the hosts then removed the damaged eggs, thus reducing the clutch size. In four parasitized nests of the Carrion Crow (*Corvus corone*), Soler found that all Great Spotted Cuckoo eggs were incubated effectively despite the Carrion Crow eggs being 29.5% wider than cuckoo eggs. This study may also support the incubation efficiency hypothesis because the smaller cuckoo eggs hatched from reduced clutches, albeit they were reduced in a different manner. This could be tested experimentally by comparing the incubation efficiency of cuckoo eggs in nests from which a host egg has been removed with those in nests without removal.

Interestingly, the average incubation length we recorded for the House Sparrow eggs incubated by grackles closely matched that of the grackle eggs, at approximately 13.5 days. This is similar to reported incubation lengths for grackles (13.2 d, Maxwell and Putnam 1972), but is dramatically different from values reported for House Sparrow eggs (10.7 d, Murphy 1978; 11.5 d, Anderson 1994). They are different probably because the smaller House Sparrow eggs received less heat than they normally would if they were incubated in a clutch of similar-sized House Sparrow eggs. As a result, they required more time to hatch. The average incubation length that we recorded for the cowbird eggs (12.1 d) is within its reported range of 10–13 days (Nice 1937, 1953; Rothstein 1975; Briskie and Sealy 1990).

Our results also suggest that the wide range of incubation lengths reported for Brown-headed Cowbird eggs may be a consequence of the eggs being in nests of different hosts with varying clutch sizes. Incubation lengths in our study varied with clutch volume, so they would also be expected to vary with the different clutches of different-sized hosts. Unfortunately, it is not yet possible to test this prediction because data in the literature usually present only the mean incubation lengths for all clutch sizes combined (but see Briskie and Sealy 1990). Furthermore, there are few data concerning the incubation lengths of cowbird eggs in the nests of larger species because many of these species are rejectors.

Egg Removal in Relation to Host Size

Removal of eggs from smaller hosts' nests might not appear to be as adaptive as egg removal from larger species' nests. The larger cowbird egg presumably would have no problem coming into contact with the host's brood patch (Hofslund 1957, Mayfield 1960, Friedmann 1963, Rothstein 1975). Hofslund (1957) found that fewer Common Yellowthroat (*Geothlypis trichas*) eggs hatched in the presence of cowbird eggs, which suggests that cowbird eggs actually caused the smaller host's eggs to be inadequately incubated. In addition, cowbirds may remove more eggs from nests of larger hosts (Elliott 1978, Blankespoor et al. 1982, Røskaft et al. 1990, but see Scott 1977) than from nests of smaller hosts (Nice 1937, Clark and Robertson 1981, Sealy 1992, Zimmerman 1983, but see Nolan 1978). However, the strength of the relationship between host size and frequency of egg removal is only moderate, perhaps because egg removal within a host species varies with cowbird density. Where cowbirds are common, egg removal by multiple females is common (e.g., Elliott 1978).

Decreasing the number of eggs in a smaller host's nest may also increase the chance that the parasite's egg will hatch. However, this benefit may be negligible compared to the potential costs. Cowbirds, unlike cuckoos, typically do not remove an egg at the time they parasitize a nest (Sealy 1992). As a result, the female cowbird must make another visit to the nest (if she removes an egg at all). If a host detects a brood parasite, the host may be more likely to reject its egg (see Davies and Brooke 1988, Moksnes and Røskaft 1989, but see Sealy 1995). In the case of larger hosts, the potential for injury is also greater (Leathers 1956, see also Davidson 1994), yet cowbirds still remove eggs from nests of large species regularly. Cowbirds are capable of removing eggs as large as those of grackles; Blincoe (1935) and S. G. Sealy (unpubl. data) have both observed cowbirds removing the eggs of American Robins (*Turdus migratorius*), which are the same size as grackle eggs.

Hatching before a host's nestlings is not always required for cowbirds to fledge from the nests of smaller species (Clark and Robertson 1981, Marvil and Cruz 1989, Petit 1991). However, for the cowbird to fledge successfully from the nests of larger hosts, hatching earlier seems to be imperative. Mason (1986a) and Fraga (1985) found that the shorter incubation period of the Shiny Cowbird was offset by the larger size and faster growth rate of the Chalk-browed

Mockingbird (*Mimus saturninus*). Carter (1986) found that nestling Bronzed Cowbirds were much more likely to fledge from the nests of moderate-to-large hosts when they hatched in synchrony with or before the hosts' nestlings. Peer and Bollinger (1997a) found that the only Brown-headed Cowbirds that fledged from nests of the larger Common Grackle hatched at least two days earlier than the grackles.

Larger species are the optimal hosts for cowbird nestlings because they can supply them more food (Fraga 1985, Carter 1986, Mason 1986a, Wiley 1986) and they are more capable of defending their nests against predators compared to smaller hosts (Fretwell in Rothstein 1975, Gottfried 1979). The aggressive begging behavior displayed by cowbirds (Nice 1939, Gochfeld 1979) may be another adaptation that allows the cowbird to benefit from parasitizing larger species. Gochfeld (1979) and Wiley (1986) both suggested that exaggerated begging may help Shiny Cowbirds to compete against larger hosts. Persistent and loud begging may help the smaller cowbird to acquire food in the nests of larger species. Although strident begging behavior may draw the attention of nest predators (Hudson 1920, Mayfield 1960, Gochfeld 1979), this may be offset by the enhanced ability of larger hosts to defend their nests.

Evolution of Egg Removal

Brown-headed Cowbirds may have initially preferred larger hosts as the Bronzed and Shiny Cowbirds do today. In support of this idea, most rejectors of Brown-headed Cowbird eggs are larger species (Rothstein 1975, Mason 1980, Peer and Bollinger 1997a). Incubation efficiency does not constrain tropical cowbirds as it does Brown-headed Cowbirds (but see Wiley 1985), because clutch sizes decrease with latitude (Ricklefs 1980) and ambient temperatures increase. Therefore, the Bronzed and Shiny Cowbirds would be expected to remove eggs less often than Brown-headed Cowbirds. Tropical cowbirds do not remove eggs per se, but instead puncture eggs. The few studies that have documented this behavior indicate it occurs as often as egg removal (Fraga 1978, Mason 1980, Carter 1986). However, the function of egg puncture is unclear and requires further study (but see Nakamura and Cruz, Chapter 20, this volume).

While the incubation efficiency hypothesis explains the function (and perhaps lack of function in small hosts) of egg removal in Brown-headed Cowbirds, it may not explain this behavior in other brood parasites (Lerkelund et al. 1993). Brood parasitism has evolved independently in seven avian taxa (Payne 1977, Rothstein 1990) with egg removal occurring in most of these groups (see Lerkelund et al. 1993). For example, species of both cuckoos and honeyguides eliminate nestling competitors, yet they still remove host eggs (Friedmann 1955, 1968). Here, egg removal clearly does not function to reduce competition with host nestlings. Therefore, egg removal in these parasites may also

function to enhance incubation. Cuckoos have eggs that mimic those of their hosts, and cuckoo and host's eggs are thus fairly similar in size (but see Soler 1990, and above). The incubation efficiency hypothesis is not relevant in this situation because it addresses differences in host and parasitic egg sizes. Therefore, the host incubation limit hypothesis appears to be a more appropriate explanation in the case of cuckoos. However, honeyguides parasitize hosts with larger eggs (Friedmann 1955). Thus, the incubation efficiency hypothesis may apply in this taxon.

Acknowledgments

We thank the various landowners who gave us access to their properties. We also thank Roger W. Jansen for the dedicated field assistance he provided. D. Glen McMaster, Stephen I. Rothstein, and Spencer G. Sealy offered many suggestions to help improve the manuscript. Financial support was provided in part by the Council for Faculty Research at Eastern Illinois University.

References Cited

Anderson, T. R. 1994. Breeding biology of House Sparrows in northern lower Michigan. Wilson Bull. 106:537–548.

Baltz, M. E., and C. F. Thompson. 1988. Successful incubation of experimentally enlarged clutches by House Wrens. Wilson Bull. 100:70–79.

Blankespoor, G. W., J. Oolman, and C. Uthe. 1982. Eggshell strength and cowbird parasitism of Red-winged Blackbirds. Auk 99:363–365.

Blincoe, B. J. 1935. A cowbird removes a robin's egg. Wilson Bull. 47:158.

Briskie, J. V., and S. G. Sealy. 1990. Evolution of short incubation periods in the parasitic cowbirds, *Molothrus* spp. Auk 107:789–794.

Carter, M. D. 1986. The parasitic behavior of the Bronzed Cowbird in south Texas. Condor 88:11–25.

Clark, K. L., and R. J. Robertson. 1981. Cowbird parasitism and evolution of anti-parasite strategies in the Yellow Warbler. Wilson Bull. 93:249–258.

Davidson, A. H. 1994. Common Grackle predation on adult passerines. Wilson Bull. 106:174–175.

Davies, N. B., and M. De L. Brooke. 1988. Cuckoos versus Reed Warblers: Adaptations and counter-adaptations. Anim. Behav. 36:262–284.

Dunning, J. B., Jr. 1993. CRC handbook of avian body masses. CRC Press, Ann Arbor, MI.

Elliott, P. F. 1978. Cowbird parasitism in the Kansas tallgrass prairie. Auk 95:161–167.

Fraga, R. M. 1978. The Rufous-collared Sparrow as a host of the Shiny Cowbird. Wilson Bull. 90:271–284.

———. 1985. Host-parasite interactions between Chalk-browed Mockingbirds and Shiny Cowbirds. Ornithol. Monogr. 36:607–620.

Friedmann, H. 1955. The honeyguides. U.S. Natl. Mus. Bull. 208.

———. 1963. Host relations of the parasitic cowbirds. U.S. Natl. Mus. Bull. 233.

———. 1968. The evolutionary history of the avian genus *Chrysococcyx*. U.S. Natl. Mus. Bull. 265.

Gochfeld, M. 1979. Begging by nestling Shiny Cowbirds: Adaptive or maladaptive? Living Bird 17:41–48.

Gottfried, B. M. 1979. Anti-predator aggression in birds nesting in old field habitats: Experimental analysis. Condor 21:251–257.

Hofslund, P. B. 1957. Cowbird parasitism of the Northern Yellowthroat. Auk 74:42–48.

Howe, H. F. 1978. Initial investment, clutch size, and brood reduction in the Common Grackle (*Quiscalus quiscula* L.). Ecology 59:1109–1122.

Hudson, W. H. 1920. Birds of La Plata. J. M. Dent and Sons, London.

Jenner, E. 1788. Observations on the natural history of the cuckoo. Phil. Trans. Roy. Soc. Lond. 78:219–235.

Leathers, C. L. 1956. Incubating American Robin repels female Brown-headed Cowbird. Wilson Bull. 68:68.

Lerkelund, H. E., A. Moksnes, E. Røskaft, and T. H. Ringsby. 1993. An experimental test of optimal clutch size of the Fieldfare, with a discussion on why brood parasites remove eggs when they parasitize a host species. Ornis Scand. 24:95–102.

Marvil, R. E., and A. Cruz. 1989. Impact of Brown-headed Cowbird parasitism on the reproductive success of the Solitary Vireo. Auk 106:476–480.

Mason, P. 1980. Ecological and evolutionary aspects of host selection in cowbirds. Ph.D. Thesis, University of Texas, Austin.

———. 1986a. Brood parasitism in a host generalist, the Shiny Cowbird: 1. The quality of different species as hosts. Auk 103:52–60.

———. 1986b. Brood parasitism in a host generalist, the Shiny Cowbird: 2. Host selection. Auk 103:61–69.

Mayfield, H. 1960. The Kirtland's Warbler. Cranbrook Institute of Science, Bloomfield Hills, MI.

Maxwell, G. R. II, and L. S. Putnam. 1972. Incubation, care of young, and nest success of the Common Grackle (*Quiscalus quiscula*) in northern Ohio. Auk 89:349–359.

Moksnes, A., and E. Røskaft. 1989. Adaptations of Meadow Pipits to parasitism by the Common Cuckoo. Behav. Ecol. Sociobiol. 24:25–30.

Moreno, J., L. Gustafsson, A. Carlson, and T. Pärt. 1991. The cost of incubation in relation to clutch-size in the Collared Flycatcher *Ficedula albicollis*. Ibis 133:186–193.

Murphy, E. C. 1978. Breeding ecology of House Sparrows: Spatial variation. Condor 80:180–193.

Nice, M. M. 1937. Curious ways of the cowbird. Bird Lore 39:196–201.

———. 1939. Observations on the behavior of a young cowbird. Wilson Bull. 51:233–239.

———. 1953. The question of ten-day incubation periods. Wilson Bull. 65:81–93.

Nolan, V. 1978. The ecology and behavior of the Prairie Warbler *Dendroica discolor*. Ornithol. Monogr. 26:1–595

Ortega, C. P., and A. Cruz. 1991. A comparative study of cowbird parasitism in Yellow-headed Blackbirds and Red-winged Blackbirds. Auk 108:16–24.

Payne, R. B. 1977. The ecology of brood parasitism in birds. Annu. Rev. Ecol. Syst. 8:1–28.

Peer, B. D., and E. K. Bollinger. 1997a. Explanations for the infrequent cowbird parasitism on Common Grackles. Condor 99:151–161.

———. 1997b. Common Grackle (*Quiscalus quiscula*). *In* The birds of North America 271 (A. Poole and F. Gill, eds). Academy of Natural Sciences, Philadelphia; American Ornithologists' Union, Washington, DC.

Petit, L. J. 1991. Adaptive tolerance of cowbird parasitism by Prothonotary Warblers: A consequence of nest-site limitation? Anim. Behav. 41:425–432.

Post, W., and J. W. Wiley. 1977. Reproductive interactions of the Shiny Cowbird and Yellow-shouldered Blackbird. Condor 79:176–184.

Ricklefs, R. E. 1980. Geographical variation in clutch size among passerine birds: Ashmole's hypothesis. Auk 97:38–49.

Røskaft, E., G. H. Orians, and L. D. Beletsky. 1990. Why do Redwinged Blackbirds accept eggs of Brown-headed Cowbirds? Evolutionary Ecol. 4:35–42.

Rothstein, S. I. 1975. An experimental and teleonomic investigation of avian brood parasitism. Condor 77:250–271.

———. 1977. Cowbird parasitism and egg recognition of the Northern Oriole. Wilson Bull. 89:21–32.

———. 1990. A model system for coevolution: Avian brood parasitism. Annu. Rev. Ecol. Syst. 21:481–508.

Scott, D. M. 1977. Cowbird parasitism on the Gray Catbird at London, Ontario. Auk 94:18–27.

Sealy, S. G. 1992. Removal of Yellow Warbler eggs in association with cowbird parasitism. Condor 94:40–54.

———. 1995. Burial of cowbird eggs by parasitized Yellow Warblers: An empirical and experimental study. Anim. Behav. 49:877–889.

Smith, H. G. 1989. Larger clutches take longer to incubate. Ornis Scand. 20:156–158.

Soler, M. 1990. Relationships between the Great Spotted Cuckoo *Clamator glandarius* and its corvid hosts in a recently colonized area. Ornis Scand. 21:212–223.

Wiley, J. W. 1985. Shiny Cowbird parasitism in two avian communities in Puerto Rico. Condor 87:165–176.

———. 1986. Growth of Shiny Cowbird and host chicks. Wilson Bull. 98:126–131.

Wolf, L. 1987. Host-parasite interactions of Brown-headed Cowbirds and Dark-eyed Juncos in Virginia. Wilson Bull. 99:338–350.

Zimmerman, J. L. 1983. Cowbird parasitism of Dickcissels in different habitats and at different nest densities. Wilson Bull. 95:7–22.

PART IV

*Environmental
Correlates of
Cowbird Parasitism
at Multiple
Spatial Scales*

22. Introduction

SCOTT K. ROBINSON AND JAMES N. M. SMITH

This largest part of the book comprises chapters that deal with the effects of spatial scale and landscape heterogeneity (Forman 1995) on levels of cowbird parasitism. As these chapters collectively show, parasitism may vary locally with several factors: the vegetation structure of nest sites and territories, proximity to edge, vertical strata, tract size, and habitat type. Parasitism also varies at larger scales with landscape composition and biogeographic context. All of these scales must be considered when designing management plans to reduce cowbird parasitism.

Before discussing the patterns revealed by the studies below, we consider briefly the nature of the evidence available to examine environmental influences on cowbird parasitism. Nearly all the data presented here consist of correlations between (a) descriptions of habitats and landscapes and (b) cowbird abundance and/or parasitism levels in hosts. Such data may suggest that particular factors cause high levels of parasitism, particularly when similar correlations appear at the same spatial scale in different landscapes or across spatial scales within a biogeographic region, but the causal relationships may be hidden and actually due to other unmeasured factors. To confirm that a factor plays a causal role, experiments at the appropriate spatial scale are needed. For example, manipulation of vegetation can affect parasitism levels locally (Larison 1996), or creation of forest patches of varying size and degree of isolation can affect the composition of passerine communities (Bierregard and Lovejoy 1989, Schmiegelow et al. 1997). Such experimental studies are expensive and logistically difficult, particularly in the human-altered landscapes where cowbirds are common. Little such evidence is available now, nor is much of it likely to be available in the near future. There has, however, been a recent and radical shift in conservation biology away from managing populations of single species toward managing habitats and ecosystems (Meffe et al. 1997). This shift may well lead to more frequent experiments at the large spatial scales that are most likely to improve our understanding of the causes of frequent and intense cowbird parasitism.

There are, however, two types of ecological experiments that are already contributing useful data. Habitat restoration efforts, e.g., burning to encourage grasslands (Howe and Knopf Chapter 23), are ecological experiments that may affect host breeding success and the frequency of parasitism. Forest and agricultural management practices are a second type of experiment that can be used to explore the causes of high levels of parasitism. For example, evidence from two biogeographic regions suggests that clearcut logging does not elevate parasitism levels greatly in newly fragmented forested landscapes (boreal mixed wood in central Alberta, S. J. Hannon unpubl. data; montane spruce-pine in southern British Columbia, C. J. Fonnesbeck and J. N. M. Smith unpubl. data).

Thus, there are limits to the certainty of the conclusions that can be drawn about the ecological causes of high or low levels of parasitism. Despite these limits, the descriptive studies in this part reveal many common patterns that increase our confidence that causal hypotheses based on these results may be robust.

We now synthesize the thirteen chapters in the part by examining correlates of parasitism, beginning with local spatial scales and progressing upward to the biogeographic scale. Each spatial scale has several featured chapters, but some chapters deal explicitly with multiple scales (e.g., Robinson et al. Chapter 33, Donovan et al. Chapter 30, Thompson et al. Chapter 32); these chapters will be mentioned in several spatial contexts. Similarly, we will also mention several chapters that have a spatial component from other parts of the book.

Local Scales

We begin this part with a discussion of the effects on parasitism of within-habitat features, by which we mean features such as vegetation structure at nest sites and on territories, position of nests within a habitat (vertical strata), forest structure, and host species identity. We then proceed to the within-landscape scale, which includes effects on parasitism levels of proximity to edge, forest tract size, and differences among habitats. These scales are of greatest relevance to managers of particular sites or constellations of sites within a region.

Within-habitat Features

Because female cowbirds have to invest time and energy in searching for host nests (Norman and Robertson 1975, Wiley 1988), it seems reasonable that concealment by dense vegetation will often affect parasitism levels. Chapter 23 by Howe and Knopf and Chapter 24 by Uyehara and Whitfield both find evidence that vegetation structure around the nest site is correlated with parasitism levels in western riparian habitats (see also Larison 1996). Burning also generated an interaction between parasitism and nest depredation in Howe and Knopf's study. These results support Martin's (1992) contention that management of nesting microhabitat may be crucial to increasing nesting success. Uyehara and Whitfield go one step farther by adding a measure of heterogeneity of cover within territories to their analysis. Although their results are not quite statistically significant, there is an intriguing possibility that cowbirds have an easier time finding nests in territories in which suitable cover is patchily distributed. In a more general review, Robinson et al. (1995a), however, failed to find a general effect of nest cover on parasitism levels, possibly because cowbirds often use host behavior as a cue for locating nests rather than searching for nests directly (Norman and Robertson 1975).

Within forested habitats, there is conflicting evidence on the effects of nesting stratum on parasitism levels. In Chapter 13, Hahn and Hatfield (also 1995) found much higher parasitism levels on ground nesters. In contrast, Robinson et al. (Chapter 33) found high levels of parasitism in all strata, but levels were highest among canopy nesters, which Hahn and Hatfield generally did not sample. In a general review, Robinson et al. (1995a) found no consistent preferences for any particular vegetation stratum within a habitat. As more studies become available, we can address the question of how cowbird abundance affects preferences for particular strata. In the Illinois forests where cowbirds are very abundant, they may be forced to parasitize all suitable hosts. In forests where they are less abundant, cowbirds may show stronger preferences for a particular stratum.

There are few data on the effects of forest composition, stand age, and forest structure on parasitism. Dowell et al. (Chapter 29) report lower parasitism levels in old-growth forests in Maryland, but not for all species. Although Robinson et al. in Chapter 33 report a trend toward lower parasitism levels in floodplain than in upland forests, a general review (Robinson et al. 1995a) found few consistent effects of forest structure on parasitism levels.

Within habitats and nesting strata, virtually all studies show substantial interspecific variation in parasitism levels

across species. Such variation occurs even among species that are often considered to be especially suitable cowbird hosts. Dowell et al. (Chapter 29), for example, find that Hooded Warblers were parasitized nearly eight times more often than Wood Thrushes in Maryland woodlots. In grasslands, Chapter 26 by Davis and Sealy and Chapter 27 by Koford et al. report substantial intraspecific variation. Within forest habitats in Illinois, intensity of parasitism of neotropical migrants was strongly positively correlated with host size (see also Robinson et al. Chapter 33), a relationship that did not exist for shrubland habitats (see also Ward and Smith Chapter 25). Whether these interspecific differences reflect habitat or host preferences by cowbirds, or differences in nest defense or other host behaviors, remains to be shown.

Within-landscape Features

Effects of habitat fragmentation at the local (within-landscape) scale can vary from subtle to profound. For example, Winslow et al. (Chapter 34) find somewhat higher parasitism levels near agricultural and silvicultural openings in the forest, although they do not find greater cowbird abundance near edges. Thompson et al. (Chapter 32) find that edge effects vary regionally; they may be absent where cowbirds are abundant and saturate available habitat. In contrast, in landscapes where cowbirds are less abundant, they may search for nests preferentially along the edges of forest openings or in forests adjacent to agricultural feeding areas (see also Thompson and Dijak Chapter 10 and Raim Chapter 9). Edge effects may also be pronounced in grasslands. Koford et al. (Chapter 27) found higher parasitism levels in a small tract in which most nests were close to fencerows and trees where cowbirds could perch (see also Johnson and Temple 1990).

The chapters in this volume provide some of the first strong evidence that tract size influences parasitism levels. Petit and Petit demonstrated strong negative correlations between area and parasitism levels in one species, the Acadian Flycatcher. These correlations remained even when regional effects and landscape context were controlled. In Chapter 31, however, Petit and Petit find no significant correlations with area in the Wood Thrush. Dowell et al. (Chapter 29) found evidence of a negative correlation between area and parasitism levels, but there were possible confounding effects of forest age and landscape context. In an agricultural landscape in Illinois, there was a negative correlation between forest area and parasitism levels (Robinson et al. Chapter 33), but most of the relationship was caused by very high parasitism levels in small woodlots (under 200 ha) in which cowbirds were abundant and hosts were relatively scarce (i.e., a high cowbird-to-host ratio). Medium-sized tracts (200–3,500 ha) in Illinois also had high but variable levels of parasitism. Although there were few replicate sites in their study, Davis and Sealy (Chapter 26) found that parasitism levels also were lower in small grasslands than in larger ones in Manitoba. Donovan et al. (Chapter 30) found significantly higher community-wide levels of parasitism in small than in large forest tracts; in very large tracts (20,000 ha or greater), parasitism levels were close to zero (see also Thompson et al. Chapter 32). Taken together, these results show that tract size is generally negatively correlated with parasitism levels but that most of the effect comes at the extremes of the distribution, with very high parasitism in the smallest woodlots (under 200 ha) and virtually no parasitism in the interior of the largest tracts (over 10,000 ha). Within the wide range of habitat patches that fall between these extremes, there is a great deal of interspecific and intersite variation.

Habitat Effects

Within landscapes, parasitism levels also vary among habitats. In grasslands, Koford et al. (Chapter 27) found higher parasitism levels in natural grasslands and crops than in seeded grasslands. They argue that the shrubs in natural grasslands give cowbirds perches from which they can search for nests and that the lack of places to hide nests in cropland makes it easier for cowbirds to find nests. In the Okanagan Valley of British Columbia, Ward and Smith (Chapter 25) found that parasitism levels were significantly higher in riparian than in upland and sagebrush areas. Riparian habitats also seem to be preferred by cowbirds elsewhere in western North America (Uyehara and Whitfield Chapter 24, Howe and Knopf Chapter 23, Griffith and Griffith Chapter 38, and Whitfield Chapter 40). Robinson et al. (Chapter 33) also found strong habitat effects in Illinois. Parasitism was rare in grasslands, common in shrubland-edge habitats, and frequent in forests. Winslow et al. (Chapter 34) also found that birds of early successional habitats in Indiana were much less heavily parasitized (see also Hahn and Hatfield 1995). These data suggest that the need to manage for cowbird parasitism varies greatly among habitats. We know little, however, about why cowbirds often prefer one habitat over another within a landscape, although it seems likely that such habitat preferences are driven by the local abundance of preferred hosts.

Regional and Biogeographic Effects

We now review how larger-scale spatial variation in landscapes is correlated with cowbird abundance and parasitism levels. We begin by examining studies that show effects of different landscape composition on parasitism levels within the same geographic region. We then discuss some of the larger-scale patterns of biogeographic variation revealed by comparing parasitism levels in different regions.

Variation among Landscapes within Regions

Thompson et al. (Chapter 32) provide the strongest evidence to date that cowbird abundance and parasitism are correlated with landscape composition. In areas of the American Midwest that were mostly (over 80%) forested within a 10-km radius of their study site (their operational definition of a landscape), parasitism levels were so low for most species that cowbird parasitism must have had little impact on host populations. In contrast, in areas that were largely (over 50%) agricultural, parasitism frequencies were much higher for virtually all species, even reaching 90–100% for some species. Agricultural landscapes are characterized by frequent parasitism for most host species, high incidences of multiple parasitism for these hosts (see also Trine Chapter 15, Robinson et al. Chapter 33, Donovan et al. Chapter 30), and high nest predation rates (Robinson et al. 1995b). Thompson et al. argue that agricultural landscapes may be population sinks (sensu Pulliam 1988) for some species, because reproduction is insufficient to compensate for adult mortality. Even though we have much to learn before we can confidently identify population sources and sinks (see Grzybowski and Pease Chapter 16), the existence of such strong gradients in nesting success must be considered in any conservation plans. It is possible that bird populations in such heavily agricultural regions as the American Midwest may depend on the surpluses produced by a few heavily forested areas (see also Brawn and Robinson 1996). Donovan et al. even argue that agricultural areas may be population sinks for cowbirds because of the high levels of nest parasitism that also characterize agricultural landscapes (Robinson et al. 1995b).

Thompson et al. further argue that landscape composition affects patterns of cowbird abundance and parasitism within landscapes. They found that edge effects were more pronounced in mostly forested landscapes than in agricultural landscapes where cowbirds appeared to saturate all forest tracts and occurred in proportion to host availability rather than proximity to edge (see also Robinson et al. Chapter 33). Even second-growth and edge species experience heavier parasitism in more agricultural landscapes (Robinson et al.). Thompson et al. concluded by hypothesizing that cowbirds are limited by the availability of hosts in fragmented agricultural landscapes and by the availability of feeding areas in mostly forested landscapes.

Thompson et al.'s hypothesis is strongly supported by two chapters concerning cowbird abundance and distribution in heavily forested landscapes in northern New England. In the Green Mountains of Vermont (Coker and Capen Chapter 28) and the nearby White Mountains of New Hampshire (Yamasaki et al. Chapter 35), cowbird abundance is positively correlated with agricultural cover. In this region, cowbirds are almost entirely restricted to areas adjacent to small agricultural openings and are absent from most national forest land. In these areas, cowbirds do not have to travel far from the few remaining feeding areas to find hosts to parasitize. Given the near absence of parasitism from the White Mountains (e.g., Holmes et al. 1992), it is unlikely that cowbirds have a significant effect on New England forest birds, especially considering that cowbird numbers are also declining in this region (Peterjohn et al. Chapter 2, Wiedenfeld Chapter 3).

Biogeographic Effects

Cowbird numbers vary greatly across North America (Peterjohn et al. Chapter 2, Wiedenfeld Chapter 3). Cowbird abundance (as indexed by the roadside Breeding Bird Survey) peaks in the northern Great Plains and declines in virtually every direction away from this region. In light of this pattern, it is interesting that neither of the studies from the northern Great Plains (Davis and Sealy Chapter 26, Koford et al. Chapter 27) show parasitism levels comparable to those of forest birds in the lower Midwest (Thompson et al. Chapter 32, Robinson et al. Chapter 33), where cowbirds are less abundant according to the Breeding Bird Survey. Grassland birds appear to be more heavily parasitized in the Great Plains (Davis and Sealy, Koford et al.) than in the lower Midwest (Robinson et al.), as predicted by patterns of cowbird abundance. There are, however, virtually no forests in the northern Great Plains to absorb cowbird eggs in the way that midwestern forests do. Perhaps the roadside Breeding Bird Survey exaggerates the abundance of cowbirds in the northern Great Plains (see Chapter 1); roadsides, for example, may be associated with fencerows and telephone lines that provide cowbirds with their only perches in otherwise treeless landscapes. Alternatively, the two sites reported on here may not be representative of parasitism levels throughout the region.

Biogeographic variation in cowbird abundance is correlated with parasitism levels of forest birds in the eastern United States. Wood Thrushes, for example, are much less heavily parasitized in eastern woodlots (Dowell et al. Chapter 29, Petit and Petit Chapter 31) than in midwestern woodlots (Robinson et al. Chapter 33, Thompson et al. Chapter 32; see also Hoover and Brittingham 1993). Even though forest area is negatively correlated with parasitism levels in both the East (Dowell et al., Petit and Petit) and the Midwest (Robinson et al.), the frequencies of parasitism are much lower for most species in the East. Dowell et al. concluded that for some species such as the Wood Thrush, nest predation has a much greater impact on reproductive success than cowbird parasitism in the East (see also Stutchbury 1997). A few species such as the Hooded Warbler, however, appear to be parasitized just as heavily in the East (e.g., Dowell et al. Chapter 29, Stutchbury 1997) as they are in the Midwest (Robinson et al.).

Even though cowbirds are generally less abundant in the West than they are farther east (Peterjohn et al. Chapter 2), they may pose a greater threat to western host populations. Compared with the East, where forest cover is extensive, many western birds live in habitats that are being rapidly destroyed or degraded. Stable or increasing cowbird popula-

tions in areas undergoing rapid habitat loss may mean increased cowbird pressure on remaining host populations and habitat patches. For many species of riparian habitats, for example, there may no longer be large populations in unfragmented habitats that can act as source populations. As a result, the combined effects of increasing cowbird parasitism and decreasing habitat availability threaten several western species or subspecies (Uyehara and Whitfield Chapter 24, Whitfield Chapter 40, Griffith and Griffith Chapter 38, Hayden et al. Chapter 39). Several small populations at the edges of their range in Canada may also be threatened (Ward and Smith Chapter 25). In such areas, local cowbird control combined with habitat protection and restoration may offer the only hope to these populations.

Summary

In summary, the chapters in this part reveal variable correlations between local habitat factors and parasitism levels but generally strong correlations at regional and landscape scales. Edge effects, which are commonly viewed as a major impact of cowbirds (e.g., Brittingham and Temple 1983), are often weak or absent, particularly where cowbirds are very abundant. Parasitism levels are typically high when female cowbirds make up a substantial fraction (over 7% of individuals) of local communities of hosts (Robinson et al. Chapter 33). Cowbird abundance and parasitism levels are consistently low in heavily forested landscapes regardless of the degree of local disturbance (Winslow et al. Chapter 34). Thus, cowbirds are seldom a conservation threat in heavily forested areas, unless a vulnerable host is highly preferred by cowbirds, as may be the case for the Kirtland's Warbler (DeCapita Chapter 37, K. De Groot unpubl. data). Where forest fragmentation by agriculture or urbanization is widespread, cowbirds are often abundant, and host populations may be at risk. At the biogeographic scale, the continental patterns of cowbird abundance (Peterjohn et al. Chapter 2, Wiedenfeld Chapter 3) interact with influences of regional and local landscapes to determine whether cowbirds are a conservation problem.

One consequence of the dominance of influences at large spatial scales is that local management efforts to reduce impacts of parasitism will not always be effective. For example, enhancement of host nesting microhabitat (Martin 1992) will not improve breeding performance in a landscape flooded with cowbirds, unless numbers of cowbirds are reduced simultaneously. On the other hand, source-sink relationships among host populations (Pulliam 1988) may keep local host populations stable despite high levels of parasitism and consequent poor local reproductive success (Brawn and Robinson 1996, Rogers et al. 1997). At a minimum, we need to consider the issue of spatial scale carefully when designing conservation strategies to reduce cowbird parasitism.

References Cited

Bierregard, R. O., and T. E. Lovejoy. 1989. Effects of forest fragmentation on Amazonian understory bird communities. Acta Amazonica 19:215–241.

Brawn, J. D., and S. K. Robinson. 1996. Source-sink population dynamics may complicate the interpretation of long-term census data. Ecology 77:3–12.

Brittingham, M. C., and S. A. Temple. 1983. Have cowbirds caused forest songbirds to decline? BioScience 33:31–35.

Forman, R. T. T. 1995. Land mosaics: The ecology of landscapes and regions. Cambridge University Press, Cambridge.

Hahn, D. C., and J. S. Hatfield. 1995. Parasitism at the landscape scale: Cowbirds prefer forests. Conserv. Biol. 9:1415–1424.

Holmes, R. T., T. W. Sherry, P. P. Marra, and K. E. Petit. 1992. Multiple brooding and productivity in a neotropical migrant, the Black-throated Blue Warbler, in an unfragmented temperate forest. Auk 109:321–333.

Hoover, J. P., and M. C. Brittingham. 1993. Regional variation in cowbird parasitism of Wood Thrushes. Wilson Bull. 105:228–238.

Johnson, R. G., and S. A. Temple. 1990. Nest predation and brood parasitism of tallgrass prairie birds. J. Wildl. Manage. 54:106–111.

Larison, B. 1996. Avian responses to riparian restoration. M.S. Thesis, San Francisco State University, San Francisco, CA.

Martin, T. E. 1992. Breeding productivity considerations: What are the appropriate habitat features for management? Pp. 455–473 *in* Ecology and conservation of neotropical migrant landbirds (J. M. Hagan III and D. W. Johnston, eds). Smithsonian Institution Press, Washington, DC.

Meffe, G. K., C. R. Carroll, and contributors. 1997. Principles of conservation biology, 2nd ed. Sinauer Associates, Sunderland, MA.

Norman, R. F., and R. J. Robertson. 1975. Nest searching behavior in the Brown-headed Cowbird. Auk 92:610–611.

Pulliam, H. R. 1988. Sources, sinks, and population regulation. Amer. Natur. 137:550–556.

Robinson, S. K., S. I. Rothstein, M. C. Brittingham, L. J. Petit, and J. A. Grzybowski. 1995a. Ecology and behavior of cowbirds and their impact on host populations. Pp. 428–460 *in* Ecology and management of neotropical migratory birds (T. E. Martin and D. M. Finch, eds) Oxford University Press, Oxford.

Robinson, S. K., F. R. Thompson III, T. M. Donovan, D. R. Whitehead, and J. Faaborg. 1995b. Regional forest fragmentation and the nesting success of migratory birds. Science 267:1987–1990.

Rogers, C. M., M. J. Taitt, J. N. M. Smith, and G. J. Jongejan. 1997. Nest predation and cowbird parasitism create a population sink in a wetland breeding population of Song Sparrows. Condor 99:622–633.

Schmiegelow, F. K. A., C. S. Machtans, and S. J. Hannon. 1997. Are boreal birds resilient to forest fragmentation? An experimental study of short-term community responses. Ecology 78:1914–1932.

Stutchbury, B. M. 1997. Effects of female cowbird removal on reproductive success of Hooded Warblers. Wilson Bull. 109:74–81.

Wiley, J. W. 1988. Host selection in the Shiny Cowbird. Condor 90:289–303.

23. The Role of Vegetation in Cowbird Parasitism of Yellow Warblers

WILLIAM H. HOWE AND FRITZ L. KNOPF

Abstract

We studied the effects of fire on parasitism of Yellow Warbler nests by Brown-headed Cowbirds. Half of a 67-ha plot of riparian habitat in northern Colorado burned in the spring of 1989. The fire reduced the abundance of Yellow Warbler territories by nearly half, while numbers remained unchanged on the unburned area. Contrary to our expectation, we found that parasitism rates were lower on the burned area in 1989 and 1990 than in 1988, even though there was little cover for warbler nests there, and numbers of cowbird detections there did not decline after the burn. Nest predation increased on both areas after the burn, and Yellow Warblers reproduced poorly in both 1989 and 1990. We suggest that cowbirds switched from using Yellow Warblers to ground-nesting hosts with better-concealed nest sites after the burn. We speculate that cowbirds first chose host habitats by the condition of the local vegetation and then searched selectively for host nests there.

Introduction

Nest parasitism by the Brown-headed Cowbird may pose a major threat to western riparian birds (Goldwasser et al. 1980, Laymon 1987). Brood parasitism by cowbirds may be higher in western riparian habitats than in more extensive woodlands elsewhere, because of their narrow, linear, or patchy configuration (Bleitz 1956, Brittingham and Temple 1983, Sedgwick and Knopf 1988). Cowbird populations in these habitats may also be maintained at high levels by virtue of surrounding habitats being open and often grazed by cattle, thus providing ideal foraging areas for the cowbird (Laymon 1987, Sedgwick and Knopf 1988).

Much research has been directed toward the selection of host species by cowbirds (Friedmann and Kiff 1985) and their impacts on hosts' breeding success (e.g., McGeen 1972, Goldwasser et al. 1980, Brittingham and Temple 1983,

Hobson and Sealy 1989). Although some authors have speculated that cowbirds tend to parasitize the species that raised them (Walkinshaw 1949, McGeen and McGeen 1968, McGeen 1972), it has been widely assumed that a cowbird's selection of host species and nests is opportunistic. This assumption is supported by the wide range of hosts that cowbirds are known to parasitize. More than 220 host species have been used, including such inappropriate birds as Blue-winged Teal (*Anas discors*), Ferruginous Hawk (*Buteo regalis*), Wilson's Phalarope (*Phalaropus tricolor*) and California Gull (*Larus californicus*) (Friedmann and Kiff 1985). Because Brown-headed Cowbirds may lay 40 or more eggs in a single breeding season (Scott and Ankney 1980, 1983), there has apparently been little need to be selective. Spreading eggs over a large number of nests will offset any costs to the cowbird of laying eggs in some "inferior" nests.

Zimmerman (1983) found a higher rate of parasitism of Dickcissels (*Spiza americana*) in prairie habitats compared with old field habitats. Few other studies to our knowledge have addressed the processes involved in selection of habitats or of host nests by individual cowbirds (but see Hahn and Hatfield Chapter 13 and other chapters in Part IV). In this chapter we address the roles of habitat quality and predation pressure on the parasitism of Yellow Warbler (*Dendroica petechia*) nests by cowbirds.

In the second year of our study of a population of Yellow Warblers in Colorado, a fire burned nearly half of the study area, transforming the habitat from lush, thick vegetation to burned skeletons of woody shrubs. In the year after the fire, the majority of live vegetation was new sprouts from the roots. With data from one year prior to the fire and two years postfire, we were able to compare choices of burned versus unburned habitats by laying cowbirds, and to make inferences on the process that cowbirds use to select nests to parasitize. We also monitored breeding success of the warblers and noted how both nest parasitism and nest predation in the burned and unburned habitats influenced it.

Study Area and Methods

The study was conducted in a 67-ha portion of a high-elevation (2,470 m) floodplain along the Illinois River at Arapaho National Wildlife Refuge, Jackson County, Colorado. The refuge is located in an intermountain basin in the rain shadow of the surrounding Rocky Mountains. Woody vegetation along the river consists almost entirely of up to eight species of shrubby willows (*Salix* spp.) growing in narrow stringers and dense clumps (Cannon and Knopf 1984). The willows represented the only woody vegetation suitable for placement of Yellow Warbler nests and covered 18.2% of the study area. A minor component of the woody community was scattered stands of Wood's rose (*Rosa woodsii*). Vegetation covering the remaining 81% of the floodplain consisted of grasses, sedges (*Carex* spp.), and various forbs, whereas the surrounding uplands were dominated by sagebrush (*Artemisia* spp.) shrublands.

To describe the avian community in the study area we censused all species of riparian birds in the summers of 1988 to 1990 using point-count surveys following the method of Reynolds et al. (1980). Fifty census points 100 m apart were censused twice each year in June. Twenty-seven census points became the burn in 1989; 23 points were located in the unburned area.

Searches for Yellow Warbler nests commenced in late May each year, soon after the warblers arrived at the site and about the time when nest construction was initiated (W. Howe, J. Sedgwick, and F. Knopf unpubl. data). Each year, nearly all of the warblers nesting on the site were mist netted and color banded to allow individual identification of pair members. Nests were visited at intervals of 2–3 days from the day of discovery to termination of that nesting event. We recorded the incidence of brood parasitism by cowbirds and nesting success as affected by parasitism or predation.

Nest searches in 1988 were conducted as part of a greater community-wide study (Knopf et al. 1988) of high-elevation riparian birds. Yellow Warblers were not the main focus of our research at the time, and nest searches were less thorough than in 1989 or 1990. The area searched in 1988 was entirely contained within the burn of spring 1989. Coverage of the 67-ha site increased each year through 1990, but in no year were nests found on all warbler territories in the study area.

To quantify willow recovery following the fire, we measured the height and width of live and dead vegetation on five willows at each census point in 1989 and 1990. The willow nearest to each census stake was measured, as was the nearest willow in each quadrant defined by the four cardinal directions from that central willow. Willows were measured similarly on the unburned site in 1989 only.

Results

Effects of the Fire on Willows

The fire burned 32 ha of the study site in April 1989, one month prior to the arrival of Yellow Warblers for the breeding season. The aboveground portions of 66% of the willows were reduced to dead skeletons. With the exception of a few small willow stands that were killed entirely, underground portions of most of the burned willows remained alive. Willows that were still alive below ground produced root sprouts that attained an average height of 1.2 m by the end of the 1989 growing season. By the end of the 1990 growing season the root sprouts attained an average height of 1.7 m. Only 0.5 m of new growth was added on average in 1990 due to heavy browsing by moose (*Alces alces*). Mean height of live vegetation in the willows unaffected by the burn was 2.9 m.

Territories and Nests

In 1988 we counted 37 Yellow Warbler territories in the area that later became the burn. Following the fire this number decreased to 20 territories in 1989 and then increased slightly to 22 territories in 1990. Several warbler territories in 1989 were abandoned after only one or two nesting attempts, a pattern not observed in either 1988 or 1990, when three or more renesting attempts were typical. On the unburned site the number of warbler territories from 1988 to 1990 remained stable at 43–44 per year.

Including renesting attempts, we located 34 warbler nests on the burn in 1988, 25 in 1989, and 34 in 1990. The unburned area was not searched for nests in 1988. Roughly half of it was searched in 1989, and nearly all of it in 1990. Including renesting attempts, we located 29 nests in 1989 and 44 nests in 1990.

All of the nests in the unburned area were cryptic and difficult to locate. Many nests in the burn were placed in willows that had not burned, and as usual were well concealed. In 1989, however, 7 of 25 nests located in the burn were placed among burned, dead willow stems with little to no concealing vegetation and were easy to find. Two nests in particular were visible from distances up to 20 m.

Yellow Warbler Fledging Success

Yellow Warbler nesting success in the unburned portion of the study area was high in 1989; 22 of the 28 territories (78.6%) for which we had complete data successfully fledged at least one young. This percentage decreased to 19 of 35 territories (54.3%) in 1990, primarily because of higher predation that year (see below).

Before the fire, data for the burn site were incomplete for all but a few territories, and we did not get a good estimate of fledging success in 1988. After the fire in 1989, young fledged from only 2 of the 20 territories successfully. Seven of 22 territories (31.8%) produced fledging young in 1990, reflecting the improved condition of the vegetation in the second year following the fire.

Cowbird Numbers and Nest Parasitism

The number of cowbirds detected on censuses was too small to estimate density on the burned and unburned sites. Raw numbers of cowbirds detected are therefore reported. On the burn, a mean of 13.0 cowbirds was detected per census in 1988, 13.5 in 1989, and 28.0 in 1990 (Figure 23.1). The mean number detected per census on the unburned site was similar to the burn mean in 1988 (13.5) but was only 4.5 in 1989 and 10.0 in 1990. The higher numbers detected on the burn in 1989 and 1990 were most likely due to the greater detectability of birds following the fire, as there was little vegetation present to interfere with visual or auditory detections.

Although the number of cowbirds detected on the burn site after the fire first remained stable (1989) and then increased (1990), the proportion of Yellow Warbler nests parasitized by cowbirds declined (Figure 23.1). Nine of 34 nests (28.6%) were parasitized before the fire in 1988, but only 3 of 19 nests (15.8%) in 1989 and 7 of 34 nests (20.6%) in 1990 were parasitized. In contrast, 5 of 24 nests (20.8%) were parasitized on the unburned site in 1989, and 17 of 40 nests (42.5%) were parasitized in 1990. No cowbirds were known to have fledged from the burn in either 1989 or 1990. At least four cowbirds fledged from the unburned site in 1989 and one in 1990.

Nest Predation

Several nest predators were present in the study area in all years. Primary avian predators included Black-billed Magpies (*Pica pica*) and Common Grackles (*Quiscalus quiscula*), both of which were observed depredating nests of the Yellow Warbler. Each year a family of magpies was observed systematically searching the study area, apparently for nests, from late May into June at a time when egg laying was at its peak. Mammalian predators probably included striped skunk (*Mephitis mephitis*), long-tailed weasel (*Mustela frenata*), and perhaps Richardson's ground squirrel (*Spermophilus richardsonii*). No mammals were seen depredating nests, but a number of nests each year were found torn out of the supporting branches, with eggshell fragments or bands and body parts from banded nestlings present on the ground below the nests. We attribute these failures to mammals.

Nest predation accounted for the greatest loss of nests in both 1989 and 1990 (Figure 23.1). In marked contrast to nest parasitism by cowbirds, nest predation on the burn site skyrocketed in the years after the fire, almost certainly due to the greater visibility of nests to predators. In 1989, 14 of 25 nests were depredated on the burn (56.0%), but only 8 of 29 nests (27.6%) were depredated on the unburned site. Nest predation was higher in 1990 than in 1989, but the pattern remained similar: 23 of 34 nests on the burn (67.6%) and 25 of 44 nests (56.8%) on the unburned site.

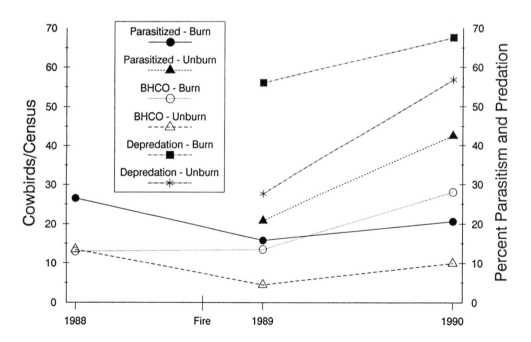

Figure 23.1. Parasitism and predation rates on Yellow Warbler nests, and number of cowbird detections on point-count censuses on burned and unburned sites from 1988 to 1990.

Discussion

Censuses confirmed the continued presence of cowbirds on the burn in years following the fire, although the increase in observations may be due to the greater visibility of cowbirds among the burned willows rather than to an actual increase in their numbers. In addition, adult warblers were fairly easy for us to observe and follow throughout their territories in most cases. We assume that female cowbirds should also have had little difficulty locating female warblers and nests on the burn because of their increased visibility. The ease with which avian and mammalian predators located nests on the burn supports this view. The decline of cowbird parasitism of Yellow Warbler nests after the fire was therefore contrary to our expectations.

The reduction of parasitism on Yellow Warbler nests may reflect a shift in host species by the cowbirds from species nesting in the willows to other bird species on the site. Our impressions were that the grasslands and meadows were less affected by the fire and recovered relatively quickly. Ground-nesting species, such as Song Sparrow (*Melospiza melodia*) and the abundant Savannah Sparrow (*Passerculus sandwichensis*), were each observed feeding fledgling cowbirds on the burn in both postfire years. The extent to which the fire affected parasitism of those and other species of open habitats, however, is not known. A community-wide look at shifts in parasitism rates among species would shed light on the ability of cowbirds to respond to microhabitat changes by adjustments in host or microhabitat selection.

The reduced parasitism of exposed Yellow Warbler nests in this study suggests that the availability of host nests is only one of several stimuli employed by searching cowbirds. We suggest that cowbirds use a hierarchical approach similar to that used by shrub-nesting passerines to select habitat (Knopf et al. 1990). In the case of the cowbird, birds seem initially to select specific features of the vegetation and landscape prior to initiating specific searches for host nests. We speculate further that the choice of landscape and vegetation for laying is influenced by the natal habitat of the individual cowbird.

Management Implications

Besides providing insights into how cowbirds may search for hosts, this opportunistic study further emphasizes that cowbird parasitism should not be viewed independently of the other principal source of reproductive loss, predation. Management of vegetation to provide good nesting cover for host species may in some situations favor increased rates of cowbird parasitism. Certain management practices, such as prescribed burns, may reduce cowbird parasitism locally but at a cost of lowered host reproductive success due to increased levels of predation. We find it interesting that predators located all parasitized nests in the burn in this study, suggesting that parasitism increased the probability that predators would detect host nests.

Acknowledgments

We thank E. C. Patten and the staff at Arapaho National Wildlife Refuge for their support during the study. Field assistance was provided by J. Sedgwick, C. Miller, and K. Stone.

References Cited

Bleitz, D. 1956. Heavy parasitism of Blue Grosbeaks by cowbirds in California. Condor 58:236–238.

Brittingham, M. C., and S. A. Temple. 1983. Have cowbirds caused forest birds to decline? BioScience 33:31–35.

Cannon, R. W., and F. L. Knopf. 1984. Species composition of a willow community relative to seasonal grazing histories in Colorado. Southwest. Natur. 29:234–237.

Friedmann, H., and L. F. Kiff. 1985. The parasitic cowbirds and their hosts. Proc. Western Found. Vert. Zool. 2:226–304.

Goldwasser, S., D. Gaines, and S. R. Wilbur. 1980. The Least Bell's Vireo in California: A de facto endangered race. Amer. Birds 34:742–745.

Hobson, K. A., and S. G. Sealy. 1989. Responses of Yellow Warblers to the threat of cowbird parasitism. Anim. Behav. 38:510–519.

Knopf, F. L., J. A. Sedgwick, and R. W. Cannon. 1988. Guild structure of a riparian avifauna relative to seasonal cattle grazing. J. Wildl. Manage. 52:280–290.

Knopf, F. L., J. A. Sedgwick, and D. B. Inkley. 1990. Regionwide correspondence among shrub-steppe bird habitats. Condor 92: 45–53.

Laymon, S. A. 1987. Brown-headed Cowbirds in California: Historical perspectives and management opportunities in riparian habitats. Western Birds 18:63–70.

McGeen, D. S. 1972. Cowbird-host relationships. Auk 89:360–380.

McGeen, D. S., and J. J. McGeen. 1968. The cowbirds of Otter Lake. Wilson Bull. 80:84–93.

Reynolds, R. T., J. M. Scott, and R. A. Nussbaum. 1980. A variable circular plot method for estimating bird numbers. Condor 82: 309–313.

Scott, D. M., and C. D. Ankney. 1980. Fecundity of the Brown-headed Cowbird in southern Ontario. Auk 97:677–683.

———. 1983. The laying cycle of Brown-headed Cowbirds: Passerine chickens? Auk 100:583–592.

Sedgwick, J. A., and F. L. Knopf. 1988. A high incidence of Brown-headed Cowbird parasitism of Willow flycatchers. Condor 90: 253–256.

Walkinshaw, L. H. 1949. Twenty-five eggs apparently laid by a cowbird. Wilson Bull. 61:82–85.

Zimmerman, J. L. 1983. Cowbird parasitism of Dickcissels in different habitats and at different nest densities. Wilson Bull. 95:7–22.

24. Association of Cowbird Parasitism and Vegetative Cover in Territories of Southwestern Willow Flycatchers

JAMIE C. UYEHARA AND MARY J. WHITFIELD

Abstract

We investigated the association of brood parasitism by Brown-headed Cowbirds and vegetation structure in 31 territories of the endangered Southwestern Willow Flycatcher in central California. These flycatchers nest between 1 and 3 m high in riparian forest. We tested whether parasitism varied with concealment on different spatial scales within territories. We predicted that unparasitized flycatchers ($N =$ 13) would have more vegetative cover throughout their territory. We measured vegetation at the nest (nest site), in a 0.04-ha plot around the nest (nest plot), and in a random 0.04-ha plot within the territory (random plot). Cover was characterized using principal component analysis (PCA). Cover at the nest was similar in territories of parasitized and unparasitized birds. Unparasitized flycatchers, however, had more cover in both random and nest plots than parasitized flycatchers. Our results suggest that while all Willow Flycatcher territories had variable densities of understory vegetation and closed canopy, unparasitized flycatchers had larger patches of dense understory. Southwestern Willow Flycatchers may therefore benefit from management activities that increase the density of vegetation below 3 m in height.

Introduction

When female Brown-headed Cowbirds lay their eggs in nests of particular host species, only some nests are parasitized (but see Trine, Chapter 15, this volume). It is therefore of interest to ask, do some nests or hosts share features that affect the likelihood of parasitism? We hypothesize here that concealment from vegetation around the nest and within the territory reduces the likelihood of parasitism.

The effect of vegetative cover on parasitism risk probably depends on cowbirds' nest-searching methods. If cowbirds rely only on visual cues provided by the nest itself, then parasitized nests should have less cover around them than unparasitized nests. It seems likely, however, that cowbirds use both visual and acoustic cues provided by the host to locate nests (Uyehara 1996). Then, parasitism risk could be lower if vegetation concealed activities of the host pair as well as the nest. Masking of such behavioral cues might require dense cover throughout the territory of the nesting pair, not just at the host nest.

We studied a host that nests in riparian forest, the Southwestern Willow Flycatcher (*Empidonax traillii extimus*). In our study area along the South Fork Kern River, 50–80% of all nests have been parasitized in recent years (Harris 1991; Whitfield 1990, Chapter 40 this volume). Nests of different pairs are clustered (Whitfield Chapter 40) with both parasitized and unparasitized nests in each area. The vegetation structure in flycatcher territories is heterogeneous, with open areas containing exposed song perches and dense patches of shrubs, where the flycatchers nest (e.g., Brown 1988, McCabe 1991, Sedgwick and Knopf 1992, Uyehara and Whitfield pers. obs.). In our study area, using univariate analyses, Whitfield (1990) did not find any differences in 15 vegetation variables measured at parasitized and unparasitized nest sites and within 0.04-ha circular plots centered at the nest. We now test the hypothesis that concealment within the whole territory is associated with parasitism events.

Methods

Description of Study Area

Our study site was located along the South Fork Kern River in central California (35° 40' N, 118° 20' W, elevation 762–805 m). The study area lies in a broad, braided-river valley that drains the southern Sierra Nevada. The riparian forest along the river is bordered by pastures and cultivated fields. The forest is narrow, with the interior no farther than 200 m from an edge or opening, and is dominated by red willow (*Salix laevigata*), black willow (*S. gooddingii*) and Fremont cottonwood (*Populus fremontii*). The understory includes young willow in thickets, mule fat (*Baccharis viminea*), and hoary nettle (*Urtica dioica holosericea*). Open areas of grass or freshwater marsh are interspersed throughout the forest. More detailed descriptions are found in Harris et al. (1987:28–29), Whitfield (1990 and Chapter 40), Harris (1991), and Uyehara and Narins (1995).

At our study site, Willow Flycatchers begin arriving in early May. By late May, territorial boundaries are stable and discrete, as seen in other Willow Flycatcher populations (e.g., Sedgwick and Knopf 1988). A pair's territory was defined as the area enclosed by perches used by the nesting pair. Pairs were determined by nesting phenology, vocalizations during nest searches and checks, observed flight patterns, and locations of banded birds (ca. 25% of adults in the population).

Brown-headed Cowbirds arrived in large numbers from mid- to late April. In 1989 and 1990, cowbirds were abundant (pers. obs.). In 1991, 60 15-min point counts separated by 200 m averaged 1.55 female cowbirds/point and 2.67 male cowbirds/point from May through June.

From late May through July, 1989–1990, we monitored Willow Flycatcher nests. After nests were no longer in use, we measured the vegetation. The delay in measuring vegetation probably had only small effects on the data because (1) trees and forbs changed little from June through August, and (2) delays were short and equal for parasitized and unparasitized pairs. Grass height may have changed over time but did so only below 1 m.

We took three sets of vegetation measurements in each territory. We measured habitat variables at the nest site, in a 0.04-ha circular nest plot (radius = 11.3 m) around the nest, and in a 0.04-ha random plot. Random plots were selected by blindly picking numbers and directions from a bag. Random plots were 23–50 m from the nest but within the territory of the pair.

Measurements taken at the nest site were percentage of nest cover, nest height, distance to top of canopy from the nest, and percentage of canopy cover. The seven variables measured in nest plots were (1) canopy height, (2) percentage of ground cover, (3) percentage of canopy cover, and (4–7) percentage of foliage cover at height intervals of 0–1 m, 1–2 m, 2–3 m, and 3–4. Measurements on random plots were identical to those on the nest plot. Our vegetation sampling methods are modified from James and Shugart (1970) and Noon (1981).

Percentage of cover at the nest site was estimated using a 1- x 0.5-m grid marked with 50 0.1-m cells. The grid was positioned behind the nest, centered on it, and observed from 5 m in four cardinal directions. The percentage of nest cover was estimated as 2x the number of cells that were at least 50% concealed by vegetation.

All height measurements were estimated using a 4-m pole marked at 10-cm intervals. Ground and canopy cover were measured by peering through the objective lens of a 7x, 6.2° monocular with a 10-cell grid attached 7.6 cm from the eyepiece (details in Laymon 1988). Percentage of cover was estimated as 10x the number of cells in which 50% or more of the cell was obscured by vegetation. For nest plots and random plots, eight measurements were taken, at 5 m and 11 m from the plot center in the four cardinal directions.

Percentage of foliage cover was measured using a 1-x-0.5-m grid subdivided into 50 10-cm cells. We calculated the estimate as 10x the number of cells in which 50% or more of the cell was obscured by vegetation when viewed from the center to the edge of the circular plot at the appropriate heights.

Data Analysis

Because vegetation features were correlated within plots, scores obtained by principal component analysis (PCA) were used to describe vegetation on territories. PCA reduced the vegetation variables to independent axes, with a few variables contributing significantly to each.

Each territory received one score for each axis. Scores for each axis were normally distributed with a mean close to zero. A score of 0 indicated that a territory had average values, especially for the vegetation variables that loaded heavily on the axis. An extreme score indicated that the territory had extreme values for the most heavily weighted variables. We used only the first two axes, noting which variables had component loadings with values greater than 0.5. Territory scores were divided into parasitized and unparasitized groups and tested for differences in cover by a Mann-Whitney U-test.

Results

Data from 17 variables for 19 parasitized territories and 14 unparasitized territories were used for the PCA (Table 24.1). The first two axes explained 30.05% and 18.13%, respectively, of the total variance in the data. The component loadings for the first axis indicated that high-scoring territories had dense understory vegetation (1–4 m) and relatively tall and continuous canopy in the random plot (Table 24.2, Figure 24.1). Canopy cover values were also high on the nest plot (Table 24.2). The mean score of unparasitized territories (0.24) was higher than that of parasitized territories (-0.16), but not significantly so (Mann-Whitney U-test, $U = 88.5$, $P = .13$, one-tailed).

Loadings for the second axis indicated that high-scoring territories had dense understory vegetation at the nest site itself, at 0–3 m height within 0.04 ha of the nest, and in the nest plot (Table 24.2, Figure 24.1). The mean score for unparasitized nests was significantly higher than scores for

Table 24.1. Mean Vegetation Measurements for 19 Parasitized (P) and 14 Unparasitized (NP) Territories from 1989 and 1990

Parameter	P/NP	Nest Site	Nest Plot	Random Plot
Nest height	NP	2.23 (0.32)	—	—
(m)	P	2.21 (0.29)	—	—
Canopy height	NP	7.11 (0.88)*	4.79 (0.57)	5.04 (0.83)
(m)	P	6.76 (0.39)	5.16 (0.31)	3.45 (0.45)
Ground cover	NP	—	28.29 (11.34)*	—
(%)	P	—	47.22 (9.52)*	—
Canopy cover	NP	93.67 (5.31)	76.63 (4.91)	63.22 (8.40)
(%)	P	94.53 (2.22)	68.40 (4.22)	49.15 (5.94)
Foliage cover	NP	92.27 (4.41)	—	—
(%)	P	89.37 (2.26)	—	—
0–1 m	NP	—	97.46 (1.59)	86.16 (3.92)
	P	—	96.12 (2.36)	84.84 (3.75)
1–2 m	NP	—	95.92 (2.42)	83.48 (4.50)
	P	—	92.26 (2.08)	68.31 (5.24)
2–3 m	NP	—	92.79 (2.56)	74.81 (6.63)
	P	—	86.53 (2.85)	67.06 (5.68)
3–4 m	NP	—	91.34 (3.09)	72.51 (6.96)
	P	—	88.23 (2.96)	63.46 (6.94)

Note: Measurements were taken at the nest (Nest Site), within a 0.04-ha plot centered around the nest (Nest Plot) and in a randomly selected 0.04-ha plot within the territory of the pair (Random Plot). Values within parentheses are one SE.
*Sample size reduced by 1, $N = 18$ for parasitized plots and $N = 13$ for unparasitized plots.

Table 24.2. Habitat Measurements of Axes 1 and 2 for Which Eigenvector Values were Greater than 0.5

Type of Plot	Habitat Variable	Value
Axis 1		
Random plot	% foliage cover, 2–3 m	0.882
Random plot	% canopy cover	0.861
Random plot	% foliage cover, 3–4 m	0.836
Random plot	% foliage cover, 1–2 m	0.818
Random plot	Canopy height	0.743
Nest plot	% canopy cover	0.616
Axis 2		
Nest plot	% foliage cover, 0–1 m	0.741
Nest plot	% foliage cover, 1–2 m	0.703
Nest plot	% foliage cover, 2–3 m	0.688
Nest site	% foliage cover, at nest	0.648

Note: Habitat measurements that had component loadings less than 0.5 are not listed.

parasitized nests, 0.21 and -0.15, respectively ($U = 74.0$, $P = .04$, one-tailed).

Discussion

Southwestern Willow Flycatchers differed in the amounts of cover at different spatial scales within their territories. Territories varied most in the amount of cover at randomly selected sites within the territory, but cover on random plots did not differ greatly between territories of parasitized and unparasitized pairs. Parasitized and unparasitized territories did, however, differ in cover provided by understory vegetation at and near the nest (nest site and nest plot). Willow Flycatchers typically placed nests in denser vegetation than was found near the nest or on random plots. Nests were usually 1–3 m high (e.g., Whitfield 1990, Harris 1991, McCabe 1991) and, again, in denser vegetation than was available at this height on random plots in the territory. Dense vegetation at the nest site is typical of Willow Flycatchers (e.g., Brown 1988, McCabe 1991, Sedgwick and Knopf 1992).

Well-concealed nests alone did not reduce parasitism for Willow Flycatchers (Whitfield 1990 and Chapter 40). Song Sparrows (*Melospiza melodia*) on Mandarte Island also showed no correspondence between nest concealment and parasitism (Smith 1981). Yellow Warblers (*Dendroica petechia*) breeding in an open burned area did not suffer increased parasitism compared to a control site or preburn year (Howe and Knopf, Chapter 23, this volume). Taken together, these studies suggest that vegetation characteristics at the nest site itself are unreliable predictors of parasitism (but see Sealy et al. 1995 for an alternative hypothesis).

Parasitism frequency was associated with vegetation features on a larger scale than at the nest site. The vegetation characteristics in the rest of the territory differ from those at the nest site (Sedgwick and Knopf 1992, this study). This variation in cover was associated with parasitism in Southwestern Willow Flycatchers. Unparasitized nests tended to be in territories with more cover in both the random and nest plots than in parasitized territories, although the difference in scores was significant only at the scale of the nest plot (Axis 2). In contrast, parasitized flycatchers occupied territories that had more heterogeneous vegetation with dense cover at the nest site, but less cover overall in the territory. Therefore, to avoid being parasitized in areas with high numbers of cowbirds, Willow Flycatchers may need to nest in territories with large patches of thick, dense understory merging with closed canopies, as indicated by the component loadings of Axis 2.

Few studies have associated territorial forest structure with parasitism frequencies. Our study indicates that in a riparian forest with many edges or openings, cover is associated with parasitism. Wolf (1987) categorized habitats as open or closed canopy and found significant differences in

the frequencies of parasitism in Dark-eyed Juncos (*Junco hyemalis*) in these habitats in one of two years. Winslow et al. (Chapter 34, this volume) reported that nests of some species near forest openings had higher parasitism rates than interior nests. Woodworth (1995) found no significant edge effect in parasitism events of Shiny Cowbirds in an open-canopy forest in Puerto Rico.

Thus, lack of canopy cover does not always correlate with higher frequencies of parasitism. Mixed results might occur if cowbirds predominantly search for and parasitize nests below the canopy (Norman and Robertson 1975, Wiley 1988, Briskie et al. 1990, Hahn and Hatfield Chapter 13; but see Robinson et al. Chapter 33) or if canopy cover is correlated with other factors affecting parasitism.

The association of parasitism with vegetation on territories may be due to the effect of dense vegetation on the searching behavior of female cowbirds. Cowbirds use elevated perches from which they scan for potential hosts (e.g., Friedmann 1963, Norman and Robertson 1975, Payne 1973, Darley 1983, Wiley 1988, Uyehara 1996; but see Freeman et al. 1990 regarding available perches). Female cowbirds probably use visual and acoustic cues from hosts to find nest sites (Hann 1941; Robertson and Norman 1976, 1977; Smith 1981; Smith et al. 1984; Wiley 1988; Uyehara and Narins 1995; Uyehara 1996). For example, Wiley

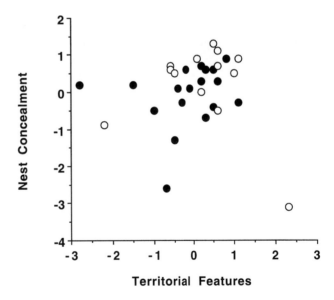

Figure 24.1. Scores of parasitized (solid circles) and unparasitized (open circles) territories of Southwestern Willow Flycatchers derived from principal component analysis. For Axis 1 (territorial features), high positive scores indicated more dense vegetation 1–4 m high and taller forest with greater canopy cover in random plots. For Axis 2, high scores indicated more dense vegetation 0–3 m high and more foliage cover in the nest plot. Scores close to 0 indicate average vegetation cover for the territory.

(1988) observed Shiny Cowbirds flying to sites where hosts had been building nests or actively defending territories. It is therefore likely that dense vegetation will impede their ability to observe host behavior and to discover nest sites from a distance.

Searching cowbirds have been observed beating shrubs (Norman and Robertson 1975, Wiley 1988) and walking in forests while looking up at passing birds (Norman and Robertson 1975; Hahn and Hatfield, Chapter 13, this volume). Because of the presence of water and dense vegetation below 4 m in height in our study area, cowbirds would have found it difficult to walk in this habitat.

In Song Sparrows, older birds are more likely to be parasitized because they are more aggressive toward cowbirds (Smith and Arcese 1994). If both aggressive behavior and habitat use are correlated with age in flycatchers, an alternative interpretation of our results is that parasitized birds were older, more aggressive, and occupied areas with more heterogeneous vegetation. In support of this idea, Willow Flycatchers at seven parasitized nests were noisier than flycatchers at six unparasitized nests (Uyehara and Narins 1995) at Kern River Preserves 1 and 2 (Whitfield, Chapter 40, this volume). Unfortunately, we have few data on the age structure in our population that could be used to compare the habitat use and behavior of older and yearling flycatchers.

Management Implications

Southwestern Willow Flycatchers nested in shrubs or understory in riparian forest that provided relatively high amounts of cover (Brown 1988, Whitfield 1990, this study, M. Sogge pers. comm.). The riparian forest in our study area is generally narrow and linear with many openings. In this area, unparasitized pairs nested in territories with larger patches of dense understory and contiguous canopy, although there were still many forest openings.

Our results suggest that forest management to enhance dense understory and canopy cover may benefit Southwestern Willow Flycatchers. However, we did not address whether similar benefits would prevail in unfragmented forests or if enhancing shrub and canopy cover might adversely affect other species nesting in riparian habitats. These issues also need to be considered before action is taken.

Acknowledgments

We are grateful to R. Tiller, The Nature Conservancy, Audubon California's Kern River Preserve, for logistic support that greatly facilitated our study. We also thank S. Laymon, Kern River Research Center, for generous loans of equipment. For recommending multivariate statistics, we thank A. Møller. For constructive critiques of an earlier draft, we thank S. Sealy, S. Robinson, R. Tollefson, and R. Gibson.

References Cited

Brown, B. T. 1988. Breeding ecology of a Willow Flycatcher population in Grand Canyon, Arizona. Western Birds 19:25–33.

Briskie, J. V., S. G. Sealy, and K. A. Hobson. 1990. Differential parasitism of Least Flycatchers and Yellow Warblers by the Brown-headed Cowbird. Behav. Ecol. Sociobiol. 27:403–410.

Darley, J. A. 1983. Territorial behavior of the female Brown-headed Cowbird (Molothrus ater). Can. J. Zool. 61:65–69.

Freeman, S., D. F. Gori, and S. Rohwer. 1990. Red-winged Blackbirds and Brown-headed Cowbirds: Some aspects of a host-parasite relationship. Condor 92:336–340.

Friedmann, H. 1963. Host relations of the parasitic cowbirds. U.S. Natl. Mus. Bull. 233.

Hann, H. W. 1941. The cowbird at the nest. Wilson Bull. 53: 211–221.

Harris, J. H. 1991. Effects of brood parasitism by Brown-headed Cowbirds on Willow Flycatcher nesting success along the Kern River, California. Western Birds 22:13–26.

Harris, J. H., S. D. Sanders, and M. A. Flett. 1987. Willow Flycatcher surveys in the Sierra Nevada. Western Birds 18:27–36.

James, F. C., and H. H. Shugart. 1970. A quantitative method of habitat description. Audubon Field Notes 24:727–736.

Laymon, S. A. 1988. Ecology of the Spotted Owl in the central Sierra Nevada, Calif. Ph.D. Dissertation, University of California, Berkeley.

McCabe, R. A. 1991. The little green bird: Ecology of the Willow Flycatcher. Rusty Rock Press, Madison, WI. 171 pp.

Noon, B. R. 1981. Techniques for sampling avian habitats. USDA For. Serv. Gen. Tech. Rept. RM-87, pp. 42–52.

Norman, R. F., and R. J. Robertson. 1975. Nest-searching behavior in the Brown-headed Cowbird. Auk 92:610–611.

Payne, R. B. 1973. The breeding season of a parasitic bird, the Brown-headed Cowbird, in central California. Condor 75:80–99.

Robertson, R. J., and R. F. Norman. 1976. Behavioral defenses to brood parasitism by potential hosts of the Brown-headed Cowbird. Condor 78:166–173.

———. 1977. The function and evolution of aggressive host behavior towards the Brown-headed Cowbird (Molothrus ater). Can. J. Zool. 55:508–518.

Sealy, S. G., D. L. Neudorf, and D. P. Hill. 1995. Rapid laying by Brown-headed Cowbirds Molothrus ater and other parasitic birds. Ibis 137:76–84.

Sedgwick, J. A., and F. L. Knopf. 1988. A high incidence of Brown-headed Cowbird parasitism of Willow Flycatchers. Condor 90: 253–256.

———. 1992. Describing Willow Flycatcher habitats: Scale perspectives and gender differences. Condor 94:720–733.

Smith, J. N. M. 1981. Cowbird parasitism, host fitness, and age of the host female in an island Song Sparrow population. Condor 83:152–161.

Smith, J. N. M., and P. Arcese. 1994. Brown-headed Cowbirds and an island population of Song Sparrows: A 16-year study. Condor 96:916–934.

Smith, J. N. M., P. Arcese, and I. G. McLean. 1984. Age, experience, and enemy recognition by wild Song Sparrows. Behav. Ecol. Sociobiol. 14:101–106.

Uyehara, J. C. 1996. Correlates and field experiments of nest searching behavior in a brood parasite, the Brown-headed Cowbird. Ph.D. Thesis, University of California, Los Angeles.

Uyehara, J. C., and P. M. Narins. 1995. Nest defense by Willow Flycatchers (*Empidonax traillii*) to brood-parasitic intruders. Condor 97:361–368.

Whitfield, M. J. 1990. Willow Flycatcher reproductive response to Brown-headed Cowbird parasitism. M.S., California State University, Chico.

Wiley, J. W. 1988. Host selection by the Shiny Cowbird. Condor 90: 289–303.

Wolf, L. 1987. Host-parasite interactions of Brown-headed Cowbirds and Dark-eyed Juncos in Virginia. Wilson Bull. 99:338–350.

Woodworth, B. L. 1995. Ecology of the Puerto Rican Vireo and the Shiny Cowbird in Guánica Forest, Puerto Rico. Ph.D. Thesis, University of Minnesota, St. Paul, MN.

25. Interhabitat Differences in Parasitism Frequencies by Brown-headed Cowbirds in the Okanagan Valley, British Columbia

DAVID WARD AND JAMES N. M. SMITH

Abstract

We compared levels of brood parasitism by Brown-headed Cowbirds among habitats and host species in the south Okanagan Valley, British Columbia, from 1992 through 1994. There were significant differences among habitats in the level of parasitism. Riverine woodland in the valley bottom was the habitat most affected by parasitism. More than 50% of the nests of three host species breeding there (Warbling Vireo, Song Sparrow, and Yellow Warbler) were parasitized. Three of the four species in riverine woodland for which historical records were available (ca. 1900–1978) showed significant increases in parasitism in our study over levels recorded previously. The levels of brood parasitism in riverine woodland may act in tandem with high human impacts in this habitat, resulting in a high probability of local extinction for heavily parasitized species.

Introduction

There has been considerable recent focus on the purported impacts of Brown-headed Cowbirds on populations of their songbird hosts (e.g., Brittingham and Temple 1983, Terborgh 1989). These studies have largely emphasized the impact of cowbirds on forest songbirds because of large-scale forest removal in North America and observations that cowbirds preferentially parasitize species that nest at forest edges (Brittingham and Temple 1983, but see Robinson et al., Chapter 33, this volume). Because deforestation necessarily increases the forest edge : interior ratio, the impact of cowbirds on their hosts is likely to increase in tandem with deforestation. Although such concerns may be well founded, it is also valuable to consider the impacts of cowbirds in other habitats. Brown-headed Cowbirds are habitat generalists (Rothstein 1975), and levels of parasitism are spatially and temporally variable. In some hosts, parasitism varies considerably with habitat; e.g., in Red-winged Blackbirds, *Agelaius phoeniceus*

(Facemire 1980, Linz and Bolin 1982), and Dickcissels, *Spiza americana* (Zimmerman 1983). Thus, we wished to test whether the severe negative impacts of cowbirds on their host populations extend to nonforest habitats.

We studied cowbird parasitism in the south Okanagan Valley, British Columbia. The terrain and conditions vary greatly in this steep-sided valley, and habitat diversity is exceptionally high in a small area. Because of this unusual habitat diversity and a biogeographical link with the fauna of the Great Basin, the vertebrate community is exceptionally rich and of regional and national significance in Canada. Several passerine species of the south Okanagan are found nowhere else in Canada, and the valley is currently experiencing two major changes in land use. First, the Okanagan is one of Canada's favorite vacation and retirement centers, and thus much land is being lost to urbanization. Second, recent changes in agricultural use, driven by a Canada-U.S. trade agreement, are changing a once-stable agricultural landscape. Many Okanagan farmers, faced with increased competition from American farmers, have changed crops, and this is altering the amount and character of land under cultivation.

The habitats occupied by cowbirds in the Okanagan Valley range from sage grasslands through dry ponderosa pine and Douglas fir forests to moister high-elevation forests of western larch, lodgepole pine, and Engelmann spruce. Altitude ranges from 300 m in the valley bottom to about 1,500 m on the extensive plateaus on each side of the valley. Above these plateaus, mountaintops occasionally exceed 2,000 m. The temperature range over the altitudinal gradient in summer is enormous; it is possible to experience 40°C at midday in the valley bottom and to have snow the same day at high altitudes (pers. obs.). A moisture gradient parallels the temperature gradient. In the valley bottom, average annual rainfall is about 300 mm, while at 1,500 m it is about 1,000 mm (Cannings et al. 1987).

We also examine historical records of parasitism because

there is a fairly good record of cowbird parasitism in the valley beginning at the turn of the 20th century (Cannings et al. 1987).

Methods

Host and Cowbird Censuses

In 1992, we attempted to conduct censuses of hosts and cowbirds in as many habitats as possible. This necessarily constrained the number of repeated censuses per site. We recorded virtually no cowbirds in certain habitats (e.g., sage) in 1992. Thus, in 1993, we restricted our censuses to a few habitats that had many cowbirds the previous year and censused these habitats every three weeks through the breeding season. No censuses were conducted in 1994. In riverine woodlands, pine-fir forests, and brushy draws, we used six census sites each in 1993 and repeated censuses at these sites through the season. All site-to-site distances (interhabitat counts) were greater than 1 km, while point-to-point distances (intrahabitat counts) were greater than 500 m. As with all other data presented in this study, our study sites were between the towns of Okanagan Falls (north) and Osoyoos (south).

We restricted our analyses to host species that accept cowbird eggs and are known to raise cowbird young successfully (Rothstein 1975, Rohwer and Spaw 1988). We recorded the number of potential host species in each of the above-mentioned habitats using point counts with distance limited to 100 m (Blondel et al. 1981). All individuals seen or heard calling during the survey period were recorded, and the numbers of individuals per species were estimated. We initially used 10-min counts in which we recorded all birds each minute. We found that almost the same amount of information was obtainable in 5 min. Hence, we restricted all subsequent censuses to 5 min. Cowbird counts included male and female vocalizations. In most instances, cowbird vocalizations were verified by sightings, as habitats were open and visibility was good.

Nest Searching

We searched for nests in eleven different habitat types from 1992 through 1994 (Table 25.1). We attempted to spend approximately equal amounts of time in each habitat surveyed, except that we devoted much more effort to marsh habitat in 1992 than in the following years. Only a few nests were found in several of these habitats. In these cases, data were combined across years to calculate habitat-specific rates of parasitism or nest failure. When a nest was discovered, it was checked carefully for the presence of cowbird and host eggs and young. Nests were revisited 0–8 times to assess their fates. Nests in trees were usually checked using bicycle mirrors mounted on extendable aluminum poles. Some nest checks were made while standing on ladders, but other nests

were so high and far from supporting trunks that they could not be checked successfully. The presence of the parents at these last-mentioned nests and the calling of chicks were used to indicate whether the nest was successful or not.

Levels of Parasitism

Nests were scored as failing if they were preyed on or deserted before the expected fledging time for the species. We used Ehrlich et al. (1988) as a guide to nestling and incubation periods. Nests were scored as successful if they contained host nestlings 1 or 2 days before the expected fledging time, or if the parents were seen feeding fledglings or carrying food near the nest after the young had fledged. Nests that contained young on the last check but were not checked near the expected fledging time were not scored as either successes or failures. When calculating daily rates of nest failure, we considered nests that raised only cowbirds as successes. Nests were scored as parasitized if they contained cowbird eggs or young at any stage and as unparasitized if they never contained cowbird eggs when checked at the egg stage. Nests that contained only host young when discovered were not assigned a parasitism status.

Each nest was assigned an exposure score, the number of days it was observed and active. When a nest failed between checks, we added half the number of days between the final and penultimate check to the exposure score. From these data, we calculated habitat- and species-specific nest failure rates following Hensler and Nichols (1981).

We compared the parasitism frequencies recorded by Cannings et al. (1987) for the Okanagan Valley with our data to determine whether there were significant temporal changes in parasitism for certain species. The Cannings data were mostly gathered in the 1960s at a low-elevation site about 20 km north of our study area in pine-fir and orchard habitats.

Results

Cowbird and Host Censuses

The number of host individuals was similar in the riverine woodlands, brushy draws, and pine-fir forests (1992: $F_{5,44} = 1.49$, $P = .214$; 1993: $F_{2,91} = 0.04$, $P = .958$; Figure 25.1), indicating that these habitats presented similar opportunities for parasitism to cowbirds. There were significant differences among habitats in 1992 in the number of host species ($F_{5,44} = 2.76$, $P = .03$; Figure 25.2), with more species in riverine woodland, pine-fir forest, and brushy draw than in the other habitats. In 1993, however, there was no significant difference among habitats in the number of host species present ($F_{2,91} = 0.44$, $P = .64$; Figure 25.2).

Although the numbers of cowbirds per count were similar in riverine woodlands and pine-fir forests in 1993 (Figure 25.3), the highest ratio of cowbirds to hosts occurred in riv-

Table 25.1. Habitat Types Where Cowbird Parasitism Surveys Were Conducted

Habitat Type	Description	Elevation (m)
Riverine woodland	Patchy willow, rose, and cottonwood scrub woodland, near water	300–400
Pine-fir forest	Ponderosa pine–dominated dry forest	300–500
Sage grassland	Open grassland with big sage	300–400
Purshia grassland	Grassland with tall shrubs and scattered pines	300–350
Brushy draws	Scrub-filled gullies, with some small trees	300–400
Mixed woodland	Aspen–Douglas fir forest with cleared openings	400–500
Subalpine forest	Lodgepole pine–Engelmann spruce forest with marshy openings, clear-cuts	1200–1500
Marsh	Valley-bottom cattail marsh	300–350
Suburban	Gardens and hedgerows	300–350
Agricultural	Field and pasture margins	300–350
Second-growth fir forest	Douglas fir–dominated forest, 15–20 years old; some aspen, willow, and western larch	900–1100

Plant species: aspen, *Populus tremuloides;* cottonwood *P. nigra;* willow, *Salix* spp.; rose, *Rosa* spp.; Purshia, *Purshia tridentata;* big sage, *Artemisia tridentata;* cattail, *Typha latifolia;* Douglas fir, *Pseudotsuga menziesii;* western larch, *Larix occidentalis;* lodgepole pine, *Pinus contorta;* ponderosa pine, *P. ponderosa;* Engelmann spruce, *Picea engelmannii.*

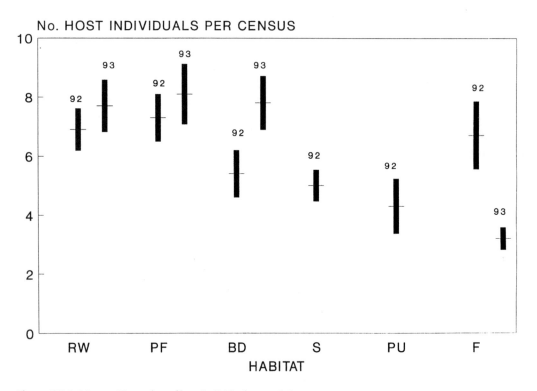

Figure 25.1. Mean ± SE number of host individuals recorded per 5-min census (acceptor species only). Habitats: RW, riverine woodland; PF, Ponderosa pine–fir forest; BD, brushy draw; S, sage; PU, *Purshia;* F, second-growth fir forest. 92 and 93 indicate the years that censuses were conducted.

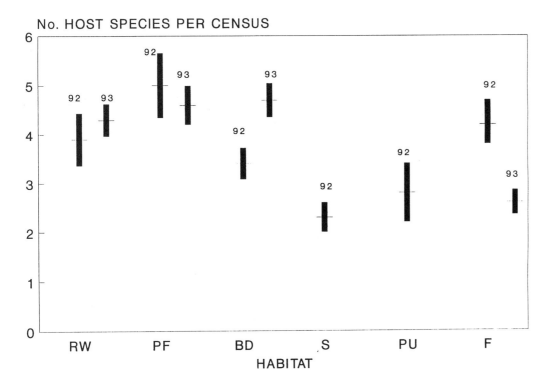

Figure 25.2. Mean ± SE number of host species recorded per 5-min census (acceptor species only). Habitat abbreviations as for Figure 25.1.

erine woodland in both 1992 ($F_{5,44}$ = 3.13, P = .017) and 1993 ($F_{2,91}$ = 4.44, P = .014), with about one-third as many cowbirds as hosts in this habitat (Figure 25.4). There were also significantly more cowbirds in riverine woodland than in the other habitats in 1992 ($F_{5,44}$ = 3.64, P = .008). In 1993, there were significantly more cowbirds in pine forest and riverine woodland than in brushy draw ($F_{2,91}$ = 4.37, P = .015). No cowbirds were found in sage habitat, and numbers of cowbirds in the high-altitude second-growth larch, fir, and spruce forests were low (Figure 25.3).

Levels of Parasitism and Nest Failure by Habitat

We found most nests in riverine woodland and marsh habitats. Parasitism was frequent in all habitats except marsh (Table 25.2). Riverine woodland, with the biggest sample of nests checked, had a consistently high rate of parasitism in all years. Levels of parasitism were slightly lower in the two habitats at mid-elevations (mixed woodland and second-growth fir forest), and no parasitism was noted in five nests in subalpine areas. Subsequent extensive work in high-elevation habitats (C. J. Fonnesbeck unpubl. data) has confirmed that parasitism is infrequent in subalpine areas. Too few nests were followed in several habitats to assess rates of parasitism with confidence. Daily rates of nest failure did not vary strongly among habitats and averaged from 2% to 6% per day.

Changes in the level of multiple parasitism may also indicate significant changes in parasitism pressure on host populations. However, levels of multiple parasitism in our study area were not significantly different from those reported by Cannings et al. (1987) (Kolmogorov-Smirnov two-sample test, P = .82, Table 25.3).

Levels of Parasitism and Nest Failure by Host Species

Levels of parasitism were moderate to high for most suitable hosts with more than five nests scored for parasitism (Table 25.4). Levels were over 48% for Yellow Warbler (*Dendroica petechia*), Warbling Vireo (*Vireo gilvus*), and Song Sparrow (*Melospiza melodia*). The first mentioned was also the host species most often noted (6 of 13 cases) feeding cowbird fledglings. Table 25.4 also presents data on parasitism levels for the Okanagan Valley reported by Cannings et al. (1987). Increased levels of parasitism were found in some species (Song Sparrow; Black-headed Grosbeak, *Pheucticus melanocephalus;* Western Wood-Pewee, *Contopus sordidulus;* and Brewer's Blackbird, *Euphagus cyanocephalus*). Overall, as many species showed an increase in rate of parasitism as showed a decrease [paired t-test, t = 0.63, P = .536, power $(1 - \beta)$ = 80.6%]. In the riverine woodland habitat, however, there was a significant increase in the rate of parasitism (paired t-test, t = 2.33, P = .04).

Levels of daily nest failure by species (Table 25.4) were

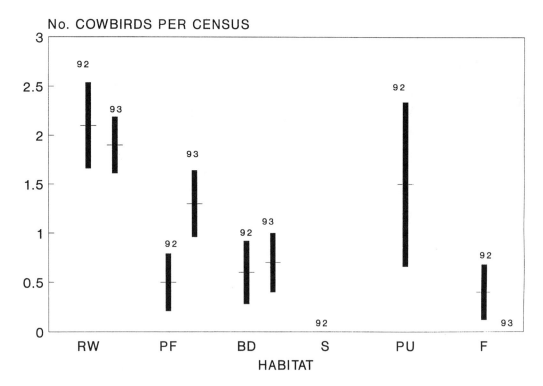

Figure 25.3. Mean ± SE number of cowbirds recorded per 5-min census. Habitat abbreviations as for Figure 25.1.

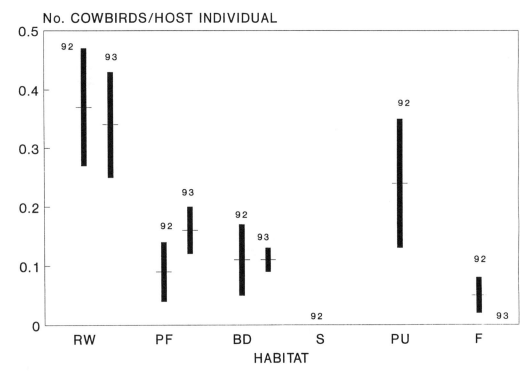

Figure 25.4. Mean ± SE ratio of the number of cowbirds per host individual per 5-min census. Habitat abbreviations as for Figure 25.1.

higher than the habitat-specific levels (Table 25.2), most notably for the Song Sparrow, Yellow Warbler, and Warbling Vireo. Comparison of our nest success data with those collected by Martin (1992), Weatherhead (1989), and Smith (1981) shows that nest success in the Okanagan is low for the seven species for which we have sufficient data to make a useful comparison (Table 25.5).

Cowbird parasitism had a significant negative impact on the survival of the young of host species for which we have sufficient data (Table 25.6). There was a mean decline of 1.6 surviving young per nest across host species.

Discussion

Habitat-specific Parasitism

Our data clearly indicate that hosts in the riverine woodland habitats in the Okanagan Valley suffer high levels of parasitism and that there are large numbers of cowbirds there. This habitat is also under the most pressure from anthropogenic sources, through direct destruction for housing to sustain the burgeoning human populations in the valley, through agriculture, and through invasions of weed species such as

poison ivy (*Toxicodendron radicans*), thistle (*Cirsium vulgare*), broom (*Cytisus scoparius*), and nettles (*Urtica dioica*).

The average levels of parasitism in all but the marsh and sage habitats in the Okanagan Valley (Table 25.2) may be sufficient to drive host populations to extinction, especially if there is additional pressure from nest predation or habitat loss (May and Robinson 1985).

Host Species of Particular Concern

Although our sample sizes (and those of Cannings et al. 1987) are modest, our data show that several host species in the Okanagan Valley are suffering increasingly high levels of parasitism. The species with the most severe parasitism are the Warbling Vireo, Yellow Warbler, Black-headed Grosbeak, and Song Sparrow. All these species are widespread in North America. Global extinction of these species is unlikely, because populations of such species are unlikely to suffer the same pressures over a wide geographic range (Terborgh 1989). However, if such levels of parasitism were universal for any species (e.g., Griffith and Griffith, Chapter 38, this volume), there would indeed be cause for concern.

Warbling Vireos in the Okanagan Valley may also be susceptible to local extinction as a result of brood parasitism. This species usually produces no surviving young when parasitized by cowbirds (Cannings et al. 1987, Ward and Smith pers. obs.). At current levels of parasitism, simple demographic models suggest that a population of 100 mated pairs of Warbling Vireos would go locally extinct in the Okanagan Valley in under 35 years (Ward and Smith unpubl. manuscript). Vireos, as a group, appear to be highly vulnerable to extinction in North America (Grzybowski et al. 1986; Griffith and Griffith, Chapter 38, Hayden et al., Chapter 39, and Chace et al., Chapter 14, this volume), perhaps because of their lack of responses to parasitism (but see Sealy 1996), long incubation periods, and high suitability as hosts. Thus, further monitoring of these species is required.

Of the four species of most concern here, two (Yellow Warbler and Warbling Vireo) are discussed in the Böhning-Gaese et al. (1993) analysis of trends in breeding bird surveys (BBS). Both species appear to have healthy populations,

Table 25.2. Parasitism and Nest Failure Rates by Habitat, 1992–1994

Habitat	% Parasitism		Daily Rate of Nest Failure	
Riverine woodland	37	(180)	0.022	(217)
Pine-fir forest	40	(10)	0.018	(10)
Brushy draw	39	(54)	0.024	(73)
Mixed woodland	27	(83)	0.021	(120)
Suburban garden	18	(11)	0.006	(10)
Second-growth fir forest	33	(13)	0.037	(10)
Marsh	4	(113)	0.057	(107)

Note: Numbers in parentheses are the numbers of nests found, all hosts combined. Too few nests were found to calculate rates in other habitats.

Table 25.3. Levels of Multiple Parasitism in Our Study Compared with Those Reported by Cannings et al. (1987)

	Number of Cowbird Eggs/Nest							
	1	2	3	4	5	6	7	8
Number of nests, this study	190	50	31	3	1	2	0	0
Number of nests, Cannings et al.	189	54	16	5	0	1	0	1

Table 25.4. Parasitism and Nest Failure Rates by Species, 1992–1994, Compared with Data from Cannings et al. (1987)

Species	% Parasitism, This Study	% Parasitism, Cannings et al.	Daily Rate of Nest Failure, This Study
American Robin	* 0 (26)	* 0 (288)	0.016 (21)
Gray Catbird	* 7 (29)	* 4 (120)	0.037 (12)
Cedar Waxwing	* 0 (8)	8 (84)	0.01 (7)
Dusky Flycatcher	16 (19)	46 (11)	0.029 (13)
Willow Flycatcher	25 (16)	49 (104)	0.04 (11)
Western Wood-Pewee	38 (26)	10 (40)	0.014 (16)
Warbling Vireo	49 (45)	79 (43)	0.045 (24)
Yellow Warbler	77 (35)	32 (28)	0.079 (21)
Song Sparrow	55 (44)	9 (44)	0.088 (26)
Chipping Sparrow	33 (6)	35 (98)	0.069 (8)
Dark-eyed Junco	0 (5)	8 (49)	0.067 (5)
Black-headed Grosbeak	35 (20)	8 (13)	0.063 (15)
Brewer's Blackbird	33 (18)	9 (155)	0.081 (16)
Red-winged Blackbird	1 (107)	11 (258)	0.057 (107)

Note: Numbers in parentheses are the numbers of nests found. Asterisks denote rejecter species (of these, only the Gray Catbird was known to be parasitized).

Table 25.5. Nest Success of Hosts in Our Study Compared with Three Other Studies

Species	Mayfield Success, This Study	Mayfield Success, Martin	Fraction Successful, Martin
Gray Catbird	0.361	0.445, 0.57	0.518, 0.591, 0.487
Willow Flycatcher	0.319	—	0.369
Warbling Vireo	0.263	0.550	0.618
Yellow Warbler	0.117	—	0.453
Black-headed Grosbeak	0.162	0.743	0.769
Red-winged Blackbird	0.205	—	0.40–0.862
Song Sparrow	0.084	—	0.74

Note: Martin (1992) gives nest success values using the Mayfield estimate (as we did) and by simply calculating the fraction of successful nests. The latter method tends to overestimate nest success. Red-winged Blackbird fraction is from Weatherhead 1989. Song Sparrow fraction is from Smith 1981.

and the abundance of Warbling Vireos has increased significantly. If the data obtained from the BBS are reliable indicators of population persistence and abundance, local extinctions in the Okanagan may be of little concern.

Martin (1992) considers predation to be a more important factor influencing nest success than cowbird parasitism. The overall low levels of nest success in our study compared with Martin's (1992) survey (Table 25.5) suggest that most species in the Okanagan Valley are reproducing poorly, which is a cause for concern.

Comparisons with Historical Data

The data on cowbird parasitism recorded by Cannings et al. (1987) come from a variety of sources and mostly represent records collected serendipitously; direct comparisons with our more uniformly collected data are therefore difficult. In addition, much of the data presented by Cannings et al. come from a single site dominated by orchards and Ponderosa pine forest, habitats for which we have little data.

Although there was little overall temporal change in the level of parasitism lumped across species, there were considerable differences in the intensity of parasitism of certain species. While these differences may indeed reflect higher levels of parasitism in the 1990s, they may also be due to differences in habitats sampled at different times or to differences in data collection. The last possibility may explain low levels of parasitism of Red-winged Blackbirds in our study. Most of our nesting records for this blackbird are from a large marsh (ca. 30 ha, see also Ward et al. 1996). Linz and Bolin (1982) have shown that the percentage of parasitism of Red-winged Blackbirds is negatively correlated with marsh size. Cowbirds preferentially parasitize individuals close to the marsh edge, and small marshes have a greater edge : interior ratio than large marshes. Because the Cannings et al. data were collected incidentally, it is unlikely that many records would be from the interiors of large marshes where access is difficult. Thus, the levels of parasitism that Cannings et al. recorded are likely to be higher than those that we recorded by systematically searching the length and breadth of marshes.

Notwithstanding the above-mentioned methodological differences, there has been a significant temporal increase in average parasitism in the most heavily parasitized habitat, riverine woodland. Three species (Yellow Warbler, Black-headed Grosbeak, and Song Sparrow) of the four acceptor species that we studied in this habitat showed large increases in parasitism (median increase of 45%, Table 25.4). Sample sizes for the Yellow Warbler and Song Sparrow are large enough that we have confidence that this pattern is robust. However, more nests of Black-headed Grosbeaks and Willow Flycatchers (where parasitism levels declined compared to historical data) need to be monitored to ascertain whether the apparent general increase in parasitism in riverine woodland is a cause for concern.

Management Implications of Cowbird-Host Coexistence

The initial colonization of new habitats by cowbirds expanding their historical range was undoubtedly a major ecological change, with potentially profound implications for vulnerable host species. However, cowbird numbers are no longer increasing in much of North America (Peterjohn et al., Chapter 2, and Wiedenfeld, Chapter 3, this volume). Cowbirds have been present in the Okanagan Valley since the beginning of historical records, although no quantita-

Table 25.6. Differences in the Mean ± SD of Surviving Young Produced by Parasitized and Unparasitized Nests

Species	Surviving Young, Unparasitized Nests	Surviving Young, Parasitized Nests	Surviving Young, Overall	Mean Change in No. of Surviving Young	N	P <
BHGR	1.4 ± 1.64	0.5 ± 0.76	1.2 ± 1.33	−0.9	15	0.05
SOSP	1.6 ± 0.48	0.3 ± 0.62	1.0 ± 1.18	−1.3	26	0.001
DUFL	1.9 ± 1.38	0	1.6 ± 1.44	−1.9	13	0.001
WIFL	2.3 ± 1.56	0	1.5 ± 1.66	−2.3	11	0.001
WWPE	2.0 ± 1.49	1.4 ± 1.18	1.7 ± 1.45	−0.6	16	0.2
YEWA	1.1 ± 1.04	1.4 ± 1.29	1.2 ± 1.15	+0.3	19	0.2
WAVI	1.7 ± 1.49	0.1 ± 0.35	1.1 ± 1.18	−1.6	24	0.01

Note: Significance values (*P*) are from t-tests comparing the numbers of surviving young in parasitized and unparasitized nests. *N* = total number of nests in sample. Abbreviations for species: BHGR, Black-headed Grosbeak; DUFL, Dusky Flycatcher; SOSP, Song Sparrow; WAVI, Warbling Vireo; WIFL, Willow Flycatcher; WWPE, Western Wood-Pewee; YEWA, Yellow Warbler.

tive data on their abundance are available. Cowbirds in the Okanagan are strongly morphologically differentiated from the two subspecies recognized in western North America (Ward and Smith 1998), implying that they have been in the Okanagan region for a long time. In the Okanagan, as in much of the continent, it is therefore likely that cowbirds and hosts coexist stably, at least where landscapes are not being altered rapidly by humans. Avian community ecologists have yet to come to a clear consensus on how stable avian community patterns are in space and time (Wiens 1989). There is, however, certainly enough temporal and spatial variation to make it difficult to detect subtle impacts of a new species arriving in a community.

We therefore caution against treating the patterns of parasitism found in our study, or similar patterns found elsewhere on the continent, as a cause for great alarm or immediate management action, such as a cowbird removal program. The highest parasitism rates in our study affected common species such as the Yellow Warbler and Warbling Vireo. These species are probably not threatened regionally and certainly not threatened continentally by cowbird parasitism. In the last-mentioned case, such levels of parasitism have persisted in the Okanagan Valley for 30 years (Cannings et al. 1987), yet the species remains common in mid-elevation mixed forests.

On a smaller spatial scale, however, some marked local population changes appear to have occurred. Parham (1937), who lived in this part of the valley during the first 20 years of this century, recorded three vireo species as common on the valley floor (Warbling; Red-eyed, *Vireo olivaceus*; and Cassin's [Solitary], *V. cassinii*). Yet, we never found Red-eyed or Cassin's Vireos nesting there, and we found only three Warbling Vireo nests on the valley floor during our study. This suggests that extinction or near-extinction has occurred on the valley floor for all three vireo species during the past century. The valley floor has suffered the highest habitat loss from anthropogenic sources. Source-sink population dynamics (Pulliam 1988) may maintain healthy populations of such species, even if local production is poor, provided the species can breed in places where cowbirds are rare or absent. In the Okanagan Valley, such places occurred for Warbling and Cassin's Vireos within 20 km of the valley bottom in higher-elevation habitats. An issue worthy of attention in the Okanagan Valley is whether increasing disturbance of such source populations by clearcut logging at higher elevations is increasing the ranges and impacts of cowbirds.

The species of greatest concern are those cowbird hosts that are restricted to habitat islands far from, or in the absence of, sources of colonists. One such species is the Yellow-breasted Chat, *Icteria virens*. There are fewer than 20 breeding pairs of chats in the Okanagan and neighboring Similkameen valleys (R. Gibbard unpubl. data). Further study of such species is needed in the south Okanagan Val-

ley. It should be noted, however, that chats are common throughout their extensive continental range in the United States and in no immediate danger of global extinction.

Management action may be warranted, however, in areas where increasing intensities of human-induced disturbance are imposed on top of a preexisting and significant level of cowbird parasitism. This situation may apply in riverine woodland in the Okanagan Valley. The human population in the valley is increasing through immigration at over 3% per annum, and the loss of riverine woodland habitat may be placing several species like chats and Willow Flycatchers in increasing jeopardy through habitat fragmentation and loss. Increasing levels of cowbird parasitism in such a situation could cause rapid population extinctions, and active management such as cowbird removal might be considered. In the absence of significant habitat restoration measures, however, cowbird control might merely postpone the extinction of such species. The southwestern subspecies of the Willow Flycatcher is currently facing this situation (Whitfield, Chapter 40, this volume). Active cowbird management has costs as well as benefits (see introduction to Part V), and both costs and benefits need to be considered when active management is being contemplated.

Acknowledgments

We thank Anna Lindholm, Christine Adkins, Kelley Kissner, and Carolina Johansson for assistance in the field and Mary Doherty and Frank Metcalfe for many favors. We thank the Hatfield family for kindly allowing us to stay on their island during the course of the study. This study was funded by a grant from NSERC (Canada) to JNMS, and by an NSERC international postdoctoral fellowship to DW.

References Cited

Blondel, J., C. Ferry, and B. Frochot. 1981. Point counts with unlimited distance. Studies in Avian Biol. 6:414–420.

Böhning-Gaese, K., M. L. Taper, and J. H. Brown. 1993. Are declines in North American insectivorous songbirds due to causes on the breeding range? Conserv. Biol. 7:76–86.

Brittingham, M. C., and S. A. Temple. 1983. Have cowbirds caused forest songbirds to decline? BioScience 33:31–35.

Cannings, R. A., R. J. Cannings, and S. G. Cannings. 1987. Birds of the Okanagan Valley, British Columbia. Royal British Columbia Museum, Victoria.

Ehrlich, P. R., D. S. Dobkin, and D. Wheye. 1988. The birder's handbook: A field guide to the natural history of North American birds. Simon and Schuster, Fireside, New York.

Facemire, C. F. 1980. Cowbird parasitism of marsh-dwelling Red-winged Blackbirds. Condor 82:347–348.

Grzybowski, J. A., R. B. Clapp, and J. T. Marshall. 1986. History and current population status of the Black-capped Vireo in Oklahoma. Amer. Birds 40:1151–1161.

Hensler, G. L., and J. D. Nichols. 1981. The Mayfield method of estimating nesting success: A model, estimators, and simulation results. Wilson Bull. 93:42–53.

Linz, G. M., and S. B. Bolin. 1982. Incidence of Brown-headed Cowbird parasitism on Red-winged Blackbirds. Wilson Bull. 94: 93–95.

Martin, T. E. 1992. Breeding productivity considerations: What are the appropriate habitat features for management? Pp. 455–473 in Ecology and conservation of neotropical migrant landbirds (J. M. Hagan III and D. W. Johnston, eds). Smithsonian Institution Press, Washington, DC.

May, R. M., and S. K. Robinson. 1985. Population dynamics of avian brood parasitism. Amer. Natur. 126:475–494.

Parham, H. J. 1937. A naturalist in British Columbia. Witherby, London.

Pulliam, H. R. 1988. Sources, sinks, and population regulation. Amer. Natur. 132:652–661.

Rohwer, S., and C. D. Spaw. 1988. Evolutionary lag versus bill-size constraints: A comparative study of the acceptance of cowbird eggs by old hosts. Evolutionary Ecol. 2:27–36.

Rothstein, S. I. 1975. An experimental and teleonomic investigation of avian brood parasitism. Condor 77:250–271.

Sealy, S. G. 1996. Evolution of host defenses against brood parasitism: Implications of puncture ejection by a small passerine. Auk 113:346–355.

Smith, J. N. M. 1981. Cowbird parasitism, host fitness, and age of the host female in an island Song Sparrow population. Condor 83:152–161.

Terborgh, J. 1989. Where have all the birds gone? Princeton University Press, Princeton, NJ.

Ward, D., A. K. Lindholm, and J. N. M. Smith. 1996. Multiple parasitism of the Red-winged Blackbird: Further experimental evidence of evolutionary lag in a common host of the Brown-headed Cowbird. Auk 113:408–413.

Ward, D., and J. N. M. Smith. 1998. Morphological differentiation of Brown-headed Cowbirds in the Okanagan Valley, British Columbia. Condor 100:1–7.

Weatherhead, P. J. 1989. Sex ratios, host-specific reproductive success, and impact of Brown-headed Cowbirds. Auk 106:358–366.

Wiens, J. A. 1989. The ecology of bird communities: Vol. 1, Foundations and patterns. Cambridge University Press, Cambridge.

Zimmerman, J. L . 1983. Cowbird parasitism of Dickcissels in different habitats and at different nest densities. Wilson Bull. 95: 7–22.

26. Cowbird Parasitism and Nest Predation in Fragmented Grasslands of Southwestern Manitoba

STEPHEN K. DAVIS AND SPENCER G. SEALY

Abstract

The loss and degradation of native grassland in the northern Great Plains has resulted in the decline of many species of grassland birds. The number of Brown-headed Cowbirds, however, has increased over the past 200 years in this region where few studies of parasitism have been conducted. We quantified the frequency of parasitism by cowbirds and nest predation on ground-nesting passerine birds in three fragments of grassland habitat in southwestern Manitoba. Cowbird parasitism was higher for hosts that nested in a 22-ha site compared with two 64-ha sites. High abundance of female cowbirds, increased edge effects from fragmentation, availability of perch sites, and close proximity of cowbird foraging areas to host nests may have promoted the higher parasitism frequencies at the smallest site. Each host species was parasitized, but at different frequencies. All hosts fledged at least one cowbird, except for Lark Buntings and Bobolinks. Western Meadowlarks were parasitized at a frequency of 44%, with an average of 3.1 cowbird eggs laid in each parasitized nest. Chestnut-collared Longspurs and Sprague's Pipits were parasitized at frequencies of 18% and 14%, respectively. Baird's, Grasshopper and Savannah Sparrows were parasitized at frequencies of 27–36% and fledged over 20% of the cowbird eggs laid. Parasitized nests of these species fledged significantly fewer young than unparasitized nests, while the impact of parasitism on host productivity per nest was negligible for the other five species. The frequencies of predation did not differ significantly between the three sites, but nests in the smallest site fledged significantly fewer young per nest than those at the two larger sites. These results suggest that parasitism was primarily responsible for the low productivity/nest in the smallest site. The combination of parasitism and nest predation greatly reduced nesting success of grassland birds in Manitoba.

Introduction

Historically, Brown-headed Cowbirds ranged throughout the Great Plains region of North America (Mayfield 1965). Over the last 200 years, however, agricultural practices and deforestation have altered the North American landscape. Cowbirds subsequently expanded their range, and their numbers have increased (Mayfield 1965, Rothstein et al. 1980) while populations of many other grassland songbirds have declined (Peterjohn and Sauer 1993). Where the grasslands were once an almost continuous expanse of habitat, now there is often only a mosaic of fragmented islands of habitat to support the native fauna (Owens and Myres 1973, Herkert 1994). In Canada, less than 2% of the original tall-grass and 24% of mixed-grass prairie remain (Trottier 1992). Fragmentation of habitats apparently has favored the lifestyle of cowbirds, because parasitism often is more frequent near habitat edges (Gates and Gysel 1978, see also Paton 1994). Studies of this relationship, however, have involved mainly forest habitats (O'Conner and Faaborg 1992, Robinson et al. 1993), and grasslands generally have been ignored (but see Johnson and Temple 1990). In this study, we fill in a gap by quantifying the frequency of nest predation and cowbird parasitism on host species in three small fragments of grassland in southwestern Manitoba.

Cowbird parasitism often reduces host productivity (e.g., Rothstein 1975, Payne 1977). Mayfield (1965) suggested that because cowbirds and potential host species on the grasslands evolved together, hosts may be more resistant to parasitism than species in areas east and west of the plains that only recently have experienced cowbird parasitism. Friedmann and Kiff (1985) supported this idea, finding few records of cowbird parasitism among old nest records of several grassland birds. More recently, however, studies have revealed that cowbirds frequently parasitize grassland birds and that parasitism reduces their nesting success (Hill 1976, Elliott 1978, Zimmerman 1983, Fleischer 1986). Here we quantify the nesting success of hosts to determine the effects of parasitism on productivity and examine factors that may influence cowbird host choices.

Methods

Study Sites

The study was conducted from 6 May to 17 August 1991 and 1 May to 25 August 1992 on three sites of typical grassland habitat in southwestern Manitoba. Site 1 (49° 04' N, 101° 14' W) was a square, 64-ha patch of idle hayland located in the Broomhill Wildlife Management Area (WMA). Graveled roads bordered the east and north sides of the site, native pasture (and fence) bordered the south and west sides, and cropland was situated to the north. *Bromus inermis* was prominent along with *Stipa spartea*, *Poa* spp., *Koeleria cristata*, *Artemisia frigida*, and *Melilotus officinalis* (see Appendix 1 in Davis 1994). Shrubs (*Salix* spp.) were present only on the southeast and northwest portions of the site. The structure and species composition of vegetation on the WMA, which continued eastward on the other side of the road, was different from that on this site. Site 2 (49° 24' N, 101° 02' W) was a square, 64-ha private cattle pasture that rolled gently with numerous depressions. A fence bordered the south, north, and east sides of the site. The pasture continued westward, and a native hay field, horse pasture, and cropland were situated along the north, south, and east sides, respectively. Elevated areas were extensively grazed by cattle and covered mostly by *Agropyron repens* and *Calamagrostis inexpansa*. Sedges (*Carex* spp.) dominated the low areas. Western snowberry (*Symphoricarpos occidentalis*) and wolf willow (*Elaeagnus commutata*) occurred in clumps throughout this site. Site 3 (49° 30' N, 100° 56' W) was an irregularly shaped, 22-ha strip of native grassland surrounded by crop and rangeland. This site had been periodically grazed in the past and was characterized by *Stipa spartea*, *Muhlenbergia richardsonis*, and *Bouteloua gracilis*. The site was bordered by stands of wolf willow, western snowberry, and *Salix* spp., with other shrubs scattered throughout.

Each site was staked out with labeled pin flags to produce a 50-m grid system, with markers kept approximately the height of the vegetation throughout the field season to avoid creating artificial perches. Hosts, but apparently not cowbirds, occasionally perched on the flags.

Nest Success

Females were flushed from nests by dragging the vegetation with a 30-m nylon rope to which aluminum cans were attached every 0.5 m. Each nest was marked with a small flag 5 m away in line with the nearest grid point (see above), and nests were inspected every 2–4 days until the young fledged or the nesting attempt ended. Several cues were used to identify successful nests: nests from which at least one host young or cowbird young fledged, including nestlings near fledging age and minimal nest disturbance with an adult(s) uttering alarm calls nearby. Nest success was quantified for each species by using Mayfield's (1975) method of calculating nest success (computer program in Krebs 1989). Results from each site and species were analyzed separately to identify differences in parasitism frequency, predation, and host and cowbird productivity.

Determination of Cowbird and Host Abundance

In 1992, five surveys were conducted between 13 May and 13 June (site 1), 11 May and 4 June (site 2), and 12 May and 12 June (site 3). Three transects were established in sites 1 and 2 (two 300-m belts, one 200-m belt) to census the entire study area, and one central transect was established for site 3. Singing males were plotted on a map for each census, and the mean abundance was calculated for each species on each site. The number of female cowbirds encountered during the surveys was recorded, and the mean abundance was computed for each site. Song surveys were conducted at sunrise at each site on days without rain and when winds were less than 20 km/hr.

The number of female cowbirds was monitored on each site during the cowbird laying period (10 May to 25 July 1992, see Davis 1994). The activity period of cowbirds was divided into three time periods (Central Standard Time): morning (0400–1000 hr), afternoon (1000–1600 hr) and evening (1600+ hr). The number of female cowbirds observed in each time period was converted into the number of cowbirds/hr to permit comparisons of time periods and sites.

Statistical Analyses

Analyses were conducted using the SAS/STAT package (SAS Institute 1989). The frequency of parasitism and predation of each host species was compared using chi-square contingency analyses or Fisher's exact test if 25% of the cells had expected counts less than 5 (PROC FREQ). Productivity of host species was analyzed using one-way ANOVA (PROC GLM) to determine whether significant differences existed between sites.

Results

Frequency of Cowbird Parasitism

The species composition of the three sites was similar, but species were not equally abundant (Table 26.1). Baird's Sparrow (*Ammodramus bairdii*), Grasshopper Sparrow (*A. savannarum*), Savannah Sparrow (*Passerculus sandwichensis*), Sprague's Pipit (*Anthus spragueii*), Brown-headed Cowbird, and Western Meadowlark (*Sturnella neglecta*) were recorded during song surveys on all sites. Chestnut-collared Longspurs (*Calcarius ornata*) were not recorded on site 3; Bobolinks (*Dolichonyx oryzivorus*) and Lark Buntings (*Cala-*

mospiza melanocorys) were detected only on sites 1 and 3, respectively. Nests of these latter two species were located only in 1992.

Parasitism frequencies varied among host species and among sites (Table 26.2). Except for Chestnut-collared Longspurs, hosts on site 3 experienced high parasitism frequencies. Parasitism frequency was highest on site 3 for Sprague's Pipit (Fisher's exact test, $P = .037$), Baird's Sparrow (X^2, df = 2, $P = .001$), Grasshopper Sparrow (Fisher's exact test, $P = .016$), and Western Meadowlark (Fisher's exact test, $P < .01$). Cowbirds multiply parasitized all host species but laid the most eggs in Western Meadowlark nests (range 1–8 eggs). The mean number of cowbird eggs laid in each host's nests did not differ significantly between sites (one-way ANOVA, df = 2, $P > .16$). Of the 183 cowbird eggs laid, 42 (23%) were laid after host clutches were completed (79% during the incubation stage, 11% during the nestling stage, and 9% after the nest was terminated).

Except for Bobolinks and Lark Buntings, which fledged only their own young (Tables 26.2 and 26.3), at least one cowbird was reared by each host species. Savannah Sparrows fledged a greater percentage of cowbird young than any other host species (Table 26.2). The mean number of host young fledged from successful unparasitized nests was significantly higher than from successful parasitized nests for Savannah, Baird's, and Grasshopper Sparrows (Table 26.4).

Table 26.1. Mean ± SE Number of Individuals Detected during Five Song Surveys on Three Study Sites in Southwestern Manitoba

	Site 1	Site 2	Site 3
Lark Bunting (LARB)	0	0	11.4 ± 2.7
Sprague's Pipit (SPPI)	12.6 ± 2.1	9.6 ± 2.9	3.8 ± 0.4
Savannah Sparrow (SAVS)	8.8 ± 2.3	20.2 ± 1.3	6.0 ± 1.3
Baird's Sparrow (BAIS)	27.4 ± 3.8	15.0 ± 1.3	8.6 ± 1.4
Grasshopper Sparrow (GRSP)	26.2 ± 2.3	7.4 ± 1.8	7.8 ± 1.5
Chestnut-collared Longspur (CCLO)	9.4 ± 1.8	14.0 ± 2.4	0
Bobolink (BOBO)	17.0 ± 4.7	0	0
Western Meadowlark (WEME)	7.0 ± 0.7	8.4 ± 1.1	7.0 ± 0.6
Brown-headed Cowbird (BHCO)	2.8 ± 1.5	7.4 ± 1.4	7.6 ± 1.1

Table 26.2. Parasitism Frequency and Cowbird Productivity for Eight Grassland Host Species, 1991 and 1992

Host Species	% Parasitism (*N*)				Cowbird Eggs Laid/ Parasitized Nest (± SE)				% Cowbirds Fledged from Eggs Laid (*N*)			
	Site 1	Site 2	Site 3	Total	Site 1	Site 2	Site 3	Total	Site 1	Site 2	Site 3	Total
SPPI	0 (10)	0 (2)	60 (5)*	18 (17)	0	0	3.0 ± 0.6	3.0 ± 0.6	0	0	11 (9)	11 (9)
LARB	—	—	100 (6)	100 (6)	—	—	2.5 ± 0.4	2.5 ± 0.4	—	—	0 (15)	0 (15)
SAVS	14 (7)	35 (20)	50 (4)	32 (31)	2.0	2.1 ± 0.4	2.0 ± 1.0	2.1 ± 0.3	0 (1)	20 (15)	50 (4)	30 (20)
BAIS	14 (29)	25 (16)	77 (22)*	37 (67)	1.7 ± 0.2	1.5 ± 0.3	2.1 ± 0.2	1.9 ± 0.2	14 (7)	50 (6)	19 (36)	22 (49)
GRSP	15 (26)	14 (7)	56 (16)*	29 (49)	1.5 ± 0.3	3.0	2.0 ± 0.2	1.9 ± 0.2	40 (6)	0 (3)	22 (18)	22 (27)
CCLO	23 (31)	4 (24)	0 (2)	14 (57)	1.3 ± 0.1	2.0	0	1.4 ± 0.2	11 (9)	0 (2)	(0)	9 (11)
BOBO	50 (6)	—	—	50 (6)	1.7 ± 0.7	—	—	1.7 ± 0.7	0 (5)	—	—	0 (5)
WEME	43 (7)	23 (31)	67 (27)*	43 (65)	1.7 ± 0.7	2.4 ± 0.7	3.6 ± 0.5	3.1 ± 0.4	40 (5)	6 (17)	18 (66)	17 (88)

*$P < 0.05$.

Table 26.3. Nest Predation Frequency, Mayfield Survival Rate, and Productivity of Eight Grassland Hosts

Host Species	% Depredated (N)	Mayfield Survival Rate (%)	Hosts Fledged/ Nest
Sprague's Pipit	53 (17)	14	1.1
Lark Bunting	71 (7)	20	1.1
Savannah Sparrow	50 (30)	7	0.8
Baird's Sparrow	39 (67)	21	1.4
Grasshopper Sparrow	46 (46)	23	1.7
Chestnut-collared Longspur	47 (57)	30	1.7
Bobolink	50 (6)	6	1.3
Western Meadowlark	55 (65)	9	0.9

Note: Values are totals of all three sites, as no significant differences were detected between sites.

Table 26.4. Comparison of Host Productivity in Parasitized and Unparasitized Nests, 1991 and 1992

Species	Host Young Fledged/ Unparasitized Nest	Host Young Fledged/ Parasitized Nest	1-way ANOVA (df = 1)	Host Young Fledged/ Successful Unparasitized Nest	Host Young Fledged/ Successful Parasitized Nest	1-way ANOVA (df = 1)
SPPI	1.36 ± 0.43 (14)	0.00 (3)	$F = 2.06, P = 0.17$	2.71 ± 0.42 (7)	—	—
LARB	0.00 (1)	1.33 ± 1.33 (6)	$F = 0.33, P = 0.59$	—	4.0 ± 1.00 (2)	—
SAVS	1.09 ± 0.33 (21)	0.11 ± 0.11 (9)	$F = 3.66, P = 0.07$	2.56 ± 0.41 (9)	0.33 ± 0.33 (2)	$F = 8.62, P = 0.02$
BAIS	1.55 ± 0.29 (42)	1.21 ± 0.31 (24)	$F = 0.58, P = 0.45$	3.25 ± 0.29 (20)	1.81 ± 0.38 (16)	$F = 9.44, P = 0.01$
GRSP	2.03 ± 0.35 (33)	0.92 ± 0.39 (13)	$F = 3.23, P = 0.08$	3.72 ± 0.24 (18)	2.40 ± 0.60 (5)	$F = 5.79, P = 0.02$
CCLO	1.73 ± 0.28 (48)	1.75 ± 0.70 (8)	$F = 0.00, P = 0.97$	3.61 ± 0.22 (23)	3.50 ± 0.50 (4)	$F = 0.04, P = 0.85$
BOBO	1.33 ± 1.33 (3)	1.33 ± 1.33 (3)	$F = 0.00, P = 1.00$	4.00 (1)	4.00 (1)	$F = 0.00, P = 1.00$
WEME	1.13 ± 0.28 (37)	0.64 ± 0.64 (28)	$F = 1.66, P = 0.20$	3.00 ± 0.38 (14)	2.25 ± 0.49 (8)	$F = 1.45, P = 0.24$

Cowbird parasitism cost these sparrows 2.2, 1.4, and 1.3 young per successful nest, respectively. However, the impact of parasitism on fledging success was negligible for the other host species (Table 26.4). On average, cowbird parasitism cost hosts 1.2 young per parasitized host nest.

Female cowbirds were more abundant in the mornings on all plots. The fewest cowbirds attended site 1, regardless of the time of day, whereas cowbirds were most abundant on site 3 (Figure 26.1). Parasitism frequency did not appear to be correlated with host abundance on site 1, but cowbirds parasitized the most abundant hosts on sites 2 and 3. On sites 2 and 3, Savannah Sparrows and Lark Buntings, respec-

tively, were the most abundant hosts and were parasitized most frequently (Figure 26.2).

Nest Success

Contingency analyses showed that the frequency of nest predation was not significantly different between sites for any of the songbird species ($P > .4$). Predation frequency differed among species, however, with predators destroying between 39% (Baird's Sparrow) and 71% (Lark Bunting) of songbird nests (Table 26.3). The number of host young fledged per nest ranged from 0.8 (Savannah Sparrow) to 1.7 (Grasshopper Sparrow and Chestnut-collared Longspur)

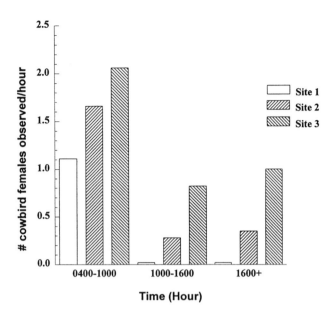

Figure 26.1. Female cowbird abundance during different times (CST) of the day at three sites in southwestern Manitoba (10 May to 25 July 1992).

and did not differ significantly between sites for any host species (one-way ANOVA, df = 2, $P > .19$).

Discussion

Cowbird Parasitism

Cowbirds parasitized more than 20% of the nests of six of the eight host species studied in southwestern Manitoba (Table 26.2). These results are in contrast with the earlier compilations of cowbird host records that revealed infrequent cowbird parasitism on grassland hosts (Mayfield 1965, Friedmann 1963). Those compilations, however, were based on small samples of nests (Friedmann and Kiff 1985). The infrequent parasitism recorded on Chestnut-collared Longspur nests does agree with results of other studies (Harris 1944, Fairfield 1968). Longspurs may be parasitized infrequently because many of their clutches are initiated before and after the cowbird laying season (Davis 1994), but this does not explain why cowbirds infrequently parasitize longspurs when their laying seasons overlap. Unlike the other hosts studied, female longspurs normally remained on or close to their nests during nest inspections, behavior that may serve to conceal nests from predators or defend them against cowbird parasitism (see Hobson and Sealy 1989). On the other hand, cowbirds may avoid longspur nests because they are exposed (Davis 1994), and presumably cowbirds would be visible when they approach the nests to parasitize them (see Neudorf and Sealy 1994).

The three sparrow species were frequent hosts of cowbirds in this study. Cowbird parasitism has rarely been recorded

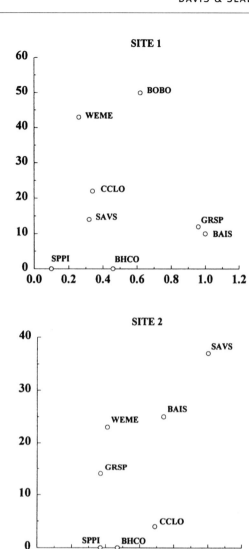

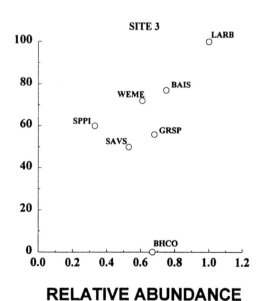

RELATIVE ABUNDANCE

Figure 26.2. Frequency of cowbird parasitism with respect to cowbird and host relative abundance during 1992.

for Baird's and Grasshopper Sparrows (Friedmann 1963, Wray et al. 1982, Friedmann and Kiff 1985; but see Hill 1976, Elliott 1978), and it appears to vary geographically in Savannah Sparrows (Lein 1968, Dixon 1978, Friedmann and Kiff 1985). Mayfield (1965) suggested that hosts that fledge more than 20% of the cowbird eggs laid could be considered "tolerant" hosts. Accordingly, the three sparrow species are important hosts for cowbirds in southwestern Manitoba (Table 26.2).

The Lark Bunting had the highest parasitism frequency, but its importance as a host cannot be assessed because only a few nests were located. Friedmann (1963) considered Lark Buntings to be infrequently victimized by cowbirds, but Hill (1976) recorded a parasitism frequency of 15.5%. Hill suggested that Lark Buntings eject cowbird eggs or desert parasitized nests. Two nests in our study contained buried cowbird eggs, but these eggs may have been laid prematurely and buried during nest construction (see Hobson and Sealy 1987).

Using Hill's (1976) criteria, the host species that is parasitized frequently and whose parasitized nests receive the most cowbird eggs indicates host preference. Accordingly, Western Meadowlarks were the preferred host in this study, because 43% of the nests were parasitized and each parasitized nest received on average three eggs (Table 26.2). Cowbirds may have synchronized their laying more effectively with meadowlarks than with the smaller, perhaps more secretive hosts. We frequently used nest-building behavior to locate meadowlark nests, and it is possible that cowbirds also used this cue. Not only are the larger Western Meadowlarks more conspicuous than the other hosts, but their nests also can hold more cowbird eggs. Elsewhere, however, parasitism on Western Meadowlark nests is infrequent (e.g., Lanyon 1957, Hergenrader 1962, Friedmann 1963, Hill 1976). This apparent geographic variation in parasitism possibly is influenced by the size of the suitable habitat fragment and the other host species in the community.

Most cowbirds were sighted in the morning on each site (Figure 26.1). Rothstein et al. (1984) found that cowbirds remained on host nesting areas in the morning, but by the afternoon they had flown up to 7 km to foraging sites. Cowbirds were consistently more abundant on site 3 throughout the day, possibly because this site was situated between two pastures where cowbirds did not have to commute far between laying and feeding areas (see also Elliott 1980). This may have given cowbirds more time to search for nests. Cowbirds, however, could also spend more time searching for nests on site 2 because they also foraged on the pasture. The combination of high host nesting density and more female cowbirds searching for nests longer each day may have contributed to the higher parasitism frequency on site 3.

Contrary to the findings of Fretwell (1977) and Zimmerman (1983), cowbirds parasitized the most abundant hosts more frequently (Figure 26.2). If cowbirds parasitized hosts only on the basis of their availability, however, frequency of parasitism should have declined as nesting densities declined. We did not observe this on all plots. Parasitism of the most abundant hosts would not be an effective strategy unless the abundant species were also high-quality hosts (Mason 1986). Neither Bobolinks nor Lark Buntings raised any cowbirds, yet they were parasitized frequently (Table 26.2). Only two parasitized bunting nests and one parasitized Bobolink nest were not depredated, however, so this result may be an artifact of small sample sizes. Cowbirds apparently lay consecutive clutches 1–2 days apart (Scott and Ankney 1983), and hence they may have chosen less appropriate nests when more suitable nests were not available. The many inappropriately timed eggs laid on the sites support this contention. Opportunistic laying may be an effective strategy for cowbirds in this habitat because females can lay up to 40 eggs during a single breeding season (Scott and Ankney 1980). These results suggest that cowbird abundance may be more strongly correlated with parasitism frequency than nest abundance (see also Fleischer 1986, Hoover and Brittingham 1993) or that nest selection is influenced by many factors.

The high levels of multiple parasitism reported in this and other studies (Hergenrader 1962, Elliott 1978, Hoover and Brittingham 1993; see also Thompson et al., Chapter 32, and Robinson et al., Chapter 33, this volume) may reflect high ratios of female cowbirds to host nests (McGeen 1972). If hosts are rare relative to cowbirds, cowbirds may be forced to lay in already parasitized nests.

The inability of cowbirds to locate enough suitable nests on our sites was also suggested by the large number of cowbird eggs laid after host clutches were complete. Freeman et al. (1990) reported that more than 20% of cowbird eggs laid in Red-winged Blackbird (*Agelaius phoeniceus*) nests were laid in inactive nests. The authors attributed these inappropriate layings to the lack of perches from which cowbirds could gain better information on the location and status of nests. In our study, particularly in site 3, perches were available that made the site especially attractive to cowbirds (see Arnold and Higgins 1986). Shrubs that bordered nearly the entire site provided perches for social displays and nest searching.

The presence of perches suitable for cowbirds may influence the frequency of brood parasitism (e.g., Gochfeld 1979, Biermann et al. 1987, Wiley 1988, Freeman et al. 1990, Davis 1994). Plot shape may account for the highest parasitism frequency recorded on site 3; the widest point there was 220 m, much narrower than the 800 m of the other two sites. Although perch distance may influence parasitism frequency, some nests in our study were parasitized even though they were far from any perch (see also Thompson et al., Chapter 32, this volume).

How do cowbirds find nests that are situated well away from perches? They may inadvertently locate some nests while foraging, particularly in pastures (see Elliott 1980). Cowbirds may also use cattle as mobile perches where shrubs are absent, although we only infrequently observed them perched on the backs of grazing and resting cattle. Wiley (1988) observed Shiny Cowbirds (*M. bonariensis*) making short flights and hops, often in the presence of a male and another female(s), and stated that this behavior was a form of active nest-searching. We did not observe cowbirds making noisy flights into nesting areas, as reported by Norman and Robertson (1975) and Wiley (1988), but females were observed frequently flying about 1 m above the ground in sites 1 and 2. Although we did not witness host aggression toward low-flying cowbirds, Robertson and Norman (1977) showed that host aggression may be used as a nest-finding cue by cowbirds where host nesting densities are low. Smith et al. (1984) believed that adult female Song Sparrows (*Melospiza melodia*) responded more strongly to passing cowbirds, which possibly facilitated the discovery of their nests.

Our results support Brittingham and Temple's (1983) contention that increased habitat fragmentation promotes higher cowbird densities and subsequently higher frequencies and intensities of parasitism. The authors believed that fragmentation increased the amount of habitat that was available for cowbirds in which to locate nests. As cowbirds generally stay within 100 m of forest edges (Brittingham and Temple 1983, but see O'Conner and Faaborg 1992; Thompson et al., Chapter 32, this volume), fragmentation apparently allows them to penetrate small patches of habitat more fully, which makes more hosts susceptible to parasitism. Fragmentation of grasslands promotes the same result because cowbirds concentrate their activities near the habitat edge and are seldom seen in the interior of large expanses of grassland (Johnson and Temple 1990). In fact, Robinson et al. (1993) stated that to reduce the impact of cowbirds on host populations, large continuous tracts of habitat, simple in shape, are preferable to smaller tracts that are irregular in shape with more edge. Our results support this claim. Sites 1 and 2 were larger, square patches of habitat compared with the irregular, fingerlike shape of the smaller site 3. The latter site was also situated between two pastures, which allowed cowbirds to spend less time commuting between foraging and laying areas. Cowbirds parasitized hosts in site 3 more frequently than they did hosts in the other two sites (Table 26.2).

Our results suggest that in order to reduce the level of parasitism on grassland birds, managers should acquire and manage large, simple-shaped tracts of habitat instead of small irregular sites. As cowbirds use perches for social displays and to synchronize their laying with their hosts, both land managers and farmers can make nesting habitat less attractive to cowbirds by reducing shrub cover using controlled burns and rotational grazing and mowing. Removing unnecessary fencelines would also limit perch sites and possibly make the nesting habitat less attractive to cowbirds.

Nesting Success

Predation frequency among the species in this study was similar on all sites (Table 26.3). Angelstam (1986) stated that predation levels are inversely related to the size of the habitat due to the increased relative amount of habitat edge. Indeed, several studies have shown that the level of predation is higher in smaller fragments (e.g., Nelson and Duebbert 1974, Gates and Gysel 1978, Wilcove 1985, Paton 1994). Site 3 was narrow, elongate, and irregular in shape, and the amount of edge was greater compared with sites 1 and 2, yet predation levels were not significantly different among the sites. Perhaps sites 1 and 2 did not differ sufficiently in size from site 3 to permit detection of a relationship between habitat size and predation frequency. Johnson and Temple (1990), for example, found that predation frequencies were lower for nests situated in fragments of tall-grass prairie larger than 130 ha.

Gates and Gysel (1978) attributed a decrease in nest predation with increasing distance from a wooded edge to a functional response by predators to higher nest densities and the greater activity of predators near the edge. Our sites were bordered by shrubs (site 3), road, crop, and/or hayland (sites 1 and 2). Nest predators may prefer areas with extensive cover along the edge, because many of them are potential prey to other animals and thus may reduce their activity in open areas where escape cover is sparse (Johnson and Temple 1990). Fritzell (1978), for example, reported that in prairie habitat raccoons (*Procyon lotor*) traveled along shelter belts. Our sites, therefore, may have been equally attractive or unattractive to nest predators, as there was no forested edge along which predators could concentrate their activities.

The combination of high parasitism and predation frequencies greatly affected the productivity of birds in the grassland community of southwestern Manitoba. Hosts suffered predation frequencies of up to 71% (Table 26.3) and parasitism frequencies of up to 100%, with many nests multiply parasitized. In some instances, nests fledged only cowbird young (four Baird's Sparrow nests and one each of Savannah Sparrow, Sprague's Pipit, and Western Meadowlark). The impact of parasitism and predation on the community, however, was difficult to assess because information was lacking on the total number of nesting attempts (see Smith 1981). Parasitism apparently reduced the productivity of hosts in site 3 because these hosts were parasitized more frequently and fledged significantly fewer young than hosts on either of the other sites, even though predation was similar among all sites. Cowbirds lowered host productivity of Savannah, Baird's, and Grasshopper Sparrow nests by 1–2 young per nest. Similar estimates of the cost of cowbird parasitism have been found in other studies (Hill 1976, Elliott 1978, Trail and Baptista 1993; but see Smith 1981, Weatherhead 1989). Although young cowbirds sometimes outcompete host young for food and impede the growth of host nestlings (Smith 1981), the productivity of five host species in this study did not appear to be influenced by cowbird parasitism. Grasslands, however, are characterized by high primary productivity that is produced in a relatively short time. The resulting superabundant food (Maher 1979, Wiens 1974) may have allowed these hosts to feed their own young adequately, along with cowbirds, and thus reduce the cost of parasitism.

Trail and Baptista (1993) calculated that Nuttall's White-crowned Sparrows (*Zonotrichia leucophrys nuttalli*) cannot maintain their numbers at parasitism frequencies greater than 20%. There is little evidence, however, to justify implicating cowbirds as the reason for the decline of certain species (Holmes 1993, see also Böhning-Gaese et al. 1993). Some species are more vulnerable than others to the negative effects of parasitism because their numbers have been reduced by previous and continuing factors, such as habitat loss on breeding and wintering grounds (see Böhning-Gaese et al. 1993). Robinson et al. (1993) stated that hosts with a limited distribution and neotropical migrants are especially vulnerable to the effects of parasitism. Sprague's Pipits, Baird's Sparrows, Grasshopper Sparrows, and Chestnut-collared Longspurs are at the northern edge of their breeding ranges in Manitoba, and their distributions have been reduced to a fraction of their former sizes (Godfrey 1986). As these species accept cowbird parasitism, they may be particularly vulnerable to high levels of parasitism. Although the host species in this study have been associated with cowbirds for a long time, it is difficult to determine if antiparasite strategies have evolved without controlled experimental studies (see Rothstein 1975, Hill and Sealy 1994). In the past, grassland birds may have experienced low parasitism frequencies because of the large, unfragmented expanse of prairie habitat was not conducive to the cowbird's mode of reproduction. Fragmentation not only may have allowed cowbirds to expand their range and increase their numbers (Brittingham and Temple 1983, Robinson et al. 1993) but also may have facilitated the use of grassland hosts that historically were inaccessible to them.

Acknowledgments

We thank B. Zarn and R. Kilfoyle for allowing us to conduct this study on their properties and L. Bidlake (Manitoba Department of Natural Resources) for providing access to the Broomhill Wildlife Management Area. We thank K. De Smet (Manitoba Department of Natural Resources) for providing logistic support and lodging and W. Harris for his generous supply of field equipment. We are indebted to our assistant, B. D. Jeske, for his dedication and enthusiasm in the field. G. McMaster, K. Mazur, and S. Hellman assisted with aspects of the fieldwork, and L. M. Armstrong (Statistical Advisory Service, University of Manitoba) provided statistical advice. Drafts of the manuscript benefited from comments made by B. J. Hann, S. R. Robinson, J. M. Shay, and J. N. M. Smith. This study was funded by grants from the Natural Sciences and Engineering Research Council of Canada to SGS and grants from the Canadian Wildlife Service of Environment Canada, Manitoba Department of Natural Resources (Wildlife Branch), and Manitoba Naturalists Society to SKD.

References Cited

Angelstam, P. 1986. Predation on ground-nesting birds' nests in relation to predator densities and habitat edge. Oikos 47:365–373.

Arnold, T. W., and K. F. Higgins. 1986. Effects of shrub coverage on birds of North Dakota mixed-grass prairie. Can. Field-Natur. 100:10–14.

Biermann, G. C., W. B. McGillivray, and K. E. Nordin. 1987. The effect of cowbird parasitism on Brewer's Sparrow productivity in Alberta. J. Field Ornithol. 58:350–354.

Böhning-Gaese, K., M. L. Taper, and J. H. Brown. 1993. Are declines in North American insectivorous songbirds due to causes on the breeding range? Conserv. Biol. 7:76–84.

Brittingham, M. C., and S. A. Temple. 1983. Have cowbirds caused forest songbirds to decline? BioScience 33:31–35.

Davis, S. K. 1994. Cowbird parasitism, predation, and host selection in fragmented grassland of southwestern Manitoba. M.S. Thesis, University of Manitoba, Winnipeg.

Dixon, C. L. 1978. Breeding biology of the Savannah Sparrow on Kent Island. Auk 95:235–346.

Elliott, P. F. 1978. Cowbird parasitism in the Kansas tallgrass prairie. Auk 95:161–167.

———. 1980. Evolution of promiscuity in the Brown-headed Cowbird. Condor 82:138–141.

Fairfield, G. M. 1968. Chestnut-collared Longspur. *In* Life histories of North American cardinals, grosbeaks, buntings, towhees, finches, sparrows, and allies (A. C. Bent and collaborators). U.S. Natl. Mus. Bull. 237:1635–1652.

Fleischer, R. C. 1986. Brood parasitism by Brown-headed Cowbirds in a simple host community in eastern Kansas. Kansas Ornithol. Soc. Bull. 37:21–29.

Freeman, S., D. F. Gori, and S. Rohwer. 1990. Red-winged Blackbirds and Brown-headed Cowbirds: Some aspects of a host-parasite relationship. Condor 92:336–340.

Fretwell, S. D. 1977. Is the Dickcissel a threatened species? Amer. Birds 31:923–932.

Friedmann, H. 1963. Host relations of the parasitic cowbirds. U.S. Natl. Mus. Bull. 233:1–276.

Friedmann, H., and L. F. Kiff. 1985. The parasitic cowbirds and their hosts. Proc. Western Found. Vert. Zool. 2:26–302.

Fritzell, E. K. 1978. Habitat use by prairie raccoons during the waterfowl breeding season. J. Wildl. Manage. 42:118–127.

Gates, J. E., and L. W. Gysel. 1978. Avian nest dispersion and fledging success in field-forest ecotones. Ecology 59:871–883.

Gochfeld, M. 1979. Brood parasite and host coevolution: Interactions between Shiny Cowbirds and two species of meadowlarks. Amer. Natur. 113:855–870.

Godfrey, W. E. 1986. The birds of Canada. National Museums of Canada, Ottawa.

Harris, R. D. 1944. The Chestnut-collared Longspur in Manitoba. Wilson Bull. 56:105–115.

Hergenrader, G. L. 1962. The incidence of nest parasitism by the Brown-headed Cowbird (*Molothrus ater*) on roadside nesting birds in Nebraska. Auk 79:85–88.

Herkert, J. R. 1994. The effects of habitat fragmentation on midwestern grassland bird communities. Ecol. Applic. 4:461–471.

Hill, D. P., and S. G. Sealy. 1994. Desertion of nests parasitized by cowbirds: Have Clay-coloured Sparrows evolved an anti-parasite defence? Anim. Behav. 48:1063–1070.

Hill, R. A. 1976. Host-parasitic relationships of the Brown-headed Cowbird in a prairie habitat of west-central Kansas. Wilson Bull. 88:555–565.

Hobson, K. A., and S. G. Sealy. 1987. Cowbird egg buried by a Northern Oriole. J. Field Ornithol. 58:222–224.

———. 1989. Responses of Yellow Warblers to the threat of cowbird parasitism. Anim. Behav. 38:510–519.

Holmes, B. 1993. An avian arch villain gets off easy. Science 262:1514–1515.

Hoover, J. P., and M. C. Brittingham. 1993. Regional variation in cowbird parasitism of Wood Thrushes. Wilson Bull. 105:228–238.

Johnson, R. G., and S. A. Temple. 1990. Nest predation and brood parasitism of tallgrass prairie birds. J. Wildl. Manage. 54:106–111.

Krebs, C. J. 1989. Ecological methodology. Harper and Row, New York.

Lanyon, W. E. 1957. The comparative biology of the meadowlarks (*Sturnella*) in Wisconsin. Publ. Nuttall Ornithol. Club 1:1–67.

Lein, M. R. 1968. The breeding biology of the Savannah Sparrow, *Passerculus sandwichensis* (Gmelin), at Saskatoon, Saskatchewan. M.S. Thesis, University of Saskatchewan, Saskatoon.

Maher, W. J. 1979. Nestling diets of prairie passerine birds at Matador, Saskatchewan, Canada. Ibis 121:437–452.

Mason, P. 1986. Brood parasitism in a host generalist, the Shiny Cowbird: 1, The quality of different species as hosts. Auk 103:52–60.

Mayfield, H. F. 1965. The Brown-headed Cowbird, with old and new hosts. Living Bird 4:13–28.

———. 1975. Suggestions for calculating nest success. Wilson Bull. 87:456–466.

McGeen, D. S. 1972. Cowbird-host relationships. Auk 82:360–380.

Nelson, H. K., and H. F. Duebbert. 1974. New concepts regarding the production of waterfowl and other gamebirds in areas of diversified agriculture. Proc. Int. Congr. Game Biol. 11:385–394.

Neudorf, D. L., and S. G. Sealy. 1994. Sunrise nest attentiveness in cowbird hosts. Condor 96:162–169.

Norman, R. F., and R. J. Robertson. 1975. Nest-searching behavior in the Brown-headed Cowbird. Auk 92:610–611.

O'Conner, R. J., and J. Faaborg. 1992. The relative abundance of the Brown-headed Cowbird (*Molothrus ater*) in relation to exterior and interior edges in forests of Missouri. Trans. Missouri Acad. Sci. 26:1–9.

Owens, R. A., and M. T. Myres. 1973. Effects of agriculture upon populations of native passerine birds of an Alberta fescue grassland. Can. J. Zool. 51:697–713.

Paton, P. W. C. 1994. The effect of edge on avian nest success: How strong is the evidence? Conserv. Biol. 8:17–26.

Payne, R. B. 1977. The ecology of brood parasitism in birds. Annu. Rev. Ecol. Syst. 8:1–28.

Peterjohn, B. G., and J. R. Sauer. 1993. North American breeding bird survey annual summary, 1990–1991. Bird Pop. 1:1–15.

Robertson, R. J., and R. F. Norman. 1977. The function of aggressive host behavior towards the Brown-headed Cowbird (*Molothrus ater*). Can. J. Zool 55:508–518.

Robinson, S. K., J. A. Grzybowski, S. I. Rothstein, M. C. Brittingham, L. J. Petit, and F. R. Thompson III. 1993. Management implications of cowbird parasitism on neotropical migrant songbirds. Pp. 93–102 *in* Status and management of neotropical migratory birds (D. M. Finch and P. W. Stangel, eds). USDA For. Serv. Gen. Tech. Rept. RM-229.

Rothstein, S. I. 1975. An experimental and teleonomic investigation of avian brood parasitism. Condor 77:250–271.

Rothstein, S. I., J. Verner, and E. Stevens. 1980. Range expansion and diurnal changes in dispersion of the Brown-headed Cowbird in the Sierra Nevada. Auk 97:253–267.

———. 1984. Radio-tracking confirms a unique diurnal pattern of spatial occurrence in the parasitic Brown-headed Cowbird. Ecology 65:77–88.

SAS Institute Inc. 1990. SAS/STAT user's guide, version 6, 4th ed., vol. 1 and 2. Cary, NC. 943 pp.

Scott, D. M., and C. D. Ankney. 1980. Fecundity of the Brown-headed Cowbird in southern Ontario. Auk 97:677–683.

———. 1983. The laying cycle of Brown-headed Cowbirds: Passerine chickens? Auk 100:583–592.

Smith, J. N. M. 1981. Cowbird parasitism, host fitness, and age of the host female in an island Song Sparrow population. Condor 83:152–161.

Smith, J. N. M., P. Arcese, and I. G. MacLean. 1984. Age, experience, and enemy recognition by wild Song Sparrows. Behav. Ecol. Sociobiol. 14:101–106.

Trail, P. W., and L. E. Baptista. 1993. The impact of Brown-headed Cowbird parasitism on populations of Nuttall's White-crowned Sparrow. Conserv. Biol. 7:309–315.

Trottier, G. C. 1992. Conservation of Canadian prairie grasslands: A landowner's guide. Environment Canada, Edmonton, AB.

Weatherhead, P. J. 1989. Sex ratios, host-specific reproductive success, and impact of Brown-headed Cowbirds. Auk 106:358–366.

Wiens, J. A. 1974. Climatic instability and the "ecological saturation" of bird communities in North American grasslands. Condor 76:385–400.

Wilcove, D. S. 1985. Nest predation in forest tracts and the decline of migratory songbirds. Ecology 66:1211–1214.

Wiley, J. W. 1988. Host selection by the Shiny Cowbird. Condor 90:289–303.

Wray, T. II, K. A. Strait, and R. C. Whitmore. 1982. Reproductive success of grassland sparrows on a reclaimed surface mine in West Virginia. Auk 99:157–164.

Zimmerman, J. L. 1983. Cowbird parasitism of Dickcissels in different habitats and at different nest densities. Wilson Bull. 95:7–22.

27. Cowbird Parasitism in Grassland and Cropland in the Northern Great Plains

ROLF R. KOFORD, BONNIE S. BOWEN,

JOHN T. LOKEMOEN, AND ARNOLD D. KRUSE

Abstract

The landscape of the Great Plains has been greatly altered by human activities in the past century, and several grassland passerines have experienced significant population declines in recent decades. We explore here whether brood parasitism by Brown-headed Cowbirds, which are abundant in the Great Plains, has contributed to these declines. We measured the frequency of cowbird parasitism of passerine species in seeded grassland, natural grassland, and cropland in studies conducted in North Dakota during 1981–1993. The proportions of parasitized nests were 25%, 34%, and 39% in seeded grassland, natural grassland, and cropland, respectively. We speculate that much of the variation in parasitism rate among these habitats is related to the local abundance of cowbirds, to nest visibility, and to the presence of suitable perches for female cowbirds. Local abundance of cowbirds may be high in areas with cattle pastures. Nests and nesting behavior are probably more visible to female cowbirds in cropland than in grassland. Female cowbirds may use shrubs as perches while searching for host nests, and shrubs are more common in natural grasslands than in the other habitats we examined. Experimental work on the determinants of cowbird abundance in grasslands is needed.

Introduction

Significant population declines have been noted recently in populations of several grassland bird species of the northern Great Plains, including the Clay-colored Sparrow (*Spizella pallida*), Lark Bunting (*Calamospiza melanocorys*), Grasshopper Sparrow (*Ammodramus savannarum*), and Bobolink (*Dolichonyx oryzivorus*) (Robbins et al. 1986, Johnson and Schwartz 1993, Peterjohn and Sauer 1993). In general, population declines of migratory species have been attributed to increased mortality from habitat destruction on the wintering grounds or poor reproduction on the breeding grounds (Hagan and Johnston 1992).

Habitat loss and fragmentation on the breeding grounds of grassland birds are known to contribute to poor reproductive success (Best 1978, Gates and Gysel 1978, Graber and Graber 1983, Johnson and Temple 1986, 1990). Johnson and Temple (1986) found that birds that nested in remnants of tall-grass prairie near wooded edges produced fewer young than birds that nested far from wooded edges.

Brood parasitism by Brown-headed Cowbirds may also contribute to poor nesting success in grassland habitats. The frequency of brood parasitism among grassland species in the northern Great Plains is not well documented, but examples of heavy parasitism are known (Table 27.1; also Knapton 1979, Johnson and Temple 1990). Breeding populations of the Brown-headed Cowbird reach peak abundance in the northern Great Plains (Robbins et al. 1986; Peterjohn et al., Chapter 2, and Wiedenfeld, Chapter 3, this volume). Brown-headed Cowbirds bred on the Great Plains long before European settlement (Mayfield 1965), but their populations have probably been enhanced by human activities. Since settlement, most of the prairie has been converted to cropland and pasture; tree plantings and agriculture have fragmented the landscape, domestic livestock have been introduced, and fires have been suppressed. Significant increases in cowbird numbers occurred between 1966 and 1991 in the northern Great Plains (Peterjohn et al. Chapter 2, Wiedenfeld Chapter 3).

Additional research is needed to understand the effects of brood parasitism on reproductive success of grassland birds and the interactions (if any) of parasitism with habitat alteration. We examined parasitism in three habitats on the northern Great Plains by using data from three independent studies. The study fields were in a matrix of cultivated lands and mixed-grass prairie in south central North Dakota. Because the abundance of cowbirds relative to their hosts may affect the frequency of parasitism, we also examined the ratio of female cowbirds to hosts based on counts of birds in the breeding season.

Table 27.1. Frequency of Parasitism of Selected Species from Previous Studies in the Great Plains

Species	Nests	Frequency (%)	Location	Authors
Horned Lark	22	5	North Dakota	T. L. George (pers. comm.)
Horned Lark	31	45	Kansas	Hill (1976)
Dickcissel	17	53	Nebraska	Hergenrader (1962)
Dickcissel	19	95	Kansas	Elliot (1978)
Dickcissel	23	65	Kansas	Fleischer (1986)
Dickcissel	65	91	Kansas	Hatch (1983)
Dickcissel	28	50	Kansas	Hill (1976)
Dickcissel	55	78	Kansas	Zimmerman (1966)
Dickcissel	620	70	Kansas	Zimmerman (1983)
Dickcissel	14	7	Oklahoma	Ely in Wiens (1963)
Dickcissel	61	31	Oklahoma	Overmire (1962)
Dickcissel	15	33	Oklahoma	Wiens (1963)
Clay-colored Sparrow	24	17	Alberta	Salt (1966)
Clay-colored Sparrow	9	89	Saskatchewan	Fox (1961)
Clay-colored Sparrow	232	36	Manitoba	Knapton (1979)
Clay-colored Sparrow	135	10	Minnesota	Johnson and Temple (1990)
Vesper Sparrow	93	5	North Dakota	T. L. George (pers. comm.)
Lark Bunting	142	15	Kansas	Hill (1976)
Lark Bunting	77	21	Kansas	Wilson (1976)
Savannah Sparrow	46	37	Minnesota	Johnson and Temple (1990)
Grasshopper Sparrow	44	7	Minnesota	Johnson and Temple (1990)
Grasshopper Sparrow	18	50	Kansas	Elliot (1978)
Grasshopper Sparrow	18	22	Kansas	Hill (1976)
Bobolink	47	44	Minnesota	Johnson and Temple (1990)
Red-winged Blackbird	17	76	North Dakota	Houston (1973)
Red-winged Blackbird	258	42	North Dakota	Linz and Bolin (1982)
Red-winged Blackbird	59	54	Nebraska	Hergenrader (1962)
Red-winged Blackbird	73	30	Kansas	Fleischer (1986)
Red-winged Blackbird	50	22	Kansas	Hill (1976)
Red-winged Blackbird	73	3	Oklahoma	Ely in Wiens (1963)
Red-winged Blackbird	33	0	Oklahoma	Wiens (1963)
Eastern Meadowlark	40	70	Kansas	Elliot (1978)
Eastern Meadowlark	10	50	Kansas	Fleischer (1986)
E & W Meadowlarks	31	16	Nebraska	Hergenrader (1962)
Western Meadowlark	76	18	Minnesota	Johnson and Temple (1990)
Western Meadowlark	39	13	North Dakota	T. L. George (pers. comm.)
Western Meadowlark	29	7	Kansas	Hill (1976)

Methods

Study Areas

We obtained data on the frequency of nest parasitism and cowbird/host ratios in seeded grassland, natural grassland, and cropland west of the Agassiz Lake Plain in south central North Dakota. This entire area is in the glaciated prairie pothole region of North Dakota (Stewart and Kantrud 1974). The terrain is gently to moderately rolling and contains numerous temporary, seasonal, and semipermanent wetlands. About one sixth of the upland habitat is seeded grassland, one sixth is natural grassland (mostly pastureland), and two thirds is cropland.

Our seeded grassland fields were either former cropland that had been enrolled in the Conservation Reserve Program (CRP) of the U.S. Department of Agriculture or upland areas on Waterfowl Production Areas (WPA) of the U.S. Fish and Wildlife Service. We examined six CRP fields (10–16 ha) and two fields (11–17 ha) on WPAs that covered 65–105 ha of wetland and upland habitat. The CRP fields were seeded in the late 1980s with mixtures of legumes (sweet clover, *Melilotus* spp., or alfalfa, *Medicago sativa*) and cool-season grasses. They were left ungrazed and unhayed during the study (1991–1993). In the 1970s, WPAs were seeded with similar grass-legume mixtures to attract nesting ducks.

Natural grassland in WPAs included the native grasses needle-and-thread (*Stipa comata*) and green needle grass (*S. viridula*), the exotic grasses Kentucky bluegrass (*Poa pratensis*) and smooth brome (*Bromus inermis*), forbs, and shrubs (mainly western snowberry, *Symphoricarpos occidentalis*). Natural-grassland fields were examined in two studies, which we refer to as the northern study and the southern study. In the northern study, we examined all species of nesting birds on 4 fields in 1991–1993. In the southern study, conducted in 1981–1987, we examined Western Meadowlarks (*Sturnella neglecta*) on 15 fenced fields (25–47 ha). These fields were used in a study of the effects of cattle grazing on ground-nesting birds (Bowen and Kruse 1993).

The cropland we examined in 1991–1993 consisted of three types of crops and three farming systems. The crops were wheat (the principal small grain in the region), sunflower (the principal row crop), and fallow. The three farming systems were organic (synthetic chemicals were not used), minimum till (tillage was minimal), and conventional (tillage and synthetic chemicals were used). On fallow fields, organic farmers grew sweet clover that was plowed under in June; minimum-till farmers left stubble and controlled weeds with herbicide; and conventional farmers left little or no vegetation on fields that were repeatedly tilled. We studied ten 40-ha fields of each combination of crop type and farming system for a total of about 90 fields on ten different farms.

In spring, crop fields had little or no residual cover. Cover varied from small-grain stubble to bare soil. Residual cover declined in late April and May when most organic and conventional fields were cultivated and seeded. Crops typically begin to grow rapidly in late May and June and were tall and lush by early July.

Nest Searching and Monitoring

Nests in seeded grassland and northern natural grassland were located mainly by flushing birds from nests with a 25-m, hand-pulled rope. We conducted three searches on each field. In the southern study, we used a 53-m cable-chain towed between two vehicles (Higgins et al. 1977). In cropland, we conducted three searches between late April and mid-July. Two searches were conducted with a 50-m chain pulled by all-terrain vehicles (ATV) or, in rowcrop fields, a specially fitted ATV. The third search was conducted in July by walking through each field looking for nests of Red-winged Blackbirds (*Agelaius phoeniceus*).

We plotted the locations of all nests on maps and placed marker flags 4 m from each nest bowl. We recorded the number of host eggs, number of cowbird eggs, and the incubation stage or nestling stage during each nest visit. We monitored nests to determine their fates.

Any active nest that contained a cowbird egg or nestling on any visit was considered parasitized. The frequency of parasitism for a given species in a given habitat was the number of parasitized nests (× 100) divided by the total number of nests located for that species.

To report an average frequency of parasitism in each habitat, we calculated an unweighted mean of the frequencies for each species. Data from species with fewer than 10 nests were pooled. Data on Western Meadowlark nests in natural grassland in the northern study were treated separately from those in the southern study. There were fewer than 10 nests of Western Meadowlarks in the northern study; these nests were pooled with other species. Differences among habitats were not tested statistically because of possible confounding effects of year, species, and land use in different areas.

The sample sizes of nests do not necessarily reflect species abundance in the three habitats. Studies varied by searched area and number of study years. In particular, studies of natural grassland in the northern study area took place on fewer fields and were of shorter duration than in the southern study. Also, our search methods may have been more likely to locate nests of some species (e.g., Red-winged Blackbirds) because of their high nest visibility and adult vigilance.

Census Methods

Populations were censused with strip transects 100 m wide. Transects covered most or all of each field. We censused birds well after sunrise in the morning and recorded their numbers; we ignored flyovers. Single birds of unknown sex or groups of host species were assigned equally to the two sexes. We assigned a single female to each group of cowbirds because we reasoned that most groups of cowbirds probably consisted of one female and several courting males. We conducted two or three breeding-season censuses on each study field between 15 May and 15 July (the earliest cowbird eggs were found in mid-May).

We calculated the ratio of female cowbirds to male hosts in each habitat by using total counts from all censuses. Robinson et al. (1993) suggested that this ratio could be used with data derived from point counts to obtain a crude index of parasitism frequency. Data from the northern study of natural grassland were analyzed separately from the southern study. We used host species that Ehrlich et al. (1988) rated as frequent, common, or uncommon hosts. Because these hosts were the most abundant birds, the results would have been similar if we had used all possible hosts.

Results

In seeded grassland, the mean frequency of parasitism was 25% among the host species whose nests we found (Table 27.2). The frequency in Red-winged Blackbirds (47%) was distinctly higher than for other species combined (21%). Most blackbirds nested in two fields in upland vegetation.

In natural grassland, the mean frequency of parasitism was 34% (Table 27.2). Western Meadowlarks in the southern study had a higher frequency of parasitism (47%) than the mean frequency of all species in the northern study (31%). We found too few meadowlark nests to make an accurate estimate of parasitism frequency in this species in the northern study.

In cropland, the mean frequency of parasitism was 39% (Table 27.2), mainly because of a high frequency among Horned Larks (*Eremophila alpestris*) late in the breeding season. We found 24 unparasitized nests of Horned Larks before the first cowbird egg was found (15 May). Of the 60 nests found after 15 May, 83% were parasitized and received 1–6 cowbird eggs (mean = 2.3). This heavy parasitism contrasts with the moderate parasitism (26%) experienced by Vesper Sparrows (*Pooecetes gramineus*), the other common species in cropland. Our observed cowbird/host ratios corresponded roughly to observed patterns of variation in frequency of parasitism. The lowest ratio (1 : 84) was observed in seeded grassland, where the lowest overall frequency of parasitism occurred (mean = 25%). Higher ratios were observed in natural-grassland fields in the northern area (1 : 13) and in cropland (1 : 10), where birds experienced moderate to heavy parasitism (mean = 31% and 39%, respectively). In natural grassland, the ratio was higher (1 : 5) in the southern study than in the northern study (1 : 13). This higher ratio corresponded to a higher frequency of parasitism in the southern study (47%).

Table 27.2. Frequency of Parasitism in Seeded Grassland, Natural Grassland, and Cropland Fields

Species	Nests	Frequency (%)
Seeded Grassland		
Vesper Sparrow	10	20
Grasshopper Sparrow	45	22
Bobolink	12	25
Red-winged Blackbird	74	47
Western Meadowlark	26	19
Other	18	17
Natural Grassland		
Clay-colored Sparrow	49	24
Western Meadowlark (southern)	294	47
Other	35	37
Cropland		
Horned Lark (early)	24	0
Horned Lark (late)	60	83
Vesper Sparrow	84	26
Lark Bunting	23	61
Grasshopper Sparrow	13	38
Red-winged Blackbird	19	26
Other	10	40

Note: In natural grassland, data on Western Meadowlark nests in the southern study area are shown and nests in the northern area are included with other nests. Horned Lark nests were classified as early (found on or before 15 May) or late (found after 15 May). Numbers of nests located may not reflect nest densities (see Methods).

Discussion

As the region with the highest concentrations of breeding cowbirds, the northern Great Plains might be expected to show parasitism that has a major influence on the reproduction of host species. Increases in the number of cowbirds in the northern Great Plains between 1966 and 1991 (Peterjohn et al. Chapter 2, Wiedenfeld Chapter 3) may be related to increased coverage of trees and brush. Current data are insufficient to evaluate the importance of brood parasitism as a factor contributing to population declines of grassland birds. Cowbirds have coexisted with their grassland hosts in this region for a long time. Mayfield (1965) reviewed the scanty data available in the mid-1960s and found that the only species characteristic of the north-temperate grassland with over 30% parasitism was the Dickcissel (*Spiza americana*, 31%).

Recent data, however, indicate that about half of grassland species in the Great Plains have moderate frequencies in the 10–30% range, and nearly half have values above 30% (Tables 27.1 and 27.2). All grassland birds in Illinois, in contrast to the general results from the Great Plains, had parasitism frequencies below 20% (see Robinson et al., Chapter 33, this volume). Two of the lightly parasitized populations were in southwestern North Dakota (George et al. 1992, Table 27.1). Red-winged Blackbirds can be moderately or heavily parasitized in uplands but are usually lightly (under 10%) parasitized in wetlands (Friedmann 1963, Fleischer 1986).

The long period of coexistence of Brown-headed Cowbirds and their grassland hosts suggests that host species have evolved tactics, such as removal of cowbird eggs or desertion of nests, that reduce the effects of parasitism on host populations (Mayfield 1965). Among grassland birds, only the Lark Bunting has been reported to remove cowbird eggs (Hill 1976), but the overall frequency of this behavior seems to be low. Nest desertion because of parasitism has been reported in Dickcissels (Elliott 1978, Zimmerman 1983), Lark Buntings (Hill 1976), and Clay-colored Sparrows (Fox 1961, Knapton 1979), but its frequency also seems to be low. Furthermore, desertion may have multiple causes and is not an effective response to parasitism in all situations (Grzybowski and Pease, Chapter 16, this volume). The moderate or heavy parasitism among these grassland species suggests that their productivity is often reduced, but further analysis and different data are needed to estimate the magnitudes of these effects (see Grzybowski and Pease Chapter 16).

We documented moderate or high frequencies of parasitism in south central North Dakota. The effects of moderate or heavy parasitism on reproductive rates of most grassland species are poorly known, largely because they are difficult to determine (Smith 1981, May and Robinson 1985). It is thus possible that parasitism is contributing to population declines of the Dickcissel, Lark Bunting, Clay-colored Sparrow, Grasshopper Sparrow, Bobolink, and Eastern Meadowlark (*Sturnella magna*) (Peterjohn and Sauer 1993). If this is confirmed, land managers may want to explore options to reduce the frequency of parasitism. Habitat management is one option that could be used to reduce parasitism of grassland birds. We found variation in the frequency of parasitism across the three habitats we examined. We speculate that much of this variation is related to the cowbird/host ratio and to two habitat features, nest visibility and availability of perch sites for cowbirds (see also Davis and Sealy, Chapter 26, this volume). These habitat features may affect frequency of parasitism directly or indirectly through the cowbird/host ratio (Figure 27.1).

The high cowbird/host ratio in the southern study of natural grassland fields may be due to the prevalence of cattle pastures in this area. The preferred foraging habitat of Brown-headed Cowbirds is short grass, which is often in

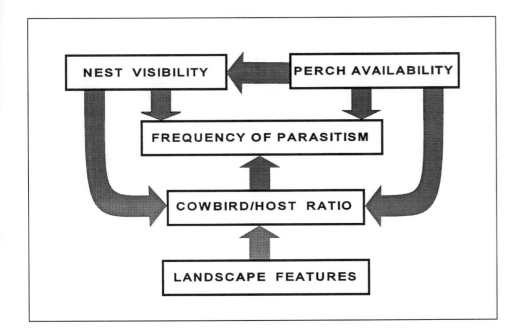

Figure 27.1. Schematic diagram of factors that may affect frequency of Brown-headed Cowbird parasitism.

pastures (Mayfield 1965, Rothstein et al. 1984). Cowbirds are capable of commuting long distances between laying sites and foraging sites if these sites are limited (Rothstein et al. 1984). In the southern area, however, laying sites were often in grassland next to or including feeding sites, making short commuting trips possible. Cowbirds may be attracted to pasture-rich landscapes and may reach higher abundance in these than in nearby landscapes with fewer pastures.

Visibility of nests and host females to female cowbirds is probably related to the height and density of the vegetation at or near the nest site. Of the habitats we examined, seeded grassland generally had the tallest and densest vegetation. Natural grassland fields were generally similar except in the southern study, in which some fields were grazed in some years. The most open habitat we examined was cropland, in which several species nested before the crops grew tall. Horned Larks, which were heavily parasitized, use open areas for nesting. Thus, a ranking of nest visibility in the three habitats corresponds to a ranking of the mean frequencies of parasitism. Furthermore, in natural grassland, birds in fields that probably had the highest visibility (that is, the grazed fields in the southern study) experienced a higher frequency of parasitism than the birds in the northern study.

The availability of suitable perch sites for female cowbirds searching for host nests (Norman and Robertson 1975) was presumably lowest in seeded grassland, which had little brush but had some sweet clover and tall forbs. The birds in seeded grassland had the lowest overall frequencies of parasitism. Availability of perch sites was highest in natural grassland, the only habitat with a substantial amount of brush. The cropland fields generally had few perches, except for the fallow fields that were farmed organically. These fields consisted entirely of sweet clover, which is sturdy enough to be used for perching and was the primary nesting habitat of Lark Buntings. This species was heavily parasitized (61%, Table 27.2).

The ratio of cowbirds to hosts, visibility of nests, and availability of perches are probably not the only factors that affect frequencies of parasitism. High nesting density of a colony of host species (e.g., Red-winged Blackbirds in wetlands) may constitute a defense against cowbird parasitism (Robertson and Norman 1976; Carello and Snyder, Chapter 11, this volume). Nesting phenology can reduce the likelihood of cowbird parasitism in some species (Middleton 1977), as indicated by our data on Horned Larks. Host-species preferences of cowbirds may also have an effect (Fleischer 1986). With the possible exception of the first, however, these factors are not amenable to management.

The correspondence between the cowbird/host ratio and frequency of parasitism suggests those effects of habitat on parasitism act in part by affecting local cowbird abundance (Table 27.2). Other studies have found that cowbirds re-

spond to changes or differences in habitat structure. Wiens (1963) reported higher frequencies of parasitism in a study area after several years of growth of low, brushy vegetation. He suggested that cowbird numbers increased relative to hosts as a result. Lowther and Johnston (1977) concluded that the prairie-shrub successional stage provided better breeding conditions for cowbirds in Kansas than do forests or extensive grasslands or prairie.

Results from these studies suggest that local densities of cowbirds in grassland habitats can be reduced by managing brush. Wiens (1963) found a lower frequency of parasitism in areas far from brushy thickets. Johnson and Temple (1990), who studied remnant tracts of tallgrass prairie, found lower frequencies of parasitism far from wooded edges and suggested that prairie fragments be as large as possible to minimize edge and that woody vegetation be removed. Managers should realize, however, that reducing brush probably affects more than cowbird densities. Reducing brush may reduce the population density of Clay-colored Sparrows (Knapton 1979) and other species that commonly nest in low brush.

Managers could also increase or maintain the amount of the landscape with seeded grassland, the habitat in which we observed the lowest frequency of parasitism. Many seeded grassland fields we studied were CRP fields. When this cropland retirement program ends, landowners and managers will have the option of returning these seeded fields to crop production. Our data suggest that such a conversion of habitat may increase the overall frequency of parasitism in the northern Great Plains, even if cowbird abundance does not change.

Another management option is to reduce the intensity of grazing by livestock. Knapton (1979) observed that parasitism declined in one study area after cattle were removed and suggested that visibility of nests decreased as vegetation gained height. In addition to decreasing visibility of nests, lower numbers of livestock may reduce cowbird abundance in the landscape by reducing the amount of short grass that cowbirds require for foraging. In addition, fewer cows will reduce foraging opportunities afforded directly by the presence of livestock (Mayfield 1965; Rothstein et al. 1980, 1984). However, reduced cowbird abundance after a reduction in livestock numbers has yet to be demonstrated, and it may only occur under extreme levels of stock reduction. A small amount of short grass in the landscape may meet the needs of many cowbirds (Mayfield 1965).

All management options have costs as well as benefits. Increased understanding of the relative importance of factors that affect parasitism in the northern Great Plains may require an experimental approach at the local and at the landscape scales.

Acknowledgments

We thank the many landowners who generously provided access to their properties to conduct these studies. The U.S. Fish and Wildlife Service allowed access to WPAs. Many fine assistants helped with the fieldwork. We thank T. L. George, L. D. Igl, D. H. Johnson, J. R. Price, S. K. Robinson, and J. N. M. Smith for comments on the manuscript.

References Cited

Best, L. B. 1978. Field Sparrow reproductive success and nesting ecology. Auk 95:9–22.

Bowen, B. S., and A. D. Kruse. 1993. Effects of grazing on nesting by Upland Sandpipers in south central North Dakota. J. Wildl. Manage. 57:291–301.

Ehrlich, P. R., D. S. Dobkin, and D. Wheye. 1988. The birder's handbook. Simon and Schuster, New York. 785 pp.

Elliott, P. F. 1978. Cowbird parasitism in the Kansas tallgrass prairie. Auk 95:161–167.

Fleischer, R. C. 1986. Brood parasitism by Brown-headed Cowbirds in a simple host community in eastern Kansas. Kansas Ornithol. Soc. Bull. 37:21–29.

Fox, G. A. 1961. A contribution to the life history of the Clay-colored Sparrow. Auk 78:220–224.

Friedmann, H. 1963. Host relations of the parasitic cowbirds. U.S. Natl. Mus. Bull. 223:1–276.

Gates, J. E., and L. W. Gysel. 1978. Avian nest dispersion and fledging success in field-forest ecotones. Ecology 59:871–883.

George, T. L., A. C. Fowler, R. L. Knight, and L. C. McEwen. 1992. Impacts of a severe drought on grassland birds in western North Dakota. Ecological Appl. 21:275–284.

Graber, R. R., and J. W. Graber. 1983. The declining grassland birds. Illinois Nat. Hist. Surv. Rep. No. 227:1–2.

Hagan, J. M. III, and D. W. Johnston (eds). 1992. Ecology and conservation of neotropical migrant landbirds. Smithsonian Institution Press, Washington, DC. 609 pp.

Hatch, S. A. 1983. Nestling growth relationships of Brown-headed Cowbirds and Dickcissels. Wilson Bull. 95:669–671.

Hergenrader, G. L. 1962. The incidence of nest parasitism by the Brown-headed Cowbird (Molothrus ater) on roadside nesting birds in Nebraska. Auk 79:85–88.

Higgins, K. F., L. M. Kirsch, H. F. Duebbert, A. T. Klett, J. T. Lokemoen, H. W. Miller, and A. D. Kruse. 1977. Construction and operation of a cable-chain drag for nest searches. U.S. Fish Wildl. Serv. Wildl. Leafl. 512. 14 pp.

Hill, R. A. 1976. Host-parasite relationships of the Brown-headed Cowbird in a prairie habitat of west-central Kansas. Wilson Bull. 88:555–565.

Houston, C. S. 1973. Northern Great Plains region. Amer. Birds 27:882–886.

Johnson, D. H., and M. D. Schwartz. 1993. The Conservation Reserve Program and grassland birds. Cons. Biol. 7:934–937.

Johnson, R. G., and S. A. Temple. 1986. Assessing habitat quality for birds nesting in fragmented tallgrass prairies. Pp. 245–249 in Modeling habitat relationships of terrestrial vertebrates (J. Verner, M. L. Morrison, and C. J. Ralph, eds). University of Wisconsin Press, Madison.

———. 1990. Nest predation and brood parasitism of tallgrass prairie birds. J. Wildl. Manage. 54:106–111.

Knapton, R. W. 1979. Breeding ecology of the Clay-colored Sparrow. Living Bird 17:137–158.

Linz, G. M., and S. B. Bolin. 1982. Incidence of Brown-headed Cowbird parasitism on Red-winged Blackbirds. Wilson Bull. 94:93–95.

Lowther, P. E., and R. F. Johnston. 1977. Influences of habitat on cowbird host selection. Kansas Ornithol. Soc. Bull. 28:36–40.

May, R. M., and S. K. Robinson. 1985. Population dynamics of avian brood parasitism. Amer. Natur. 126:475–494.

Mayfield, H. F. 1965. The Brown-headed Cowbird, with old and new hosts. Living Bird 4:13–28.

Middleton, A. L. A. 1977. Effect of cowbird parasitism on American Goldfinch nesting. Auk 94:304–307.

Norman, R. F., and R. J. Robertson. 1975. Nest-searching behavior in the Brown-headed Cowbird. Auk 92:610–611.

Overmire, T. G. 1962. Nesting of the Dickcissel in Oklahoma. Auk 79:115–116.

Peterjohn, B. G., and J. R. Sauer. 1993. North American breeding bird survey annual summary, 1990–1991. Bird Populations 1:1–15.

Robbins, C. S., D. Bystrak, and P. H. Geissler. 1986. The breeding bird survey: Its first fifteen years, 1965–1979. USFWS Resource Publ. 157. 196 pp.

Robertson, R. J., and R. F. Norman. 1976. Behavioral defenses to brood parasitism by potential hosts of the Brown-headed Cowbird. Condor 78:166–173.

Robinson, S. K., J. A. Grzybowski, S. I. Rothstein, M. C. Brittingham, L. J. Petit, and F. R. Thompson III. 1993. Management implications of cowbird parasitism on neotropical migrant songbirds. Pp. 93–102 in Status and management of neotropical migratory birds (D. M. Finch and P. W. Stangel, eds). USDA For. Serv. Gen. Tech. Rep. RM-229.

Rothstein, S. I., J. Verner, and E. Stevens. 1980. Range expansion and diurnal changes in dispersion of the Brown-headed Cowbird in the Sierra Nevada. Auk 97:253–267.

———. 1984. Radio-tracking confirms a unique diurnal pattern of spatial occurrence in the parasitic Brown-headed Cowbird. Ecology 65:77–88.

Salt, W. R. 1966. A nesting study of Spizella pallida. Auk 83:274–281.

Smith, J. N. M. 1981. Cowbird parasitism, host fitness, and age of the host female in an island Song Sparrow population. Condor 83:152–161.

Stewart, R. E., and H. A. Kantrud. 1974. Breeding waterfowl populations in the prairie pothole region of North Dakota. Condor 76:70–79.

Wiens, J. A. 1963. Aspects of cowbird parasitism in southern Oklahoma. Wilson Bull. 75:130–139.

Wilson, J. K. 1976. Nesting success of the Lark Bunting near the periphery of its breeding range. Kansas Ornithol. Soc. Bull. 27:13–22.

Zimmerman, J. L. 1966. Polygyny in the Dickcissel. Auk 83:534–546.

———. 1983. Cowbird parasitism of Dickcissels in different habitats and at different nest densities. Wilson Bull. 95:7–22.

28. Distribution and Habitat Associations of Brown-headed Cowbirds in the Green Mountains of Vermont

DANIEL R. COKER AND DAVID E. CAPEN

Abstract

We studied the distribution and habitat associations of Brown-headed Cowbirds in an extensively forested region of New England where densities of cowbirds are low. Our study area was a 620-km² portion of the Green Mountain National Forest and a 7-km buffer that included agricultural and suburban areas. We conducted roadside surveys to determine the distribution of cowbirds and to sample habitat; we also mapped all areas where livestock was present. Cowbirds were detected at 21.1% of 481 survey points. They were found most consistently at survey points where there was new suburban development and at points within 1 km of livestock areas. Cowbirds were common within the national forest, but only in areas where residential, commercial, and agricultural uses were present.

Introduction

Forests of eastern North America provide habitat for a rich diversity of birdlife (Robbins et al. 1986, Sauer and Droege 1992), and the extensively forested regions of northern New England represent a substantial portion of the remaining eastern forest (Brooks and Birch 1988). Brown-headed Cowbirds are not common in these extensively forested regions (Ellison 1985; Yamasaki et al., Chapter 35, this volume), but there is reason to believe that they might become more abundant if forested areas are converted to other habitat types and become fragmented. Fragmentation may lead to the creation of land-use types preferred by feeding cowbirds and an increase in forest edges where cowbird parasitism is often comparatively high (Wilcove and Whitcomb 1983, Temple and Cary 1988, Robinson et al. 1993). Studies by Robinson (1992) and Brittingham and Temple (1983) have documented alarmingly high rates of cowbird parasitism on forest songbirds in regions of North America where extensive forest cover has been lost.

Brown-headed Cowbirds demonstrate spatial and tem-poral separation of breeding and feeding activities in many areas. They often breed secretively in host-rich areas in the morning, then feed later in the day with other cowbirds in human-disturbed areas, especially areas where livestock concentrate (Darley 1983, Dufty 1982, Rothstein et al. 1984, Airola 1986). Cowbirds travel up to 7 km from areas where they feed to habitats where they breed (Smith 1981, Rothstein et al. 1984). This ability to separate their breeding and feeding activities is reason for concern in a mostly forested landscape that contains small patches of suitable cowbird feeding habitat. These suitable feeding areas may, in fact, serve as sources that promote cowbird penetration into surrounding forested lands. Robinson et al. (1993) suggested that any practices that increase suitable cowbird feeding areas in large tracts of forest should be avoided.

In Vermont, Brown-headed Cowbirds were detected in 95% of 179 25-km² sample blocks surveyed by the state's Breeding Bird Atlas project between 1978 and 1981 (Ellison 1985). However, the species was rare in portions of the Green Mountain physiographic region. Data from the U.S. Fish and Wildlife Service's Breeding Bird Survey routes from 1966 to 1989 showed no significant change in Vermont's Brown-headed Cowbird population. Survey data from 1980 to 1989, however, indicated an average annual decrease of 6.4% ($P < .10$) in numbers of cowbirds detected on the 21 routes used in the analysis (Office of Migratory Bird Management, U.S. Fish and Wildlife Service, unpubl. data). There is little information available on cowbird-host interactions in our study area or Vermont in general. Ellison (unpubl. data) reported 26 known hosts of cowbirds in Vermont. This number is likely an underestimate of the actual number of species that serve as cowbird hosts. Of the species listed by Ellison, 11 nest in the forest interior. The most common host reported was the Red-eyed Vireo (*Vireo olivaceus*).

We examined cowbird distribution and habitat associations in and around an extensive, mostly continuous forest bordered by agricultural land uses. Our results provide a basis for future studies of cowbird populations and distribution in this and similar landscapes. The landscape under study is 94% forested and is in the very early stages of fragmentation, providing an opportunity to study cowbird distribution and make recommendations concerning future land management practices.

Our objectives were (1) to map agricultural areas that may serve as sources of cowbirds and to determine the distribution and abundance of cowbirds in these areas, (2) to assess distribution of cowbirds along roadsides in the study area, (3) to evaluate relationships between the occurrence of cowbirds and various habitat types at roadside survey stops; and (4) to assess the relationship between cowbird presence at roadside survey stops and the proximity of livestock areas.

Methods

The study area was a 620-km^2 portion of the Green Mountain National Forest and a 7-km surrounding region. This portion of the national forest contains the spine of the north-south trending Green Mountains with peaks approaching 1,320 m. It is dominated by extensive northern hardwood cover with a scattering of relatively small openings created by forest harvesting, residential development, ski-area development, hayfields, and small livestock operations (Figure 28.1). The valley on the eastern side of the mountains is narrow and contains ski areas, a few small horse stables, and residential development. In contrast, the western valley is quite broad, extending to Lake Champlain, and is dominated by agriculture, especially dairy farming.

We classified an area as a cowbird feeding area if livestock or signs of recent livestock use were present. Signs of recent livestock use included fresh manure, stables in good repair, and the presence of waste grain. To map these areas, we drove all roads within the study area and the 7-km surrounding region and recorded locations of livestock areas on 7.5-min USGS quadrangle maps. All mapped livestock areas were then digitized from maps, and 47 were selected randomly and surveyed for cowbirds twice during afternoons in June 1992. Surveys were 10-min, unlimited-distance counts conducted between 1400 and 1700 hr. All cowbirds seen or heard during counts were recorded.

From a Geographic Information System (GIS) road coverage (developed from 1 : 5000 orthophotographs), we randomly chose starting points for 25 survey routes. Each route had approximately 20 stops within the core study area or surrounding 7-km region. Survey stops were separated by 1-km driving distance, and surveys were 5-min point counts, similar to those on Breeding Bird Survey routes. Surveys were run in the early morning, twice a summer (1 June–7 July), between 0545 and 1030 hr. Twenty routes were surveyed in 1992, and 5 new routes were added in 1993. Survey routes were mapped from the existing digital road coverage of the area, and stops were subsequently mapped using dynamic segmentation, a GIS procedure for rapidly relating tabular data to spatial features.

After survey stops were mapped, their distances to livestock areas were determined using GIS proximity analysis. Each stop was classified as to whether it had livestock areas within a 1-km radius. Subsequently, chi-square analysis was used to determine whether cowbird presence at a stop was independent of land use by livestock nearby.

Cover types at each survey point were quantified using a procedure developed by Adamus et al. (1992). Observers classified cover types into 1 of 78 categories within a 100-m circle and estimated the percentage of each cover type in the circle around the survey point. Cover types included general categories such as suburban, hayfield, and cropland and more detailed categories for forest cover. Forest cover was broken into 27 combinations of type, height, and closure. For our analysis, we combined all forest types into a single category.

We used chi-square analyses and the method of Neu et al. (1974) to generate simultaneous confidence intervals to determine if stops with certain cover types were used disproportionately to their availability. These confidence intervals are inflated to account for multiple comparisons conducted with the same data set. Only cover types present at 100 stops or more were included in the simultaneous confidence interval analyses; remaining cover types were grouped into the "other" category. Assessment of preference or avoidance in this manner has been widely reported in recent years (see Manly et al. 1993 for review). Our approach differed only slightly from the common model, Design 1 with known proportions of available resource units (Manly et al. 1993:40–46), where availability of a habitat category is exclusive of another category at a particular sampling location. In our study, more than one habitat type often was recorded at each survey point. Thus, we tested for differences between proportions of survey points where a habitat category occurred and proportions of those points where we found cowbirds and also recorded the habitat type in question.

For example, if we sampled 200 roadside survey points and found that 50 points had habitat type A present, then $P_a = 0.25$ (P_a = proportion available). If cowbirds were detected at 100 of the roadside points and if 25 of these points had habitat A present, then $P_u = 0.25$ (P_u = proportion used). Therefore, the proportional occurrence of habitat A (P_a) at all survey points is the same as the proportion of habitat A (P_u) at the 100 points where cowbirds were detected. Thus, $P_a = P_u$, and there is no evidence that habitat A is preferred or avoided.

Results

We mapped 217 livestock areas, illustrating a marked difference between eastern and western portions of the study area (Figure 28.2). The wider western valley, with a concentra-

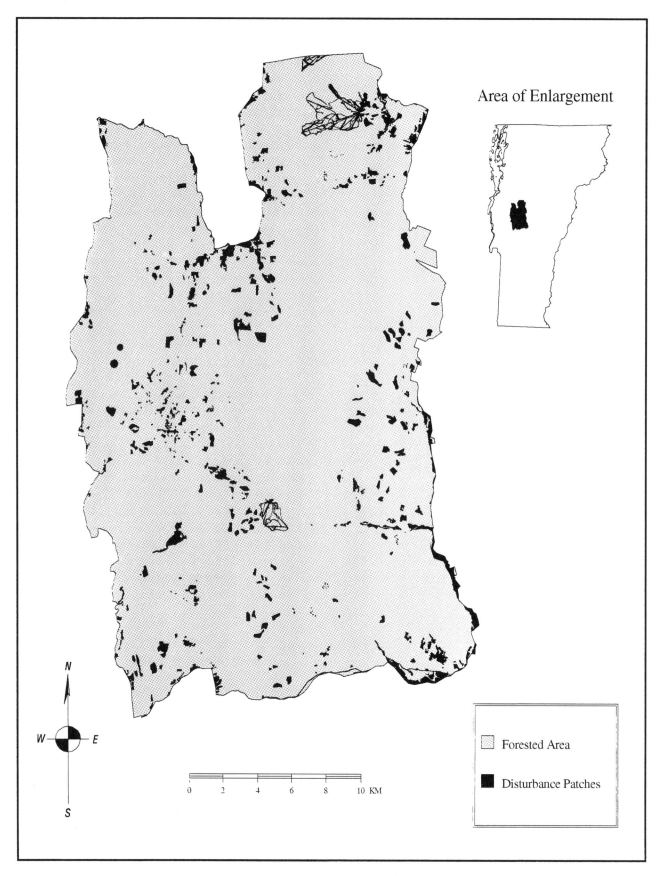

Area of Enlargement

Forested Area

Disturbance Patches

N
W—E
S

0   2   4   6   8   10 KM

Figure 28.1. Green Mountain National Forest study area in central Vermont, showing forested area and disturbance patches.

tion of dairy farms, contained 0.2 livestock areas/km road, whereas the eastern valley contained 0.07 livestock areas/km road. Of the 47 livestock areas surveyed, cowbirds were observed feeding at 29 (62%). At sites where cowbirds were observed, an average of only 6 cowbirds was recorded (range, 1–25). The overall sex ratio was 2.2 males : 1 female, similar to other skewed sex ratios previously reported for cowbirds (Darley 1983, Dufty 1982). Most (86%) of these livestock areas were outside the national forest boundary, but there was a scattering of small livestock pastures within the national forest.

Cowbirds were present at 100 (21.1%) of 481 survey stops (Figure 28.3). They appeared to occur most consistently around survey points outside the national forest (71 of 317, or 22%) and less often within the national forest boundaries (29 of 164, 18%), but these detection rates were not different ($X^2 = 1.44$, $P > .10$). The proportion of survey points where cowbirds were detected ranged from 0 to 0.37 per route (mean = 0.14, SE = 0.018). A one-sample runs test (Daniel 1978) of the data combined across replicates showed that 20 of the 23 survey routes where cowbirds occurred were random series of cowbird presence/absence stops ($P > .05$). We therefore included all survey stops as independent data points in our analyses.

At least some forest cover was present at 424 (88%) survey stops; 252 (52%) stops had more than 50% forest cover within 100 m of the survey point. Cowbirds showed strong preference for stops with new suburban development (Table 28.1). In addition, cowbirds occurred at 62 of 225 (27%) morning survey points within 1 km of an active livestock area and only 38 of 256 (15%) morning survey points not within 1 km of an active livestock area. The presence of a livestock area within 1 km of the survey stop was a significant indicator of cowbird presence ($X^2 = 11.03$, $P = .001$). Cowbirds showed neither a preference for nor an aversion to survey stops that had at least some forest cover (Table 28.1). They were found regularly along roadsides within the national forest boundary but preferred survey stops with less than 50% forest cover (Table 28.1). Pastures occurred at 65 (13.5%) of the 481 survey stops; cowbirds were detected at 17 (26.2%) of the 65 survey stops with pasture and at 83 (20%) of the stops without pasture. They did not show a preference for morning stops that had pasture cover versus those that did not $X^2 = 1.31$, $P = .25$). Because pasture cover occurred at only 65 stops, it was not included in the Bonferroni confidence interval generation (Table 28.1).

Discussion

The relatively small numbers of cowbirds detected at individual livestock areas suggest either that the overall population of cowbirds in the study region is not high or that feeding areas do not function as central gathering places for cowbirds. Because of the large number (217) of active live-

stock areas within the study area and surrounding region, cowbirds may not have to travel far from breeding areas to reach suitable feeding sites. Cowbirds are likely feeding at a large number of sites; therefore, the local population is widely distributed in the afternoon. Rothstein et al. (1980) found much higher concentrations of cowbirds in afternoon feeding gatherings in the Sierra Nevada. In contrast to the situation in our study area, however, suitable feeding areas in their study area were quite scarce.

Short grass lawns that provide feeding sites for cowbirds (Darley 1983, Verner and Ritter 1983, Airola 1986) and an abundance of bird feeders may explain the preference cowbirds showed for new suburban habitat. New suburban areas may function like livestock areas in aiding cowbird penetration into the forest. In addition to its possible role as a feeding area, new suburban habitat in a predominantly forested region provides abundant forest edges, which are known for high concentrations of songbird nests (Gates and Gysel 1978). Cowbird parasitism rates also are greater at edges, probably in response to the increase in diversity and abundance of nesting bird species (Wilcove and Whitcomb 1983). But many other habitat types that produced forest edge in our study area were not favored by cowbirds. Thus, new suburban areas may produce an optimal habitat situation for this species: high-quality breeding areas (forest edge) in close proximity to short grass feeding areas that are often supplemented with bird feeders.

We acknowledge, however, that roadside surveys were conducted in the morning hours when cowbirds normally are engaged in reproductive functions rather than feeding activities. Thus the apparent lack of preference for survey stops where pasture was present may not accurately reflect the importance of this habitat type during afternoon hours. Survey stops within 1 km of an agricultural operation were preferred, when compared to those more than 1 km away, and cowbirds were regularly found feeding at mapped livestock areas in the mid- to late afternoon.

What was most obvious from our study was that Brown-headed Cowbirds were relatively common in and around the Green Mountain National Forest, especially in areas where feeding and breeding habitat were in close proximity. This finding was expected. Of more significance, however, was the detection of cowbirds on 29 survey points within the national forest boundary. Not surprisingly, points where cowbirds were detected concurred with forest openings along roads where our surveys were conducted. It is clear that even an area of extensive forest cover, such as this portion of the Green Mountain National Forest, is subject to use by cowbirds when residential, commercial, and agricultural development is present.

Our findings differ somewhat from those of Yamasaki et al. (Chapter 35), who conducted similar studies in and around the White Mountain National Forest in New Hampshire. They found virtually no Brown-headed Cowbirds on

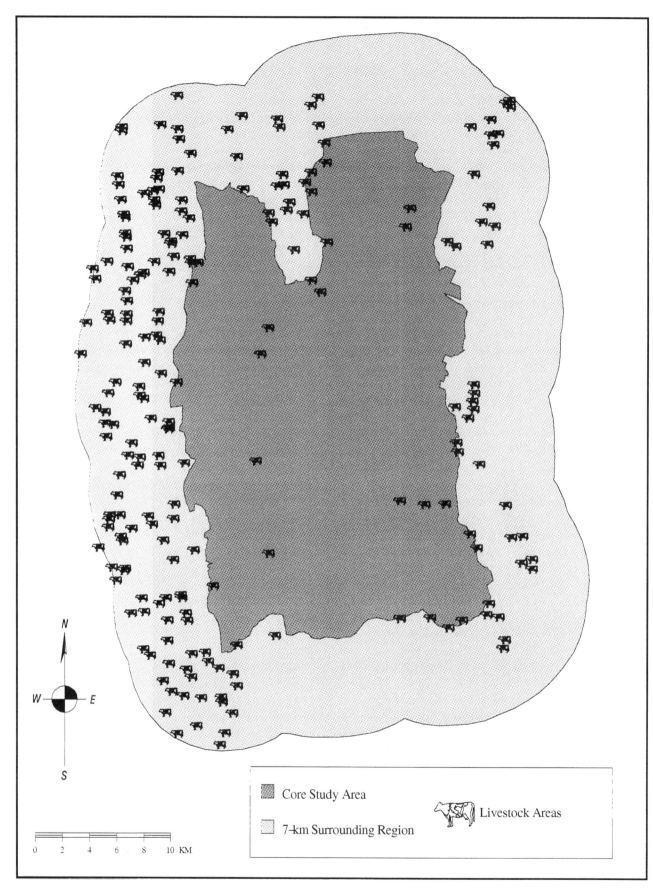

Figure 28.2. Green Mountain National Forest study area with
7-km surrounding region showing livestock areas.

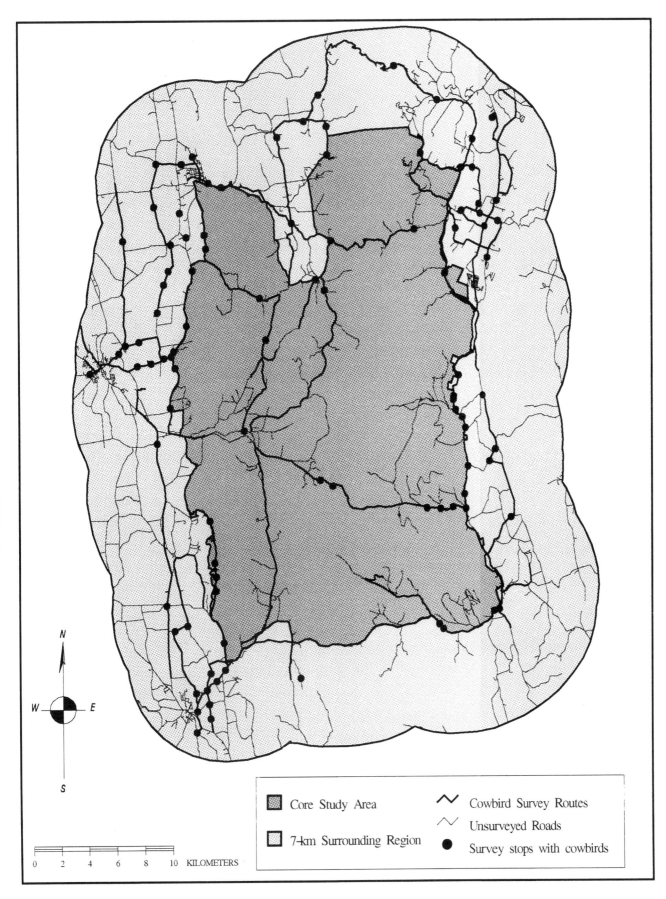

Figure 28.3. Green Mountain National Forest study area and 7-km surrounding region showing roadside survey stops where cowbirds were detected.

Table 28.1. Occurrence of Brown-headed Cowbirds According to Habitat Categories on Roadside Survey Routes in Vermont, 1991–1992

Habitat Category	Proportion of Survey Stops with Habitat Category (P_a)	Proportion of Survey Stops Used by Cowbirds (P_u)[a]	95% Bonferroni Confidence Intervals[b]
Forest	0.88 (424/481)	0.92 (92/100)	0.85–1.00
Hayfield	0.41 (195/481)	0.52 (52/100)	0.38–0.65
New suburban	0.37 (180/481)	0.59 (59/100)	0.45–0.72[c]
Old field/grass	0.25 (120/481)	0.25 (25/100)	0.13–0.37
Old field/shrub	0.24 (116/481)	0.26 (26/100)	0.14–0.38
Transportation/communication	0.35 (169/481)	0.38 (38/100)	0.24–0.52
Less than 50% forest cover	0.42 (201/481)	0.57 (57/100)	0.44–0.71[c]
Other	0.64 (308/481)	0.68 (68/100)	0.54–0.82

[a]The denominator used to calculate these proportions is 100 in each case because this is the number of stops (out of 481) at which cowbirds were detected. The sum of the numbers in this column exceeds the total number of stops because some stops had more than one habitat category.

[b]Confidence interval around P_u. If P_a falls below interval, preference for the habitat category is indicated.

[c]Preference for habitat category ($P < .05$).

roadside survey points in the national forest. These differences may be due to a larger local population of cowbirds in central Vermont than in northern New Hampshire or to a more favorable juxtaposition of breeding and feeding habitat in the forested region of our study area.

It is also interesting to compare our findings to those of Hahn and Hatfield (Chapter 13) in east central New York, a region where forests occupy a minor portion of the landscape. They found that Brown-headed Cowbirds routinely used edge habitat and forest interior sites for reproduction. In contrast, Germaine (1993) surveyed forest-interior birds on a portion of our study area in 1991 and 1992 and found no Brown-headed Cowbirds on 102 survey points that were visited four times each. His data were collected in areas remote from roads and human development. Germaine's findings are consistent with other surveys of forest-interior plots on our study area (D. E. Capen unpubl. data).

Management implications seem straightforward. Although Brown-headed Cowbird populations appear to be declining in northern New England and forest cover is increasing, cowbirds are abundant enough in regions where feeding habitat is suitable to penetrate forested areas if even mild forest fragmentation has occurred. If cowbirds are believed to threaten the well-being of forest-nesting birds in this region, then careful planning of the extent and location of future forest openings is recommended.

Acknowledgments

Our research was funded by Cooperative Forest Research Funds from the School of Natural Resources at the University of Vermont and by the forest bird monitoring program of the Green Mountain National Forest. We thank Peter Norton and Stephen S. Germaine for their assistance in the field. B. K. Williams and D. Wang offered constructive comments on our study design and data analyses. S. I. Rothstein and an anonymous reviewer made helpful comments that improved the manuscript.

References Cited

Adamus, P. R., R. J. O'Connor, and A. A. Whitman. 1992. Quality assurance of BBS habitat assessment project. University of Maine, Orono. Unpubl. report. 61 pp.

Airola, J. A. 1986. Brown-headed Cowbird parasitism and habitat disturbance in the Sierra Nevada. J. Wildl. Manage. 50:571–575.

Brittingham, M. C., and S. A Temple. 1983. Have cowbirds caused forest songbirds to decline? BioScience 33:31–35.

Brooks, R. T., and T. W. Birch. 1988. Changes in New England forests and forest owners: Implications for wildlife habitat resources and management. Trans. N. Amer. Wildl. and Nat. Resour. Conf. 53:78–87.

Daniel, W. W. 1978. Applied nonparametric statistics. Houghton Mifflin Co., Boston, MA. 503 pp.

Darley, J. A. 1983. Territorial behavior of the female Brown-headed Cowbird (*Molothrus ater*). Can. J. Zool. 61:65–69.

Dufty, A. M., Jr. 1982. Movements and activities of radio-tracked Brown-headed Cowbirds. Condor 84:15–21.

Ellison, W. G. 1985. The Brown-headed Cowbird. Pp. 368–369 in The atlas of the breeding birds of Vermont (S. B. Laughlin and D. P. Kibbe, eds). University Press of New England, Hanover, NH. 456 pp.

Gates, J. E., and L. W. Gysel. 1978. Avian nest dispersion and fledging success in field-forest ecotones. Ecology 59:871–883.

Germaine, S. S. 1993. The effects of small patch clearcuts on the forest-interior bird community in the Green Mountain National Forest, Vermont. M.S. Thesis, Bowling Green State University, Bowling Green, OH. 75 pp.

Manly, B. F., L. L. McDonald, and D. L. Thomas. 1993. Resource selection by animals. Chapman and Hall, London. 177 pp.

Neu, C. W., C. R. Byers, and J. M. Peek. 1974. A technique for analysis of utilization-availability data. J. Wildl. Manage. 39:541–545.

Robbins, C. S., D. Bystrak, and P. H. Geissler. 1986. The breeding bird survey: Its first fifteen years, 1965–1979. U.S. Fish and Wildl. Serv. Resource Publ. 157. 196 pp.

Robinson, S. 1992. Population dynamics of breeding neotropical migrants in a fragmented Illinois landscape. Pp. 408–418 in Ecology and conservation of neotropical migrant landbirds (J. M. Hagan III and D. W. Johnston, eds). Smithsonian Institution Press, Washington, DC.

Robinson, S. K., J. A. Grzybowski, S. I. Rothstein, M. C. Brittingham, L. J. Petit, and F. R. Thompson. 1993. Management implications of cowbird parasitism on neotropical migrant songbirds. Pp. 93–102 in Status and management of neotropical migratory birds (D. M. Finch and P. W. Stangel, eds). USDA For. Serv. Gen. Tech. Report RM-229.

Rothstein, S. I., J. Verner, and E. Stevens. 1980. Range expansion and diurnal changes in dispersion of the Brown-headed Cowbird in the Sierra Nevada. Auk 97:253–267.

———. 1984. Radio-tracking confirms a unique diurnal pattern of spatial occurrence in the parasitic Brown-headed Cowbird. Ecology 65:77–88.

Sauer, J. R., and S. Droege. 1992. Geographical patterns in population trends of neotropical migrants in North America. Pp. 26–42 in Ecology and conservation of neotropical migrant landbirds (J. M. Hagan III and D. W. Johnston, eds). Smithsonian Institution Press, Washington, DC.

Smith, J. N. M. 1981. Cowbird parasitism, host fitness, and age of the host female in an island Song Sparrow population. Condor 83:152–161.

Temple, S. A., and J. R. Cary. 1988. Modeling dynamics of habitat-interior bird populations in fragmented landscapes. Conserv. Biol. 2:340–347.

Verner, J., and L. V. Ritter. 1983. Current status of the Brown-headed Cowbird in the Sierra National Forest. Auk 100:355–368.

Wilcove, D., and R. F. Whitcomb. 1983. Gone with the trees. Natural History 41:82–91.

29. Impacts of Cowbird Parasitism on Wood Thrushes and Other Neotropical Migrants in Suburban Maryland Forests

BARBARA A. DOWELL, JANE E. FALLON, CHANDLER S. ROBBINS,

DEANNA K. DAWSON, AND FREDERICK W. FALLON

Abstract

During 1988–1993, we monitored nests of neotropical migrant birds in seven suburban Maryland forests to compare parasitism and predation rates in forests of different areas. Of 1,122 nests monitored, 672 were of Wood Thrush, the most commonly found nesting species. Study sites were forests that ranged in size from 21 ha to more than 1,300 ha in the Piedmont and Coastal Plain regions of Maryland within 50 km of Washington, D.C. Parasitism rates of Wood Thrush nests varied greatly among sites, ranging from 0% (29 nests in 1990–1992) in a site in extensive forest to 68% (31 nests 1992–1993) in a 21-ha, selectively logged old-growth forest. A sudden increase in parasitism from 9% (102 nests 1990–1991) to 35% (125 nests 1992–1993) in a 23-ha old-growth forest was noteworthy. The surrounding environment at this site is changing from rural to residential. Wood Thrush parasitism rates dropped as the breeding season progressed, but peaks of parasitism coincided with peaks of nesting activity. Parasitism rates for Hooded Warblers (88% of 17 nests—all sites) were most alarming. High predation rates were a much greater factor in low productivity for Wood Thrushes than parasitism.

Introduction

During the mid-1980s, fieldwork for the Maryland-D.C. Breeding Bird Atlas (Robbins et al. 1996) indicated the presence of "forest interior" neotropical migrant birds such as Hooded and Kentucky Warblers in some very small suburban woodlots in the Washington, D.C., metropolitan area. Previous research during the breeding seasons of 1979–1983 in Maryland and adjacent states (Robbins et al. 1989) showed that forest-dwelling neotropical migrants occurred infrequently in small isolated woodlots compared with larger stands. Although a high correlation existed between forest area and composition of the avifauna, the methods used in our previous studies were based on counts of singing males and did not reveal whether the males were mated or raising young. Results of other studies (Brittingham and Temple 1983, Wilcove 1985, Andrén and Angelstam 1988) suggest that both nest parasitism by Brown-headed Cowbirds and nest predation are often higher in small woodlots and along forest edges than in the interior of extensive forests. Therefore, the smallest woodlots in which neotropical migrants occur might be below the size in which they can successfully nest. Also, studies of the Wood Thrush in forest fragments in Illinois (Robinson 1989) reported alarming levels of predation and parasitism. Additionally, the North American Breeding Bird Survey (Robbins et al. 1986) indicated that cowbird populations were increasing in many sections of the United States. Yet, there were insufficient recent nest data to show trends in cowbird parasitism in our area.

Forest loss and fragmentation from suburban development in Maryland and Virginia continue at a rapid rate. Some counties have experienced as much as a 50% loss in forest cover since 1950, resulting in efforts to pass legislation to preserve trees. Therefore, in 1988, with concerns about increasing cowbird populations and the dependency of birds on smaller and more isolated forest fragments for breeding habitat, we undertook a study of nesting birds to assist ongoing conservation and land-use planning efforts. In this chapter, we report on rates of nest parasitism by Brown-headed Cowbirds and nest success rates, with emphasis on Wood Thrushes, in relation to tract size, distance from edge, and surrounding landscapes. We also compare historical parasitism rates for Maryland with parasitism rates from this study for Wood Thrush and other species.

Study Sites

Using previously collected information on the occurrence of neotropical migrants, we selected for study seven forested sites in Maryland ranging in size from 21 to more than 1,300 hectares. Table 29.1 shows study plot size, forest size, percentage of forest within 1 and 2 km of plot center, and characteristics of the surrounding landscapes. Sites chosen were in the Piedmont and Coastal Plain regions of Maryland, in an area dominated by suburban development within 50 km of Washington, D.C. Six sites were in Prince Georges County and one in adjoining Howard County.

Two of the sites, North and South Belt Woods, are of special interest because they contain sections of old-growth forest. Stewart and Robbins (1947) described South Belt as one of the finest examples of mature upland hardwood forest remaining on the Coastal Plain. It has rolling topography and is dominated by mature oaks and tulip trees (*Liriodendron tulipifera*) with a dogwood (*Cornus florida*) understory that supports a high density of Wood Thrushes (83 males/40 ha, O'Brien and Dowell 1990a). North Belt was similar in topography and tree composition until selective logging for oaks in 1971 and 1981 opened the upper canopy (Whitcomb et al. 1977b, Robbins 1989). This resulted in a very dense shrub layer and a low closed canopy with few remaining large trees. Neither of these sites contains streams, but both are adjacent to a seepage forest and are noted for their highly fertile greensand soils.

The Clarksville site (Robbins 1990) comprises two irregular connected fragments (12.5 and 8.9 ha) of mature Piedmont forest. Selectively logged about 28 years ago for oaks and walnuts, it is now dominated by red maples (*Acer rubrum*) and tulip trees. No point in the forest is more than 150 m from an edge. The topography is rolling, with a small seasonal stream in each half.

Woodmore, 2 km north of the two Belt Woods sites, is part of a recreational ballpark complex surrounded by an even-age forest of rolling topography (O'Brien and Dowell 1990b). Our study plot was a forest of mature beech (*Fagus grandifolia*) and tulip trees with mixed understory and small intermittent streams and swampy areas.

Watkins is a county-owned regional park with playgrounds, ballfields, and picnic areas clustered in clearings in the northern portion. Our study site was in a mature beech-oak forest of uneven age in the southern portion of the park and was bordered on the north and west by streams.

The remaining two sites were within the last extensive forest in central Maryland, which stretches unbroken for 8 km along the Patuxent River within the Patuxent Wildlife Research Center (PWRC). Site 1 was mature beech-oak forest in the Patuxent floodplain and terrace (Robbins 1991). Site 2, 1 km away, was in upland forest adjacent to the floodplain; roughly one third of the plot was in mature beech forest, the rest being a mixture of tulip trees and oaks about 60 years old. Both Patuxent sites were flat and contained numerous streams and vernal pools. Heavy deer browsing in recent years has eliminated the understory and shrub vegetation from all except the swampy areas in both sites.

Methods

Sites were searched for nests two or three times a week, beginning in early May. Methods for locating nests varied by species and included locating females by call notes and fol-

Table 29.1. Area of Study Plots and Sites, Percentage of Forest within 1 km and 2 km of Plot Center, and Adjacent Nonforest Habitats

| Site | Area (ha) | | % Forest within | | Adjacent Habitats |
	Plot	Forest	1 km	2 km	
North Belt[a]	21	21	57	50	Agricultural fields, ballfields, housing
South Belt[b]	21	23	44	45	Fallow fields, housing
Clarksville[c]	21	21	27	20	Plant nursery, mowed fields, housing
Woodmore[d]	18	45	37	41	Pasture, powerline, ballfields, abandoned fields
Watkins	16	183	65	38	Park, powerline, housing, agricultural fields
Patuxent 1[e]	42	> 1300	70	79	Fallow fields, lawns, powerlines
Patuxent 2	18	> 1300	70	72	Lawns, fallow fields, powerline, housing

[a]Robbins 1989.
[b]Whitcomb et al. 1977a, O'Brien and Dowell 1990a.
[c]Robbins 1990.
[d]O'Brien and Dowell 1990b.
[e]Robbins 1991.

lowing them to nests, locating adults carrying nesting or feeding material, or simply searching for the nest. When a nest was located, the number of host and cowbird eggs or nestlings was determined by inspection with an extension pole and mirror (for nests up to approximately 7 meters above ground level). Active nests were generally checked at 2- to 4-day intervals, especially during egg-laying (to detect cowbird eggs) and near the time of fledging (to determine outcome). After egg-laying was complete, checks to determine the presence of adult activity at the nest were made from a distance with binoculars to minimize predation risk. If no adults were present during a short observation period, the nest contents were examined to determine their status.

For nests with known contents, we calculated the percentage with one or more cowbird eggs or young. For Wood Thrush, we used logistic regression (Hosmer and Lemeshow 1989) to identify factors that influenced the probability that a nest was parasitized. In this analysis, the dependent variable had a value of 1 if a nest was parasitized and 0 if it contained no cowbird eggs or young. The independent variables were distance of the nest from the forest edge, time of the season (days from 1 May) when the nest was found, forest size, and study site in which a nest was located. The interactions of all these factors were also included in an initial regression model. Factors, beginning with the highest-order interactions, were eliminated ($\hat{A} = 0.1$) from the regression based on chi-square statistics until all remaining factors were deemed to be significant.

Using a geographic information system of the forests of Prince Georges County (1990), which was supplied to us by the Maryland–National Capital Park and Planning Commission, we determined the amount of forest within 1 and 2 km of each study site. In addition, we measured the area of forest within 100 meters of the edge (in 25-m intervals) and beyond 100 meters from the edge for two sites, South Belt and North Belt. The average number of nests per hectare and the percentage parasitized in each of these segments were calculated. The Jonckheere-Terpstra test (Lehman 1975, Statxact 1989) was used to determine whether there was a trend in the proportion of parasitized nests and distance to edge, and the Fisher exact test was used to determine the significance of the distance-from-edge categories and parasitism rates. Simple linear regression, using the midpoint of each interval as the independent variable, tested the significance of nest density in relation to distance from the edge.

We examined the success of nests in two ways. For nests for which we were certain of the outcome, we calculated the success rate as the percentage of nests that fledged at least one host young. We also used the Mayfield method (Mayfield 1975, Bart and Robson 1982) to calculate daily survival rates and estimated success rates for nests. For Wood Thrush, daily survival rates of parasitized and unparasitized nests were compared using the program CONTRAST (Hines

and Sauer 1989). Common and scientific names of all species mentioned in this chapter are listed in the Appendix.

Results

During the study we monitored 1,122 nests; we report on 743 of these for this chapter. We did not include nests that were too high for their contents to be observed, those of unsuitable cowbird hosts such as the Gray Catbird (a rejector), Mourning Dove, or cavity nesters, and nests of species with small (fewer than 5) samples.

Wood Thrushes

The most commonly found nests in all seven sites were those of Wood Thrushes, with a total of 467 nests with known contents. The number of observable nests in a year in any of the sites ranged from 8 to 74. Half of all Wood Thrush nests, 233, were in South Belt, the unlogged old-growth forest.

The overall parasitism rate for Wood Thrush nests for all years at the seven sites was 20%. The rates varied greatly among years and among sites (Table 29.2, Figure 29.1). The two sites within the extensive forest at the Patuxent Wildlife Research Center had very low rates: 0% of 29 nests at site 2 during 1990–1992 and 8% (2 of 26 nests) at site 1 in 1990–1991. The Clarksville site also had a low rate of 7% (4 of 58 nests) during 1988–1990. This contrasts with North Belt, which had a parasitism rate of 68% for 31 nests in 1992–1993. The rate at South Belt Woods climbed from 9% of 102 nests during 1990–1991 to 35% of 125 nests in 1992–1993. However, graphing the parasitism rates by site and year shows no clear trend (Figure 29.1).

In our study sites the Wood Thrush arrives in late April. Nesting begins the first week of May and continues into August, with the peak in the second half of May. Although the parasitism rate generally decreased as the season progressed, peaks of parasitism clearly occurred during the peaks of nesting activity (Figure 29.2). Of the 94 parasitized Wood Thrush nests, only 12 contained 2 Brown-headed Cowbird eggs and 1 contained 3. Four nests containing only a cowbird egg and no host eggs were either abandoned or depredated.

The logistic regression analysis indicated that nests were significantly more likely to be parasitized if they occurred early in the season ($P = .001$) or were located in either North Belt ($P = .000$) or South Belt Woods ($P = .004$). A small effect from the size of site was not statistically significant ($P = .10$).

A second logistic regression included all the same factors except the study-site variable to determine whether the site effect might dominate the analysis. Based on this model the significant factors were size of site ($P = .002$) and time of the nesting season ($P = .003$). The model that included the site provided the better fit (Akaike's information criteria = 404.120 compared to 438.509 for the second regression).

Table 29.2. Parasitism Rates of Wood Thrush Nests by Site

Site	Years	% Parasitized (N)[b]	Cowbird Eggs per Nest (N)[c]
North Belt[a]	1992–1993	68 (31)	1.10 (21)
South Belt[a]	1990–1991	9 (102)	1.00 (9)
	1992–1993	35 (125)	1.20 (44)
Clarksville	1988–1990	7 (58)	1.25 (4)
Woodmore[a]	1990–1991	10 (30)	1.00 (3)
	1992–1993	21 (23)	1.00 (5)
Watkins	1990–1992	19 (31)	1.33 (6)
Patuxent 1	1990–1991	8 (26)	1.00 (2)
Patuxent 2	1990–1992	0 (29)	— —

[a]Does not include 12 nests found in 1988 and 1989.

[b]N = total nests.

[c]N = number of parasitized nests.

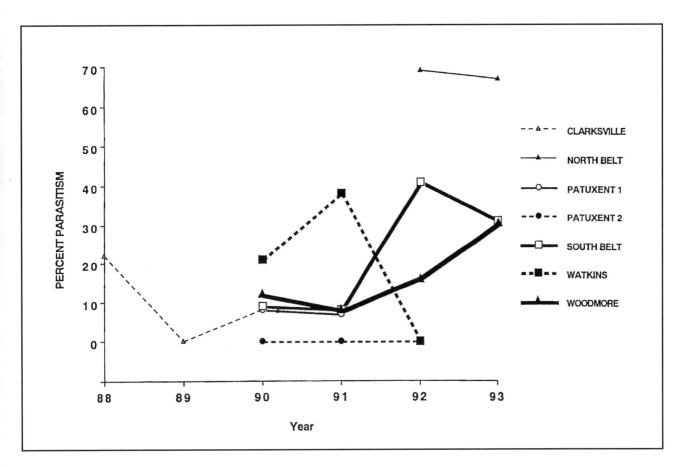

Figure 29.1. Percentage of Wood Thrush nests that were parasitized by Brown-headed Cowbirds at seven study sites, 1988–1993. Number of nests for each site and year: Clarksville (1988–1990) 9, 24, 25; North Belt (1992–1993) 13, 18; South Belt (1990–1993) 65, 37, 51, 74; Watkins (1990–1992) 14, 8, 9; Woodmore (1990–1993) 17, 13, 13, 10; Patuxent 1 (1990–1991) 12, 14; Patuxent 2 (1990–1992) 8, 13, 8.

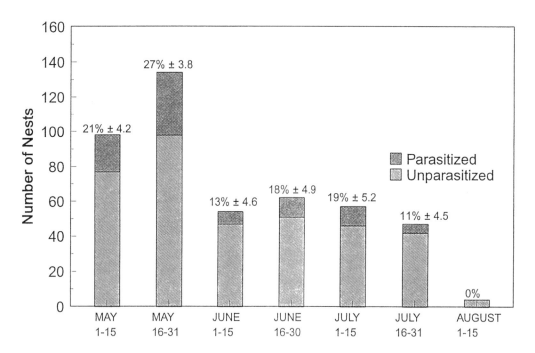

Figure 29.2. Number of Wood Thrush nests and percentage parasitized by Brown-headed Cowbirds at seven study sites, 1988–1993. Number of nests: 1–15 May, 98 nests; 16–31 May, 134 nests; 1–15 June, 54 nests; 16–30 June, 62 nests; 1–15 July, 57 nests; 16–31 July, 47 nests; 1–15 August, 5 nests.

Although distance from edge was not a significant variable in the regression analyses, edge effects were detected in South Belt Woods when we related numbers of Wood Thrush nests to the number of hectares of habitat available in each 25-m band (Figure 29.3). The density of nests increased significantly with distance from edge ($P = .002$), and parasitism rates declined beyond 75 m from the edge (but not significantly, $P = .182$). For North Belt, the site with the highest parasitism rate but with many fewer nests, a different pattern emerges: the density of nests increased with distance from edge (not significantly, $P = .236$), but parasitism rates were not related to distance from edge. In this site the maximum possible distance from edge was 175 m.

Unlike smaller hosts such as Hooded Warblers, Red-eyed Vireos, and Acadian Flycatchers (which if parasitized raised only cowbird young in our study), Wood Thrushes are generally able to raise both their young and the cowbird young. We compared the clutch sizes of parasitized and unparasitized nests that were observed during incubation (Table 29.3). Nests with fewer than 4 eggs, the typical clutch size for Wood Thrush, were included when we were able to confirm that the clutch was complete. We grouped the nests by half-month intervals starting 1 May and used a log-linear model to test for association among clutch size, month, and parasitism. The test showed that clutch size was significantly smaller ($P = .006$) in parasitized nests, and that the pattern

of association between clutch size and cowbird parasitism is not affected by month ($P = .999$). The mean clutch size was 3.0 for parasitized nests and 3.4 for unparasitized nests.

Nests in the three largest forest sites were not necessarily the most likely to fledge young (Table 29.4). The overall success rate for Wood Thrush for all sites combined was 42%, with an individual site high of 72% and a low of 24%. The daily survival rates of unparasitized nests (0.9686, standard error 0.0024, exposure days 5048) and parasitized nests (0.9636, standard error 0.0053, exposure days 1,244) were not significantly different. When daily survival rates of the parasitized and unparasitized thrush nests (all sites) were compared for four different time periods (two biweekly periods in May, and the months of June and July), we also found no significant difference. Thrush nests built during the first two weeks of the season had the highest daily survival rate (0.9752, standard error 0.0038, exposure days 1,674).

Other Species

Parasitism and success rates for other species are given in Table 29.5. Hooded Warblers (with 5 or fewer pairs per site) were the most heavily parasitized, with 15 of the 17 nests found containing cowbird eggs or young. The Red-eyed Vireo was probably the most abundant species (O'Brien and Dowell 1990a,b; Robbins 1989, 1990, 1991) in our study plots, but its nests were difficult to find (only 40 nests all years all sites) or too high (8) to be certain of contents. It was the species most often observed feeding only fledged cowbird young. Therefore, 32 nests with a 19% parasitism rate may be misleading. Scarlet Tanagers, whose nests were generally too high (7) to examine, were frequently observed feeding only cowbird young. Acadian Flycatcher nests were the second most numerous nests observable (124 of 187 could be examined), with an overall parasitism rate of 4%.

The observed success rate was lowest (21%) for Hooded Warblers (Table 29.5). The sample size of Hooded Warbler nests was too small for the calculation of a Mayfield estimate. For other species the Mayfield estimate of success ranged from 26% for Northern Cardinal to 42% for Wood Thrush.

Discussion

Trends

The Breeding Bird Survey for 1966–1992 (National Biological Service unpubl. data) indicates that cowbird numbers in Maryland are increasing at a rate of 1.4% per year ($P < .10$). Hooded Warbler numbers are increasing by 1.9% per year ($P > .10$), while Wood Thrushes are decreasing by 2.1% per year ($P < .01$). However, Breeding Bird Censuses conducted

in South Belt Woods (1947, 1975, 1976, and 1989) reflect little change over 40 years in density of the cowbird (1 pair/40 hectares in 1947, 2 pairs/40 hectares in 1989). Other sites for which Breeding Bird Census data are available, including North Belt Woods (1975, 1976, 1988, and 1989) and Patuxent 1 (1959–1990), did not show marked changes in cowbird density. Thus, the high parasitism rates we observed for

Table 29.3. Number of Wood Thrush Eggs in Parasitized versus Unparasitized Nests, All Sites, 1988–1993

Clutch Size	Unparasitized Nests		Parasitized Nests	
	N	(%)	N	(%)
0	0	(0.0)	1	(1.6)
1	1	(0.3)	3	(4.8)
2	33	(12.8)	14	(22.2)
3	115	(41.1)	23	(36.5)
4	129	(46.1)	21	(33.3)
5	2	(0.7)	1	(1.6)
Total nests	280		63	

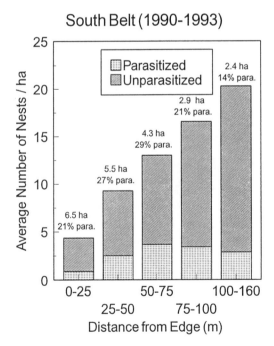

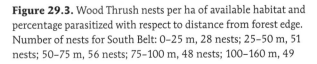

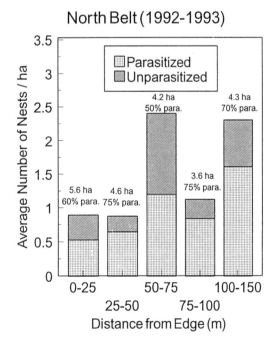

Figure 29.3. Wood Thrush nests per ha of available habitat and percentage parasitized with respect to distance from forest edge. Number of nests for South Belt: 0–25 m, 28 nests; 25–50 m, 51 nests; 50–75 m, 56 nests; 75–100 m, 48 nests; 100–160 m, 49 nests. Number of nests for North Belt: 0–25 m, 5 nests; 25–50 m, 4 nests; 50–75 m, 10 nests; 75–100 m, 4 nests, 100–150 m, 10 nests.

Table 29.4. Success Rate by Site for Wood Thrush Nests

Site (Forest Size)	% Successful	(N)[a]	MES (%)[b]	(N)[c]
Watkins Park (183 ha)	73	(22)	72	(542)
South Belt (23 ha)	49	(215)	46	(3344)
North Belt (21 ha)	45	(29)	34	(410)
Patuxent 2 (> 1300 ha)	38	(21)	48	(462)
Patuxent 1 (> 1300 ha)	35	(23)	35	(352)
Clarksville (21 ha)	34	(44)	33	(675)
Woodmore (45 ha)	34	(50)	24	(579)

[a]N = number of nests for which outcome was known.
[b]MES = Mayfield Estimate of Survival.
[c]N = exposure days; survival estimate based on 26 days.

Table 29.5. Parasitism and Success Rates by Species

Species	% Parasitized (N)[a]		Intensity of Parasitism (N)[b]		% Successful (N)[c]		Mayfield Nest Success (N)[d]	
Hooded Warbler	88	(17)	1.27	(15)	21	(14)	—	—[e]
Wood Thrush	20	(467)	1.15	(94)	45	(404)	42	(6364)
Red-eyed Vireo	19	(32)	1.00	(6)	37	(27)	41	(332)
Ovenbird	9	(23)	1.00	(2)	47	(17)	39	(218)
Northern Cardinal	6	(80)	1.00	(5)	28	(57)	26	(694)
Acadian Flycatcher	4	(124)	1.00	(5)	46	(114)	37	(1632)

[a]N = total nests for which the contents were known.
[b]Number of cowbird eggs/parasitized nest; N = number of parasitized nests.
[c]N = nests for which outcome was known.
[d]N = exposure days; survival estimate based on 23 days for Hooded Warbler, Cardinal, and Ovenbird; 25 days for Red-eyed Vireo; 26 for Wood Thrush; 28 for Acadian Flycatcher.
[e]Sample size too small to calculate Mayfield estimate.

Hooded Warblers or the temporal changes in parasitism rates in South Belt Woods were not due solely to local increases in cowbird abundance.

Thanks to the nest record cards of the Maryland Ornithological Society, which contain data from the turn of the century to the present time, we can now compare cowbird parasitism rates in Maryland over different periods (Table 29.6). To the historic records we have added our observations for the period 1988–1993 for comparison. The four species shown are representative of the increase in parasitism rates in Maryland. Parasitism rates for Wood Thrush and especially Hooded Warbler in our small study sites are higher than nest records from earlier dates. However, the number of cases of multiple parasitism is still small and may be a

problem only for the Hooded Warblers and in North Belt, where Wood Thrush density is low. The rates at our two large Patuxent sites continue to be very low, as in earlier years, even though cowbirds feed frequently along nearby roadsides and mowed areas. We repeated the survey at the Patuxent 1 site in 1994 and again found no parasitized nests, but observed cowbird fledglings with Red-eyed Vireos, Scarlet Tanagers, and an Ovenbird.

Effects of Parasitism and Predation on Nest Success

There has been a long-standing misunderstanding of the severity of the cowbird problem for some of the scarcer host species. Friedmann's classic work, *The Cowbirds* (1929), which has been widely cited, described cowbird impacts on the basis of number of nests reported parasitized without reference to the total number of nests examined. Thus, a species whose nest is rarely found was assumed to be an uncommon victim, whereas the rate of parasitism may have been very high. Table 29.7 compares cowbird parasitism rates from Ontario with Friedmann's comments for the same species. Most species in Table 29.7 actually suffer much more from parasitism than suggested by Friedmann's assessments.

For Wood Thrush the direct effect of the cowbird is to reduce the number of host fledglings (Trine, Chapter 15, this volume). For smaller birds such as Hooded Warbler the effect is much more serious, as parasitized nests usually produce only cowbird fledglings. The Maryland Breeding Bird Atlas (Robbins et al. 1996) shows Prince Georges County to have the highest occurrence rate of the Hooded Warbler in Maryland, yet our study and data from an adjacent county (D. Niven pers. comm.) indicate that parasitized Hooded Warblers rarely raise their own young.

Predation was our greatest source of nest failure. Nests in four of the seven sites experienced greater than 60% predation, including one of the extensive forest sites at Patuxent. A few nests were lost to bad weather each summer, and one or two were abandoned for unknown reasons. We never actually observed a nest being depredated and therefore have no conclusive information on sources of predation. However, we suspect that most nest loss resulted from avian predators such as Blue Jays, Common Grackles, and American Crows. Other potential predators included raccoons (a

few nests were torn apart), opossums, chipmunks, and black rat snakes. The combination of parasitism and predation is greatly reducing productivity for these forests. Few Wood Thrush young are being produced in any except the South Belt old-growth forest.

Variability among Sites and Years

In an analysis of nest records submitted to the Cornell Laboratory of Ornithology, Hoover and Brittingham (1993) found that rates of cowbird parasitism of Wood Thrush nests differed significantly among geographic regions. Our results indicate that even within a local area rates of nest parasitism can vary among sites and among years. For example, parasitism rates at Watkins dropped from 38% in 1991 ($N = 8$) to 0% in 1992 ($N = 9$), while 3 km away rates at South Belt rose from 8% ($N = 37$) to 41% ($N = 51$). Because we did not monitor nests in all sites through all years of the study, it is not possible to separate directly the effects of site and year on parasitism rates. The highest rates of parasitism in our study were observed in the North Belt site in 1992 and 1993 (Table 29.2), and parasitism rates at two sites, South Belt and Woodmore, were higher during these years than in earlier years of the study. However, additional years of study are required to determine whether these higher rates represent a trend or simply reflect among-year variation within sites. Data from South Belt for 1994 ($N = 49$) indicate that parasitism rates dropped back to 18%, with no nests parasitized after 25 May. In other years parasitism continued into late July. There may be an association be-

Table 29.6. Historical and Recent Parasitism Rates in Maryland from the Maryland Ornithological Society Nest Records

Species	Time Period	% Nests Parasitized	(N)	Location
Wood Thrush	1882–1900	0	(109)	Baltimore County
	1901–1930	6	(31)	Statewide
	1943–1973	0	(23)	PWRC
	1988–1993	20	(467)	Suburban Maryland
Worm-eating Warbler	1885–1968	15	(13)	Statewide
	Mid-1980s	71	(17)[a]	Statewide
Hooded Warbler	1932–1968	11	(19)	Statewide
	1973–1991	55	(40)	Statewide
	1988–1993	88	(17)	Suburban Maryland
	1988–1991	43	(30)[b]	Suburban Maryland
Chipping Sparrow	Pre-1950	1	(90)	Statewide
	1950–1990	12	(151)	Statewide

[a]R. Greenberg, 1980s (unpubl. data from Maryland).
[b]D. Niven, 1994 (unpubl. data from Maryland).

Table 29.7. Historical Records of Parasitism in Ontario and Maryland in Relation to Comments by Friedmann

Species	% Parasitized (N)[a]		Comments by Friedmann (1929)
Golden-winged Warbler	40.6	(32)	Very uncommon victim
Yellow-rumped Warbler	31.1	(122)	Very rarely imposed upon
Cerulean Warbler	18.0	(39)	Very uncommon victim
Northern Waterthrush	13.3	(83)	Rarely victimized
Hooded Warbler	67.0	(6)	Rather uncommon victim
Canada Warbler	20.0	(25)	Very uncommon victim
Prairie Warbler	33.3	(24)	Very uncommon host
Black-throated Green Warbler	34.4	(32)	Very seldom bothered
Black-throated Blue Warbler	13.0	(23)	Decidedly uncommon victim
Worm-eating Warbler	70.6	(17)[b]	Rather uncommonly imposed upon
Black-and-white Warbler	20.9	(43)	Rather uncommon victim
Pine Warbler	50.0	(8)	Rare victim
Yellow-throated Vireo	50.0	(44)	Common victim

[a]Peck and James 1987.
[b]R. Greenberg, 1980s. (Unpubl. data from Maryland).

tween cowbird fledging success and parasitism rates the following year, which could explain in part the among-year variability in parasitism rates at sites. More years of nest monitoring are necessary to determine if a significant correlation exists. Also needed are color-banding or genetic studies that establish the relationships between cowbird hatchlings and the adult cowbirds that parasitize the same study sites in subsequent breeding seasons (see Raim, Chapter 9, this volume).

Edge Effects

There is a strong indication (Figure 29.3) that nesting Wood Thrushes tend to avoid edges. In the study site with the largest sample of nests (South Belt, N = 233) there was a high correlation between nest density and distance from edge. In the North Belt tract (N = 33 nests) the relation was less clear, as would be expected from the smaller sample.

Parasitism rates declined beyond 75 m from edge at South Belt but not at North Belt. In the two largest sites (Patuxent 1 and 2) the number of parasitized nests (2 and 0, respectively) was too small to permit a meaningful separation with respect to distance from edge. In view of the differences among sites, a larger sample of nests will be necessary to establish whether distance from edge is a consistent factor affecting parasitism of Wood Thrush nests.

Within-site Characteristics and Surrounding Habitats

Vegetation structure may influence the placement and visibility of nests and, as a result, their vulnerability to parasi-

tism and predation. Most Wood Thrush nests in South Belt were in dogwood, an understory tree that is less abundant in other sites. The appearance in recent years of dogwood anthracnose, which reduces vitality and causes mortality in dogwoods, could be an important factor in the increase in cowbird parasitism in this site. In addition, the high predation rates at the two Patuxent sites may be attributable to the openness of the forest resulting from the removal of all the vegetation below the deer browse line.

Furthermore, cowbird abundance and rates of nest parasitism in forests are likely to be influenced by extrasite characteristics. All of our sites are located in a suburban-rural matrix, which is increasingly becoming dominated by residential developments. As development continues in this and other metropolitan regions in the eastern United States, much of the remaining forest will be restricted to isolated tracts smaller in area than the forests we studied. Maintenance of breeding populations of forest birds in landscapes such as these will require not only the preservation of suitable nesting habitat but also an understanding of the distribution and dynamics of cowbird populations. Of particular interest is the extent to which suburban environments can support populations of cowbirds, which are generally associated with agricultural habitats (e.g., Thompson 1994). Knowledge of whether lawns and bird feeders provide adequate food sources for cowbirds and whether any bird species adapted to nesting in suburban habitats provide suitable alternative hosts is critical to the development of conservation plans for forest-nesting birds.

Management Recommendations

We reiterate some of the excellent recommendations of Robinson et al. (1993) for land planners and forest managers and add our own:

(1) Land planners and managers should strive to prevent fragmentation of forests, especially when this results in a close admixture of agricultural land, suburban developments, and forest remnants, as these conditions increase cowbird populations. The increase in parasitism in South Belt Woods coincided with conversion of adjacent forest to a suburban housing development. Also, our highest parasitism rates were in North Belt Woods, which is primarily surrounded by agricultural fields. Maintenance of breeding populations of forest birds in urban-suburban areas may require management to make the surrounding habitats less attractive as feeding areas for cowbirds. Thompson (1994) stated that most cowbirds feed within 2 km of their nesting sites and recommended providing core areas more than 2 km from potential feeding habitats. Reducing the amount of pastureland and row crops within 2 km of key forest areas may be an effective way of discouraging cowbirds during the nesting season.

(2) Within forests, avoid any practice that creates cowbird feeding opportunities, such as mowing roadsides or along trails.

(3) Where possible, managers should seek to maintain and establish large areas of contiguous forest cover that include core areas of forest interior. In areas where large forested areas cannot be acquired, the quality of the habitat may be improved by allowing the forest to mature to promote higher structural diversity. The high carrying capacity and productivity of the old-growth forest in South Belt Woods appears to compensate in part for its small size. The two episodes of selective logging in the North Belt Woods may have contributed to the very high parasitism and predation rates and greatly reduced the density of nesting Wood Thrushes. Deer populations should be controlled before overbrowsing reduces the structural diversity of the forests. Overbrowsing may also become a problem in small forests as open space continues to be reduced by development.

(4) The health of forests needs to become a higher priority. Loss of oaks from gypsy moths, elms from Dutch elm disease, dogwoods and sycamores from anthracnose, hemlocks from woolly adelgid, and damage to oaks, sycamores, red maples, elms, and other trees from the bacterium *Xylella fastidiosa* may seriously reduce the ability of our forests to support the present diversity of nesting birds. In South Belt Woods, the dogwood anthracnose is eliminating dogwoods, the nesting substrate for over 90% of the Wood Thrush nests.

Future Research Needs

To help develop sound management guidelines more information is needed on (1) the effects of surrounding land-

scapes on parasitism and predation, (2) the sources of predation, (3) the number of young produced and survival of young, and (4) cowbird breeding dynamics.

Acknowledgments

We thank Dan and Lisa Petit for the 1993 Woodmore data, Michael O'Brien for field assistance, and Grey Pendleton and Jeff Hatfield for statistical advice. We also greatly appreciate the editorial comments from Scott Robinson and Jamie Smith.

APPENDIX
Common and Scientific Names of Birds, Mammals, and Reptiles

Mourning Dove, *Zenaida macroura*
Acadian Flycatcher, *Empidonax virescens*
Blue Jay, *Cyanocitta cristata*
American Crow, *Corvus brachyrhynchos*
Wood Thrush, *Hylocichla mustelina*
Gray Catbird, *Dumetella carolinensis*
Yellow-throated Vireo, *Vireo flavifrons*
Red-eyed Vireo, *V. olivaceus*
Golden-winged Warbler, *Vermivora chrysoptera*
Black-throated Blue Warbler, *Dendroica caerulescens*
Yellow-rumped Warbler, *D. coronata*
Black-throated Green Warbler, *D. virens*
Pine Warbler, *D. pinus*
Prairie Warbler, *D. discolor*
Cerulean Warbler, *D. cerulea*
Black-and-white Warbler, *Mniotilta varia*
Worm-eating Warbler, *Helmitheros vermivorus*
Ovenbird, *Seiurus aurocapillus*
Northern Waterthrush, *S. noveboracensis*
Kentucky Warbler, *Oporornis formosus*
Hooded Warbler, *Wilsonia citrina*
Canada Warbler, *W. canadensis*
Scarlet Tanager, *Piranga olivacea*
Northern Cardinal, *Cardinalis cardinalis*
Chipping Sparrow, *Spizella passerina*
Common Grackle, *Quiscalus quiscula*
Brown-headed Cowbird, *Molothrus ater*
Eastern Raccoon, *Procyon lotor*
Opossum, *Didelphis virginiana*
Eastern Chipmunk, *Tamias striatus*
Black Rat Snake, *Elaphe obsoleta obsoleta*

References Cited

Andrén, H., and P. Angelstam. 1988. Elevated predation rates as an edge effect in habitat islands: Experimental evidence. Ecology 69:544–547.

Bart, J., and D. S. Robson. 1982. Estimating survivorship when the subjects are visited periodically. Ecology 63:1078–1090.

Brittingham, M. C., and S. A. Temple. 1983. Have cowbirds caused forest songbirds to decline? BioScience 33:31–35.

Friedmann, H. 1929. The cowbirds: A study in the biology of social parasitism. C. C. Thomas, Springfield, IL.

Hines, J. E., and J. R. Sauer. 1989. Program CONTRAST, a general program for the analysis of several survival or recovery rate estimates. U.S. Fish and Wildlife Service Tech. Rep. 24:1–7.

Hoover, J. P., and M. C. Brittingham. 1993. Regional variation in cowbird parasitism of Wood Thrushes. Wilson Bull. 105:228–238.

Hosmer, D. W., Jr., and S. Lemeshow. 1989. Applied logistic regression. John Wiley and Sons, New York.

Lehman, E. L. 1975. Nonparametrics: Statistical methods based on ranks. Holden-Day, San Francisco, CA.

Mayfield, H. F. 1975. Suggestions for calculating nest success. Wilson Bull. 87:456–466.

O'Brien, M., and B. A. Dowell. 1990a. Breeding bird census: Mature tuliptree-oak forest. J. Field Ornithol. 61 (Suppl.): 32–33.

———. 1990b. Breeding Bird Census: Isolated tuliptree-oak forest. J. Field Ornithol. 61 (Suppl.):33–34.

Peck, G. K., and R. D. James. 1987. Breeding birds of Ontario: Nidiology and distribution vol. 2, Passerines. Royal Ontario Museum, Toronto.

Robbins, C. S. 1989. Breeding Bird Census: Selectively logged mature tuliptree-oak forest. J. Field Ornithol. 60 (Suppl.):71–72.

———. 1990. Breeding Bird Census: Isolated moist tuliptree-red maple upland forest. J. Field Ornithol. 61 (Suppl.):35–36.

———. 1991. Breeding Bird Census: Mature beech- maple-oak bottomland forest. J. Field Ornithol. 62 (Suppl.):37–38.

———, Sr. Editor. 1996. Breeding Bird Atlas of Maryland and the District of Columbia. University of Pittsburgh Press, Pittsburgh, PA.

Robbins, C. S., D. Bystrak, and P. H. Geissler. 1986. The breeding bird survey: Its first 15 years, 1965–1979. USDI Fish and Wildlife Service Res. Publ. 157.

Robbins, C. S., D. K. Dawson, and B. A. Dowell. 1989. Habitat area requirements of breeding forest birds of the middle Atlantic states. Wildlife Monogr. 103:1–34.

Robinson, S. K. 1989. Population dynamics of breeding neotropical migrants in a fragmented Illinois landscape. Pp. 408–415 in Ecology and conservation of neotropical migrant landbirds (J. M. Hagan III and D. W. Johnston, eds). Smithsonian Institution Press, Washington, DC.

Robinson, S. K., J. A. Grzybowski, S. I. Rothstein, M. C. Brittingham, L. J. Petit, and F. R. Thompson. 1993. Management implications of cowbird parasitism on neotropical migrant songbirds. Pp. 93–102 in Status and management of neotropical migratory birds (D. M. Finch and P. W. Stangel, eds). USDA Forest Serv. Gen. Tech. Rep. RM-229.

Statxact 1989, 1991. CYTEL Software Corporation, Cambridge, MA 02139.

Stewart, R. E., and C. S. Robbins. 1947. Breeding bird census: Virgin central hardwood deciduous forest. Audubon Field Notes 1:211–212.

Thompson, F. R. III. 1994. Temporal and spatial patterns of breeding Brown-headed Cowbirds in the midwestern United States. Auk 111:979–990.

Whitcomb, B. L., D. Bystrak, and R. Whitcomb. 1977a. Breeding bird census: Selectively logged mature tuliptree oak forest. Amer. Birds 31:92–93.

Whitcomb, B. L., R. F. Whitcomb, and D. Bystrak. 1977b. Long-term turnover and effects of selective logging on the avifauna of forest fragments. Amer. Birds 31:17–23.

Wilcove, D. S. 1985. Nest predation in forest tracts and the decline of migratory songbirds. Ecology 66:1211–1214.

30. Cowbird Distribution at Different Scales of Fragmentation:

Trade-offs between Breeding and Feeding Opportunities

THERESE M. DONOVAN,

FRANK R. THOMPSON III, AND JOHN R. FAABORG

Abstract

The distribution of Brown-headed Cowbirds should reflect the distribution of their feeding (agricultural or grassy areas) and breeding (host) resources. Because an increase in one resource (e.g., agricultural areas) is often at the expense of the second resource (forest hosts), relationships between cowbird abundance, forest area, and number of hosts may reflect this trade-off. We studied cowbird distribution and abundance in the extensively forested Missouri Ozarks and in fragmented central Missouri. Cowbirds were more abundant on fragments than on unfragmented Ozark study areas, even though hosts were more abundant in the Ozarks. In the Ozarks, there was no relationship between cowbird and host abundance, possibly because cowbirds there were limited more by feeding habitat than by hosts. In contrast, cowbird abundance on fragments was positively related to host abundance, possibly because cowbirds there were limited more by hosts than by food.

Although cowbirds frequently occur in fragmented landscapes, their reproductive success ultimately depends on the ability of hosts to fledge cowbirds successfully. We examined the nesting success of cowbirds in habitats of varying sizes and shapes. The number of cowbird eggs per nest increased as forest size decreased, but daily survival of cowbirds in host nests increased as forest size increased. Thus, a second type of trade-off occurred in that more eggs were laid per nest in fragments where fledging success was relatively low, and fewer eggs were laid per nest in unfragmented forests where fledging success was relatively high.

We examined habitat characteristics at varying spatial scales (1-, 3-, 5-, and 10-km radius circles) to determine what habitat scale best explains cowbird distribution and abundance. Abundance was most strongly related to percentage of forest cover and forest perimeter-to-area ratio at the 3–5-km radius scale. These results suggest that in addition to local-scale factors, Brown-headed Cowbirds may be regulated by habitat characteristics at the landscape scale and that any future cowbird population control should incorporate land management at spatial scales of more than 3 km.

Introduction

A species' distribution and abundance often reflects the distribution of its resources (Brown 1984). For Brown-headed Cowbirds, food resources are distributed in open grassy or agricultural areas whereas breeding resources (hosts) are often distributed in forested areas (Rothstein et al. 1984, Thompson 1994; Thompson and Dijak, Chapter 10, this volume). The probability that a cowbird occurs in a forest therefore depends at least partly on the probability that a feeding area is nearby. As areas become more forested, cowbird breeding opportunities may increase but feeding opportunities may decline. Conversely, as forest habitat is converted to agricultural habitat, feeding opportunities may increase but breeding opportunities may decrease because cowbirds parasitize grassland and shrubland hosts less frequently than forest hosts (Robinson et al., Chapter 33, this volume). The occurrence of cowbirds within landscapes that vary in the amount of forest and agricultural areas may reflect this apparent trade-off between breeding and feeding resources.

A second type of trade-off is also expected to occur within landscapes of varying forest cover. Cowbird production ultimately depends on the nesting success of host species (Lowther 1993). Cowbird hosts on small, fragmented forests often experience higher nest predation than hosts on large, unfragmented forests (Wilcove 1985, Askins et al. 1990, Robinson 1992, Donovan et al. 1995, Robinson et al. 1995). In fragmented landscapes, the distribution of breeding and feeding resources may be optimal for cowbirds, but these landscapes are also suitable for many nest predators (Dijak 1996, Donovan et al. 1997). Thus, cowbirds that lay their eggs in nests within a highly fragmented landscape may ex-

perience low nesting success if host nests are frequently depredated. Such fragments may constitute cowbird population
sinks, where reproduction does not compensate for adult
mortality (Pulliam 1988). In contrast, cowbirds in large, unfragmented forests may be food limited, but depredation in
large forest tracts is low, and cowbirds that lay eggs there may
be more successful than cowbirds in fragmented forests.

Our goal was to document cowbird distribution and compare cowbird reproductive success in landscapes that vary in
the amount of forest and nonforest cover. Our objectives
were to (1) compare cowbird abundance in fragmented central Missouri forests with that in the extensively forested
Ozark region, (2) determine if cowbird abundance is related
to the number of hosts (breeding resources) in fragmented
and unfragmented habitats, (3) compare the number of
cowbirds per host nest and the nesting success of cowbirds
in forests of varying size, and (4) examine the relationship
between cowbird abundance and the distribution of forested
(breeding) and nonforested (feeding) areas at several landscape scales.

Methods

Cowbird Distribution and Abundance

We studied Brown-headed Cowbird distribution and parasitism on seventeen study plots within seven forest tracts of
varying size and shape in Missouri from 1991 to 1993 (Table
30.1). A total of nine fragmented study plots were situated
in the highly fragmented central Missouri landscape, and
eight study plots were situated within the heavily forested
Ozarks in southeastern Missouri (Figure 30.1). In 1991,
nine fragmented plots and six contiguous forest plots were
studied. In 1992, two additional plots were added within
contiguous forest, and one fragmented plot was dropped
from the study. Study plots were approximately 22 ha. Forest
tracts containing study plots ranged from 7.4 to 18,258 km²
(Table 30.1). Plots were located within mature oak-hickory
forest and appeared to be homogeneous in forest structure
(Wenny et al. 1993). The nonforested portion of these landscapes was predominantly cool-season pasture.

We gridded each study plot in 150-m intervals. Grids
were established by randomly selecting a point within the
forest and situating a grid around that point. Fifteen points
along grid intersections were designated "counting points."
All counting points were located more than 70 m from an
ecotonal edge between forest and nonforest habitat.

Within each plot, we surveyed abundances of Brown-
headed Cowbirds and potential hosts by 10-min point
counts (Verner 1988) at the 15 counting points. Each point
was counted four times during the breeding season. Three
to four different observers conducted counts each year according to a protocol described in detail in Donovan et
al. (1997) that minimized effects of observer variability
(Verner and Milne 1989). Counts began after most territories were established (after 5 May) and ended by mid-June
when most nests had fledged young.

In each 10-min count, bird detections were recorded
within 50-m and 70-m fixed-radius circles, as well as total
(unlimited distance) detections. The mean number of detections in each distance class was computed for each plot in
each year. Mean detections at a plot were based on 15 counting points that were censused 4 times within a season
($N = 60$ counts). We selected the appropriate distance class
(50 m, 70 m, or unlimited distance) for analysis based on
univariate F-tests. In this chapter, we used the unlimited
distance class in all analyses because it yielded the highest
F-value in discriminating Brown-headed Cowbirds on fragments and contiguous forests (Bradley and Schumann
1957). However, results were similar for the 50-m and 70-m
distance classes.

Female cowbirds were differentiated from males based on
rattle or chatter calls (Darley 1968). Because female cowbirds are responsible for parasitism and its consequences for
host species, we present point-count results for female cowbird detections alone and for all cowbird (male and female)
detections. Caution must be used in interpreting female detections based solely on rattle calls because the social context in which these calls are given and how they influence
detectability are poorly understood (Lowther 1993; Rothstein et al., Chapter 7, this volume).

We compared cowbird abundance on fragmented ($N = 9$
plots) and contiguous forests ($N = 8$ plots) using a repeated
measures analysis of variance, with landscape (fragmented
or contiguous) as a main effect and year as a repeated effect.
Replicate plots within a landscape were used as the error
term.

Relationship of Host Abundance and Cowbird Abundance

Host abundance was surveyed in the same manner as for
Brown-headed Cowbirds, and records consisted predominantly of singing males. We identified hosts as those species
that bred during the time Brown-headed Cowbirds were
censused and received Brown-headed Cowbird eggs in over
10% of their nests (based on nests located within the study
plots and parasitism rates in the literature). Possible errors
in host detection may have occurred because hosts vary in
their detectability and because females of some species may
sing (e.g., Northern Cardinal, *Cardinalis cardinalis*). We
summed the number of potential hosts at a given plot in a
year over all host species, and used all observations of hosts
to compute mean host abundance for each plot.

We compared host abundance on fragmented ($N = 9$
plots) and contiguous forests ($N = 8$ plots) using a repeated
measures analysis of variance, with landscape (fragmented
or contiguous) as a main effect and year as a repeated effect.
Plots within a landscape were used as the error term.

Simple linear regressions were used to determine the relationship between cowbird abundance and host abundance

Table 30.1. Description of the Seven Forest Tracts Containing 17 Study Plots

Forest	Plots[a]	Sites[b]	Type[c]	Area (km²)	Perimeter (km)	Perimeter-to-Area Ratio	Forest Area[d] $r = 1$ km	Forest Area[d] $r = 3$ km	Forest Area[d] $r = 5$ km	Forest Area[d] $r = 10$ km
Ashland	2	1	frag	40.6	142.5	3.5	2.5	18.9	41.4	132.3
Bennitt	1	1	frag	20.2	51.9	2.6	3.1	17.8	41.4	114.9
County J	2	1	frag	7.9	37.1	4.7	1.4	7.7	22.4	72.7
Fulton	1	1	frag	7.4	32.7	4.4	2.3	6.3	15.0	63.2
Hungry Mother	1	1	frag	24.9	82.2	3.3	2.5	16.8	37.1	103.1
Whetstone	2	1	frag	39.0	150.8	3.9	2.6	14.0	28.6	105.9
Ozarks	8	4	cont	18,258.0	16,842.9	0.9	3.1	27.3	75.7	289.4

[a]Number of study plots situated within a particular forest tract.
[b]Number of sites within a forest tract.
[c]Forest types are fragmented plots (frag) within central Missouri or contiguous plots (cont) within the Ozarks region.
[d]Areas are in square kilometers.

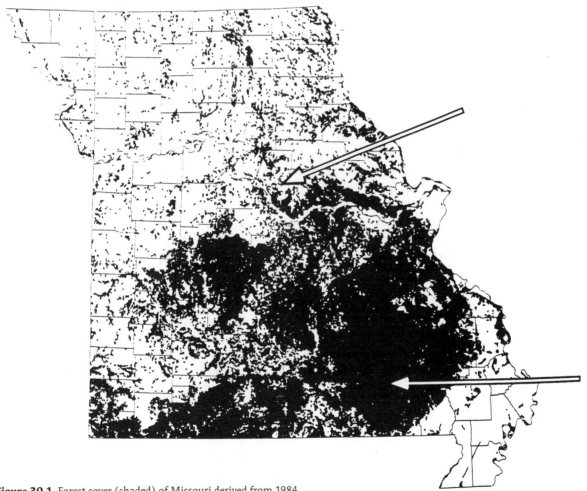

Figure 30.1. Forest cover (shaded) of Missouri derived from 1984 Transverse Mercator scenes showing general locations of study plots within fragmented central Missouri ($N = 9$ study plots) and unfragmented Ozarks in southeastern Missouri ($N = 8$ study plots).

at the plot level for each landscape. In these analyses, cowbird and host abundance at a plot were considered separately for each year because abundance of each varied between years (thus, $N = 17$ for the fragmented landscape, $N = 14$ for the contiguous landscape). Separate analyses were used because host abundance differed between the two landscapes examined.

Reproductive Success of Cowbirds in Landscapes of Varying Forest Cover

Nests that contained cowbird eggs were located in 1991–1993 on four forest tracts of varying size and were monitored every three to five days. These forest tracts included the Ashland, Hungry Mother, and Bennitt fragments and the unfragmented Ozarks (Table 30.1). For each parasitized nest, we recorded the number of cowbird eggs, chicks, and fledglings. We determined the mean number of cowbirds per nest in each forest, and used Spearman's rank correlation to determine if there was an association between the number of cowbird eggs per nest and log forest area.

We calculated daily survival of cowbirds on a per-egg basis (Mayfield 1975) and used those estimates to determine the probability that a cowbird egg would survive to fledging (nesting success). In multiply parasitized nests, these estimates may be upwardly biased because cowbirds may remove conspecific eggs. Because the incidence of multiple parasitism is related to forest size (see below), we considered this potential bias as unimportant because it would bias *against* showing an effect of forest size on cowbird nesting success. We calculated survival days across host species within a site to determine cowbird daily survival for each forest. This approach was necessary because sample sizes of parasitized nests on a per-host basis were limited. However, all hosts were forest-interior species and have similar responses to parasitism. In addition, the composition of parasitized hosts was fairly consistent among the four forests we studied for this analysis, with the large Wood Thrush being the most frequently parasitized host in three of the four plots (Table 30.2). Nevertheless, caution should be used in interpreting cowbird survival data from a combination of host nests, because hosts vary in their ability to fledge cowbird young; future studies should include effects of host quality. We used Spearman's rank correlation to determine if there was an association between cowbird daily nest survival and log forest area.

Landscape Patterns and Cowbird Abundance at Selected Spatial Scales

Because cowbird distribution may represent a trade-off between feeding and breeding resources, we examined landscape characteristics at different spatial scales to determine the habitat scale that best explains cowbird distribution. We used an existing forest-cover GIS database developed by the Missouri Department of Conservation and the Geographic

Resources Center at the University of Missouri–Columbia (Giessmann et al. 1986) to compute landscape statistics at 1-, 3-, 5-, and 10-km radius spatial scales. The distribution of forestland in Missouri was determined from sixteen Landsat Satellite Thematic Mapper photographic images obtained during the 1984 growing season (Giessmann et al. 1986). Areas of more than 2 ha with at least 10% canopy cover were considered forested habitat and were digitized using an analytical mapping system. Files were incorporated into a GIS, and Map Overlay Statistical System (MOSS) was used to manage digitized data to produce forest area estimates and maps.

We located all seventeen study plots within the GIS database. Some fragmented study plots were situated within the same forest polygon; all plots within the heavily forested Ozarks were situated within the same forest polygon (Table 30.1). For fragments, we averaged cowbird detections among plots that were located within the same forest polygon for analyses. As a result, six forest tracts were evaluated in the fragmented landscape. Although all eight contiguous plots were located within the unfragmented Ozarks, these plots were spatially located as four paired plots that were separated by more than 5 km, and thus the four pairs were considered as independent. We averaged cowbird detections of paired plots for analyses ($N = 4$). For clarity, these ten forests ($N = 6$ fragments and 4 contiguous) will be called sites; sites consisted of 1–2 study plots (Table 30.1).

We calculated the percentage of forest cover and perimeter-to-area ratio within 1-, 3-, 5-, and 10-km radius circles on each site (Figure 30.2 and Table 30.1). We evaluated the relationship between cowbird abundance at a site ($N = 10$) and percentage of forest cover and perimeter-to-area ratios at these scales. This scale of evaluation is appropriate because female cowbirds move up to 10 km between feeding and breeding areas within a breeding season in this region (Thompson 1994; Thompson and Dijak, Chapter 10, this volume).

We used simple, univariate regression models to determine if linear relationships existed between cowbird abundance, cowbird survival, and percentage of forest cover and perimeter-to-area ratio across sites at the 1–10-km radius scales. We examined the adjusted r^2 for each model to determine the scale (1-, 3-, 5-, or 10-km radius circles) at which the linear relationship was strongest.

Results

Cowbird Distribution and Abundance

Brown-headed Cowbirds occurred much more frequently on fragmented central Missouri plots than on extensively forested Ozark plots (Table 30.3). Results were similar for analyses based solely on female cowbird detections (Table 30.3). Additionally, cowbirds increased between 1991 and 1992 (repeated measures analysis of variance main effect of year,

Table 30.2. Numbers of Parasitized Nests Located for Each Host Species in Each Forest

Forest	ACFL	BAWW	INBU	KEWA	NOCA	OVEN	REVI	WEWA	WOTH	Total
Ashland	1 (1)	0 (0)	1 (1)	4 (5)	1 (3)	2 (6)	0 (0)	2 (4)	11 (31)	22 (51)
Bennitt	1 (1)	0 (0)	0 (0)	1 (3)	1 (1)	0 (0)	0 (0)	3 (6)	2 (6)	8 (17)
Hungry Mother	1 (1)	0 (0)	0 (0)	0 (0)	1 (2)	1 (2)	1 (3)	4 (10)	9 (24)	17 (42)
Ozarks	3 (3)	1 (1)	0 (0)	0 (0)	0 (0)	1 (1)	3 (3)	2 (2)	1 (1)	11 (11)

Note: Total number of cowbird eggs or young detected in parasitized nests of each host species is shown in parentheses. Abbreviations: ACFL, Acadian Flycatcher. BAWW, Black-and-white Warbler. INBU, Indigo Bunting. KEWA, Kentucky Warbler. NOCA, Northern Cardinal. OVEN, Ovenbird. REVI, Red-eyed Vireo. WEWA, Worm-eating Warbler. WOTH, Wood Thrush.

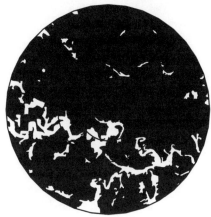

Figure 30.2. Images of a fragmented central Missouri study site (top) and a contiguous forest southeastern Missouri study site (bottom) showing the forest cover (shaded) within 1-, 3-, 5-, and 10-km radius circles.

$F = 18.00$, $P = .001$), and this increase was greater on fragments than in contiguous forests (landscape × year interaction, $F = 15.98$, $P = .002$).

Relationship of Host Abundance and Cowbird Abundance

Although cowbirds were more abundant on fragments, hosts were more abundant in unfragmented habitat (Table 30.3). Cowbird abundance at a plot was related to the abundance of potential hosts at a plot, but this relationship depended on the forest landscape in question. On fragmented study plots, cowbirds (males and females combined) were positively related to the number of hosts (Figure 30.3). This trend was not significant when examining female cowbirds alone ($F = 2.70$, $P = .12$, $r^2 = 0.153$). In contrast, cowbird (males and females combined) and host abundances were not related in unfragmented Ozark plots ($F = 3.11$, $P = .10$, $r^2 = 0.21$).

Reproductive Success of Cowbirds in Landscapes of Varying Forest Cover

We analyzed the incidence of multiple parasitism and nest survival on four forests of varying size. In the Ozarks, parasitism was almost nonexistent; of more than 500 host nests located in the Ozarks, only 11 nests (six host species) were parasitized (Faaborg and Clawson unpubl. data). On the three fragments, 47 total nests (eight host species) containing 110 cowbird eggs or young were located (Table 30.2). All nests in the Ozarks were singly parasitized, whereas most nests in fragments were multiply parasitized (mean = 2.55 cowbird eggs/parasitized nest, SD = 1.35; Wilcoxon two-sample test, $Z = -3.76$, $P = .000$). Of the eight host species on fragments, nests of six hosts were multiply parasitized. In the three fragments studied, parasitized Wood Thrush nests contained the majority (> 50%) of the cowbird eggs located. Figure 30.4 shows that the number of cowbird eggs per parasitized nest increased as forest size decreased (Spearman's rank correlation, $r_s = -1.00$, $P = .000$). In contrast, cowbird nesting success (the probability that a cowbird egg would survive to fledging) increased as forest size increased ($r_s = 1.00$, $P = .000$).

Table 30.3. Repeated Measures ANOVA at a Study Plot Showing Main Effect
of Landscape

Detections	Fragments			Contiguous			
	Mean	SD	N	Mean	SD	N	P
Mean BHCO[a]	0.95	0.32	17	0.27	0.12	14	.000
Mean female[b]	0.20	0.12	17	0.04	0.02	14	.000
Mean host	5.66	1.32	17	6.88	0.78	14	.004

[a]Female and male cowbird observations.
[b]Female cowbird observations only.

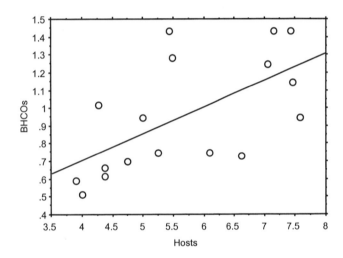

Figure 30.3. Relationship between cowbirds (male and female)
and hosts on fragmented study plots in central Missouri. Cowbird
abundance = 0.15 (number of hosts) + 0.096; $F = 9.59$, $P = .007$, r^2
= 0.39. Because cowbird abundance varied with years, the relationship of cowbirds to hosts was calculated separately for each year
($N = 9$ plots in 1991 and 8 plots in 1992).

Landscape Patterns and Cowbird Abundance at Selected Spatial Scales

Cowbird abundance (male and female) was not related to
forest area at the 1-km scale ($P = .073$) but was negatively related to forest area at the 3-, 5-, and 10-km scales ($P = .002$,
.001, and .000, respectively), as shown in Table 30.4. The variation in cowbird abundance was best explained by the variation in percentage of forest cover at the 10-km scale (adj.
$r^2 = 0.854$). The abundance of female cowbirds was negatively related to percentage of forest cover at all spatial scales
analyzed (Table 30.4). The variation in female cowbirds was

best explained by variation in percentage of forest cover at
the 3-km and 5-km scales (adj. $r^2 = 0.93$ for both scales).

Cowbird abundance was also related to the perimeter-to-
area ratio of the landscape in question (Table 30.4). The variation in cowbird abundance was positively related to the
variation in perimeter-to-area ratio at the 3-, 5-, and 10-km
scales ($P = .009$, .002, and .001, respectively) but not at the
1-km scale ($P = .117$). Female cowbirds showed a similar pattern but were significantly related to perimeter-to-area ratios
at all scales examined. As with forest area, the relationship
between cowbird abundance and habitat edge was best explained at spatial scales greater than 3 km (Table 30.4).

Host abundance was positively related to percentage of forest cover and negatively related to perimeter-to-area ratio at the 3-, 5-, and 10-km scales ($P = .04, .03, .04$, and $P = .03, .03, .04$, respectively), as shown in Table 30.4. Host abundance was not related to percentage of cover or perimeter-to-area ratio at the 1-km scale ($P = .08$ and $.18$, respectively).

Discussion

Cowbird Distribution and Abundance

Cowbirds evolved in open grasslands where their breeding and feeding resources overlapped spatially (Lowther 1993), but presettlement populations were potentially limited because many sympatric hosts evolved strategies against parasitism (Briskie et al. 1992). In the past 200 years, however, cowbirds have benefited tremendously by the clearing of forests for agricultural purposes (Brittingham and Temple 1983). These changes in landscape have increased feeding resources (agriculture) and introduced new breeding resources (naive hosts) that were previously inaccessible to cowbirds.

Telemetry studies in Missouri and New York have shown that although feeding and breeding resources can overlap spatially, cowbirds often move considerable distances between breeding and feeding areas (Thompson 1994; Hahn and Hatfield, Chapter 13, this volume). In Missouri, female cowbirds tend to breed in host-rich forests in the early morning and move to open grassy or agricultural areas to feed later in the day (Thompson 1994). In the Ozarks, feed-

ing areas are limited and there may be costs (energetic, behavioral) in moving longer distances to sites with abundant hosts. In contrast, mixed forest and agricultural landscapes in central Missouri possess both breeding and feeding resources for cowbirds; the distribution of these two resources in fragmented landscapes may be more favorable for cowbirds than in heavily forested landscapes.

At some point, however, severely fragmented forests within agricultural landscapes may lack sufficient hosts. There may be costs in traveling between breeding and feeding areas in such landscapes. The fragmented landscape in central Missouri averaged approximately 32% forest cover at the 10-km scale (8% core habitat, Donovan et al. 1995). Thus, the fragmented landscape we studied was perhaps too forested to show an effect of host limitation. We suggest that future studies focus on determining the distribution of cowbirds across an even broader spectrum of forest-field landscapes.

Trade-offs in Host Abundance, Cowbird Abundance, and Reproductive Success

Although cowbirds breed in a wide variety of habitats, several studies suggest that cowbirds select habitats with high host densities (Rothstein et al. 1986, Verner and Ritter 1983; Thompson et al., Chapter 32, this volume). However, this relationship may depend on whether feeding resources are nearby. In our study, cowbird abundance was not related to host abundance in large unfragmented forests, despite an abundance of breeding opportunities, possibly because cowbirds were limited more by feeding resources than by host resources. Our study plots were situated well within forest core habitat, buffered from cowbird feeding habitats. In heavily forested landscapes such as the Ozarks, positive associations between host and cowbird abundance may be evident in portions of the forest that are located near agricultural openings.

In many forest habitats throughout the United States, the total number of forest hosts and host densities decrease as forest area decreases (Askins et al. 1990, Wenny et al. 1993), creating an interplay of trade-offs between forest size, host abundance, and cowbird abundance in fragmented landscapes. In our study, as forest size decreased, hosts were less common but cowbirds increased. As forest size increased, host abundance increased but cowbird numbers decreased. Additionally, host nests in fragments were often multiply parasitized and had low nesting success, whereas host nests in the Ozarks were singly parasitized and had higher nesting success.

We suggest that habitat distribution patterns in fragmented landscapes contribute to the incidence of multiple parasitism by increasing the number of female cowbirds that seek limited host resources. We do not believe that increased incidence of multiple parasitism in fragments is a sampling artifact due to differences in host susceptibility to parasitism. First, although we combined host species to examine

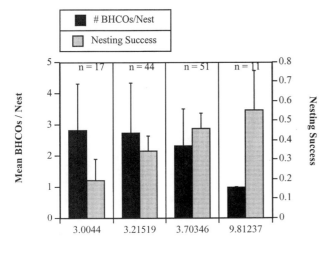

Figure 30.4. Solid bars show the mean number and standard deviation of cowbird eggs per parasitized nest in forests of different areas; stippled bars show nesting success ± SE of cowbirds in forests of different areas.

Table 30.4. Univariate regressions of Cowbird and Host Abundance against Percentage of Forest Cover and Perimeter-to-Area Ratios at Selected Spatial Scales

Dependent Variable	Independent Variable	Scale	Slope	Adj. r^2	P
BHCO abundance	% forest cover	1 km	−1.503	0.266	.073
		3 km	−1.059	0.670	.002
		5 km	−1.106	0.773	.001
		10 km	−1.463	0.854	.000
	Perimeter/area ratio	1 km	+0.154	0.189	.117
		3 km	+0.126	0.547	.009
		5 km	+0.127	0.685	.002
		10 km	+0.129	0.773	.001
Female BHCO abundance	% forest cover	1 km	−0.534	0.678	.002
		3 km	−0.303	0.939	.000
		5 km	−0.296	0.933	.000
		10 km	−0.366	0.850	.000
	Perimeter/area ratio	1 km	+0.053	0.480	.016
		3 km	+0.039	0.900	.000
		5 km	+0.036	0.905	.000
		10 km	+0.033	0.839	.000
Host abundance	% forest cover	1 km	+3.899	0.258	.077
		3 km	+2.148	0.351	.042
		5 km	+2.170	0.378	.034
		10 km	+2.033	0.357	.042
	Perimeter/area ratio	1 km	−0.353	0.117	.177
		3 km	−0.288	0.383	.033
		5 km	−0.273	0.415	.026
		10 km	−0.251	0.370	.037

Note: N = 10 sites.

multiple parasitism and nest survival in forests of varying size, the composition of host nest samples was similar on all fragments, with cowbird eggs being laid primarily in Wood Thrush nests. Second, Wood Thrush nests show the least variation in predation rates in relation to forest size compared to other hosts (Robinson et al. 1995, Chapter 33, this volume). Third, the relationship between multiple parasitism and forest size is the same even when analyzed on a per-species basis (Thompson et al., Chapter 32, this volume). Thus, multiple parasitism appeared to increase as forest size decreased, even though cowbird nesting success was lower there.

Although we have suggested that landscape habitat patterns contribute to multiple parasitism and high predation rates in fragmented landscapes, alternative explanations should be evaluated. An alternative to the landscape hypothesis is that the number of parasites in a nest may influence nesting success more than forest area per se. In our study areas, however, nesting success of cowbirds in singly versus multiply parasitized nests did not differ (Donovan unpubl. data; also see Trine, Chapter 15, this volume). Thus, we suggest that landscape features influence the trade-off between cowbird occurrence, host occurrence, and cowbird nesting success. A clear understanding of such trade-offs requires studying host and cowbird distribution, cowbird philopatry, and cowbird nesting success across many landscapes that vary in their distribution of cowbird breeding and feeding areas. Additionally, comparisons of annual and lifetime fitness of females that parasitize hosts on fragments with females that parasitize hosts on larger forest tracts would be illuminating.

Management Implications

Cowbird Abundance and Habitat Features at Selected Spatial Scales

Our analyses suggest that cowbird occurrence is best explained by forest cover and perimeter-to-area ratio at spatial scales greater than 3 km. Many studies of the distribution of cowbirds have focused on local-scale factors such as distance to edge, vegetation characteristics, size and shape of the study plot, etc. (Robinson et al. 1993). Although local habitat characteristics may influence cowbird distribution (Uyehara and Whitfield, Chapter 24, this volume), we emphasize that cowbird populations may also be regulated by landscape-scale habitat patterns. For female cowbirds, relationships between cowbird distribution and forest cover were the strongest when habitat characteristics were examined 3-5 km from the study site. For all cowbirds, these relationships gained strength as scale increased (Table 30.4). These results suggest that in managing cowbird numbers, habitat characteristics should be evaluated beyond the local scale (Robinson et al., Chapter 33, and Thompson et al., Chapter 32, this volume).

Strong biological reasons may underlie the patterns we

detected. Telemetry studies of female cowbirds in fragmented central Missouri revealed that cowbirds move on average 1-2 km between breeding and feeding areas (range 0.03-7.34 km; Thompson 1994; Thompson and Dijak, Chapter 10, this volume). In these movements, suitable feeding and breeding habitat is apparently bypassed. Because cowbirds can move great distances to optimize their resource use, managing for forest area or perimeter-to-area ratios at scales of less than 3 km may have little impact on cowbird distribution if the surrounding landscape provides optimal cowbird feeding and breeding resources.

Acknowledgments

We are grateful to a number of people who assisted in data collection, including A. Anders, D. Burhans, C. Freeman, K. Loraff, D. Novinger, J. Porath, P. Porneluzi, and K. Winter. R. Clawson and the Missouri Ozark Forest Ecosystem Project provided the Ozark nest data. S. Muller of the Geographic Resources Center at the University of Missouri provided maps and GIS expertise for spatial analyses. G. Krause and R. Semlitsch offered many insights regarding data analysis and interpretation. D. Burhans, C. Galen, P. Jones, M. Puterbaugh, M. Ryan, and R. Semlitsch critically reviewed and improved the manuscript. The USDA Forest Service North Central Forest Experiment Station and NBS Global Change Program, Breeding Biology Research Database (BBIRD), funded this project.

References Cited

Askins, R., J. F. Lynch, and R. Greenberg. 1990. Population declines in migratory birds in eastern North America. Curr. Ornith. 7: 1-57.

Bradley, R. A., and D. E. Schumann. 1957. The comparison of the sensitivities of similar experiments: Applications. Biometrics 13:496-510.

Briskie, J. V., S. G. Sealy, and K. A. Hobson. 1992. Behavioral defenses against avian brood parasitism in sympatric and allopatric host populations. Evolution 46:334-340.

Brittingham, M. C., and S. A. Temple. 1983. Have cowbirds caused forest songbirds to decline? BioScience 33:31-35.

Brown, J. H. 1984. On the relationship between abundance and distribution of species. Amer. Natur. 124:255-279.

Darley, J. A. 1968. The social organization of breeding Brown-headed Cowbirds. Ph.D. Thesis, University of Western Ontario, London, Ontario.

Dijak, W. D. 1996. Landscape characteristics affecting the distribution of mammalian predators. M.S. Thesis, University of Missouri, Columbia.

Donovan, T. M., P. W. Jones, E. M. Annand, and F. R. Thompson. 1997. Variation in local-scale edge effects: Mechanisms and landscape context. Ecology 78:2064-2075.

Donovan, T. M., F. R. Thompson III, J. Faaborg, and J. Probst. 1995. Reproductive success of migratory birds in habitat sources and sinks. Conserv. Biol. 9:1380-1395.

Giessman, N. F., T. W. Barney, T. L. Haithcoat, J. W. Myers, and R. Massengale. 1986. Distribution of forestland in Missouri. Trans. Missouri Acad. Sci. 20:5–20.

Lowther, P. 1993. Brown-headed Cowbird (*Molothrus ater*). *In* The birds of North America, no. 47 (A. Poole and F. Gill, eds). Academy of Natural Sciences, Philadelphia; American Ornithologists' Union, Washington, DC.

Mayfield, H. F. 1975. Suggestions for calculating nest success. Wilson Bull. 87:456–466.

Pulliam, H. R. 1988. Sources, sinks, and population regulation. Am. Nat. 132:652–661.

Robinson, S. K. 1992. Population dynamics of breeding neotropical migrants in a fragmented Illinois landscape. Pp. 408–418 *in* Ecology and conservation of neotropical migrant landbirds (J. H. Hagan III and D. W. Johnston, eds). Smithsonian Institution Press, Washington, DC.

Robinson, S. K., J. A. Grzybowski, S. I. Rothstein, M. C. Brittingham, L. J. Petit, and F. R. Thompson. 1993. Management implications of cowbird parasitism for neotropical migrant songbirds. Pp. 93–102 *in* Proceedings of a workshop on the status and management of neotropical migratory birds (D. M. Finch and P. W. Stangel, eds). USDA For. Serv. Gen. Tech. Rep. RM-229.

Robinson, S. K., F. R. Thompson III, T. M. Donovan, D. R. Whitehead, and J. Faaborg. 1995. Regional forest fragmentation and the nesting success of migratory birds. Science 267:1987–1990.

Rothstein, S. I., J. Verner, and E. Stevens. 1984. Radio-tracking confirms a unique diurnal pattern of spatial occurrence in the parasitic Brown-headed Cowbird. Ecology 65:77–88.

Rothstein, S. I., D. A. Yokel, and R. C. Fleischer. 1986. Social dominance, mating and spacing systems, female fecundity, and vocal dialects in captive and free-ranging Brown-headed Cowbirds. Curr. Ornithol. 3:127–185.

Thompson, F. R. III. 1994. Temporal and spatial patterns of breeding Brown-headed Cowbirds in the midwestern United States. Auk 111:979–990.

Verner, J. 1988. Optimizing the duration of point counts for monitoring trends in bird populations. Pacific Southwest Forest and Range Experiment Station, Forest Service, USDA Res. PSW-395: 1–4.

Verner, J., and K. A. Milne. 1989. Coping with sources of variability when monitoring population trends. Ann. Zool. Fennici 26:191–199.

Verner, J., and L. V. Ritter. 1983. Current status of the Brown-headed Cowbird in the Sierra National Forest. Auk 100:355–368.

Wenny, D. G., R. L. Clawson, J. Faaborg, and S. L. Sheriff. 1993. Population density, habitat selection and minimum area requirements of three forest-interior warblers in central Missouri. Condor 95:968–979.

Wilcove, D. S. 1985. Nest predation in forest tracts and the decline of migratory songbirds. Ecology 66:1211–1214.

31. Brown-headed Cowbird Parasitism of Migratory Birds:
Effects of Forest Area and Surrounding Landscape

LISA J. PETIT AND DANIEL R. PETIT

Abstract

Fragmentation of eastern deciduous forests has decreased available habitat for forest-breeding birds, and it may have increased exposure of individuals to predators and brood parasites. Although predation and parasitism by Brown-headed Cowbirds are commonly cited as explanations for the population declines of many bird species, little empirical support exists for these contentions, especially for cowbird parasitism.

Nests of Acadian Flycatchers ($N = 358$) and Wood Thrushes ($N = 329$) were monitored during 1993 to determine occurrence of cowbird parasitism in seven forest sites in northeastern Ohio and eleven forest sites in central Maryland and Washington, D.C. Forest patches varied in area from 20 to more than 500 ha and were categorized as surrounded primarily by either agricultural land or urban development. Overall parasitism frequencies were similar for both species in both Ohio and Maryland, but some patterns differed between the host species. Frequency of parasitism on Acadian Flycatchers was significantly and inversely related to size of forest fragments, a trend that also was apparent within each region and in both types of landscapes. These patterns were less apparent for Wood Thrushes. Landscape type played little role in the probability of parasitism for both species. These results support the hypothesis that in some species, individuals breeding in small forest fragments are more susceptible to cowbird parasitism than individuals occupying unbroken forest.

Introduction

Removal and conversion of native vegetation types in North America has resulted in the decline of local and regional wildlife populations through direct displacement or mortality of individual animals. Whereas alteration of pristine habitats is universally recognized as one cost of economic expansion and human population growth, land-use plan-
ners, resource managers, and private citizens historically have attempted to mitigate those impacts by creating reserves and parks. Those habitat remnants have ameliorated many of the direct detrimental consequences of habitat loss on wildlife populations by providing, for example, large tracts for area-sensitive mammals, snags for cavity-nesting birds, or refugia for small prey species. However, wildlife ecologists recognize that indirect and cumulative effects of habitat and landscape fragmentation also can contribute substantially to local population extinctions. For instance, whereas small forest reserves may provide suitable breeding habitat for some species of ground- and shrub-nesting birds, extirpation of large predators from those reserves could "release" populations of small- to medium-sized predators and omnivores (e.g., Fonseca and Robinson 1990). Increased numbers of small predators could reduce the nesting success of ground- and shrub-nesting birds (Wilcove and Robinson 1990). Understanding such indirect effects on population viability may help to manage wildlife in disturbed landscapes.

Direct loss of habitat alone cannot account for shrinking populations of neotropical migratory landbirds during the past 25 years (Holmes et al. 1986, Sauer and Droege 1992, Whitham and Hunter 1992). Declines of many of these species may be a consequence of cumulative degradation of habitats and indirect effects of habitat alteration such as increased nest predation. Another indirect effect of habitat disruption, elevated access of the parasitic Brown-headed Cowbird to open and fragmented eastern forests, has been implicated in local and regional declines of many migratory bird species (Mayfield 1965, Brittingham and Temple 1983, Böhning-Gaese et al. 1993).

Although Brown-headed Cowbirds have clearly increased the probability of extinction for several endangered songbirds with highly restricted ranges (Mayfield 1977, Grzybowski et al. 1986), definitive evidence of such an effect for migratory birds as a group is lacking. In fact, the widely presumed relationship between parasitism frequency and forest fragmentation was generally unsubstantiated when the authors of this volume assembled to report on their research in 1993. Furthermore, the loss of eggs or nestlings to predators is usually several times greater than that attributable to cowbird parasitism (Martin 1993), raising doubts over the contribution of cowbird parasitism to declines of songbirds. On the other hand, range expansion of cowbirds since the early 20th century (Part I, this volume) could have upset the population dynamics of migratory birds, stimulating the declines observed in the latter half of the century (May and Robinson 1985). In response to dwindling migratory bird populations, federal, state, and nongovernmental agencies have launched research and management initiatives to explore the ecological effects of cowbird parasitism.

In 1993, we commenced a study in Ohio, Maryland, and the District of Columbia to investigate the influences of forest size, nesting microhabitat, landscape features, and geographic region on reproductive success of forest-breeding birds. Brown-headed Cowbird nest parasitism is a major aspect of the study. The purpose of this chapter is to assess the relationship between two aspects of the physical environment (forest patch size and composition of the surrounding landscape) and nest parasitism levels in Acadian Flycatchers (*Empidonax virescens*) and Wood Thrushes (*Hylocichla mustelina*) breeding in forest fragments within highly managed landscapes.

Study Sites and Methods

Eighteen forest fragments were selected in the Coastal Plain and Western Shore physiographic regions of Maryland (see Figure 1 in Robbins et al. 1989) and District of Columbia (hereafter, Maryland; $N = 11$), and in the Allegheny Plateau and Ohio Hills regions (see Figure 1 in Bystrak 1981) of northeastern Ohio ($N = 7$). Fragments ranged from 20 to more than 500 ha and were located in either predominantly agricultural-rural ($N = 10$) or urban ($N = 8$) landscapes. Landscapes were defined according to the land uses within 5 km of each site. Urban sites were located within metropolitan areas surrounding Washington, D.C., and Cleveland and Akron, Ohio, where urban and suburban development are extensive (often over 50% of the matrix). Agricultural-rural landscapes contained less than 1% of the numbers of buildings found in urban areas and were dominated by agricultural fields and forest patches.

Generally, forests had not been harvested for approximately 80–150 years. Canopy trees in several fragments were at least 200 years old. Nearly all plots were located in parks and reserves, but human impacts were restricted to small foot trails in areas where we searched for nests. Based on tree species composition, each plot was loosely characterized as either a beech-maple or oak-hickory association. Ohio sites, located in Cuyahoga, Medina, Summit, and Ashland counties, were composed predominantly of beech (*Fagus grandifolia*), sugar maple (*Acer saccharum*), red maple

(*A. rubrum*), white oak (*Quercus alba*), red oak (*Q. rubra*), wild cherry (*Prunus serotina*), and white ash (*Fraxinus americana*). Maryland sites were located in Montgomery, Prince Georges, and Anne Arundel counties, as well as in the District of Columbia. Vegetation composition in Maryland was similar to that of Ohio, although tulip trees (*Liriodendron tulipifera*) and oaks (e.g., chestnut oak, *Q. velutina,* and red oak) were often more dominant in Maryland.

On each site, searches for nests were restricted to a predefined plot covering 20–40 ha (determined by fragment size and homogeneity of vegetation). Typically, at least one side of a plot abutted a forest-edge interface. Nests were located between May and July during searches conducted every 3–5 days per plot. Active nests (containing live eggs or young, or with obvious signs of adult activity such as nest building) were also monitored at similar intervals. Plastic flagging was placed more than 10 m away from active nests in cases where nests would be difficult to relocate because of a lack of natural landmarks.

Analyses in this chapter were restricted to Acadian Flycatchers and Wood Thrushes because of small sample sizes for other species. Five sites were discarded in analyses for Wood Thrushes because nest contents were positively known for fewer than five nests. In all statistical analyses, each site (as opposed to individual nests) was treated as an independent replicate. Difference in parasitism levels between Wood Thrushes and Acadian Flycatchers was assessed with the Wilcoxon matched-pairs signed-ranks statistic (T), which controlled for species-specific differences in distribution across sites. For each species, nonparametric Mann-Whitney U-tests (Z approximation) were used to evaluate differences between parasitism frequency in large (> 400 ha) and small (< 130 ha, mean = 65 ha) fragments, in urban and agricultural landscapes, and in Ohio and Maryland. The relationship between fragment area and parasitism frequency was assessed with Spearman's rank correlation coefficients (r_s). A critical probability level of .10 rather than .05 was set for significance in all statistical tests because of the value of minimizing Type II errors in conservation-related hypothesis testing (Askins et al. 1990).

Results

Acadian Flycatchers and Wood Thrushes nested on 18 and 17 plots, respectively. However, numbers of Wood Thrush nests were sufficient for statistical analyses on only 13 sites. Totals of 358 Acadian Flycatcher and 329 Wood Thrush nests were located, but only 70% of the nests of each species were used in statistical analyses because of uncertainty over the occurrence of parasitism in the other nests. In the following analyses, numbers of nests with known contents for each plot averaged 17 for Wood Thrush (median = 12, range = 5–74) and 14 for Acadian Flycatcher (median = 10, range = 5–49).

Overall, 10.7% of all Acadian Flycatcher and 21.7% of all

Wood Thrush nests were parasitized. Parasitism frequencies did not vary by region for either species (Acadian Flycatcher, $Z = 0.56$, $P = .57$; Wood Thrush, $Z = 0.63$, $P = .53$), although for both species the average parasitism frequency was slightly greater in Maryland (Figure 31.1).

Effects of Surrounding Landscape

Although parasitism was slightly more prevalent in urban areas compared with agricultural areas for both Acadian Flycatchers ($Z = 1.26$, $P = .22$) and Wood Thrushes ($Z = 0.43$, $P = .67$), landscape type played only a minor role in overall probability of parasitism in 1993 (Figure 31.1).

When analyses were restricted by region, parasitism of Acadian Flycatchers was slightly more frequent in urban landscapes in both Maryland (mean parasitism frequency in agricultural areas = 13%, urban = 22%; $Z = 0.84$, $P = .40$) and Ohio (agricultural = 10%, urban = 15%; $Z = 0.55$, $P = .58$). For Wood Thrushes, however, this relationship was not evident in either Maryland (agricultural = 19%, urban = 22%; $Z = 0.13$, $P = .90$) or Ohio (agricultural = 14%, urban = 0%: 0 of 5 nests from 2 plots).

Effects of Forest Fragment Size

Parasitism frequencies were five times greater on Acadian Flycatchers breeding in small forest fragments compared with large forests ($Z = 2.82$, $P < .01$; Figure 31.1a). Wood

Thrushes, however, displayed a much weaker relationship with fragment size ($Z = 0.73$, $P = .47$; Figure 31.1b). However, the overall correlation between fragment area and parasitism frequency was significant for both Acadian Flycatchers ($r_s = -0.725$, $P < .01$, $N = 18$) and Wood Thrushes ($r_s = -0.506$, $P = .08$, $N = 13$).

In both Ohio ($Z = 2.07$, $P = .03$) and Maryland ($Z = 1.65$, $P = .10$), Acadian Flycatchers suffered greater cowbird parasitism in small forest fragments than in large forests. That trend was apparent, but nonsignificant, for Wood Thrushes in Ohio ($Z = 1.22$, $P = .22$); thrushes in Maryland were parasitized nearly equally often in large and small fragments ($Z = 0.12$, $P = .90$). Correlation analysis supported the relationship between parasitism and fragment area for Acadian Flycatchers in both Ohio ($r_s = -0.675$, $P = .09$, $N = 7$) and Maryland ($r_s = -0.811$, $P < .01$, $N = 11$; Figure 31.2a). Negative relationships between those two variables were also observed for Wood Thrushes in Ohio ($r_s = -0.778$, $P = .22$, $N = 4$) and, to a much lesser extent, in the 9 forest fragments in Maryland ($r_s = -0.247$, $P = .52$; Figure 31.2b).

The negative relationship between fragment size and parasitism frequencies in Acadian Flycatchers was apparent in both agricultural ($Z = 2.24$, $P = .02$) and urban ($Z = 2.18$, $P = .03$) landscapes. For Wood Thrushes breeding in agricultural areas, parasitism frequencies in small fragments (23%) were twice as high as those in large fragments (11%),

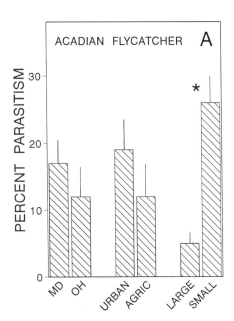
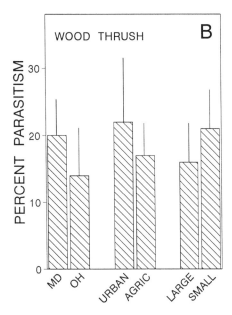

Figure 31.1. Average percentage of parasitism of (A) Acadian Flycatchers and (B) Wood Thrushes in different regions, landscapes, and fragment sizes. Bars represent mean values from study sites, vertical lines signify standard errors, and the asterisk indicates a significant ($P < .10$) difference.

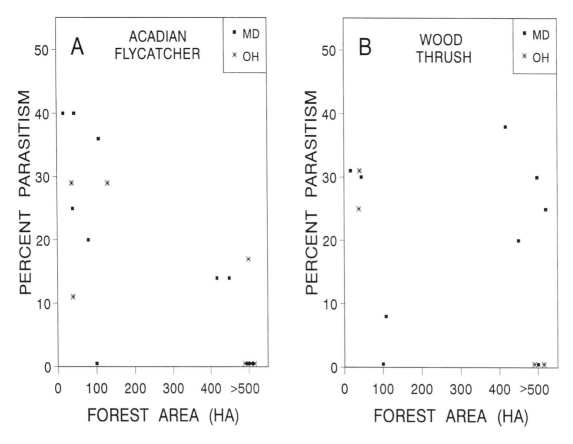

Figure 31.2. Relationship between frequency of parasitism and fragment size for (A) Acadian Flycatchers and (B) Wood Thrushes breeding in forest remnants in Ohio and Maryland.

although this difference was not significant ($Z = 1.41$, $P = .16$). Sample sizes for Wood Thrushes in urban fragments were too small for statistical analyses.

Discussion

Frequencies of parasitism in both Acadian Flycatchers and Wood Thrushes here were generally consistent with levels of parasitism observed elsewhere for these species in the eastern U.S. (e.g., Walkinshaw 1966, Hoover and Brittingham 1993, Roth and Johnson 1993; Dowell et al., Chapter 29, this volume). However, results from this study should be interpreted cautiously because local parasitism frequencies can vary substantially among years (Petit 1991, Roth and Johnson 1993).

The slightly higher frequencies of parasitism in Maryland than Ohio for both species contrast with the conclusion of Hoover and Brittingham (1993), who suggested that there is a negative relationship between parasitism level and distance from the Great Plains. However, parasitism levels in both areas were lower overall than levels found in fragmented landscapes in the midwestern U.S. (Robinson et al.

1995; Donovan et al., Chapter 30, this volume).

Parasitism levels for both Acadian Flycatchers and Wood Thrushes were clearly influenced more by the area of a forest fragment used for breeding than by either geographic region or surrounding landscape. Thus, dissection of forest habitat results in increased brood parasitism of forest-dwelling birds. Increased parasitism levels in relatively small forest fragments have important ramifications for design of forest reserves. The question of the value of small versus large reserves was initially evaluated during the 1970s SLOSS (single large or several small) debate over reserve design (Simberloff and Abele 1976, Whitcomb et al. 1976). Reserve size is now incorporated into algorithms used for identification of critical areas for preservation of species diversity (Margules et al. 1988), using simple presence or absence of species as a function of forest area. The fact that the presence of a species may not necessarily indicate a suitable local environment for reproduction has only recently begun to be incorporated into management and conservation plans (Martin 1992). Acadian Flycatchers and Wood Thrushes do not exhibit strong forest-area sensitivity (Robbins et al. 1989). Breeding populations of such species in small woodlots and forest fragments in eastern North America may experience low reproductive success due to area- and edge-related factors, including cowbird parasitism (e.g., Roth and Johnson 1993). Such potential sink populations (Pulliam 1988) may be maintained only by immigration from areas with successful reproduction. Although the impact of cowbird parasitism on Acadian Flycatchers and Wood Thrushes was minor in relation to losses from nest predation (L. Petit and D. Petit unpubl. data), larger forest reserves may enhance reproductive success of some forest birds through both reduced predation and reduced parasitism (Robinson et al. 1995).

Nevertheless, the effect of forest fragment area on probability of parasitism was weak for Wood Thrushes breeding in eastern Maryland. Likewise, in an investigation conducted over several years in Maryland, Dowell et al. (Chapter 29, this volume) also found that forest area apparently had little influence on parasitism frequencies of Wood Thrushes. Other factors, such as microhabitat surrounding a nest (Petit and Petit unpubl. manuscript), distance to a canopy opening (Brittingham and Temple 1983), or relative abundances of different hosts, may influence probability of parasitism more strongly. These factors deserve further attention.

Landscape type had little effect on parasitism frequencies, although they were generally higher in urban areas. However, when all nests of Acadian Flycatchers (as opposed to replicate plots) were summed by landscape type, nests in urban areas suffered much higher parasitism (21%) than nests in agricultural areas (6%; log-likelihood ratio test, $G = 12.65$, $P < .01$). Cowbirds are attracted to feeding sites in agricultural fields and human settlements (Robinson et al. 1993), so that parasitism frequencies in habitats near these areas are inflated. Within urban areas, grassy lawns, golf courses, recreation areas, and high densities of songbird feeders probably provide good feeding sites for cowbirds. Our results indicate that one cost of a large cowbird population in highly urbanized areas is reduced nesting success of the avian hosts in urban parks and reserves.

The impact of Brown-headed Cowbird parasitism on small birds has been suspected for more than a century (Friedmann 1929, Mayfield 1965). However, organized avian monitoring programs and the resulting discoveries of shrinking migratory bird populations and expanding numbers of Brown-headed Cowbirds (Part I, this volume) have only recently led wildlife ecologists to study those trends. Although it is easy to presume that anthropogenic activity is harmful to wildlife populations, well-designed scientific study is indispensable for advancement of viable and long-term conservation strategies. In our study, the presumed relationship between fragment area and frequency of cowbird parasitism was confirmed for the Acadian Flycatcher and, to a lesser extent, for the Wood Thrush. For several other species breeding on these study sites, however, fragment size had little apparent influence on probability of

parasitism (L. Petit and D. Petit unpubl. data). Thus, as for many ecological phenomena, explanations are neither obvious nor simple. Wildlife and resource managers, as well as researchers, should not be content with current knowledge of cowbird-host relationships. Rather, we need to know the impact of Brown-headed Cowbirds relative to other sources of mortality to understand songbird population dynamics in fragmented landscapes.

Acknowledgments

This research could not have been conducted without cooperation and assistance from a large number of federal, state, and local land management agencies and individual landowners. In Maryland and Washington, D.C., we thank R. Ford, W. Shields, and N. Howell-Streeter (National Park Service); R. Gibbs, G. Lewis, G. Kearns, R. Dolesh, M. Brooks, and L. Straw (Maryland National Capitol Parks and Planning Commission); E. Denike (Maryland Department of Natural Resources); J. Lynch and D. Correll (Smithsonian Environmental Research Center); S. Briggs, S. Droege, and R. Pittman. In Ohio, we thank J. O'Brien (Mohican State Park), W. Starcher (Metroparks serving Summit County), P. Saldutte (Medina County Metroparks), and especially T. Stanley, E. Kuilder, and K. Smith (Cleveland Metroparks). Assistance with fieldwork was provided by G. Lowe, C. Tate, J. Jones, and J. Bridgham. Special thanks also go to K. Petit, B. Dowell, and J. Fallon for conducting a major portion of fieldwork. S. Sealy and S. Robinson provided helpful comments on the manuscript.

References Cited

Askins, R. A., J. F. Lynch, and R. Greenberg. 1990. Population declines in migratory birds in eastern North America. Curr. Ornithol. 7:1–57.

Böhning-Gaese, K., M. L. Taper, and J. H. Brown. 1993. Are declines in North American insectivorous songbirds due to causes on the breeding range? Conserv. Biol. 7:76–86.

Brittingham, M. C., and S. A. Temple. 1983. Have cowbirds caused forest songbirds to decline? BioScience 33:31–35.

Bystrak, D. 1981. The North American breeding bird survey. Studies in Avian Biol. 6:34–41.

Fonseca, G. A. B., and J. G. Robinson. 1990. Forest size and structure: Competitive and predatory effects on small mammal communities. Biol. Conserv. 53:265–294.

Friedmann, H. 1929. The cowbirds: A study in the biology of social parasitism. C. C. Thomas, Springfield, IL.

Grzybowski, J. A., R. B. Clapp, and J. T. Marshall. 1986. History and current population status of the Black-capped Vireo in Oklahoma. Amer. Birds 40:1151–1161.

Holmes, R. T., T. W. Sherry, and F. W. Sturges. 1986. Bird community dynamics in a temperate deciduous forest: Long-term trends at Hubbard Brook. Ecol. Monogr. 50:201–220.

Hoover, J. P., and M. C. Brittingham. 1993. Regional variation in cowbird parasitism of Wood Thrushes. Wilson Bull. 105:228–238.

Margules, C. R., A. O. Nicholls, and R. L. Pressey. 1988. Selecting networks of reserves to maximize biological diversity. Biol. Conserv. 43:63–76.

Martin, T. E. 1992. Breeding productivity considerations: What are the appropriate habitat features for management? Pp. 455–473 in Ecology and conservation of neotropical migrant landbirds (J. M. Hagan III and D. W. Johnston, eds). Smithsonian Institution Press, Washington, DC.

———. 1993. Nest predation among vegetation layers and habitat types: Revising the dogmas. Ecology 141:897–913.

May, R. M., and S. K. Robinson. 1985. Population dynamics of avian brood parasitism. Amer. Natur. 126:475–494.

Mayfield, H. F. 1965. The Brown-headed Cowbird, with old and new hosts. Living Bird 4:13–28.

———. 1977. Brown-headed Cowbird: Agent of extermination? Amer. Birds 31:107–113.

Petit, L. J. 1991. Adaptive tolerance of cowbird parasitism by Prothonotary Warblers: A consequence of nest-site limitation? Anim. Behav. 41:425–432.

Pulliam, H. R. 1988. Sources, sinks, and population regulation. Amer. Natur. 132:652–661.

Robbins, C. S., D. K. Dawson, and B. A. Dowell. 1989. Habitat area requirements of breeding forest birds of the middle Atlantic states. Wildl. Monogr. 103:1–34.

Robinson, S. K., J. A. Grzybowski, S. I. Rothstein, M. C. Brittingham, L. J. Petit, and F. R. Thompson. 1993. Management implications of cowbird parasitism on neotropical migrant songbirds. Pp. 93–102 in Status and management of neotropical migratory birds

(D. M. Finch and P. W. Stangel, eds). USDA Forest Serv. Gen. Tech. Rep. RM-229.

Robinson, S. K., F. R. Thompson III, T. M. Donovan, D. R. Whitehead, and J. Faaborg. 1995. Regional forest fragmentation and the nesting success of migratory birds. Science 267:1987–1990.

Roth, R. R., and R. K. Johnson. 1993. Long-term dynamics of a Wood Thrush population breeding in a forest fragment. Auk 110:37–48.

Sauer, J. R., and S. Droege. 1992. Geographic patterns in population trends of neotropical migrants in North America. Pp. 26–42 in Ecology and conservation of neotropical migrant landbirds (J. M. Hagan III and D. W. Johnston, eds). Smithsonian Institution Press, Washington, DC.

Simberloff, D. S., and L. G. Abele. 1976. Island biogeography theory and conservation practice. Science 191:285–286.

Walkinshaw, L. H. 1966. Studies of the Acadian Flycatcher in Michigan. Bird-Banding 37:227–257.

Whitcomb, R. F., J. F. Lynch, P. A. Opler, and C. S. Robbins. 1976. Island biogeography and conservation: Strategy and limitations. Science 193:1030–1032.

Whitham, J. W., and M. L. Hunter, Jr. 1992. Population trends of neotropical migrant landbirds in northern coastal New England. Pp. 85–95 in Ecology and conservation of neotropical migrant landbirds (J. M. Hagan III and D. W. Johnston, eds). Smithsonian Institution Press, Washington, DC.

Wilcove, D. S., and S. K. Robinson. 1990. The impact of forest fragmentation on bird communities in eastern North America. Pp. 319–331 in Biogeography and ecology of forest bird communities (A. Keast, ed). SPB Academic Publishing, The Hague, The Netherlands.

32. Biogeographic, Landscape, and Local Factors Affecting Cowbird Abundance and Host Parasitism Levels

FRANK R. THOMPSON III, SCOTT K. ROBINSON, THERESE M. DONOVAN,

JOHN R. FAABORG, DONALD R. WHITEHEAD, AND DAVID R. LARSEN

Abstract

Cowbird abundance and host brood parasitism levels may vary at biogeographic, landscape, and local scales. We use Breeding Bird Survey data to examine biogeographic or continental patterns in cowbird abundance, and data from our ongoing studies in the midwestern United States to demonstrate landscape and local factors affecting cowbird numbers and levels of brood parasitism. Cowbird abundance in the 48 contiguous United States was negatively related to distance from the center of the cowbird's range in the Great Plains and the amount of forest cover. In Illinois, Indiana, Missouri, and Wisconsin landscapes, cowbird abundance, cowbird/host ratio, and host parasitism level were negatively correlated with the proportion of forest cover, proportion of core-area forest, and mean forest patch size and positively correlated with edge density. At a local scale, parasitism levels and cowbird abundance were related to proximity to edge in some landscapes and to forest-host densities in others. Edge or host-density effects may be related to factors limiting cowbird numbers at the landscape scale. Management to reduce community-wide host parasitism levels should focus primarily on landscape-level constraints on cowbird numbers, such as eliminating feeding areas within forested landscapes.

Introduction

Ecological patterns are dependent on the spatial and temporal scale at which they are viewed. The continuum of possible spatial scales can be broken into (1) the space occupied by an individual, (2) a local patch occupied by many individuals and species, (3) a region that contains many patches or local populations, and (4) a biogeographic scale that encompasses different climates, vegetation formations, and assemblages of species (Wiens et al. 1986). Ecological patterns or processes at any one of these scales may be influenced by factors at the other scales. In a hierarchical system, higher levels or larger scales may even act to constrain lower levels (Allen and Starr 1982). A multiscale approach, therefore, may be needed to understand distribution patterns of species and the factors limiting population sizes (Wiens 1989).

We investigated the distribution of Brown-headed Cowbirds at biogeographic, landscape, and local scales (analogous to scales 4, 3, and 2 in Wiens et al. 1986). The Brown-headed Cowbird is an obligate brood parasite and a major threat to some populations of neotropical migratory songbirds (Mayfield 1977, Brittingham and Temple 1983, Robinson and Wilcove 1994). An increased understanding of the factors limiting the abundance of Brown-headed Cowbirds at different spatial scales may be important for the conservation of host species.

Cowbirds were limited primarily to open grasslands of central North America prior to European settlement, but their numbers increased dramatically in the East as forests were cleared by settlers in the mid- to late 1700s (Mayfield 1965). It is not clear if they were completely absent from the East or simply uncommon prior to this (Rothstein 1994). In the West, cowbird abundance has greatly increased in the Great Basin and adjoining areas east of the Cascades since the 1800s. Colonization of California began around 1900 (Rothstein 1994). Cowbirds are still most abundant in the Great Plains (Lowther 1993; Peterjohn et al., Chapter 2, and Wiedenfeld, Chapter 3, this volume). Other biogeographic patterns in cowbird distribution include the decline in levels of parasitism of Wood Thrushes (*Hylocichla mustelina*) from the Midwest to Mid-Atlantic to Northeast (Hoover and Brittingham 1993). We hypothesized that as a result of its recent range expansion and affinity for grasslands, biogeographic patterns in the cowbird's distribution should be related to distance from the center of its range and to continental patterns in land cover.

Within a particular biogeographic context, cowbird numbers may be affected by landscape patterns in feeding and breeding habitats. Because cowbirds do not rear their own young, breeding activities (nest searching and egg laying) and feeding can be separated spatially and temporally. This uncoupling of breeding and feeding allows cowbirds to select separate areas appropriate for each activity (Rothstein et al. 1984, 1986). Cowbirds usually feed in short grass habitats or with large grazing mammals (Friedmann 1929, Mayfield 1965, Dufty 1982, Rothstein et al. 1986). They breed in a variety of habitats from prairie to forest but often select habitats with high host densities (Rothstein et al. 1980, 1986). In midwestern landscapes, most cowbirds breed in forest habitats and commute up to 5 km to feed in agricultural habitats such as pasture (Thompson 1994). We therefore hypothesized that within a given biogeographic context, cowbird numbers and parasitism levels are related to the availability of feeding habitat in a landscape.

Within a given biogeographic and landscape context, cowbird numbers and levels of brood parasitism may also be related to local- or habitat-scale relationships. Numbers of cowbirds and host parasitism levels are sometimes higher near forest edges (Gates and Gysel 1978, Chasko and Gates 1982, Brittingham and Temple 1983, O'Conner and Faaborg 1992; Winslow et al., Chapter 34, this volume). Other studies, however, have found no relationship between parasitism levels and distance to edge (Hoover 1992, Robinson and Wilcove 1994). Cowbird numbers and parasitism levels have often been presumed to be higher near forest edges because cowbirds may be unwilling to fly large distances from feeding areas into forests to breed. However, cowbirds may also be responding positively to host density, either at edges (Chasko and Gates 1982, Gates and Giffen 1991) or even away from edges. We hypothesized that at a local scale, numbers of breeding cowbirds in forest habitats should be determined by host densities or distance to forest edge and that these relationships should depend on the landscape context. We predicted that cowbird densities should be more closely related to host numbers in fragmented landscapes where cowbirds are abundant and saturate forest habitats and that cowbird numbers should be more closely related to distance to edge in extensively forested landscapes with limited numbers of cowbirds.

We used data from several sources to investigate these hypotheses. We examined biogeographic patterns by relating mean cowbird abundance (as determined by the Breeding Bird Survey) in states to percentage of forest cover and a state's distance from the center of the cowbird's range. We investigated landscape-level patterns by assessing correlations among cowbird abundance, brood parasitism levels, and the degree of forest fragmentation in seven landscapes. These seven landscapes are part of our ongoing songbird demographic studies and occur within similar biogeographic contexts in Illinois, Indiana, Missouri, and Wisconsin. To investigate local-scale effects, we explored the relationships of cowbird abundance to host abundance and of parasitism levels in common hosts to distance from the forest edge in some of these landscapes.

Methods

Biogeographic Patterns

We used multiple regression analysis to relate mean statewide abundance of cowbirds to percentage of forest cover and distance from the center of the cowbird's range. We used states as the experimental unit because existing data on cowbird abundance (Breeding Bird Survey, National Biological Survey) and forest inventory data (USDA Forest Service) are easily summarized by state. We calculated percentage of forest cover for each state from total land area and total forested land area (Waddell et al. 1989). We defined the center of the cowbird's range as the two large high-density contours encompassing portions of North and South Dakota, Nebraska, Kansas, and Oklahoma on a map of cowbird abundance derived from Breeding Bird Survey data (Lowther 1993). This region also probably represents the historical center of the cowbird's range (Friedmann 1929, Mayfield 1965). We calculated the distance from the mean *x* and *y* coordinates for all Breeding Bird Survey routes in a state, weighted by cowbird abundance on each route, to the closest part of the center of the cowbird's range. We used the weighted mean coordinates for each state because they represent the center of the density distribution of cowbirds in a state (though the location may be geographically meaningless). In addition to the regression model, we calculated partial correlations for forest cover and distance to the center of the cowbird's range to determine the effects of each of these variables on cowbird abundance, while removing the effect of the other statistically.

Landscape Patterns

We investigated landscape-level patterns by relating cowbird abundance and brood parasitism levels to levels of forest fragmentation in seven study areas in Illinois, Indiana, Missouri, and Wisconsin. Estimates of cowbird and host abundance and brood parasitism levels come from separate studies reported in this volume by Robinson et al. (Chapter 33), Winslow et al. (Chapter 34), and Donovan et al. (Chapter 30) and elsewhere by Donovan (1994). These study areas are north-central Missouri (NC-MO), south-central Missouri (SC-MO), northwest Illinois (NW-IL), southwest-Illinois (SW-IL), northwest Wisconsin (NW-WI), west-central Wisconsin (WC-WI), and southern Indiana (SO-IN) (Figure 32.1). These landscapes included central hardwood forests, cropland, cool-season pasture, and urban and suburban habitats. There was some variation in methods and years of study. However, we believe our approach is robust because it is based on means from 3 to 8 sites for 2 to 4 years for each study area.

Study areas consisted of 3 to 8 study sites at which abundance of breeding forest birds was estimated by point count surveys. Levels of parasitism were estimated by monitoring nests. The Indiana and southern Illinois study sites were wa-

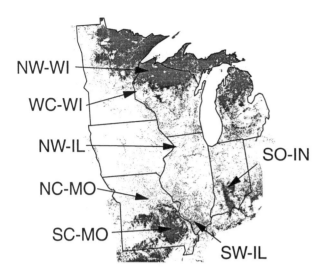

Figure 32.1. Location of study areas used to determine regional-landscape patterns in cowbird abundance, host brood parasitism levels, and the distribution of forest cover (in black) in Illinois, Indiana, Missouri, and Wisconsin.

tersheds 2 to 5 km long and 200 to 350 ha in area. The remaining sites were forest patches of 40 to 350 ha that were either isolated or part of a larger contiguous forested area, depending on the landscape context. Data were pooled from the individual study sites to calculate estimates for each study area. All data from point counts are reported as the mean numbers of detections within a 70-m radius during a 10-min count, except for Illinois sites, which are reported for 6-min counts. This difference in count length is probably not significant, as 85% of cowbirds and 88% of hosts were detected during the first 6 min of 10-min point counts in Indiana (F. Thompson unpubl. data). We calculated the mean number of detections of female cowbirds/point and the mean ratio of female cowbirds to hosts for each study area (reported as cowbird abundance or cowbirds/point and cowbird/host, respectively). We considered a species to be a host if it bred on the study area and accepted cowbird eggs. Alternatively, we could have included rejector species in this ratio. A cowbird's assessment of breeding opportunities in these habitats may include rejector species, because these species are often parasitized by cowbirds (Rothstein 1976, Scott 1977). However, rejector species, such as American Robins (*Turdus migratorius*), Gray Catbirds (*Dumetella carolinensis*), Northern Orioles (*Icterus galbula*), and Brown Thrashers (*Toxostoma rufum*), were very uncommon in the forest habitats we studied. Their inclusion or exclusion would not have significantly affected the cowbird/host ratio on our study sites.

We report brood parasitism levels for four common host species that occurred in the forest on nearly all sites, the Ovenbird (*Seiurus aurocapillus*), Wood Thrush, Acadian Fly-

catcher (*Empidonax virescens*), and Red-eyed Vireo (*Vireo olivaceus*). These species represent a ground nester, two understory–small tree nesters, and one canopy nester, respectively. Parasitism frequency was calculated as the proportion of monitored nests with cowbird eggs or young, and parasitism intensity as the number of eggs/parasitized nest on all study sites within a study area. We calculated mean parasitism frequency and intensity for each study area from the parasitism levels of these four species.

We used digital data for land use and land cover from the U.S. Geological Survey (Office of Geographic and Cartographic Research, Reston, VA) to determine patterns in forest cover for each study area. Individual scenes were pieced together to form an image for the four-state area with a 500-m resolution (pixel size). All forest land-use types were reclassified as forest and all other types as nonforest. For each study area, we used a spatial analysis program (Fragstats, Forest Science Department, Oregon State University, Corvallis, OR) to determine the proportion of the landscape in forest cover, the proportion of the landscape in forest core area, the amount of forest/nonforest edge, and the mean forest patch size within a 10-km radius of each study site. We selected a 10-km radius because cowbird movements between breeding and feeding areas are less than 10 km (Thompson 1994), and it is an appropriate scale for relating forest cover to cowbird abundance (Donovan et al., Chapter 30, this volume). We defined forest core area as forest more than 250 m from nonforest because it was similar to the 200-m core-area model proposed by Temple (1986). Also, it was easily analyzed because 250 m is half the dimension of a pixel in the land cover image. Mean values of each landscape statistic were calculated for the seven study areas from their respective sites. We calculated Pearson correlation coefficients between cowbird abundance, cowbird/host ratio, mean parasitism frequency, number of cowbird eggs/parasitized nest, the proportion of the landscape in forest cover, the proportion of the landscape in forest core area, the amount of forest/nonforest edge, and the mean forest patch size.

Local Patterns

Within study areas, we were able to examine relationships between cowbird and host abundance and between brood parasitism levels and distance to forest edge. Regression analysis was used to relate cowbird abundance to host abundance on the SW-IL study area. Mean cowbird and host detections per point were calculated for 16 transects surveyed by point counts on the SW-IL study area in 1992. Transects were more than 5 km apart and 2 to 5 km long with points every 150 m. Each transect was used as an observation in the regression analysis. Donovan et al. (Chapter 30, this volume) report results of a similar regression analysis for the NC-MO and SC-MO sites.

We determined the effects of edges on parasitism levels of hosts breeding in forest habitat on the SW-IL study area during 1989–1990. We identified forest openings greater than 0.2 ha from aerial photographs and mapped the openings and nest locations on topographic maps. Openings included wildlife openings maintained in grass and forb cover, forest stands that had been clearcut less than 10 years ago, and recreational areas with mowed grass. We measured distances from monitored nests to the nearest opening and pooled nests into classes describing the distance to edge (0–50, 51–100, 101–200, 201–400, over 400 m). We used a Mantel-Haenszel chi-square test to determine if there was a linear association between parasitism level and distance to edge. Winslow et al. (Chapter 34, this volume) present an analysis of edge effects on the SO-IN study area.

Results

Biogeographic Patterns

Mean statewide cowbird abundance was negatively related to forest cover in a state and to a state's distance from the center of the cowbird's breeding range ($r^2 = 0.67$; Table 32.1, Figure 32.2). Regression coefficients for distance to center of range and forest cover were both significant. However, the partial correlation of distance to center of range with cowbird abundance was greater than that for forest cover and cowbird abundance (Table 32.1).

Landscape Patterns

Mean cowbird abundance, cowbird/host ratio, and the frequency and intensity of parasitism varied greatly among study areas. Relative abundance of female cowbirds ranged from 0.0007/point in NW-WI to 0.675/point in NW-IL. The mean proportion of nests parasitized ranged from 0.013 in NW-WI to 0.678 in SW-IL. The mean number of cowbird eggs/parasitized nest ranged from 1.0 in NW-WI and SC-MO to 2.75 in NW-IL. Cowbird abundance and parasitism levels were positively correlated across study areas (Figure 32.3, Table 32.2). The four species had similar patterns of parasitism across study sites, but the overall magnitude of parasitism varied among species. Wood Thrushes and Red-eyed Vireos had the greatest parasitism levels (up to 88% in NW- and SW-IL), and Acadian Flycatchers had the lowest levels (Figure 32.4).

Landscape patterns varied greatly among study areas. The SC-MO and NW-WI study areas had the greatest proportions of forest cover, and NW-IL and WC-WI the least (Table 32.3). Edge density was greatest on the SW-IL and NC-MO study areas, which had intermediate levels of forest cover (Table 32.3).

Cowbird abundance, cowbird/host ratio, proportion of nests parasitized, and numbers of cowbird eggs/parasitized nest were negatively correlated with mean forest patch size and proportion of the landscape in forest and forest core area, and positively correlated with edge density (Table 32.2).

Local Patterns

Cowbird abundance was related to host abundance on the SO-IL study area ($r^2 = 0.44$, $F = 10.24$, $P < .01$; Figure 32.5). There were sufficient numbers of Wood Thrush ($N = 173$) and Acadian Flycatcher nests ($N = 145$) in all distance-to-edge classes to test for edge effects on the SO-IL study area. There was no linear relationship between distance to edge and parasitism level for Wood Thrushes ($X^2 = 1.19$, $P = .28$) or Acadian Flycatchers ($X^2 = 0.13$, $P = .72$; Figure 32.6).

Discussion

Biogeographic Patterns

The biogeographic pattern we observed in the distribution of the Brown-headed Cowbird is consistent with that expected

Table 32.1. Multiple Regression and Partial Correlation Analysis Relating Mean Statewide Relative Abundance of Brown-headed Cowbirds to Statewide Percentage of Forest Cover and Distance to the Center of Their Breeding Range

Parameter	Estimate	P value	Partial Correlation
Intercept	56.00	.0001	
Log_e % forest cover	−4.83	.0412	−0.30
Log_e distance to center of range	−2.00	.0021	−0.44

$F = 45.4$; $P < .001$; $r^2 = 0.67$.

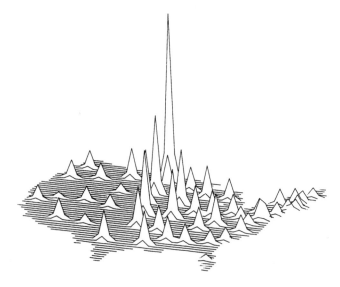

Figure 32.2. Continental patterns in the relative abundance of Brown-headed Cowbirds. Data are the mean number of cowbirds per Breeding Bird Survey route by state.

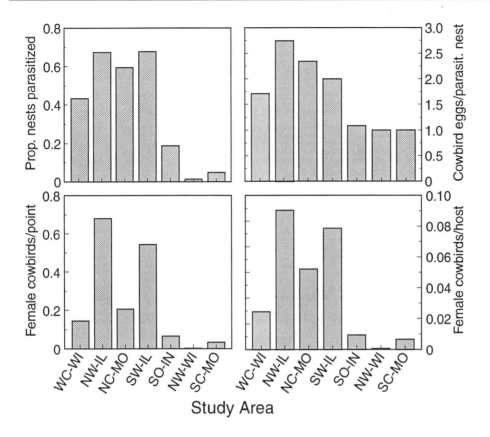

Figure 32.3. Mean proportion of nests parasitized by Brown-headed Cowbirds, number of cowbird eggs/parasitized nest, relative abundance (detections/point) of cowbirds, and ratio of cowbird abundance/host abundance on seven study areas in the midwest-ern United States. Study areas are in order of decreasing levels of forest cover along the x-axis. Study-area abbreviations as listed in Methods.

for a species expanding its range. Mayfield (1965) hypothesized that two factors facilitated the cowbird's range expansion: first, the clearing of eastern forests by settlers; second, the exceptional reproductive success achieved by cowbirds using tolerant new hosts as their range began to expand. These explanations seem plausible, given the relationships found here between cowbird abundance and forest cover, and abundance and distance to the center of its range. The grasslands of the Great Plains remain the current center of the cowbird's range (Friedmann 1929, Lowther 1993; Figure 2.2 in Peterjohn et al., Chapter 2, this volume). While both partial correlations were significant, the effect of distance to the center of the range was stronger, indicating the importance of biogeographic constraints. This biogeographic pattern is also reflected in host parasitism levels. Wood Thrush parasitism levels decrease from the Midwest to the Mid-Atlantic region and to New England (Hoover and Brittingham 1993).

Landscape Patterns

We found convincing evidence that within a particular biogeographic context, landscape patterns of forest habitat affect numbers of cowbirds and levels of brood parasit-

ism. Cowbird abundance, cowbird/host ratio, proportion of nests parasitized, and number of cowbird eggs/parasitized nest varied greatly across midwestern landscapes within a similar biogeographic context. Cowbirds were virtually absent from heavily forested landscapes such as SC-MO and NW-WI but were very abundant in landscapes composed of an agricultural matrix with forest fragments.

Although cowbird abundance and parasitism levels in hosts were significantly correlated with landscape structure, there was some variation in this relationship. Mean cowbird abundance and percentage of parasitism did not decrease monotonically across study areas ordered by decreasing habitat fragmentation (Figures 32.3 and 32.4). For instance, cowbird abundance and parasitism levels were greatest in the two Illinois study areas, while the NC-MO and WC-WI study areas had similar or greater levels of fragmentation yet many fewer female cowbirds. A likely explanation is that nonforest portions of these landscapes varied in some way important to cowbirds. For instance, the nonforested portion of the Illinois study sites was predominately cropland with some pasture and developed area, while the nonforested portions of the Missouri and Wisconsin study areas

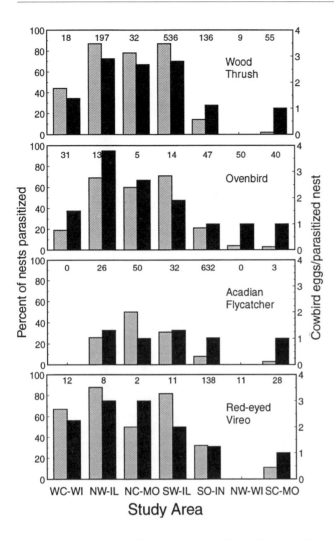

Figure 32.4. Frequency of brood parasitism (hatched bars) and numbers of cowbird eggs per parasitized nest (solid bars) for four common forest-host species on seven study areas in the midwestern United States. Study areas are in order of decreasing levels of forest fragmentation along the x-axis. Study-area abbreviations as listed in Methods.

were predominately pasture. The type of nonforest landscape matrix may influence cowbird distribution as much as forest cover.

There was nearly complete correspondence among study areas in proportion of forest, proportion of forest core area, and mean patch size. Edge density, however, was greatest at intermediate values of forest cover because it depends on the presence of both forested and unforested patches. Other studies have reported high levels of parasitism in fragmented forests (Robinson 1992) or near forest edges (Gates and Gysel 1978, Chasko and Gates 1982, Brittingham and Temple 1983; Winslow et al., Chapter 34, this volume). While we found a relationship between cowbird numbers and edge density, we do not believe that edge per se regulates cowbird numbers at the landscape scale. Rather, we suggest

that cowbird numbers depend on the interspersion of open grassland and agricultural habitat in these landscapes and that the amount of edge is a correlate of the degree of interspersion. Thompson (1994) and Thompson and Dijak (Chapter 10, this volume) radiotracked cowbirds on three of our study areas and determined that cowbirds breeding in the forest commute daily to grassland or cropland to feed in flocks and that most commuting distances were less than 5 km. We found a negative relationship between numbers of cowbirds and the amount of forest, but this was probably because the availability of feeding habitat was inversely related to the amount of forest. We believe that the interspersion of cowbird feeding habitat within the breeding habitat of forest songbirds increases a landscape's carrying capacity for cowbirds. In addition to any increase in cowbird carrying capacity, however, edge per se increases the impact of brood parasitism on a songbird population only if edges function as ecological traps. This requires not only that cowbirds are attracted to edges but also that a host species prefers to nest in edge habitats over more productive habitat (such as interior habitats). Although total bird densities and species richness may be higher near edges (Gates and Gysel 1978), few forest species appear to prefer forest edge to forest interior (Kroodsma 1984).

Local Patterns

At a local- or habitat-level scale, edge and host density were correlated with cowbird numbers and host parasitism levels. The importance of these factors appeared to vary with landscape context (Table 32.4). We found no evidence of an edge-related increase in parasitism levels on the SW-IL study area, but Winslow et al. (Chapter 34, this volume) found strong edge effects on our SO-IN study area. We hypothesize that cowbird numbers and host parasitism levels are affected by the distribution of edge when cowbird numbers are limited by the availability of feeding habitat in the landscape, or by biogeographic constraints. If cowbird numbers are limited at a regional or biogeographic scale, hosts may be readily available and cowbirds should occupy only high-quality breeding habitat. Hence, most parasitism should occur near feeding areas (to minimize commuting) or in areas of high host density (to maximize parasitism opportunities). We found this on our SO-IN study area, where overall cowbird abundance and parasitism level were low. Parasitism levels were higher near forest-agricultural edges (near potential feeding areas) and at interior edges created by timber harvest (areas of potentially high host density) (Winslow et al., Chapter 34, this volume).

In areas where cowbirds are not constrained by landscape or biogeographic factors, female cowbird distribution within the forest is more likely to be affected by host abundance than by the location of edges. In such landscapes, the forest may be saturated with cowbirds and competition for hosts may be keen. On our SO-IL study area, parasitism levels

Table 32.2. Pearson Correlations of Cowbird Data and Four Measures of Forest Fragmentation for Seven Study Areas in the Midwestern United States

	Cowbird Statistics				Landscape Statistics			
	Cowbird Abundance	Cowbird/ Host	Prop. Nests Parasitized	Cowbird Eggs/Nest	Prop. Forest	Prop. Core Area	Edge Density	Forest Patch Size
Cowbird abundance	1.000	0.801	0.865	0.725	−0.718	−0.737	0.678	−0.864
	0.000	0.030	0.012	0.065	0.069	0.059	0.094	0.012
Cowbird/host	0.801	1.000	0.878	0.747	−0.737	−0.757	0.702	−0.871
	0.030	0.000	0.009	0.054	0.059	0.049	0.079	0.011
Proportion of nests parasitized	0.865	0.878	1.000	0.922	−0.855	−0.885	0.826	−0.923
	0.012	0.009	0.000	0.003	0.014	0.008	0.008	0.003
Cowbird eggs/parasitized nest	0.725	0.747	0.922	1.000	−0.858	−0.859	0.639	−0.792
	0.065	0.054	0.003	0.000	0.013	0.013	0.122	0.034
Proportion forest	−0.718	−0.737	−0.855	−0.858	1.000	0.994	−0.736	0.913
	0.069	0.059	0.014	0.013	0.000	0.001	0.059	0.004
Proportion core area	−0.737	−0.757	−0.885	−0.859	0.994	1.00	−0.806	0.944
	0.059	0.049	0.008	0.013	0.001	0.000	0.029	0.001
Edge density	0.678	0.702	0.826	0.639	−0.736	−0.806	1.000	−0.903
	0.094	0.079	0.008	0.122	0.059	0.029	0.000	0.005
Forest patch size	−0.864	−0.871	−0.923	−0.792	0.913	0.944	−0.903	1.000
	0.012	0.011	0.003	0.034	0.004	0.001	0.005	0.000

Table 32.3. Landscape Patterns in Forest Cover for Seven Study Areas

Study Area	Proportion Forest	Proportion Core Area	Edge Density (m/ha)	Mean Forest Patch Size (ha)
SC-MO	0.95	0.75	2.7	26,768
NW-WI	0.93	0.70	3.7	29,150
SO-IN	0.78	0.52	6.2	10,962
SW-IL	0.54	0.27	7.8	1,385
NC-MO	0.31	0.07	8.3	624
NW-IL	0.23	0.06	5.9	379
WC-WI	0.22	0.04	6.7	549

Note: Estimates are means of measurements for the area within a 10-km radius of the 4–9 sites for each study area.

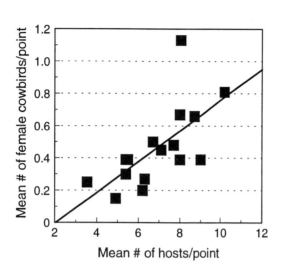

Figure 32.5. Relationship of mean relative abundance (detections/point) of female Brown-headed Cowbirds to host density on 16 transects in southwest Illinois.

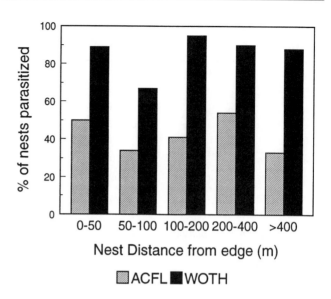

Figure 32.6. Percentages of parasitized Acadian Flycatcher (hatched bars) and Wood Thrush (solid bars) nests in southern Illinois in relation to distance to the nearest forest opening.

Table 32.4. Summary of Evidence for Edge Effect and Host Abundance Effect

Study Area	Edge Effect[a]	Host Abundance Effect[b]
Northern Missouri	Unknown (no data)	Yes (Donovan et al.)
Southwestern Illinois	No (this study)	Yes (this study)
Southern Indiana	Yes (Winslow et al.)	Unknown
South-central Missouri	Unknown (no data)	No (Donovan et al.)

Note: Study areas are listed in order of increasing forest area.
[a]Relationship between brood parasitism levels and distance to forest edge.
[b]Relationship between Brown-headed Cowbird abundance and host abundance.

were high and not related to distance to edge, but the abundance of female cowbirds in the forest in the morning was related to host abundance. The high frequency and magnitude of multiply parasitized nests on the NW-IL and SW-IL study areas is another indication that competition for hosts is high in these landscapes. Similarly, cowbird abundance was more closely related to host abundance in forest fragments in the NC-MO study area, where cowbirds are abundant, than in the SC-MO study area, where cowbirds are uncommon (Donovan et al., Chapter 30, this volume).

Conservation Implications

Conservation efforts to reduce brood parasitism levels for forest songbirds should recognize that cowbird numbers and parasitism levels are regulated at several spatial scales. If, for instance, cowbird numbers are regulated by large-scale bi-

ogeographic and regional-landscape patterns, efforts directed solely at reducing internal edges may not lower parasitism. Land managers should realize that the potential for locally high levels of brood parasitism varies greatly with different biogeographic and landscape contexts.

Within any given biogeographic context, we believe conservation efforts should focus primarily on reducing the number of cowbirds in landscapes by reducing the interspersion of cowbird feeding habitat and host-rich breeding habitats. Based on known commuting distances (Rothstein et al. 1984, Thompson 1994; Raim, Chapter 9, this volume) and relationships between landscape statistics and parasitism levels (this study and Donovan et al., Chapter 30, this volume), we believe cowbird feeding habitat should be reduced within a 2- to 10-km radius of the area in which parasitism is a concern. Vegetative management or modifications of current agricultural practices could also be used to make feeding habitats less suitable to cowbirds. There is a strong need for research on habitat preferences of feeding cowbirds and on how to modify agricultural practices to reduce the suitability of agricultural habitats for feeding cowbirds.

Secondarily, managers should consider reducing edge where there is evidence that it functions as an ecological trap. In addition to cowbird parasitism, managers should be aware of the association of edge with high levels of nest depredation (reviewed by Paton 1994).

Acknowledgments

Our landscape-level analysis was based on data from individual studies in Indiana, Illinois, Missouri, and Wisconsin. Many individuals, agencies, and organizations contributed to those studies and are acknowledged in chapters in this volume reporting on those specific studies. In addition we thank Jeff Price for providing Breeding Bird Survey data on cowbird abundance and route locations.

References Cited

Allen, T. F. H., and T. B. Starr. 1982. Hierarchy: Perspectives for ecological complexity. University of Chicago Press, Chicago.

Brittingham, M. C., and S. A. Temple. 1983. Have cowbirds caused forest songbirds to decline? BioScience 33:31–35.

Chasko, G. G., and J. E. Gates. 1982. Avian habitat suitability along a transmission-line corridor in an oak-hickory forest region. Wildl. Monogr. 82:1–41.

Donovan, T. M. 1994. The demography of neotropical migrant birds in habitat sources and sinks. Ph.D. Thesis, University of Missouri–Columbia. 180 pp.

Dufty, A. M., Jr. 1982. Movements and activities of radio-tracked Brown-headed Cowbirds. Auk 99:316–327.

Friedmann, H. 1929. The cowbirds: A study in the biology of social parasitism. C. C. Thomas, Springfield, IL.

Gates, J. E., and N. R. Giffen. 1991. Neotropical migrant birds and edge effects at a forest-stream ecotone. Wilson Bull. 103:204–217.

Gates, J. E., and L. W. Gysel. 1978. Avian nest dispersion and fledgling success in field-forest ecotones. Ecology 59:871–883.

Hoover, J. P. 1992. Factors influencing Wood Thrush (*Hylocichla mustelina*) nesting success in a fragmented forest. M.S. Thesis, Pennsylvania State University, University Park.

Hoover, J. P., and M. C. Brittingham. 1993. Regional variation in cowbird parasitism of Wood Thrushes. Wilson Bull. 105:228–238.

Kroodsma, R. L. 1984. Effect of edge on breeding forest bird species. Wilson Bull. 96:426–436.

Lowther, P. E. 1993. Brown-headed Cowbird (*Molothrus ater*). *In* The birds of North America 47 (A. Poole and F. Gill, eds).

Academy of Natural Sciences, Philadelphia; American Ornithologists' Union, Washington, DC.

Mayfield, H. F. 1965. The Brown-headed Cowbird with old and new hosts. Living Bird 4:13–28.

———. 1977. Brown-headed Cowbird: Agent of extermination? Amer. Birds 31:107–113.

O'Conner, R. J., and J. Faaborg. 1992. The relative abundance of the Brown-headed Cowbird (*Molothrus ater*) in relation to exterior and interior edges in forests of Missouri. Trans. Missouri Acad. Sci. 26:1–9.

Paton, P. W. C. 1994. The effect of edge on avian nest success: How strong is the evidence? Conserv. Biol. 8:17–26.

Robinson, S. K. 1992. Population dynamics of breeding neotropical migrants in a fragmented Illinois landscape. Pp. 408–418 *in* Ecology and conservation of neotropical migrant landbirds (J. H. Hagan III and D. W. Johnston, eds). Smithsonian Institution Press, Washington, DC.

Robinson, S. K., and D. S. Wilcove. 1994. Forest fragmentation in the temperate zone and its effects on migratory songbirds. Bird Conservation International 4:233–249.

Rothstein, S. I. 1976. Cowbird parasitism of the Cedar Waxwing and its evolutionary implications. Auk 93:498–509.

———. 1994. The cowbird's invasion of the far West: History, causes, and consequences experienced by host species. Studies in Avian Biol. 15:301–315.

Rothstein, S. I., J. Verner, and E. Stevens. 1980. Range expansion and diurnal changes in dispersion of the Brown-headed Cowbird in the Sierra Nevada. Auk 97:253–267.

———. 1984. Radio-tracking confirms a unique diurnal pattern of spatial occurrence in the parasitic Brown-headed Cowbird. Ecology 65:77–88.

Rothstein, S. I., D. A. Yokel, and R. C. Fleischer. 1986. Social dominance, mating and spacing systems, female fecundity, and vocal dialects in captive and free-ranging Brown-headed Cowbirds. Curr. Ornithol. 3:127–185.

Scott, D. M. 1977. Cowbird parasitism on the Gray Catbird at London, Ontario. Auk 94:18–27.

Temple, S. A. 1986. Predicting impacts of habitat fragmentation on forest birds: A comparison of two models. Pp. 301–304 *in* Wildlife 2000: Modeling habitat relationships of terrestrial vertebrates (J. Verner, M. L. Morrison, and C. J. Ralph, eds). University of Wisconsin Press, Madison.

Thompson, F. R. III. 1994. Temporal and spatial patterns of breeding Brown-headed Cowbirds in the midwestern United States. Auk 111:979–990.

Waddell, K. L., D. D. Oswald, and D. S. Powell. 1989. Forest statistics of the United States, 1987. USDA Forest Serv. Bull. PNW-RB-168. 106 pp.

Wiens, J. A. 1989. The ecology of bird communities, vol. 2: Processes and variations. Cambridge University Press, Cambridge.

Wiens, J. A., J. F. Addicott, T. J. Case, and J. Diamond. 1986. Overview: The importance of spatial and temporal scale in ecological investigations. Pp. 145–153 *in* Community ecology (J. Diamond and T. J. Case, eds). Harper and Row, New York.

33. Cowbird Parasitism in a Fragmented Landscape: Effects of Tract Size, Habitat, and Abundance of Cowbirds and Hosts

SCOTT K. ROBINSON, JEFFREY P. HOOVER,

AND JAMES R. HERKERT

Abstract

Brood parasitism by Brown-headed Cowbirds is a potential problem for many host species in the fragmented habitats of Illinois. We examined levels of brood parasitism at 22 sites scattered throughout Illinois in relation to five variables: (1) forest tract size (up to 3,500 ha), (2) cowbird and host abundance (estimated from point counts), (3) habitat type (forest, shrubland, and grassland), (4) host size, and (5) migration. Cowbird abundance did not vary significantly with forest tract size, but host abundance increased significantly with increasing tract size. As a result, the ratio of female cowbird to host abundance in fixed-radius point counts (the cowbird-to-host ratio) and community-wide frequencies of parasitism decreased significantly with increasing forest tract size. Tracts of 200 ha or less typically had mean frequencies of parasitism of 60–80% per species and intensities of 2.0–3.0 cowbird eggs per parasitized nest. Larger (over 200 ha) tracts had lower but variable levels of parasitism, and bottomland forests had lower levels of parasitism (44.6% parasitized, 1.6 cowbird eggs per parasitized nest) than upland forests (58.3% parasitized, 2.1 cowbird eggs per parasitized nest). The cowbird-to-host ratio was a good predictor of levels of cowbird parasitism for forest tracts. Overall, grassland and shrubland host species were parasitized significantly less often than forest species. Most grassland species were seldom parasitized (under 5% of all nests). Shrubland and edge-nesting species were significantly more likely to be parasitized in areas with less regional forest coverage.

We found few consistent differences in parasitism levels of species nesting in different vertical strata of forests. Levels of parasitism on ground-nesting hosts varied greatly among different regions of the state. Canopy nesters, for which we have only limited data, appeared to be heavily parasitized in all regions of the state. The number of cowbird eggs per parasitized nest increased significantly with increasing host mass for forest-dwelling neotropical migrants, as well as for shrubland-dwelling neotropical migrants, short-distance migrants, and year-round residents. Forest-dwelling neotropical migrants, however, typically had more cowbird eggs per parasitized nest than the other groups and may be more vulnerable to multiple parasitism, or alternatively, cowbirds may prefer them as hosts.

These results suggest several ways that managers can reduce the effect of cowbird parasitism on hosts, even in fragmented landscapes where cowbirds may saturate the available habitat. The best way to reduce parasitism levels may be to increase host populations through landscape-level management practices. Managers in Illinois, for example, should strive to create and/or protect contiguous forest tracts of 200 ha or greater. Restoration efforts should work outward from existing tracts, especially when they involve conversion of nearby pastures to forest habitat. These management practices should also reduce nest predation levels. In Illinois, however, habitats adjacent to major cowbird feeding areas will have high parasitism levels regardless of tract size and may require cowbird removal to protect hosts of special concern. Cowbird parasitism appears not to be a serious problem for most grassland and shrubland species in Illinois, which should enable managers to focus on reducing nest predation levels in these habitats. Forest-dwelling neotropical migrants appear to be the group of species most vulnerable to cowbird parasitism, which partially justifies the level of conservation concern this group of species has attracted.

Introduction

Management plans for reducing brood parasitism by the Brown-headed Cowbird require consideration of the regional context and of the host species most likely to be at risk. Thompson et al. (Chapter 32, this volume) have demonstrated the importance of considering regional and landscape scales when designing management plans to reduce parasitism. For example, in fragmented landscapes where habitat patches are generally small and isolated, cowbird populations can saturate available habitats and parasitism levels can be extremely high (Robinson 1992, Robinson and Wilcove 1994, Rogers et al. 1997). In fragmented landscapes, cowbird populations may be more limited by the availability of hosts than by the availability of nearby agricultural areas where they feed. In contrast, parasitism levels are generally low in extensively forested landscapes where cowbird populations may be more limited by the availability of feeding areas than by hosts (Robinson et al. 1995a). In these less fragmented landscapes, levels of cowbird parasitism may be highest near edges (Gates and Gysel 1978, Chasko and Gates 1982, Brittingham and Temple 1983, Temple and Cary 1988; Winslow et al., Chapter 34, and Chace et al., Chapter 14, this volume). In more fragmented landscapes, on the other hand, cowbirds may be abundant throughout habitat tracts, not just near edges (Robinson and Wilcove 1994). Managing cowbird parasitism will therefore require different strategies in different landscapes.

The purpose of this chapter is to extend the analysis of Thompson et al. to include the factors determining parasitism levels within a fragmented landscape. Our focus is on the effects of tract size, habitat, nesting strata within habitats, abundance of cowbirds and hosts, and on several life-history traits that may affect host vulnerability. Our goal is to provide data that can help managers faced with the formidable problems of minimizing parasitism levels in landscapes where cowbirds may saturate the available habitat. Specifically, we ask the five following questions: (1) Do levels of parasitism decrease with increasing tract size as predicted by the fragmentation hypothesis (Wilcove 1985, Wilcove et al. 1986, Brawn and Robinson 1996)? Cowbird parasitism has often been shown to be higher near edges (reviewed in Yahner 1988), but only recently have studies shown lower levels of parasitism in larger tracts within a geographical region (Hoover et al. 1995; Dowell et al., Chapter 29, and Petit and Petit, Chapter 31, this volume). (2) Are levels of parasitism different in forests of different structure and composition (e.g., upland versus bottomland)? Robinson et al. (1995b) found few studies of the effects of forest type on parasitism. (3) Are levels of parasitism higher in forest habitats, where hosts presumably have a shorter coevolutionary history with cowbird parasitism (Mayfield 1965), or are they higher in grassland and shrubland habitats, where cowbirds may have evolved (Robinson et al. 1995b)? Studies of the effects of habitat on levels of parasitism show few clear trends (Robinson et al. 1995b, Petit and Petit unpubl. manuscript), although cowbirds may avoid some habitats and prefer others (Hahn and Hatfield 1995). Community-wide levels of parasitism are high in forests (e.g., Brittingham and Temple 1983, Robinson 1992, Petit and Petit unpubl. manuscript) and grasslands (Elliott 1978; Davis and Sealy, Chapter 26, and Koford et al., Chapter 27, this volume). (4) Are hosts nesting in certain vertical strata within a habitat (ground, shrub, and canopy) more vulnerable to

parasitism? Petit and Petit (unpubl. manuscript) concluded that shrub nesters were more vulnerable than ground nesters, which in turn are more heavily parasitized than canopy-subcanopy nesters, although Robinson et al. (1995b) did not find universal support for this conclusion. Hahn and Hatfield (Chapter 13, this volume), however, found higher parasitism among ground nesters. (5) Do certain life-history traits (e.g., size and migratory status) affect host vulnerability to cowbirds? Petit and Petit (unpubl. manuscript) concluded that heavier birds were slightly less likely to be parasitized, and Robinson et al. (1993) argued that neotropical migrants may be especially vulnerable to cowbird parasitism.

In this chapter, we show that levels of parasitism significantly decrease with increasing forest tract size even in the fragmented landscapes of Illinois and that neotropical migrants of forest habitats are particularly vulnerable to parasitism. We argue that the cowbird-to-host ratio may be a good general predictor of levels of cowbird parasitism but that nesting stratum is not. We further show that cowbird parasitism is not a problem for most grassland and many shrubland species in Illinois. We conclude by developing management guidelines specific to fragmented landscapes where cowbird parasitism is most likely to be a problem.

Study Areas and Methods

Study Area

This study uses census data from 33 forested areas in four regions of Illinois and nesting data from 22 of these areas (Figure 33.1, Table 33.1). Most of the tracts in the Illinois Ozarks region are large (over 1,500 ha), and five have been studied intensively (Table 33.1). The Illinois Ozarks region is one of the two most heavily forested regions in Illinois (Iverson et al. 1989). Forests in these regions are topographically uniform and are dominated by narrow ridges and ravines covered by 80–130-year-old oaks (Quercus spp.) and hickories (Carya spp.). The ravines are generally younger and have species of trees characteristic of mixed-mesic stands such as sugar maple (Acer saccharum), American beech (Fagus grandifolia), and tulip tree (Liriodendron tulipifera) as well as oaks and hickories.

The tracts in the Cache River region vary enormously in size (Table 33.1) and are dominated by bottomland hardwoods, especially diverse species of oak and hickory as well as some American sycamores (Platanus occidentalis). These forests represent the northernmost extension of Mississippi River alluvial bottomland (swamp) forest. Most forest tracts in the Cache River area are located in the floodplain of the river and are surrounded by vast areas of agricultural fields. All data from the Cache River region come from a single field season (1993).

The medium to small tracts (under 500 ha) of the central region of Illinois are all along rivers on the oak- and hickory-

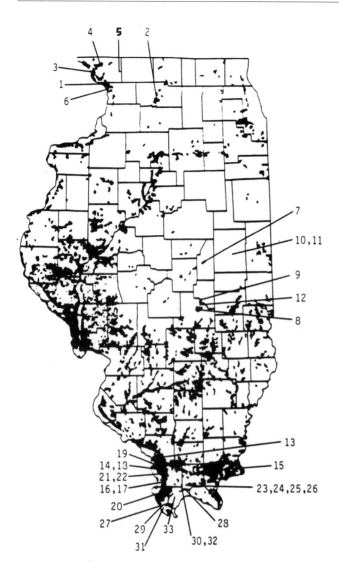

Figure 33.1. Locations of forest study sites in Illinois. Numbers correspond to study sites listed in Table 33.1.

dominated slopes that were too steep for farming or that were preserved as a source of wood for local landowners. These tracts are among the most isolated in North America and generally form habitat islands or archipelagoes of islands in a sea of corn and soybeans (Kendeigh 1982, Blake and Karr 1987, Robinson 1988). These tracts have been studied at various times during the 1985–1993 period (e.g., Robinson 1992, J. D. Brawn unpubl. data).

The tracts of northwestern Illinois are mostly medium-sized (100–450 ha) in an unglaciated landscape dominated by the bluffs of the Mississippi River (Hanover Bluff and North and South Mississippi Palisades State Park) and hilltop woodlots away from the river (Tapley Woods and Wards Grove). The forest composition of these sites is dominated by upland oak and hickory, with more mesic areas dominated by black walnut (*Juglans nigra*) and American bass-

wood (*Tilia americana*). These woodlots were all studied in 1992 and 1993.

Grassland Species

We obtained most of our data on grassland birds from sites other than those listed in Table 33.1. The only sites in Table 33.1 where we also obtained grassland data were in the Cache River area of southern Illinois, where we found 146 nests of seven grassland species. The remainder of the data on nests of grassland birds in fragmented Illinois grasslands were obtained from 13 study areas located in northeastern and east-central Illinois. Grassland fragments were 18–1,000+ ha in size and included native prairie, restored prairie, and cool-season grass fields. Dominant grass species from the native and restored prairie study areas included big bluestem (*Andropogon gerardii*), yellow Indian grass (*Sorghastrum nutans*), panic grass (*Panicum* spp.), prairie cordgrass (*Spartina pectinata*), prairie dropseed (*Sporobolus heterolepis*), and upland sedges (*Carex* spp.). Dominant grass species from the nonprairie areas included Kentucky bluegrass (*Poa pratensis*), meadow fescue (*Festuca pratensis*), smooth bromegrass (*Bromus inermis*), timothy (*Phleum pratense*), orchard grass (*Dactylis glomerata*), and redtop (*Agrostis alba*).

Additional data on grassland bird nests were obtained from the Illinois nest card files housed at the Illinois Department of Conservation. Only nest cards that described the habitat as a grassland or a grass-dominated field were used. For the grassland fragment and nest card data combined, grassland bird nesting data were included from 38 of Illinois's 102 counties.

Censusing

We used a modified version of Hutto et al.'s (1986) fixed-radius point counts to quantify cowbird and host abundance in each tract. Points were located within each tract at 150-m intervals on grids, or in transects along ravines. Each point was censused 1–4 times a season by an experienced observer for 6-min periods, during which the species, compass direction, distance, and vocalization type (song vs. call) were noted for each bird heard or observed. Censuses began within half an hour of sunrise and continued until the transect was completed (usually before 1030) between 15 May and 5 July. Censuses were not conducted on windy or rainy days or after 1100. We chose 6 min partly to maintain consistency with censuses conducted in 1985 and 1986 (Robinson 1988, 1992) and partly because we needed the full 6 min to census swamp forests where hosts were particularly abundant (typically 30–40 records per point).

For the purposes of this study, we counted all records of cowbirds and hosts within a 70-m radius of each point. We chose 70 m to minimize overlap between adjacent points, which were 150 m apart, and because all cowbird hosts were easily audible at this distance (S. K. Robinson unpubl. data).

When we judged that the same individual had moved from one 70-m radius circle to the next on consecutive census points, we counted it only once. Overall host abundance for a site was quantified by lumping all point counts for the site together and calculating the mean number of hosts counted per point. Each tract is therefore an independent sample with a single value for host abundance. We do not use these results to estimate population density, because it is unlikely that all birds within a 70-m radius vocalize during a 6-min point count and because in species such as Northern Cardinals (*Cardinalis cardinalis*) and Eastern Wood-Pewees (*Contopus virens*), both males and females sing. Our index of abundance therefore may have overestimated the abundance of some species, but without knowing how often females of such species sing, we cannot correct for this potential bias.

We used vocalizations of female cowbirds to estimate cowbird abundance. Because male cowbirds generally outnumber females and often travel alone (Robinson et al. 1995b), we decided that the abundance of females was a better predictor of local parasitism pressure. To estimate female cowbird abundance, we recorded the rattle calls separately on our censuses. Males occasionally give rattle calls (Robinson pers. obs.), but females gave over 99% of the rattle calls heard in Illinois (Robinson unpubl. data).

To control for the effects of host abundance when estimating relative abundance of cowbirds, we used the ratio of cowbird to host abundance as an index. We divided the mean number of female cowbirds detected per 70-m radius point count by the mean number of hosts to obtain this ratio. As we document below, this index is a good predictor of community-wide levels of parasitism in the forested habitats of Illinois. Hosts included all species known to accept cowbird eggs regularly. We excluded all cavity nesters except the Prothonotary Warbler (*Protonotaria citrea*), all species known to eject cowbird eggs (e.g., Gray Catbird *Dumetella carolinensis* and American Robin *Turdus migratorius*), and all species too large to be parasitized (e.g., Blue Jay *Cyanocitta cristata*, raptors, Wild Turkey *Meleagris gallopavo*). We also excluded the Blue-gray Gnatcatcher (*Polioptila caerulea*), which has usually finished nesting by the time censuses begin (15 May) and is only rarely parasitized (Robinson unpubl. data). Gray Catbirds and American Robins are parasitized occasionally (Robinson unpubl. data), which argues that they perhaps should be included. Neither species, however, was abundant enough in our study areas for its inclusion or exclusion to change our results.

Parasitism Levels of Forest and Shrubland Species

We used percentage of nests parasitized and mean number of cowbird eggs per parasitized nest by species as indices of levels of parasitism. At each site, teams of skilled fieldworkers searched for nests on forest study areas and in adjacent second-growth and edge habitats for shrubland species. Fieldworkers searched fields adjacent to these forest habitats

for nests of grassland species. Once nests were located, the cowbird and host eggs were counted. For this chapter, we used only nests located during the incubation and early nestling period (within 2–3 days of hatching) and did not include cowbird eggs laid in nests that were abandoned before incubation began. We also did not count cowbird eggs laid after hatching began. These results may slightly underestimate the species-specific intensity of parasitism by excluding some nests that may have been abandoned as a result of parasitism (Neudorf and Sealy 1994). Similarly, by excluding species that eject cowbird eggs, we are also underestimating the potential intensity of parasitism in a habitat.

Statistical Analysis

We first define all of the variables involved in the following analyses. Forest tract size is the area (ha) of forest for each study site. Cowbird abundance is the mean number of female cowbirds per 70-m radius census point per study site. Similarly, host abundance is the mean number of potential cowbird hosts per 70-m radius census point per study site. The cowbird-to-host ratio is simply the female cowbird abundance divided by the host abundance for each study site. For statistical analyses, the cowbird-to-host ratio was transformed by taking the natural log (ln) of the ratio. The cowbird-to-host ratio was transformed to normalize the variance for linear regression analysis (Neter et al. 1990). The range of values for the cowbird-to-host ratio was from 0.011 (-4.51 after being transformed) to 0.19 (-1.66 after being transformed). For graphical presentation, nontransformed values are used so that others can more readily compare their results with our data.

Levels of parasitism were determined for each study site by calculating the mean percentage of nests parasitized (parasitism frequency) and the mean number of cowbird eggs per parasitized nest (parasitism intensity, independent of its frequency). These two measures in combination provide a good indication of the severity of parasitism. The percentage of nests parasitized on each study site was determined by computing the percentage of nests that was parasitized for each species and calculating an overall mean percentage per species for each study site. Similarly, the number of cowbird eggs per parasitized nest was determined for each study site by computing the mean number of cowbird eggs per parasitized nest for each species and calculating an overall mean for the number of cowbird eggs per parasitized nest per species for each study site.

Single-factor regression analyses were used to test for effects between variables. Cowbird abundance, host abundance, the cowbird-to-host ratio, the mean percentage of nests parasitized, and the mean number of cowbird eggs per parasitized nest were each compared with forest tract size, which was the independent variable. Next, the percentage of nests parasitized and the mean number of cowbird eggs per parasitized nest were each compared with host abundance,

Table 33.1. Forested Study Sites in Illinois Where Songbird Censuses Were Conducted

Study Site[a] (Years)	Latitude Longitude	Forest Area (ha)	Habitat Type[b]
Northwestern Illinois			
1 MS Palisades North*	N 42°16'24"	525	1, 2
(1992–1993)	W 90°14'07"		
2 Castle Rock	N 42°03'19"	350	1, 2
(1992–1993)	W 89°21'04"		
3 Hanover Bluff*	N 42°14'08"	200	1
(1992–1993)	W 90°18'00"		
4 Tapley Woods*	N 42°21'25"	150	1
(1992–1993)	W 90°17'43"		
5 Wards Grove*	N 42°19'18"	140	1, 2
(1992–1993)	W 89°57'34"		
6 MS Palisades South*	N 42°13'18"	130	1, 2
(1992–1993)	W 90°14'41"		
Central Illinois			
7 Allerton Park*	N 39°59'33"	400	1, 2, 3
(1992–1993)	W 88°39'28"		
8 Lake Shelbyville Boot*	N 39°28'54"	65	1
(1985–1993)	W 88°43'45"		
9 Railroad Woods	N 39°33'25"	23	1
(1985–1993)	W 88°42'31"		
1 Trelease Woods*	N 40°07'56"	22	1
(1992–1993)	W 88°08'28"		
11 Brownfield Woods	N 40°08'11"	20	1
(1992–1993)	W 88°09'51"		
12 Pogue Woods	N 39°31'15"	13	1
(1985–1993)	W 88°44'02"		
Illinois Ozarks			
13 Cave Valley*	N 37°38'54"	3,000	3, 4
(1991–1993)	W 89°21'30"		
14 Pine Hills*	N 37°31'27"	2,900	1, 2
(1989–1993)	W 89°21'30"		
15 Lusk Creek*	N 37°27'38"	2,500	1, 2
(1989\ -1990)	W 88°33'03"		
16 Dutch Creek*	N 37°25'52"	2,400	1, 2
(1989–1993)	W 89°21'30"		
17 Atwood Bottoms*	N 37°24'25"	2,000	3, 4
(1989–1993)	W 89°22'20"		
18 Pine Hills Ridge	N 37°33'44"	2,000	1
(1990–1993)	W 89°25'52"		
19 Ranbarger Hollow	N 37°31'45"	2,000	1, 2
(1992)	W 89°22'17"		

Table 33.1. Continued

Study Site[a] (Years)	Latitude Longitude	Forest Area (ha)	Habitat Type[b]
20 South Ripple Hollow*	N 37°19'39"	1,600	1, 2
(1989–1993)	W 89°22'36"		
21 North Trail of Tears*	N 37°30'37"	1,200	1, 2
(1991–1993)	W 89°21'30"		
22 South Trail of Tears*	N 37°28'33"	700	1, 2
(1989–1993)	W 89°22'20"		
Cache River			
23 Forman Tract*	N 37°20'54"	1,800	1, 3, 4
(1993)	W 88°55'10"		
24 Wildcat Swamp*	N 37°22'06"	1,800	1, 3, 4
(1993)	W 88°55'26"		
25 Little Black Slough	N 37°21'42"	1,400	1, 3, 4
(1993)	W 88°57'25"		
26 North Cache	N 37°22'43"	1,400	3
(1993)	W 88°57'15"		
27 Section 11 Woods*	N 37°08'18"	550	3, 4
(1993)	W 89°17'58"		
28 Main Tract*	N 37°19'02"	140	3, 4
(1993)	W 88°59'32"		
29 Equestrian Woods	N 37°07'26"	70	3, 4
(1993)	W 89°20'35"		
30 Hogue Woods*	N 37°20'54"	65	3
(1993)	W 89°03'48"		
31 Roth Woodlots	N 37°06'26"	30	3
(1993)	W 89°18'38"		
32 Kessler Tract*	N 37°20'28"	22	3
(1993)	W 89°04'23"		
33 Ullin Woodlots	N 37°15'54"	20	3
(1993)	W 89°13'18"		

[a]Numbers to the left of site names correspond to site numbers on Figure 33.1. Asterisks indicate sites where nesting data were collected.
[b]1, upland oak-hickory on steep slopes and ridges; 2, mixed-mesic, usually along streams; 3, floodplain bottomland forest; 4, swamp forest (tupelo/cypress).

with host abundance being the independent variable. Then the percentage of nests parasitized and the mean number of cowbird eggs per parasitized nest were each compared with the cowbird-to-host ratio, with the cowbird-to-host ratio being the independent variable. Finally, the percentage of nests parasitized and the mean number of cowbird eggs per parasitized nest were each compared with cowbird abundance, with cowbird abundance being the independent variable.

Two-factor stepwise regression analyses were used to determine the variables that best predicted levels of parasitism. One stepwise regression analysis was completed for the percentage of nests parasitized, and one for the mean number of cowbird eggs per parasitized nest. In each analysis, the cowbird-to-host ratio and forest tract size were the independent variables. Host and cowbird abundances were not included in these analyses because they are a part of the cowbird-to-host ratio. The independent variables had to meet a minimum F-value ($P \leq .10$) to be entered into the model.

The percentage of nests parasitized and the mean number of cowbird eggs per parasitized nest was compared between bottomland and upland forest sites. Bottomland sites ($N = 8$) were typically in floodplains and sometimes included swamp forest (labeled as 3 or 4 in Table 33.1) and upland sites ($N = 14$) typically included oak-hickory ridges and mixed-mesic ravines (labeled as 1, or 1, 2 in Table 33.1). A t-test was used to determine whether the percentage of nests parasitized or the mean number of cowbird eggs per nest differed between bottomland and upland forests.

A chi-square test was used to determine if the frequency of five different levels of parasitism differed between combined shrub-sapling and canopy-nesting species and ground-nesting species.

Single-factor regression analyses were used to determine if the number of cowbird eggs per parasitized nest depended on host body mass for forest-dwelling neotropical migrants or for shrubland-dwelling neotropical migrants, short-distance migrants, and year-round residents. Values for host body mass were obtained from Dunning (1984).

Results

Effects of Forest Tract Size

Cowbird abundance was not related to forest tract size ($F = 2.16$, df = 1, 31, $P = .152$ with the outlier; $F = 1.35$, df = 1, 30, $P = .255$ without the outlier) (Figure 33.2a). Host abundance, however, increased significantly with increasing tract size ($F = 14.85$, df = 1, 31, $P = .001$) (Figure 33.2b). Tracts of 200 ha or smaller had fewer hosts per census point, but above that size, host abundance was not related to tract size ($F = 1.15$, df = 1, 17, $P = .298$). As a result, the cowbird-to-host ratio decreased significantly with increasing forest tract size (Figure 33.2c). Cowbird and host abundance, however, were not significantly correlated ($r = 0.064$, $P > .30$), a result that differs considerably from comparisons of cowbird and host abundance within study areas in the Illinois Ozarks (Thompson et al., Chapter 32, this volume).

Frequency of parasitism decreased significantly with increasing forest tract size ($F = 10.97$, df = 1, 20, $P = .003$), whereas intensity of parasitism showed a decreasing trend with increasing forest tract size ($F = 3.33$, df = 1, 20, $P = .083$) (Figure 33.3a,b). Both frequency and intensity of parasitism decreased significantly with increasing host abundance ($F = 14.45$, df = 1, 20, $P = .001$; $F = 5.30$, df = 1, 20, $P = .032$, respectively) (Figure 33.3c,d). Percentage of nests parasitized and number of cowbird eggs per parasitized nest, however, were best predicted by the cowbird-to-host ratio, and the two dependent variables increased significantly with increasing cowbird-to-host ratios ($F = 29.03$, df = 1, 20, $P < .001$; $F = 21.02$, df = 1, 20, $P < .001$, respectively) (Figure 33.3e,f). The percentage of nests parasitized

was not related to cowbird abundance ($F = 3.48$, df = 1, 20, $P = .077$), but the number of cowbird eggs per parasitized nest increased significantly with increasing cowbird abundance ($F = 7.68$, df = 1, 20, $P = .012$) (Figure 33.3g,h). We found this result largely because areas with very high cowbird abundance were also areas with high levels of parasitism regardless of host abundance.

From a management perspective, the key result is that levels of parasitism decreased significantly with increasing forest tract size. Tracts of 200 ha or less tended to have very high levels of parasitism (Figure 33.3a,b), mostly as a result of the scarcity of hosts relative to the abundance of cowbirds. Managing for large tracts both increases host populations and decreases levels of parasitism by decreasing the abundance of cowbirds relative to hosts. It is the cowbird-to-host ratio alone, however, that best models both percentage of nests parasitized and number of cowbird eggs per parasitized nest when both the cowbird-to-host ratio and forest tract size are included in stepwise multiple regression analyses (see Figure 33.3e,f for models).

Habitat Effects

Both the percentage of nests parasitized and number of cowbird eggs per nest were significantly lower in bottomland forests than in upland forests (mean = 44.6% bottomland vs. 58.3% upland, $t = 2.09$, df = 20, $P = .049$; mean = 1.58 eggs bottomland vs. 2.05 eggs upland, $t = 2.63$, df = 20, $P = .016$). These results, however, should be interpreted with caution because we have only a single year of data from each of our bottomland sites (Table 33.1). On a broader scale, levels of parasitism differed greatly among birds nesting in forest, shrubland, and grassland habitats in Illinois (Figure 33.4).

Facing page

Figure 33.2. Forest tract size and bird numbers for the 33 study sites with census data. Forest tract size is compared with (a) the abundance of female cowbirds, (b) the abundance of potential cowbird hosts, and (c) the cowbird-to-host ratio.

Overleaf

Figure 33.3. Relationships for the 22 study sites with both census and nesting data. Effect of forest tract size on (a) the percentage of nests parasitized and (b) the mean number of cowbird eggs per parasitized nest. Effect of the abundance of potential cowbird hosts on (c) the percentage of nests parasitized and (d) the mean number of cowbird eggs per parasitized nest. Effect of the cowbird-to-host ratio on (e) the percentage of nests parasitized and (f) the mean number of cowbird eggs per parasitized nest. Effect of the abundance of female cowbirds on (g) the percentage of nests parasitized and (h) the mean number of cowbird eggs per parasitized nest.

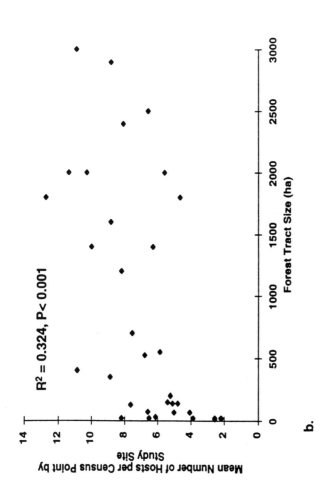

a.

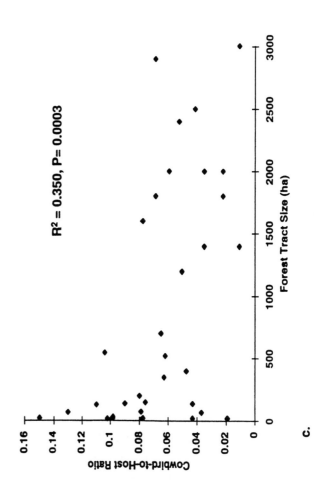

c.

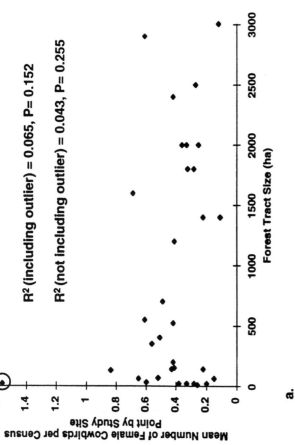

b.

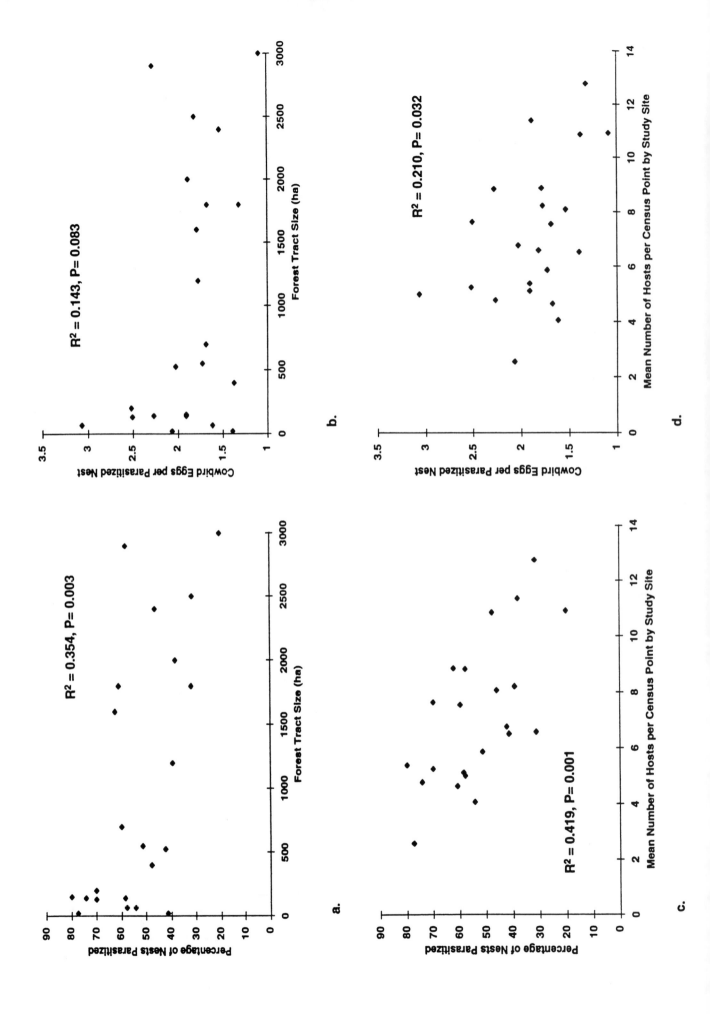

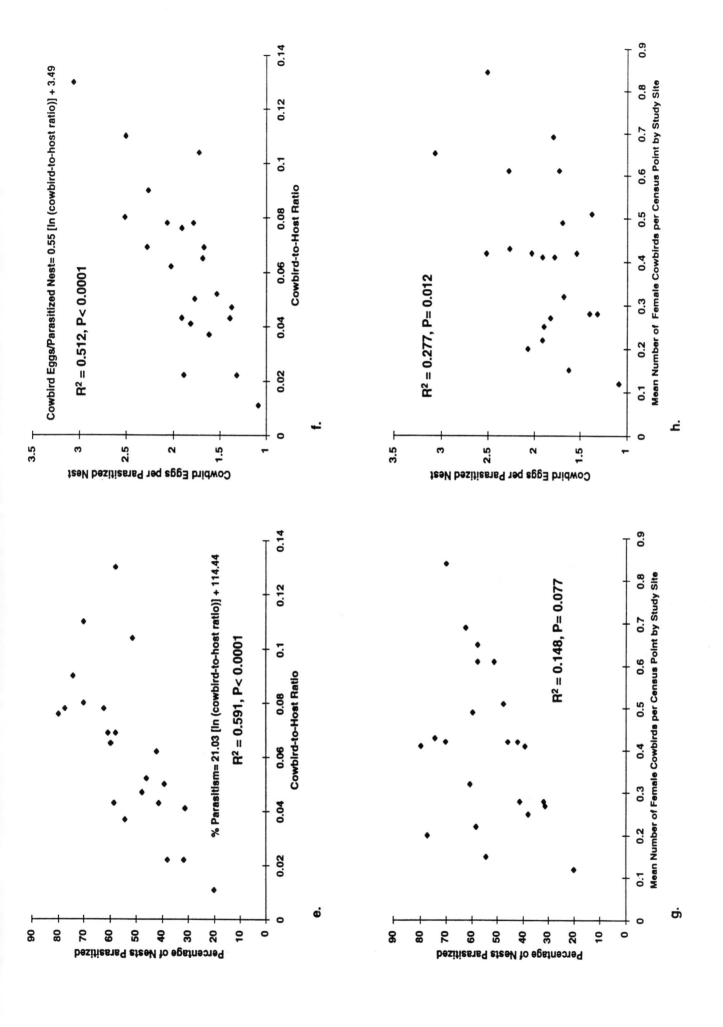

None of the 54 nests found in 13 grassland fragments in east-central and northeastern Illinois were parasitized, and only the Common Yellowthroat (*Geothlypis trichas*) was heavily parasitized in grassland fragments in the Cache River area (Table 33.2). Results from the nest-card program reinforce these patterns, with an overall frequency of parasitism of only 8.1% of 272 nests (Table 33.2). The low frequencies of parasitism of grassland birds in the Cache River area contrasted with adjacent forest tracts in which parasitism levels typically ranged from 50% to 80% per host species (S. K. Robinson and J. P. Hoover unpubl. data).

Frequencies of parasitism of shrubland species were intermediate between those of grassland and forest hosts (Figure 33.4) and showed considerable variation among the four regions of the state (Figure 33.5). In central Illinois, where less than 5% of the landscape was forested, levels of parasitism of shrubland birds were generally high. In contrast, none of the shrubland species nesting along the edges of extensive forests of the Illinois Ozarks were parasitized at levels of 50% or greater (Figure 33.4b). Among species characteristic of shrubland habitats, only the Orchard Oriole (*Icterus spurius*) in central Illinois had levels of parasitism approaching those of forest-nesting Wood Thrushes (*Hylocichla mustelina*) and tanagers (Robinson 1992, Robinson and Wilcove 1994).

Nesting Stratum

Nesting stratum was not significantly related to levels of parasitism for forest birds. Although shrub-sapling and canopy nesters were more likely to be heavily parasitized than ground nesters (over 80% of all nests, Figure 33.6), the difference was not statistically significant ($X^2 = 3.01$, df = 1, $P > .10$). Within forests, therefore, cowbirds appeared to be using all available hosts, regardless of nesting stratum.

Host Size and Migratory Status

Intensity of parasitism, as indexed by cowbird eggs per parasitized nest, increased significantly with increasing host mass for forest-dwelling neotropical migrants ($F = 33.50$, df = 1, 15, $P < .001$; Figure 33.7a). For neotropical migrants of shrubland habitats, short-distance migrants, and year-round residents, intensity of parasitism was also related to host mass ($F = 8.21$, df = 1, 10, $P = .017$ not including the outlier; Figure 33.7b). When the single outlier (Common Yellowthroat) was included, the trend for increasing multiple parasitism of larger species was masked ($F = 0.62$, df = 1, 11, $P = .447$). Forest-dwelling neotropical migrants, however, typically received 1.5–3.0 cowbird eggs per parasitized nest, whereas the other groups typically received only 1.0–1.5. These results suggest either that forest-dwelling neotropical migrants are more likely to accept multiple cowbird eggs or that cowbirds avoid multiply parasitizing nests of other kinds of host species. For some large hosts such as the Wood Thrush, Veery (*Catharus fuscescens*), and Scarlet (*Piranga olivacea*) and Summer (*P. rubra*) Tanagers, parasitized nests averaged more cowbird than host eggs (e.g., Robinson 1992, unpubl. data).

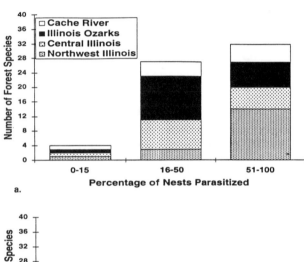

a.

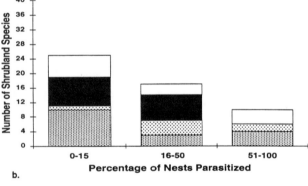

b.

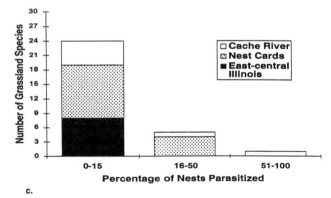

c.

Figure 33.4. The number of (a) forest, (b) shrubland, and (c) grassland bird species from each of the four study regions experiencing low (0–15%), moderate (16–50%), or high (51–100%) frequencies of brood parasitism.

Discussion

In the fragmented landscapes of Illinois, levels of cowbird parasitism were highest in small forest tracts and lowest in large forest areas and in grasslands. Shrubland habitats had intermediate levels of cowbird parasitism. Within forest tracts, cowbirds appeared to be unselective, parasitizing

Table 33.2. Levels of Brood Parasitism in Illinois Grasslands

Species	Grassland Fragments (1988–1993) Nests/ Parasitized	Cache River Grasslands (1993) Nests/ Parasitized	Nest Cards[a] (1970–1988) Nests/ Parasitized	Total % Parasitized	(N)
Horned Lark (*Eremophila alpestris*)	1/0	0/0	10/1	9.1	(11)
Sedge Wren (*Cistothorus platensis*)	0/0	0/0	1/0	0.0	(1)
Common Yellowthroat (*Geothlypis trichas*)	1/0	10/7	3/0	50.0	(14)
Dickcissel (*Spiza americana*)	0/0	22/0	7/0	0.0	(29)
Field Sparrow (*Spizella pusilla*)	2/0	40/5	44/7	14.0	(86)
Vesper Sparrow (*Pooecetes gramineus*)	3/0	0/0	4/2	28.6	(7)
Lark Sparrow (*Chondestes grammacus*)	0/0	0/0	11/2	18.1	(11)
Savannah Sparrow (*Passerculus sandwichensis*)	2/0	0/0	4/1	16.7	(6)
Grasshopper Sparrow (*Ammodramus savannarum*)	0/0	1/0	12/1	7.7	(13)
Henslow's Sparrow (*A. henslowii*)	0/0	0/0	2/0	0.0	(2)
Song Sparrow (*Melospiza melodia*)	0/0	6/3	8/1	28.6	(14)
Bobolink (*Dolichonyx oryzivorus*)	1/0	0/0	2/0	0.0	(3)
Red-winged Blackbird (*Agelaius phoeniceus*)	19/0	66/0	132/6	2.8	(217)
Eastern Meadowlark (*Sturnella magna*)	25/0	1/0	31/1	1.8	(57)
Western Meadowlark (*S. neglecta*)	0/0	0/0	1/0	0.0	(1)
Overall	54/0	146/15	272/22	7.8	(472)

[a]Only nest cards that describe the habitat as grassland or grass-dominated field are included.

hosts nesting in all strata. Neotropical migrants appeared to be particularly vulnerable to multiple parasitism, which can be costly to host species (Trine, Chapter 15, this volume; Robinson unpubl. data). In the following discussion, we compare our results with those from other studies and expand on their implications for host-cowbird coevolution and management strategies to reduce parasitism.

Effects of Tract Size

Our results strongly support the fragmentation hypothesis, which predicts an inverse relationship between tract size and levels of cowbird parasitism (Wilcove et al. 1986). Levels of parasitism were high in all of the tracts we studied, but they decreased in tracts greater than 200 ha. It is only recently that similar effects of tract size on levels of parasitism have been demonstrated in other studies (Hoover et al. 1995; Dowell et al., Chapter 29, and Petit and Petit, Chapter 31, this volume). In Illinois, the effects of forest area on levels of parasitism do not result from the increased edge-to-interior ratios of small tracts, which Brittingham and Temple (1983) hypothesized would account for higher levels of parasitism in small tracts. In Illinois forests, levels of parasitism generally do not decrease in the interior of large tracts, a possible indication of saturated cowbird populations (Robinson and

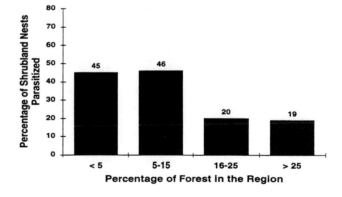

Figure 33.5. The percentage of nests parasitized in shrubland habitat in central Illinois (< 5% forested), the Cache River (5–15% forested), northwest Illinois (16–25% forested), and the Illinois Ozarks (> 25% forested).

Wilcove 1994, Brawn and Robinson 1996, Robinson unpubl. data; Thompson et al., Chapter 32, this volume). Instead, levels of parasitism appear to decrease in large tracts because host abundance increases, which decreases the cowbird-to-host ratio and reduces the intensity of parasitism for most host species. Thus, the creation and maintenance of large

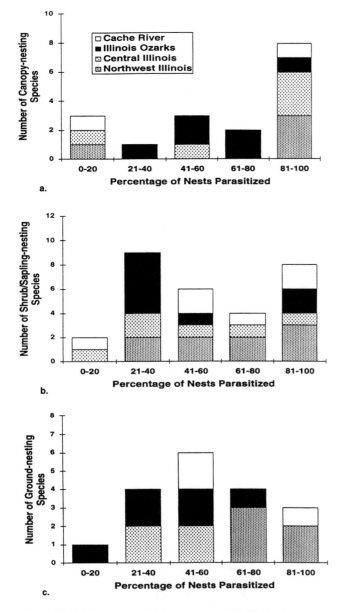

Figure 33.6. The number of (a) canopy-nesting, (b) shrub- or sapling-nesting, and (c) ground-nesting bird species from the four study regions that are parasitized at one of five different frequencies of brood parasitism.

tracts of forest benefits songbirds by providing sufficient habitat for many area-sensitive species (Hayden et al. 1985, Blake and Karr 1987, Robbins et al. 1989) and by decreasing the ratio of cowbirds to hosts within tracts. Efforts to restore entire ecosystems should therefore increase bird community richness, abundance, and nesting success. Such efforts are currently under way in the Cache River area by the Nature Conservancy, the U.S. Fish and Wildlife Service, and the Illinois Department of Conservation. There has been a recent parallel emphasis on creating large tracts of wildlife habitat (macrosites) by the Illinois Department of Conservation.

Nevertheless, there is a great deal of residual variation in

the regressions shown in Figures 33.2–33.3. Cowbird abundance, for example, varies greatly among similar-sized sites. Some of the tracts with high abundances of cowbirds are immediately adjacent to known cowbird foraging areas (e.g., Pine Hills and Forman Tract). Pine Hills, in particular, has a campground in which cowbirds forage regularly. The interiors of large tracts (more than 200 ha) of floodplain forest have unusually low cowbird-to-host ratios. These low ratios may be a result of the increased host abundance that we have documented in this habitat type and the rarity of some of the most heavily parasitized hosts there, such as tanagers and Wood Thrushes (S. K. Robinson and J. P. Hoover unpubl. data). At present, there are too few replicate plots to test for the additional effects of proximity to cowbird feeding areas and forest type (bottomland vs. upland) on levels of parasitism. Managers, however, might be able to reduce local parasitism levels within sites by not mowing areas such as roadsides and campgrounds within and adjacent to large tracts and by local cowbird trapping (see Robinson et al. 1993 for more detailed management recommendations). Similarly, conversion of pastures to forest adjacent to nature preserves might reduce cowbird abundance in the whole area. Nevertheless, very large areas may need to be restored to reduce cowbird parasitism in Illinois. We estimate that a minimum area of 8,000 ha with no agricultural inholdings where cowbirds can forage may be necessary to escape the effects of cowbird parasitism.

As a general rule for Illinois forest habitats, management efforts to benefit songbirds should be aimed at restoring and preserving tracts of 200 ha and greater (Herkert et al. 1993). Tracts smaller than 200 ha lack many area-sensitive species (Blake and Karr 1987), have low abundances of hosts, and have correspondingly high cowbird-to-host ratios and levels of brood parasitism. These management guidelines will also likely reduce levels of nest predation (Herkert et al. 1993, Robinson unpubl. data).

Habitat Effects

At least in Illinois, cowbird parasitism does not appear to be a major problem for birds of grassland habitats. These results are surprising, given the high levels of parasitism sustained by many of the same species elsewhere in their ranges (e.g., Friedmann 1929). In the prairies of North Dakota and Kansas, for example, grassland species are often heavily parasitized (Elliott 1978; Koford et al., Chapter 27, and Davis and Sealy, Chapter 26, this volume). The lack of parasitized Dickcissel (*Spiza americana*) and Eastern (*Sturnella magna*) and Western (*S. neglecta*) Meadowlark nests in our sample is particularly remarkable, given the intense parasitism on these species documented by Elliott (1978) and by Koford et al. and Davis and Sealy in this volume. It is likely that cowbirds avoid grassland habitats in Illinois, but it is also possible that the landscape of Illinois makes grassland birds less vulnerable to parasitism here than in other regions.

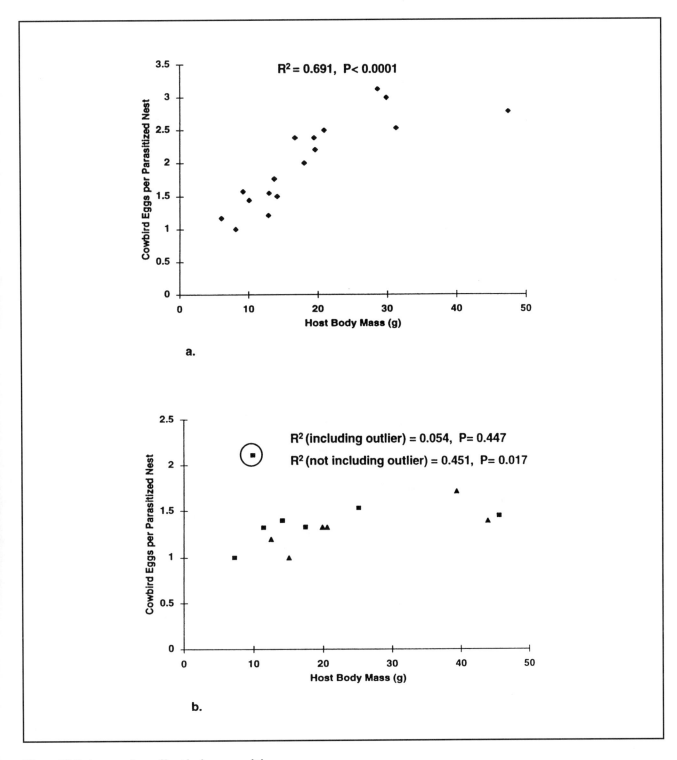

Figure 33.7. A comparison of host body mass and the mean number of cowbird eggs per parasitized nest for (a) forest-dwelling neotropical migrant songbirds and for (b) combined shrubland-dwelling neotropical migrant species (squares) and short-distance migrant or resident species (triangles).

Cowbirds are relatively rare in Illinois grassland fragments. In a study on 11 prairie fragments in Illinois (Herkert in press), cowbirds were encountered on less than 10% of 42 transects. Cowbirds may avoid grassland fragments in Illinois because of the relatively low density of potential hosts there. Bird densities in tallgrass prairie habitat are considerably lower than in eastern deciduous forests (Herkert 1991a). Although Illinois is intensively farmed, there are numerous small forest fragments and wooded riparian strips throughout the state (Iverson et al. 1989) that may provide higher host densities than grassland fragments. Additionally, the most common breeding bird species in Illinois grassland fragments is the Red-winged Blackbird (*Agelaius phoeniceus*), which accounts for over one-fourth of the breeding birds in grassland fragments in Illinois (Herkert 1991b). Although Red-winged Blackbirds accept cowbird eggs (e.g., Weatherhead 1989), they also can be aggressive toward cowbirds (Friedmann 1929), which may limit the success of the cowbirds in these areas. Other grassland birds may benefit from the defensive behavior of Red-winged Blackbirds (see also Clark and Robertson 1979).

The location of suitable feeding areas may also limit cowbird densities in Illinois grassland fragments, many of which occur in areas surrounded by intensive rowcrop agriculture. Therefore, cowbirds would need to travel relatively long distances between grassland fragments and pastures used for feeding. Additionally, nests of grassland birds can be very difficult to locate, because the birds often land on the ground several meters from their nest before proceeding to the nest under the cover of the surrounding tall grasses. This behavior presumably makes it more difficult for a cowbird to locate a nest of a grassland bird, compared with nests of forest birds, because forest birds typically cannot hide as they approach their nests. We therefore suggest that the combination of low densities of hosts, distant cowbird feeding areas, the difficulties in finding host nests in tall grasslands, and the availability of more host-rich habitats nearby makes searching for and parasitizing grassland bird nests in Illinois less efficient for cowbirds than in other regions.

Managers of grassland sites in Illinois therefore appear to have little need to make special efforts to reduce cowbird populations. Levels of nest predation, however, are very high in Illinois grassland birds (Herkert 1991b) and may be related to tract size and proximity to edge (Johnson and Temple 1986, 1990; Burger 1988). Many of the same management guidelines that should reduce cowbird parasitism (e.g., creating and restoring large tracts) might also reduce nest predation (Herkert et al. 1993).

Brood parasitism may also be less of a problem for species of shrubland and edge habitats in Illinois. In extensively forested landscapes such as the Illinois Ozarks and the Mississippi River bluffs of northwestern Illinois, levels of parasitism on shrubland species were generally lower than they were in the less forested landscapes of central Illinois and the Cache River area (Figure 33.6). In areas where many forest hosts are available, cowbirds may bypass second-growth habitats (see Hahn and Hatfield 1995). In landscapes where forest hosts are scarce, cowbirds may have little choice but to parasitize second-growth species.

Cowbirds might avoid shrubland habitats because of the high degree of resistance to parasitism shown by many species there. In Illinois, many shrubland species eject cowbird eggs (e.g., Gray Catbird, Northern Oriole (*Icterus galbula*), American Robin, Rothstein 1975) or often abandon parasitized nests (e.g., Field Sparrow [*Spizella pusilla*], Burhans, Chapter 18, this volume; Indigo Bunting [*Passerina cyanea*] and White-eyed Vireo [*Vireo griseus*], Robinson unpubl. data; Prairie Warbler [*Dendroica discolor*], Nolan 1978; Yellow Warbler [*D. petechia*], Sealy 1992). Because cowbirds may not be able to discriminate between hosts that accept and reject cowbird eggs, cowbird breeding success might be much lower in shrubland habitats where there are many more species that reject cowbird eggs or abandon parasitized nests than there are in forest habitats. Thus, the lower levels of parasitism in shrubland habitats may reflect the combined effects of avoidance of this habitat by cowbirds and the greater resistance of hosts there.

The low levels of parasitism of grassland and shrubland species in Illinois are consistent with Mayfield's (1965) hypothesis that hosts with long histories of exposure to cowbird parasitism are more likely to have evolved antiparasite defenses. Closed-canopy forests in Illinois may once have been largely restricted to the floodplains of major rivers (Iverson et al. 1989). With fire suppression, closed-canopy forests may have replaced the more open oak and shrub savannahs that some think may have covered much of the state that was not open tallgrass prairie (Iverson et al. 1989). Thus, the structure of the forests of Illinois and their bird communities today might be much different from those before European settlement. Many of the forest birds that now occupy upland habitats may be recent colonists from areas where cowbird parasitism was not much of a problem historically. The grassland and shrubland species of Illinois, on the other hand, may have been exposed to cowbird parasitism for a long time in this part of the Midwest, which gave them a longer time to evolve defenses. The lack of effective defenses of many grassland and shrubland species in the Great Plains (Koford et al., Chapter 27, this volume), where cowbirds have long been abundant (Mayfield 1965), however, argues against this interpretation. Studies of the composition and susceptibility to cowbird parasitism of savannah bird communities might shed light on the coevolution of cowbirds and their hosts in the American Midwest.

Management efforts that increase regional coverage of forest and increase tract size may also indirectly benefit many of the shrubland species that occur along shrubby forest edges and in disturbed areas within forests. In areas of abundant forest, cowbirds may avoid shrubland habitats in favor of the more vulnerable hosts of forests. Managers responsible for very small (under 200 ha) forest tracts might reduce overall levels of cowbird parasitism by promoting shrubland (or grassland-savannah) habitats near these small tracts, because hosts in shrubland habitats tend to be less vulnerable to cowbird parasitism.

Nesting Strata

The lack of a clear relationship between nesting strata and levels of parasitism in Illinois contrasts with the results of Hahn and Hatfield (Chapter 13, this volume) in New York. Perhaps in the fragmented forests of Illinois, where cowbirds saturate most forest habitats, cowbirds are more likely to search for and find nests in all strata. Just as cowbirds are forced deep into the interior of forest tracts in Illinois (Robinson and Wilcove 1994), they may also be forced to search all strata to find nests that have not already been parasitized. Alternatively, cowbirds may not show strong preferences for nesting strata within some regions.

Host Size and Migratory Status

The remarkably strong correlation between multiple parasitism and host mass in neotropical migrants of forest habitats provides a further indication of just how susceptible forest birds are to cowbird parasitism. Cowbirds appear to adjust the number of eggs they lay to the size of the host and/or its nest when parasitizing forest-dwelling neotropical migrants. Large hosts such as Wood Thrushes can raise several cowbirds as well as several of their own young, whereas smaller hosts can generally raise only a single cowbird (Trine et al. 1998, Robinson unpubl. data; Trine, Chapter 15, this volume). Multiple parasitism of larger hosts is therefore adaptive for cowbirds. Elliott (1978) even found that cowbirds replaced each other's eggs in the nests of multiply parasitized species. Tanagers, Wood Thrushes, and Veeries will incubate clutches that consist entirely of cowbird eggs (Robinson 1992). In contrast, the lack of a strong relationship between host size and multiple parasitism in shrubland-dwelling neotropical migrants, short-distance migrants, and year-round residents suggests that larger hosts in these groups have evolved defenses against the costs of multiple parasitism.

The Cowbird-to-Host Ratio as an Index of Levels of Parasitism

The results in this chapter provide strong evidence of the usefulness of the cowbird-to-host ratio as an index of community-wide levels of parasitism. Used properly, this index can provide managers with a quick and inexpensive way to assess whether cowbird parasitism is likely to be a problem in an area of concern. The ratio is preferable to absolute cowbird abundance because it controls for host abundance.

Several caveats should be mentioned before adopting the cowbird-to-host ratio for general use. First, the cowbird-to-host ratio is not necessarily a good predictor of parasitism levels for all species (Robinson unpubl. data). Hosts that are preferred by cowbirds may be heavily parasitized even in areas where cowbirds are scarce relative to abundance of hosts. Wood Thrushes, for example, suffer 65–80% parasitism frequencies even in the sites with the lowest cowbird-to-host ratios (0.01, Robinson unpubl. data). Second, it is unclear how well the ratio would work in nonforest habitats where many species may be resistant to cowbird parasitism. We did not include known ejector species (which are rare in forest habitats), but it might be more appropriate to include them because cowbirds do sometimes lay eggs in their nests. Third, the presence of feeding areas such as campgrounds and logging roads may inflate estimates of cowbird abundance, especially in areas where census takers use these as census routes. Fourth, we have not yet tested whether estimates using male cowbird abundance are as good a predictor of levels of parasitism as female cowbird abundance. Fifth, we do not yet know how different census durations and fixed radii would affect our results, although there is little reason to believe that the ratio should depend on any of these factors (Thompson et al., Chapter 32, this volume).

Conclusions

Within fragmented landscapes where cowbirds are abundant, managers can reduce levels of cowbird parasitism by promoting and creating large forest tracts or by managing for grassland and shrubland habitats. For many forest species, parasitism will still remain a serious problem even in the largest habitat patches unless intensive cowbird removal efforts are initiated or managers seek opportunities to consolidate and restore very large contiguous habitats (over 8,000 ha). For some species, the best we may be able to do is to reduce the extent to which fragmented landscapes act as a drain on populations, an effect analogous to reducing the size of the drain in population sinks (Pulliam 1988). Alternatively, a network of large preserves scattered throughout Illinois might create local population sources (*sensu* Pulliam 1988) for some species and reduce the cowbird-to-host ratio by increasing the regional abundance and diversity of host species. The high vulnerability of neotropical migrants to cowbird parasitism in fragmented landscapes provides further justification for the widespread concern over the conservation problems of neotropical migrants (e.g., Hagan and Johnston 1992). The lack of significant population declines during the six years of censuses (Robinson unpubl. data) suggests that many of the populations experiencing very low nesting success might be maintained by immigration from unfragmented sources outside the region (Thompson et al., Chapter 32, this volume).

Future research directions in Illinois and elsewhere should include the following kinds of studies. First, we need data on the composition and parasitism levels of birds nesting in the savannah habitats that are currently being intensively restored in Illinois. Second, experimental studies of the community-wide effects of cowbird removal have not yet been conducted. Third, we need more replicates of tracts representing different forest types (e.g., bottomland vs. upland forest) and positions relative to cowbird feeding areas (near vs. far). Fourth, studies of geographical variation in host defenses against cowbird parasitism might shed light on the remarkable variation in levels of parasitism of grassland species. Fifth, we need more data on the interactions of nest predation and brood parasitism in fragmented landscapes. And sixth, we need better demographic data to identify sources and sinks (e.g., Smith 1981, May and Robinson 1985, Pease and Grzybowski 1995).

Acknowledgments

We thank the nest searching and monitoring crews who found virtually all of the nests used in this study, especially Steve Bailey, Caleb Morse, Lonny Morse, Steve Daniels, Robb Brumfield, Rhetta Jack, Leslie Jette, Rob Lu-Olendorf, Steve Amundsen, John Knapstein, Shane Heschel, Rebecca Whitehead, Suzanne Kercher, Jason Weber, Tara Robinson, Doug Robinson, Glendy Vanderah, Miguel Marini, Mike Ward, Todd Fink, Peg Gronemeyer, Andy Suarez, Karin Congdon, Tony Berto, Mindy Bailey, Johannes Foufopoulos, Troy Jesse, John Gmitro, Laura Johnson, Terry Chesser, Todd Freeberg, Liz Day, Kris Bruner, Kevin Ellison, Greg Fizzell, Robert Katz, Kevin Redvay, Steve Buck, Jeff Brawn, Tony Schreck, Chris Caruso, Barb Pilolla, Scott Saffer, Rhonda Andrews, and John Andrews. We also thank the Illinois Department of Conservation Avian Ecology Program for allowing us access to Illinois nest card data. J. Brawn and E. J. Heske reviewed the manuscript and provided many useful comments. Funding for this research was provided by the Illinois Department of Energy and Natural Resources, the Illinois Department of Conservation (Divisions of Natural Heritage and Wildlife Resources), the U.S. Fish and Wildlife Service, the U.S. Army Construction Engineering Research Laboratory, the U.S. Forest Service, the National Fish and Wildlife Foundation, the Illinois Chapter of the Nature Conservancy, the Department of Ecology, Ethology, and Evolution of the University of Illinois, the Illinois Natural History Survey, the Audubon Council of Illinois, and the Champaign County Audubon Society.

References Cited

Blake, J. G., and J. R. Karr. 1987. Breeding birds in isolated woodlots: Area and habitat relationships. Ecology 68:1057–1068.

Brawn, J. D., and S. K. Robinson. 1996. Source-sink dynamics may complicate the interpretation of long-term census data. Ecology 77:3–12.

Brittingham, M. C., and S. A. Temple. 1983. Have cowbirds caused forest songbirds to decline? BioScience 33:31–35.

Burger, L. D. 1988. Relations between forest and prairie fragmentation and depredation of artificial nests in Missouri. M.S. Thesis, University of Missouri, Columbia.

Chasko, G. G., and J. E. Gates. 1982. Avian habitat suitability along a transmission-line corridor in an oak-hickory forest region. Wildl. Monogr. 82:1–41.

Clark, K. L., and R. J. Robertson. 1979. Spatial and temporal mul-

tispecies nesting aggregations in birds as antiparasite and antipredator defenses. Behav. Ecol. Sociobiol. 5:359–371.

Dunning, J. B., Jr. 1984. Body weights of 686 species of North American birds. Western Bird Banding Monogr. 1.

Elliott, P. F. 1978. Cowbird parasitism in the Kansas tallgrass prairie. Auk 99:719–724.

Friedmann, H. 1929. The cowbirds: A study in the biology of social parasitism. C. C. Thomas, Springfield, IL.

Gates, J. E., and L. W. Gysel. 1978. Avian nest dispersion and fledging success in field-forest ecotones. Ecology 59:871–883.

Hagan, J. M. III, and D. W. Johnston (eds). 1992. Ecology and conservation of neotropical migrant landbirds. Smithsonian Institution Press, Washington, DC.

Hahn, D. C., and J. S. Hatfield. 1995. Parasitism at the landscape scale: Cowbirds prefer forests. Conserv. Biol. 9:1415–1424.

Hayden, T. J., J. Faaborg, and R. L. Clawson. 1985. Estimates of minimum area requirements for Missouri forest birds. Trans. Missouri Acad. Sci. 19:11–22.

Herkert, J. R. 1991a. Prairie birds of Illinois: Population response to two centuries of habitat change. Illinois Natural History Survey Bull. 34:393–399.

———. 1991b. An ecological study of the breeding birds of grassland habitats within Illinois. Ph.D. Thesis, University of Illinois, Urbana-Champaign.

———. In press. Breeding bird communities of midwestern prairie fragments: The effects of prescribed burning and habitat area. Natural Areas J.

Herkert, J. R., R. E. Szafoni, V. M. Kleen, and J. E. Schwegman. 1993. Habitat establishment, enhancement, and management for forest and grassland birds in Illinois. Nature Heritage Tech. Publ. 1.

Hoover, J. P., M. C. Brittingham, and L. J. Goodrich. 1995. Effects of forest patch size on nesting success of Wood Thrushes. Auk 112: 146–155.

Hutto, R. L., S. M. Pletschet, and P. Hendricks. 1986. A fixed-radius point count method for nonbreeding and breeding season use. Auk 103:593–602.

Iverson, L. R., L. Oliver, D. P. Tucker, P. G. Risser, C. D. Burnett, and R. G. Rayburn. 1989. Forest resources of Illinois: An atlas and analysis of spatial and temporal records. Illinois Natural History Survey Special Publ. 2. 181 pp.

Johnson, R. G., and S. A. Temple. 1986. Assessing habitat quality for birds nesting in fragmented tallgrass prairie. Pp. 245–250 in Wildlife 2000: Modeling habitat relationships of terrestrial vertebrates (J. Verner, M. L. Morrison, and C. J. Ralph, eds). University of Wisconsin Press, Madison.

———. 1990. Nest predation and brood parasitism of tallgrass prairie birds. J. Wildl. Manage. 54:106–111.

Kendeigh, S. C. 1982. Bird populations in east central Illinois: Fluctuations, variations, and development over a half-century. Illinois Biol. Monogr. 52.

May, R. M., and S. K. Robinson. 1985. Population dynamics of avian brood parasitism. Amer. Natur. 126:475–494.

Mayfield, H. F. 1965. The Brown-headed Cowbird with old and new hosts. Living Bird 4:13–28.

Neter, J., W. Wasserman, and M. H. Kutner. 1990. Applied linear statistical models. R. D. Irwin, Boston. 1,181 pp.

Neudorf, D. L., and S. G. Sealy. 1994. Sunrise nest attentiveness in cowbird hosts. Condor 96:162–169.

Nolan, V., Jr. 1978. The ecology and behavior of the Prairie Warbler (*Dendroica discolor*). Ornithological Monogr. 26, American Ornithologists' Union, Washington, DC. 595 pp.

Pease, C. M., and J. A. Grzybowski. 1995. Assessing the consequences of brood parasitism and nest predation on seasonal fecundity in passerine birds. Auk 112:343–363.

Pulliam, H. R. 1988. Sources, sinks, and population regulation. Amer. Natur. 132:652–661.

Robbins, C. S., D. K. Dawson, and B. A. Dowell. 1989. Habitat area requirements of breeding forest birds of the Middle Atlantic states. Wildl. Monogr. 103.

Robinson, S. K. 1988. Reappraisal of the costs and benefits of habitat heterogeneity for nongame wildlife. Trans. N. A. Wildl. Nat. Res. Conf. 53:145–155.

———. 1992. Population dynamics of breeding neotropical migrants in a fragmented Illinois landscape. Pp. 408–418 *in* Ecology and conservation of neotropical migrant landbirds (J. M. Hagan III and D. W. Johnston, eds). Smithsonian Institution Press, Washington, DC.

Robinson, S. K., J. A. Grzybowski, Jr., S. I. Rothstein, M. C. Brittingham, L. J. Petit, and F. R. Thompson III. 1993. Management implications of cowbird parasitism on neotropical migrant songbirds. Pp. 93–102 *in* Status and management of neotropical migratory birds (D. M. Finch and P. W. Stangel, eds). USDA Forest Serv. Gen. Tech. Rep. RM-229.

Robinson, S. K., Rothstein, S. I., M. C. Brittingham, L. J. Petit, and J. A. Grzybowski, Jr. 1995a. Ecology and behavior of cowbirds and their impact on host populations. Pp. 428–466 *in* Ecology and management of neotropical migratory birds (T. E. Martin and D. M. Finch, eds). Oxford University Press, New York.

Robinson, S. K., F. R. Thompson III, T. M. Donovan, D. R. Whitehead, and J. Faaborg. 1995b. Regional forest fragmentation and the nesting success of migratory birds. Science 267:1987–1990.

Robinson, S. K., and D. S. Wilcove. 1994. Forest fragmentation in the temperate zone and its effects on migratory songbirds. Bird Conserv. Int. 4:233–249.

Rogers, C. M., M. J. Taitt, J. N. M. Smith, and G. Jongejan. 1997. Nest predation and cowbird parasitism create a demographic sink in wetland-breeding Song Sparrows. Condor 99:622–633.

Rothstein, S. I. 1975. An experimental and teleonomic investigation of avian brood parasitism. Condor 77:250–271.

Sealy, S. G. 1992. Removal of Yellow Warbler eggs in association with cowbird parasitism. Condor 94:40–54.

Smith, J. N. M. 1981. Cowbird parasitism, host fitness, and age of the host female in an island Song Sparrow population. Condor 83:152–161.

Temple, S. A., and J. R. Cary. 1988. Modeling dynamics of habitat-interior bird populations in fragmented landscapes. Conserv. Biol. 2:340–347.

Trine, C. L., W. D. Robinson, and S. K. Robinson. 1998. Effects of cowbird parasitism on host population dynamics. Pp. 273–295 *in* Brood parasites and their hosts: Essays in honour of Herbert Friedmann (S. I. Rothstein and S. K. Robinson, eds). Oxford University Press, New York.

Weatherhead, P. J. 1989. Sex ratios, host-specific reproductive success, and impact of Brown-headed Cowbirds. Auk 106:358–366.

Wilcove, D. S. 1985. Nest predation in forest tracts and the decline of migratory songbirds. Ecology 66:1211–1214.

Wilcove, D. S., C. H. McClellan, and A. P. Dobson. 1986. The impact of forest fragmentation of bird communities in eastern North America. Pp. 319–331 *in* Biogeography and ecology of forest bird communities (A. Keast, ed). SPB Academic Publishing, The Hague, The Netherlands.

Yahner, R. 1988. Changes in wildlife communities near edges. Conserv. Biol. 2:333–339.

34. Within-landscape Variation in Patterns of Cowbird Parasitism in the Forests of South-central Indiana

DONALD E. WINSLOW, DONALD R. WHITEHEAD, CAROLYN FRAZER WHYTE,

MATTHEW A. KOUKAL, GRANT M. GREENBERG, AND THOMAS B. FORD

Abstract

Forest-breeding neotropical migrant birds are heavily parasitized by Brown-headed Cowbirds in many fragmented landscapes in the Midwest. Levels of parasitism are lower in more heavily forested landscapes such as south-central Indiana. Local patterns of disturbance within such areas, however, may increase cowbird densities and thus elevate levels of parasitism. We use data from a study of avian community composition in south-central Indiana to evaluate the hypothesis that cowbirds occur in higher densities in forest near internal and external edges than in forests farther from such ecotones. We use nest records from this landscape to evaluate the hypothesis that parasitism levels are also elevated in proximity to external and internal edges. We monitored 1,293 nests of various species during four breeding seasons (1990–1993) in six landscape contexts: (1) interior forest, (2) exterior edges, (3) forest near clearcuts in Hoosier National Forest, (4) forest near forest openings in Hoosier, (5) within small timbercuts in Yellowwood State Forest, and (6) within an old field. Cowbird densities were not significantly higher at internal and external forest edges than in interior forest. The overall level of parasitism was 14.6%, summed across all species and field sites, and much lower than in more fragmented midwestern forests. At a smaller scale, frequency of parasitism for several species varied with patterns of human land use such as timber and wildlife management on public lands. Levels of parasitism were higher at sites adjacent to both exterior and interior edges than in the forest interior. Different host species experienced different frequencies of parasitism, as well as different patterns of parasitism within landscapes. Acadian Flycatchers were more heavily parasitized in forests near clearcuts than in either the forest interior or near exterior edges. Red-eyed Vireos, however, were more heavily parasitized in both forests near clearcuts and near exterior edges than in the forest interior. These data suggest that minimizing the risk of brood parasitism to neotropical migrants will require management at the land-scape level. Increasing the area of forest core habitat and decreasing the extent of interior edge should benefit forest-breeding neotropical migrants.

Introduction

Recent studies indicate that many midwestern populations of forest-breeding neotropical migrant birds are heavily parasitized by Brown-headed Cowbirds (Brittingham and Temple 1983, Robinson 1992, Hoover and Brittingham 1993, Robinson et al. 1993; Thompson et al., Chapter 32, and Robinson et al., Chapter 33, this volume). These studies indicate that high levels of parasitism are associated with patterns of forest fragmentation and disturbance over both large and small spatial scales. On a regional scale, parasitism frequencies are high in fragmented forests in Illinois (Robinson et al., Chapter 33, this volume), southern Wisconsin, and north-central Missouri (Thompson et al., Chapter 32, this volume) but quite low in the heavily forested Ozark Plateau in southeastern Missouri and in northern Wisconsin (Thompson et al., this volume). On a smaller scale (within forested landscapes), levels of parasitism are higher close to openings within forest (Gates and Gysel 1978, Chasko and Gates 1982, Brittingham and Temple 1983) than at distances farther from such disturbances (but see Thompson et al., this volume).

These regional and landscape patterns of brood parasitism probably result from the behavior and ecology of Brown-headed Cowbirds. Cowbirds feed primarily in short grass and on bare ground and are attracted to concentrations of large ungulates such as cattle (Rothstein et al. 1986, Thompson 1994). The clearing of midwestern forests may have allowed cowbird populations to increase in this region, drawing recruits from high population densities in the Great Plains (Robinson et al. 1993). High levels of parasitism in forest fragments within agricultural landscapes probably result largely from the availability of extensive cowbird feeding habitat.

Cowbirds are active in breeding areas in the early morning and fly to feeding sites in the late morning and afternoon (Rothstein et al. 1986, Thompson 1994; Raim, Chapter 9, and Thompson and Dijak, Chapter 10, this volume). They can thus affect the reproductive success of forest-breeding birds, despite their dependence on open areas for feeding. Cowbirds appear to be attracted to openings within forests during the morning. O'Conner and Faaborg (1992) showed that density of cowbirds decreases with distance to both exterior and interior edges in large forest tracts in the Missouri Ozarks. Cowbirds may preferentially use internal edge habitats because of high densities of breeding birds near the ecotones (Gates and Gysel 1978), or because trees and snags near such edges provide perches for nest-searching and courtship display (Gates and Gysel 1978, Mayfield 1965).

These observations of cowbird behavior and ecology may help to explain patterns of parasitism over large and small spatial scales. Over regional scales, cowbird density and frequency of parasitism may depend on the relative proportions of various land-use types (Thompson et al., Chapter 32, this volume). Within a forested landscape, parasitism levels may be elevated near edges. Evidence for such small-scale effects is equivocal and may depend on regional cowbird densities (discussed in Thompson et al., this volume). Also, it is difficult to interpret published data sets, because results are typically expressed in terms of totals pooled across host species. Because levels of parasitism vary with species, differences in species composition of samples may lead to biases. For this reason, it is necessary to examine spatial patterns of parasitism for individual host species.

Heavily forested regions in the midwestern United States may be important for forest-breeding birds by providing habitat with relatively low parasitism pressure. Also, rates of nest predation may be lower in such areas (e.g., Chasko and Gates 1982, Wilcove 1985). An area with great potential significance for forest-dependent neotropical migrants lies in south-central Indiana. The forests in this region are extensive and relatively continuous at large spatial scales (Figure 34.1) and may therefore constitute a population source for species of forest-nesting neotropical migrants. A source is a patch in which annual recruitment to a breeding population exceeds annual mortality (Pulliam 1988). Examination of the landscape pattern at a finer resolution, however, reveals a mosaic of different-aged deciduous forest stands and pine plantations on which is superimposed a pattern of disturbed patches (clearcuts, forest openings, waterholes, roads, and utility rights-of-way; Figure 34.2). These disturbances reflect the timber and wildlife management activities of the Hoosier National Forest and the Indiana Department of Natural Resources. Agricultural areas and lawns around and within the forest provide potential feeding sites for cowbirds (Figure 34.3).

These internal and external disturbances may increase levels of parasitism, and thus reduce the breeding success of forest-dependent neotropical migrants. For this reason, we initiated a long-term demographic study of neotropical migrant populations in the Pleasant Run Unit of the Hoosier National Forest and surrounding state and privately owned forest lands. In this chapter we present data on spatial patterns of cowbird distribution and brood parasitism in this area. We use these data to test two hypotheses: (1) The density of cowbirds within forests is elevated near exterior and interior edges. (2) Levels of cowbird parasitism are higher in forests adjacent to edges (both external and internal) than in forests with less surrounding disturbance. Our ultimate objective is to provide the information needed to manage viable populations of forest-dependent neotropical migrants at all spatial scales.

For this chapter, we define the following terms: "Landscape" is an area of land (e.g., the Pleasant Run Unit and surrounding forests) in which patches within the landscape vary in successional state and species composition. "Landscape pattern" refers to a pattern of heterogeneity (e.g., in vegetative structure, parasitism level, etc.) within this landscape. "Landscape context" refers to the vegetative structure of lands surrounding a study site.

Study Sites and Methods

Census Methods

As part of a study of the effect of edge on forest avian community structure (Frazer 1992), cowbird numbers were monitored at 50 points in 1990, 54 points in 1991, and 53 points in 1992. In 1990 and 1991, 20 points were located in the forest interior, defined as being at least 200 m from a forest edge. Ten points were located at five forest-field interfaces. Each interface had one point 50 m into the forest from the edge and another point 150 m into the forest from the edge. Ten points were at the interfaces of forests and clearcuts (5–8 years after cutting, 12 points in 1991), and 10 points were at interfaces between forests and utility corridors (12 in 1991). Points used in 1990 were reused in 1991. In 1992, cowbird numbers were monitored at 19 of the interior forest points and at 12 points at six interfaces between forests and forest openings. (Forest openings are small disturbances maintained by the U.S. Forest Service, described in detail under "Landscape Patterns of Parasitism" below.) All points were located at least 50 m from the nearest road.

Birds were censused using a modification of the variable circular plot method (DeSante 1981). Each point was censused for 10 min on three different days between mid-May and late June. Each of the three counts at a given point was conducted at a different time between 0530 and 0930. No counts were taken on windy or rainy days. The same observer conducted all counts, thus eliminating between-observer variability.

In 1990 only male cowbirds were censused. In 1991 and 1992 both male and female cowbirds were included. For each

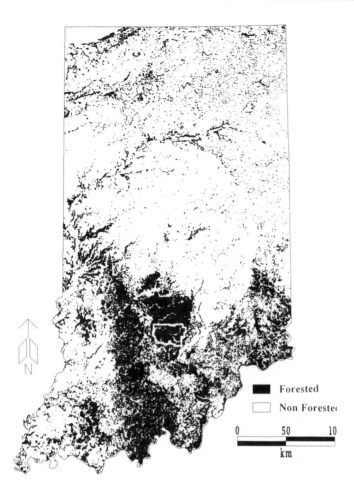

Figure 34.1. Forest cover of Indiana, adapted from Landsat imagery taken during the period 1973–1981. The Pleasant Run Unit of the Hoosier National Forest is outlined in white. Prepared by Jeffrey L. Ehman

bird heard or seen during a census, the compass direction and estimated distance to the bird from the point were recorded, as were its activity and flight direction, if necessary.

Although all birds seen or heard were recorded, the analysis was performed on birds recorded within 50 m of each monitoring point. This way, birds recorded close to one edge point were not included in the data for the other point at the same edge, and birds recorded within the openings themselves were not included unless they actually moved into the forest edge areas. The number of birds per point was determined as the maximum number each of males and females seen or heard at one time within 50 m of the point.

Numbers of cowbirds were compared among landscape contexts (counting different distances from the same interface type as the same context) by one-way analysis of variance of transformed values, using the transformation $y' = (y + 0.5)^{0.5}$ (Steel and Torrie 1981, SYSTAT 1992).

Landscape Patterns of Parasitism

We monitored nests in four mature-forest landscape contexts and two early successional habitats in the Pleasant Run Unit area from 1991 to 1993 (Figure 34.3). The forest contexts are (1) interior forest (studied 1991–1993), (2) exterior edge (1992–1993), (3) forests near clearcuts in Hoosier (1992–1993), and (4) forests near forest openings in Hoosier (1993). The successional habitats are (5) small timbercuts in Yellowwood State Forest (1992) and (6) an old field owned by Indiana University (1993).

The vegetation at our mature forest sites is classified as western mesophytic forest (Braun 1950) and is dominated by sugar maple (*Acer saccharum*) and American beech (*Fagus grandifolia*) on moister sites and oaks (*Quercus* spp.) and hickories (*Carya* spp.) on drier sites. Broad ridgetops and ravine bottoms are often planted in shortleaf pine (*Pinus echinata*) or white pine (*Pinus strobus*).

Interior Forest

The interior forest landscape context consisted of three contiguous watersheds located 6 km south of Belmont, Indiana, in the Pleasant Run Unit of the Hoosier National Forest. The three watersheds embrace an area of 170 ha. Over 80% of the area within a 3-km radius of the study site has complete forest cover; the nearest agricultural fields are 5 km to the north. There are some internal disturbances near these ravines. A gravel road traverses the ridgetop between two of the watersheds; there is a small (1 ha) field north of the site, a clearcut to the west (cut in 1981), and an older clearcut (cut in 1976) at the northern end of one of the ravines. There is also an old pine plantation at the northern end of one of the ravines. Nevertheless, this degree of internal disturbance is low compared with the Pleasant Run Unit as a whole.

Exterior Edge

The exterior context consisted of three contiguous watersheds adjacent to agricultural fields, located near the T. C. Steele State Memorial (Figure 34.3). This study site is about 4.5 km north of the interior forest site, just south of Belmont, Indiana, in Brown County. The three exterior watersheds total 184.2 ha. Internal disturbances include a paved road running along the ridgetop between two of the watersheds, a power-line cut in one watershed, extensive development around the T. C. Steele Memorial, and small openings and logging roads to the west in Yellowwood.

Forests near Clearcuts

We monitored nests in the forests adjacent to five clearcuts in the Pleasant Run Unit of Hoosier. All of these clearcuts are south of Lake Monroe (Figure 34.3). Two of them are just west of State Road 446 in Monroe County, and three of them are at more isolated locations along gravel roads east

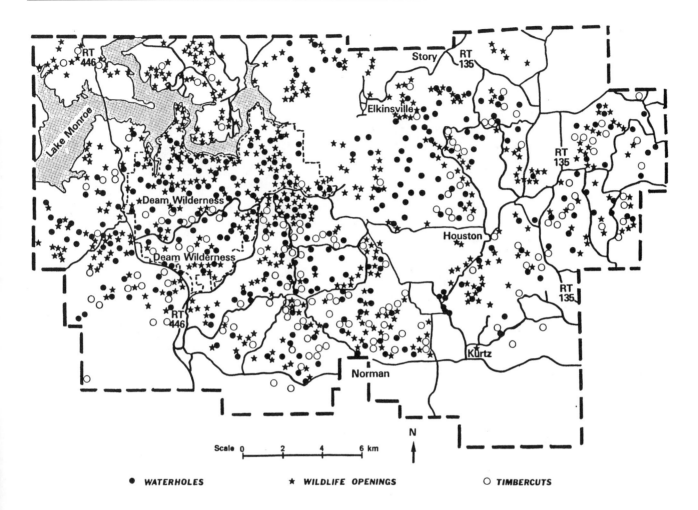

Figure 34.2. The Pleasant Run Unit of the Hoosier National Forest, showing the location of internal disturbances. Adapted from a map produced by the Indiana Department of Natural Resources, Division of Fish and Wildlife.

of State Road 446 and south of the Charles C. Deam Wilderness in Jackson and Lawrence counties. All were harvested in 1984 and/or 1985, thus the regrowth was about 7 years old in 1992. The clearcuts ranged in size from 6.6 to 11.1 ha.

The area of forest surrounding each clearcut in which nests were monitored ranged from 17.1 to 33.4 ha in 1992; the total area searched was 132.9 ha. The areas monitored in 1993 were somewhat different but similar in size. All of these clearcuts are close to other clearcuts and internal disturbances, and thus some nests were actually much closer to other disturbances than to the focal clearcut. Some of the forest surrounding these clearcuts consists of pine plantations, but most of it is deciduous forest. A few nests within the clearcuts themselves were also monitored.

Forests near Forest Openings

Forest openings are small (typically under 1 ha) disturbances that are maintained by the U.S. Forest Service. These

openings are managed by cutting or burning vegetation every few years and by girdling trees. Sections of each opening are left uncut in most years, resulting in patches of regrowth of varying successional ages along the margins and within the openings. The purpose of this management practice is "to provide early successional vegetation beneficial to some wildlife species, provide habitat for rare native plant communities, add visual variety, and provide for associated recreation opportunities such as hunting, berry picking, and wildlife observation" (USFS 1991).

We monitored nests in forests adjacent to four forest openings in the Pleasant Run Unit. These openings range in size from 0.3 to 1.0 ha and are clustered together on neighboring ridgetops in Lawrence County, Indiana (Figure 34.3). Each of these was cut in late summer in 1992, and they all contain patches of older vegetation of various ages. Two of these are adjacent to young (9–12 years) clearcuts, one is adjacent to a mature pine plantation, and one contains an ar-

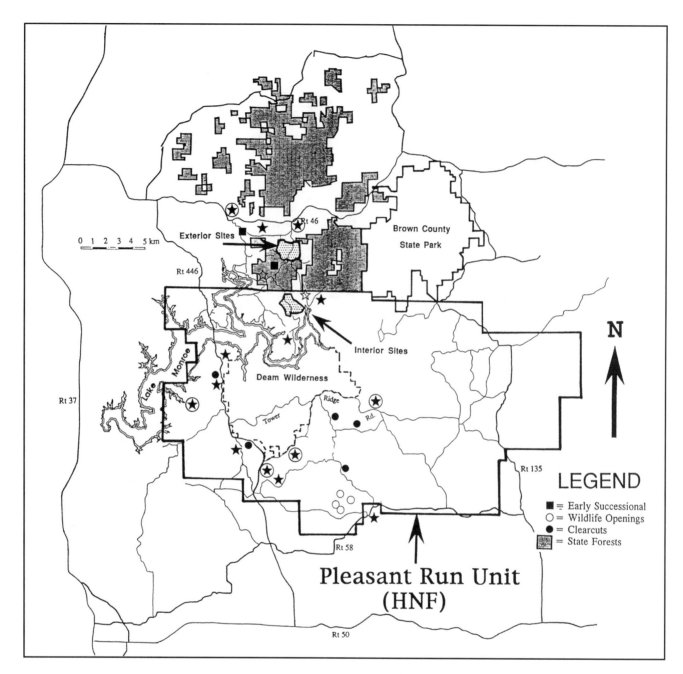

Figure 34.3. Map of study sites in and around the Pleasant Run Unit of the Hoosier National Forest. See text for descriptions of site types. Also shown are locations where we have observed cowbirds feeding in the afternoon: open star < 5 individuals, solid star 5–15 individuals, circled star > 25 individuals.

tificial pond. We monitored nests from the edges of the openings to the bottoms of the ravines immediately surrounding them. Because the openings are on adjacent ridgetops, the study area for each opening was therefore contiguous with the study area for one other opening. Thus there were effectively two study sites for this context, each containing two forest openings. We also followed a few nests within the openings themselves and along the trails between the openings.

Timbercuts in State Forest

We monitored nests within three small (0.47–1.56 ha) timbercuts in Yellowwood (Figure 34.3). All three sites are dominated by shrubby vegetation including sassafras (*Sassafras albidum*), greenbrier (*Smilax rotundifolia*), red maple (*Acer rubrum*) saplings, and saplings of other deciduous trees. Unused timber roads border each of the cuts.

Old Field

In 1993 we monitored nests at an old-field site at Kent Farm, Monroe County, Indiana, on land owned by Indiana University (Figure 34.3). This field complex is part of the agricultural corridor that is exterior to our forest sites. The site was seeded with several species of grass in 1962, and the eastern half of the field was deep plowed in 1972. A gravel road transects this old field. As of 1993, this site is still dominated by grasses and forbs, but there is also a high diversity of invading shrubs and saplings.

Methods for Determining Nesting Success

In 1990, Peter S. McKinley (unpubl. data) monitored nests of Wood Thrushes at various locations in the vicinity of the Pleasant Run Unit. Between 1991 and 1993, we monitored reproductive success of birds breeding within the habitat and landscape contexts described above. We focused our study on the Acadian Flycatcher but also monitored nests of other neotropical migrants and some resident species. (See Table 34.1 for common and Latin names of hosts.) We attempted to find every Acadian Flycatcher nest within our study sites and probably succeeded in monitoring 80–90% of the nests of this species, although some were too high to monitor. We checked nests using mirrors and poles at intervals of three days or less.

We used the G statistic in contingency table analyses (Gokhale and Kullback 1978, SYSTAT 1992) of patterns of parasitism with regard to host species, landscape context, and year. Data were pooled among sites within the same context. Because we did not study all landscape contexts in each year, we concentrated our analysis on data from 1992 and 1993 in the interior forest, exterior edge, and forest near clearcuts in Hoosier National Forest.

Proportions of parasitized nests did not differ significantly between 1992 and 1993 in these three contexts for Acadian Flycatcher (Mantel-Haenszel $\chi^2 = 0.005$, $P = .946$, controlling for the effect of context), Red-eyed Vireo ($\chi^2 = 0.671$, $P = .413$), or Worm-eating Warbler ($\chi^2 = 0.002$, $P = .965$). Low expected values precluded analysis of Wood Thrush data. For comparisons among landscape contexts, we therefore pooled data from 1992 and 1993.

We measured distances from nests to various types of interior and exterior edges. Some of these measurements were taped in the field, and others were estimated from maps. In this chapter, we consider parasitism as a function of distance to the edge of clearcuts for Acadian Flycatcher nests followed in 1992. The relationship between parasitism frequency and distance to edge was examined using logistic regression (SPSS 1990).

Results

Spatial Distribution of Cowbirds

In 1990, male cowbirds were counted more frequently at points 50 m from clearcuts and 50 m from utility corridors than in interior forest, although there were no statistically significant differences among landscape contexts (pooling among distances from edge within contexts, $F = 1.464$, $P = .237$; Figure 34.4). In 1991 male cowbirds were counted nearly as frequently in interior forest as at points 50 m from clearcuts and in utility corridors; again the overall pattern was not statistically significant ($F = 2.139$, $P = .107$; Figure 34.4). Density of female cowbirds in 1991 did not differ significantly among landscape contexts ($F = 0.259$, $P = .855$; Figure 34.4). In 1992, cowbirds were counted more frequently at sites near forest openings than in interior forest, but differences were again not significant (males, $F = 2.994$, $P = .094$; females, $F = 0.832$, $P = .369$; Figure 34.4).

Landscape Patterns of Brood Parasitism

We monitored 1,278 nests from 29 species from 1990 to 1993. Table 34.1 shows the proportion of nests parasitized for each species in each year (including species parasitized only rarely and species that consistently reject cowbird eggs). Of these nests, 186 (14.6%) were parasitized by cowbirds. The level of parasitism differed considerably among species. Of the four species for which we have large samples (> 50 nests), the most heavily parasitized was the Red-eyed Vireo (31.9%). The Worm-eating Warbler was also heavily parasitized (28.6%). Lower levels of parasitism were seen among nests of the Acadian Flycatcher (8.4%) and the Wood Thrush (14.0%). The proportion of parasitized nests differed significantly among these four species ($G = 58.467$, df = 3, $P < .001$).

In general, the proportion of nests parasitized was lower in interior forest than in the forest edge contexts. The dramatic variation in level of parasitism among species, however, precludes meaningful statistical analysis of samples pooled across all species, because samples varied in the relative proportions of each species. For this reason, we concentrated our analysis on those species for which we have large samples of nest records (Table 34.2).

Landscape patterns of parasitism differed among host species (Table 34.2, Figure 34.5). Three-way loglinear analysis of parasitism frequency with respect to species and landscape context for Acadian Flycatchers and Red-eyed Vireos shows that the proportion of parasitized nests varied with

Table 34.1. Percentage of Brood Parasitism for Each Host Species for Each Year Summed over All Landscape Contexts

Common Name	Species Name	AOU Code	1990 N	1990 % Parasitism	1991 N	1991 % Parasitism	1992 N	1992 % Parasitism	1993 N	1993 % Parasitism	Total N	Total % Parasitism
Mourning Dove	*Zenaida macroura*	MODO							2	0.0	2	0.0
Yellow-billed Cuckoo	*Coccyzus americanus*	YBCU			1	0.0	5	0.0	8	0.0	14	0.0
Black-billed Cuckoo	*Coccyzus erythropthalmus*	BBCU							1	0.0	1	0.0
Ruby-throated Hummingbird	*Archilochus colubris*	RTHU					1	0.0	4	0.0	5	0.0
Eastern Wood-pewee	*Contopus virens*	EAWP					1	0.0	2	0.0	3	0.0
Acadian Flycatcher	*Empidonax virescens*	ACFL			52	11.5	255	8.2	325	8.0	632	8.4
Blue Jay	*Cyanocitta cristata*	BLJA							3	0.0	3	0.0
Carolina Chickadee	*Parus carolinensis*	CACH					1	0.0	1	0.0	2	0.0
Carolina Wren	*Thryothorus ludovicianus*	CARW					2	0.0			2	0.0
Wood Thrush	*Hylocichla mustelina*	WOTH	12	8.3	10	0.0	40	10.0	74	18.9	136	14.0
White-eyed Vireo	*Vireo griseus*	WEVI							3	0.0	3	0.0
Red-eyed Vireo	*Vireo olivaceus*	REVI	1	100.0	9	11.1	58	31.0	70	32.9	138	31.9
Blue-winged Warbler	*Vermivora pinus*	BWWA					1	0.0	3	0.0	4	0.0
Black-and-white Warbler	*Mniotilta varia*	BAWW					2	50.0	1	100.0	3	66.7
Chestnut-sided Warbler	*Dendroica pensylvanica*	CSWA							1	0.0	1	0.0
Prairie Warbler	*Dendroica discolor*	PRAW							2	0.0	2	0.0
Kentucky Warbler	*Oporornis formosus*	KEWA			1	0.0	8	25.0	21	19.0	30	16.7
Hooded Warbler	*Wilsonia citrina*	HOWA					5	60.0	13	61.5	18	61.1
Worm-eating Warbler	*Helmitheros vermivorus*	WEWA			2	0.0	22	27.3	53	30.2	77	28.6
Ovenbird	*Seiurus aurocapillus*	OVEN			1	0.0	15	26.7	32	18.8	48	21.3
Louisiana Waterthrush	*Seiurus motacilla*	LOWA					4	25.0	2	0.0	6	16.7
Common Yellowthroat	*Geothlypis trichas*	COYE							1	0.0	1	0.0
Yellow-breasted Chat	*Icteria virens*	YBCH							15	0.0	15	0.0
Northern Cardinal	*Cardinalis cardinalis*	NOCA			2	0.0	14	0.0	33	24.2	49	16.3
Indigo Bunting	*Passerina cyanea*	INBU			2	0.0	14	14.3	32	9.4	48	10.4
Rufous-sided Towhee	*Pipilo erythrophthalmus*	RSTO					4	0.0	10	0.0	14	0.0
Song Sparrow	*Melospiza melodia*	SOSP							1	0.0	1	0.0
Field Sparrow	*Spizella pusilla*	FISP							14	7.1	14	7.1
Scarlet Tanager	*Piranga olivacea*	SCTA			1	100.0	1	100.0	4	75.0	6	83.3
TOTAL			13	15.4	81	9.9	453	13.9	731	15.5	1278	14.6

N = total nests.

both species and context (Table 34.3). Also, the three-way interaction term, although not quite significant (*P* = .067), suggests that the landscape pattern of parasitism differed between these two species.

Pairwise comparisons between contexts for each species reveal these differences in landscape pattern. Acadian Flycatchers were parasitized more heavily in forests near clearcuts (15.7%) than in either exterior edge (5.4%, Fisher exact probability = .001) or interior forest (1.5%, *P* < .001). This species was parasitized more heavily in exterior edges than in interior forest, although the difference was not quite significant (*P* = .087). Red-eyed Vireos, however, were parasitized as heavily in exterior edges (60.7%) as in forests near clearcuts (54.8%, *P* = .793), and nests of this species were parasitized more heavily in each of these edge contexts than in interior forest (9.4%, *P* < .001 for both exterior and clearcuts). The results of these pairwise comparisons were qualitatively similar when examined for each year separately.

The landscape pattern of parasitism for the Wood Thrush appeared similar to that seen for the Acadian Flycatcher (Figure 34.5), but the differences among contexts were not statistically significant (*G* = 1.822, df = 2, *P* = .402, pooling data from 1992 and 1993).

The landscape pattern for the Worm-eating Warbler was similar to that seen for the Red-eyed Vireo, and this pattern was significant (*G* = 6.219, df = 2, *P* = .045, pooling data from 1992 and 1993). Pairwise comparisons of contexts, pooling both years, show that parasitism was higher in exterior edge (43.3%) than in the interior (13.0%) for this species (Fisher exact *P* = .033). The frequency of parasitism near clearcuts for this species was intermediate (25.0%) but not significantly different from either of the other contexts (interior, *P* = .415; exterior, *P* = .338).

We monitored nests of several other species of ground-nesting warblers, but we do not have large samples for any one species. We therefore pooled data from all ground-nesting warblers in order to look for landscape patterns of parasitism in this suite of species. The species included in this sample are the Worm-eating Warbler, Ovenbird, Kentucky Warbler, Louisiana Waterthrush, Blue-winged Warbler, Black-and-white Warbler, and Common Yellowthroat. Proportions of parasitized nests did not differ between 1992 and 1993 for these species (Mantel-Haenszel χ^2 = 0.001, *P* = .975). Proportions of parasitized nests for these species varied with landscape context (*G* = 13.130, df = 2, *P* = .001, pooling data from 1992 and 1993 and comparing interior, exterior, and forests near clearcuts). Parasitism was higher in the exterior (36.7%, *P* = .001) and clearcut edge (32.5%, *P* = .008) contexts than in the interior (10.0%) for these species. The level of parasitism did not differ significantly between exterior and forests near clearcuts (*P* = .823).

The overall level of parasitism in forest adjacent to forest openings in 1993 was high (13 of 78 nests, all species pooled, including nests in successional areas) relative to the

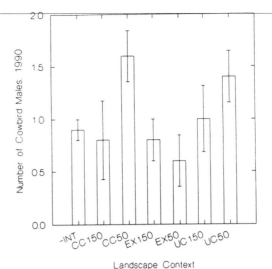

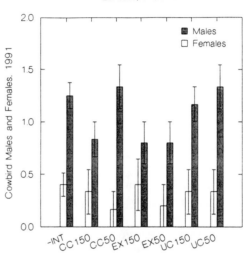

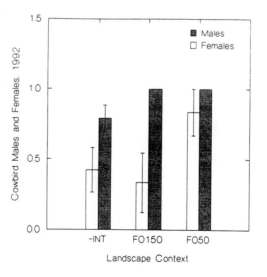

Figure 34.4. Numbers of cowbirds detected during morning censuses in 1990, 1991, and 1992, in interior forest, 150 m from edge, and 50 m from edge. INT, interior; CC, clearcut edge; EX, exterior (field) edge; UC, utility corridor edge; FO, forest opening edge. Females were not noted in 1990.

Table 34.2. Levels of Parasitism for Selected Species, Summarized by Landscape Context

Landscape Context	Interior Forest		Exterior Edge		Forest near Large Clearcuts		Forest near Forest Openings		Within Small Clearcuts	
Species	N	% Parasitism	N	% Parasitism	N	% Parasitism	N	% Parasitism	N	% Parasitism
ACFL	183	4.4	203	5.4	210	15.7	36	2.8	0	—
WOTH	52	7.7	37	16.2	29	20.7	4	50.0	2	0.0
REVI	73	9.6	28	60.7	31	54.8	5	20.0	0	—
KEWA	9	11.1	4	25.0	14	28.6	2	0.0	1	0.0
HOWA	0	0.0	0	—	11	63.6	7	57.1	0	—
WEWA	25	12.0	30	43.3	16	25.0	6	33.3	0	—
OVEN	25	8.0	11	27.3	6	50.0	6	33.3	0	—
YBCH	1	0.0	0	—	3	0.0	0	—	6	0.0
NOCA	13	30.8	14	7.1	12	8.3	4	25.0	0	—
INBU	16	6.3	16	6.3	10	30.0	2	0.0	1	0.0
RSTO	0	0.0	0	—	7	0.0	1	0.0	1	0.0
SCTA	2	50.0	2	100.0	2	100.0	0	—	0	—

Note: See Table 34.1 for species names. Dates: interior forest, 1991–1993; exterior edge, 1992–1993; forest near large clearcuts in the Hoosier National Forest, 1992–1993; forest near forest openings in Hoosier, 1993; within small clearcuts in Yellowwood, 1993.

parasitism level in interior forest for that year. Of 33 Acadian Flycatcher nests near forest openings, however, only 1 was parasitized.

We have a small sample of nests from within timbercuts. In 1992, 3 of 10 nests (30%) within cuts in Hoosier and 34 of 150 (22.7%) in forests near clearcuts were parasitized. In 1993, 2 of 25 nests (8%) within Hoosier clearcuts and 42 of 184 nests (22.8%) near clearcuts were parasitized. We also monitored 13 nests within timbercuts and 2 nests in forest near small timbercuts in Yellowwood in 1993, and none of these nests were parasitized. Pooling data from both types of cut and controlling for the effect of year, a test of independence of parasitism rate and habitat shows a nonsignificant trend toward higher parasitism in the forest near timbercuts than in the cuts themselves (Mantel-Haenszel $X^2 = 3.416$, $P = .065$). This difference reflects variation in species composition between forest and successional samples. For instance, 35.4% of the nests in successional areas (pooled over all sites and years) were of species that are only rarely parasitized or commonly reject cowbird eggs. Only 2.1% of the nests in mature forest were of such species.

The rate of parasitism in the old-field site monitored in 1993 was low (2 of 38 nests, all species pooled, including 8 from rejectors and species that were rarely parasitized).

Parasitism and Proximity to Edges

Logistic regression of parasitism of Acadian Flycatcher nests monitored in 1992 on distance to clearcut edge yielded the fitted model equation

$$par = \frac{exp\,(-1.0424 - 0.0047 \times distance)}{1 + exp\,(-1.0424 - 0.0047 \times distance)}$$

where *par* is the probability that a nest is parasitized, and *distance* is the distance in meters to the edge of the focal clearcut (Figure 34.6). Inspection of Figure 34.6 and the positive value for the coefficient of *distance* suggest that the incidence of parasitism decreased with increasing distance to clearcut edge; however, this estimate was not significantly different from 0 (SE = 0.0039, $P = .227$).

Discussion

O'Conner and Faaborg (1992) showed that cowbird densities increased near interior and exterior edges in the Missouri Ozarks. Our census data do not strongly support the hypothesis that cowbird densities are elevated near edges but do suggest that male cowbirds occur more frequently within 100 m of forest openings, in clearcuts, and in utility corridors than at points farther from edges. Morning densities of cowbirds were not higher near field edges than in interior forest. The numbers of sites and the numbers of cowbirds detected at each site were low, and there was substantial variation among sites and years. These factors make it difficult to detect trends. Our data indicate that cowbirds are found throughout the forest in and around the Pleasant Run Unit. The high degree of internal disturbance within these forests may allow cowbirds to range throughout the landscape.

Subjective observations suggest that cowbirds are more abundant near edges, although the data presented here do not support this claim strongly. The census plots did not include the disturbances themselves but only the forests surrounding them. Cowbirds were commonly present in clearcuts in the early mornings in censuses done in 1992 to determine daily patterns of habitat use (unpubl. data). We often saw and heard them in clearcuts and forest openings,

chasing and displaying from prominent perches. Additional censuses conducted at forest openings in 1995 (Doran et al. unpubl. data) indicated that male cowbirds occurred in higher numbers within forest openings than in nearby forest. These observations could, however, be deceptive, as cowbirds may simply be more conspicuous in openings and thus more likely to be detected. Cowbirds were absent from these interior openings by the afternoon, when they were feeding in nearby agricultural fields and lawns. This daily activity pattern has been documented for other populations of cowbirds through radiotelemetry (Rothstein et al. 1986, Thompson 1994; Raim, Chapter 9 and Thompson and Dijak, Chapter 10, this volume).

In future field seasons, we plan to census cowbirds and host species at the sites where we are investigating nesting success. This will provide better estimates of cowbird and host densities and enable us to relate these variables to landscape patterns and nesting success.

Levels of parasitism in our study area are low relative to other areas of the Midwest. Robinson and coworkers have documented extremely high parasitism in sites throughout Illinois (Robinson 1992; Robinson et al., Chapter 33, this volume). Faaborg and coworkers have found high levels of parasitism in fragmented sites in northern Missouri (Thompson et al., Chapter 32, this volume). Hoover and Brittingham (1993) analyzed data from Wood Thrush nests compiled by the Cornell Nest Record Program and reported frequencies of parasitism of over 40% for this species in In-

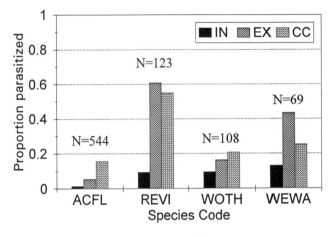

Figure 34.5. Proportion of nests parasitized in 1992 and 1993 among nests of four species of forest-breeding neotropical migrants in interior forest (IN), exterior edge (EX), and forests near clearcuts (CC; includes 6 nests within clearcuts). See Table 34.2 for species names.

Table 34.3. Loglinear Modeling of Proportion of Parasitized Nests with Respect to Landscape Context and Host Species

Model	G	df	P
context × species + par	103.82	5	0.000
context × species + context × par	74.58	3	0.000
context × species + species × par	61.51	4	0.000
context × species + context × par + species × par	5.40	2	0.067
Effect of context	29.24	2	< 0.001
Effect of species	42.31	1	< 0.001

Note: For Acadian Flycatcher and Red-eyed Vireo nests in continuous forest, at exterior edges, and in forests near clearcuts in 1992 and 1993, including three nests within clearcuts at exterior edges. The effect of context is tested by subtracting the log-likelihood ratio (*G*) and degrees of freedom for the model *species × context + context × par* from the respective statistics for the model *species × context + par* (Gokhale and Kullback 1978). The effect of species is tested in a similar manner. The degree of non-independence from the model *species × context + context × par + species × par* represents the effect of the interaction between species and context on parasitism.

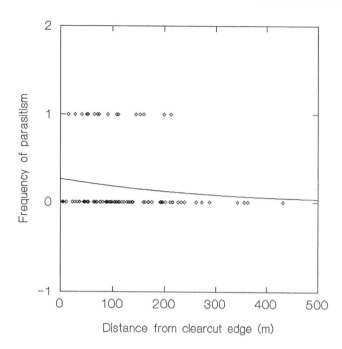

Figure 34.6. Parasitism as a function of distance from clearcut edge for Acadian Flycatcher nests in forests near clearcuts in 1992, all sites pooled. Points with values on the *y*-axis of 1 represent parasitized nests; points with values of 0 represent nonparasitized nests. The curve shows logistic model of parasitism with respect to distance from clearcut edge. The fitted model equation is given in the text.

diana and in the Midwest region as a whole. On the other hand, parasitism levels in the heavily forested Ozarks of Missouri are as low as 3% (Thompson et al., this volume). Parasitism pressure at regional scales is probably influenced heavily by landscape factors, such as percentage of forest cover and habitat heterogeneity. The landscape within which we work is heavily forested relative to areas in the Midwest where cowbird parasitism is high; this may explain the low levels of parasitism we observed (Robinson et al. 1995).

In our study area, Red-eyed Vireos and ground-nesting warblers were more heavily parasitized than Acadian Flycatchers and Wood Thrushes. This contrasts with patterns in Illinois, where Wood Thrushes often suffer parasitism frequencies close to 100% and ground-nesting species usually exhibit lower levels of parasitism than understory- and canopy-nesting species (Robinson 1992; Robinson et al. Chapter 33 and Trine, Chapter 15, this volume). Such regional differences in host-specific parasitism frequencies may result from regional differences in the behavior of cowbirds, host species, or both. Cowbirds may actively choose certain host species. Which species are preferred in a given area may be a function of local community structure or species-specific levels of nest predation. Alternatively, variables

such as host defense behavior or nest height may differ for a given species over regional scales, and this may affect the ability of cowbirds to find or their propensity to parasitize nests of that species. More information addressing the mechanisms of cowbird nest-searching and host defense behavior is needed.

Our data indicate that levels of parasitism are lower in interior forest than in forests adjacent to exterior or interior edges. This provides support for the hypothesis that levels of parasitism are elevated in proximity to internal and external edges (e.g., Brittingham and Temple 1983). Such local landscape patterns of parasitism probably result to a large extent from the habitat selection behavior of cowbirds. Host species differ in landscape patterns of parasitism over local scales (Figure 34.5). For the Acadian Flycatcher and perhaps the Wood Thrush, parasitism was higher near clearcuts than in the interior forest or exterior edge ravines. For the Red-eyed Vireo and perhaps ground-nesting warblers, parasitism was as high or higher in the exterior as in forests near clearcuts.

Levels and within-landscape patterns of parasitism were quite similar in 1992 and 1993, but data from 1994 (unpubl.) indicate that annual variation may be important. In the two Hoosier clearcut sites that were monitored in that year, only 2 of 50 Acadian Flycatcher nests were parasitized, whereas 6 of 66 were parasitized in interior forest. This may also reflect variation among sites within a landscape context. Data from the five clearcuts were pooled in this analysis; however, these sites vary in levels of parasitism and in landscape variables that may be related to parasitism intensity (Doran et al. unpubl. data).

Our initial results suggest that levels of parasitism may be lower within successional areas than in surrounding forest. The data are difficult to interpret, however, because sampling effort has varied between years. A larger sample of nests from small timbercuts and adjacent forest in Yellowwood in 1994 showed this same pattern (0 of 26 nests within cuts and 27 of 170 in forest near cuts parasitized; Ford unpubl. data). This pattern may result from differences in host species composition between these two habitats. A large proportion of the sample of nests from timbercuts consisted of species that are only rarely parasitized and of species that reject cowbird eggs. Early successional species may be more likely to have evolved defenses against parasitism, such as egg rejection (Mayfield 1965). Also, cowbirds may prefer to parasitize forest species. Several recent studies have found parasitism rates to be low in successional areas (e.g., Robinson et al., Chapter 33, this volume).

Data from 1992 suggest that parasitism of Acadian Flycatcher nests declines with distance from clearcut edge, although the result is not statistically significant. We have information that will allow us to calculate distances from nests monitored in other years and in other edge contexts, which should enable us to detect trends (if they exist) of the magnitude suggested by the model. Data from forest openings suggest that parasitism decreases with distance to forest opening edge (Doran unpubl.).

Comparison of these results with those of other studies shows that parasitism levels vary between species and over both large and small spatial scales. Furthermore, spatial patterns of parasitism at large and small scales vary with species. This may result from variation in the behavior of cowbirds and/or hosts. Because cowbird females have large breeding ranges and are often territorial (Rothstein et al. 1986, Thompson 1994; Thompson and Dijak, Chapter 10, and Raim, Chapter 9, this volume), it is conceivable that the behavior of individual females could greatly influence patterns of parasitism at local scales.

These results have important implications for both research and management. In terms of research, it is clear that the results of field studies at one location cannot be used to predict landscape patterns of parasitism for individual species in other areas. Coordinated intensive research efforts over large spatial scales will be necessary to understand the factors that determine the frequency and pattern of parasitism.

In terms of management, it is evident that disturbance and fragmentation at all spatial scales can increase levels of parasitism in forest-breeding neotropical migrants. Parasitism levels are elevated in proximity to both internal and external edge. It is not yet clear how parasitism pressure affects the population dynamics of neotropical migrants. When combined with other deleterious effects of forest fragmentation such as reduced habitat availability and increased nest predation, brood parasitism may seriously threaten neotropical migrant populations. If we are to preserve viable populations of these forest-breeding birds, it may be necessary to maintain a network of large continuous tracts of forest and to minimize the amount of internal edge. The forests of south-central Indiana may constitute an important source for populations of neotropical migrants. Management activities presently occurring in state and national forests, such as timber harvest and the creation and maintenance of forest openings, increase the area of internal edge habitat. Such habitat alteration may reduce nesting success and thus detract from this landscape's value as a source for populations of neotropical migrant birds.

Acknowledgments

Funding was provided by the National Fish and Wildlife Foundation, the Nongame and Endangered Wildlife Program of the Indiana Department of Natural Resources, the Martin Foundation, Inc., the Conservation and Research Foundation, the Indiana Academy of Science, Wild Birds Unlimited, Inc., the David G. Frey Memorial Fund, many chapters of Audubon (Amos Butler Audubon Society, Sassafras Audubon Society, South Bend Audubon Society, Evansville Audubon Society, Louisville Audubon Society, Indiana Audubon Society, Wabash Valley Audubon Society, East Central Audubon Society, Dunes-Calumet Audubon Society, Knob and Valley Audubon Society), and many private individuals.

We are indebted to Scott K. Robinson, John Faaborg, and Frank R. Thompson III for advice and encouragement in all phases of our work. We owe special thanks to the dedicated team leaders and all the field assistants who braved the elements to locate and monitor nests. In this context, special thanks are due to Peter McKinley, Elizabeth Geils, Bernadette Slusher, Patrick Doran, and Cherie Wilson-Heuser.

References Cited

Braun, E. L. 1950. Deciduous forests of eastern North America. Free Press, New York.

Brittingham, M. C., and S. A. Temple. 1983. Have cowbirds caused forest songbirds to decline? BioScience 33:31–35.

Chasko, G. C., and J. E. Gates. 1982. Avian habitat suitability along a transmission-line corridor in an oak-hickory forest region. Wildlife Monogr. 82:1–41.

DeSante, D. F. 1981. A field test of the variable circular-plot censusing technique in a California coastal scrub breeding bird community. Studies in Avian Biol. 6:177–185.

Frazer, C. F. 1992. Factors influencing the habitat selection of breeding birds in Hoosier National Forest, Indiana: Clearcutting and edges. Ph.D. Thesis, Department of Biology, Indiana University, Bloomington.

Gates, J. E., and L. W. Gysel. 1978. Avian nest dispersion and fledging success in field-forest ecotones. Ecology 59:871–883.

Gokhale, D. V., and S. Kullback. 1978. The information in contingency tables. Marcel Dekker, New York.

Hoover, J. P., and M. C. Brittingham. 1993. Regional variation in cowbird parasitism of Wood Thrushes. Wilson Bull. 105:228–238.

Mayfield, H. 1965. The Brown-headed Cowbird, with old and new hosts. Living Bird 4:13–28.

O'Conner, R. J., and J. Faaborg. 1992. The relative abundance of the Brown-headed Cowbird (*Molothrus ater*) in relation to exterior and interior edges in forests of Missouri. Trans. Missouri Acad. Sci. 26:1–9.

Pulliam, H. R. 1988. Sources, sinks, and population regulation. Amer. Natur. 132:652–661.

Robinson, S. K. 1992. Population dynamics of breeding neotropical migrants in a fragmented Illinois landscape. Pp. 408–418 in Ecology and conservation of neotropical migrant landbirds (J. M. Hagan III and D. W. Johnston, eds). Smithsonian Institution Press, Washington, DC.

Robinson, S. K., J. A. Grzybowski, S. I. Rothstein, M. C. Brittingham, L. J. Petit, and F. R. Thompson III. 1993. Management implications of cowbird parasitism on neotropical migrant songbirds. Pp. 93–102 in Status and management of neotropical migratory birds (D. M. Finch and P. W. Stangel, eds). USDA Forest Service Gen. Tech. Rep. RM-229.

Robinson, S. K., F. R. Thompson III, T. M. Donovan, D. R. White-head, and J. Faaborg. 1995. Regional forest fragmentation and the nesting success of migratory birds. Science 267:1987–1990.

Rothstein, S. I., D. A. Yokel, and R. C. Fleischer. 1986. Social dominance, mating and spacing systems, female fecundity, and vocal dialects in captive and free-ranging Brown-headed Cowbirds. Curr. Ornithol. 3:127–185.

SPSS. 1990. SPSS for VAX/VMS, version 4.1. SPSS, Inc., Chicago, IL.

Steel, R. G. D., and J. H. Torrie. 1981. Principles and procedures of statistics: A biometrical approach. 2nd ed. McGraw-Hill International Book Company, Singapore.

SYSTAT. 1992. SYSTAT for Windows, version 5. SYSTAT, Inc., Evanston, IL.

Thompson, F. R. III. 1994. Temporal and spatial patterns of breeding Brown-headed Cowbirds in the midwestern United States. Auk 111:979–990.

USFS. 1991. Amendment to the Hoosier National Forest land and resource management plan. USDA Forest Service, Eastern Region.

Wilcove, D. S. 1985. Nest predation in forest tracts and the decline of migratory songbirds. Ecology 66:1211–1214.

35. Effects of Land-use and Management Practices on the Presence of Brown-headed Cowbirds in the White Mountains of New Hampshire and Maine

MARIKO YAMASAKI, TONI M. MCLELLAN,

RICHARD M. DEGRAAF, AND CHRISTINE A. COSTELLO

Abstract

Forest managers in northern New England concerned about brood parasitism by Brown-headed Cowbirds need information on whether openings in this forested region increase cowbird abundance. We evaluated the effects of vegetation management and changing land use on cowbird presence on 10 transects 12.87 km long on roads into the White Mountain National Forest in New Hampshire and Maine from surrounding nonindustrial and private forestland and agricultural and suburban areas. We counted singing males or visible females weekly from late April through early July 1991. We classified land use and vegetation within 0.4 km of each listening station and verified habitat composition with aerial photography and ground-truthing procedures. Cowbirds were observed 365 times during the 1991 breeding season; 361 of these observations were on the 10 transect segments outside the national forest boundary. Four cowbird observations were recorded on 2 transect segments within the boundary. Land-use composition on plots outside the boundary averaged 65% forest, 18% agricultural, 9% residential, and 7% other uses. Land use on transect segments within the national forest was almost fully forested (mean = 97%). We grouped all silvicultural age-classes into a single land-use type. New clearcuts (under 5 years old) covered 5% of the area of national forest transect segments and less than 1% of the area of off-forest transect segments. Landscape composition is a major influence on the degree and extent of cowbird presence in these heavily forested landscapes.

Introduction

Much recent research on Brown-headed Cowbirds has focused on the midwestern United States, where forest cover is often a minor landscape component (e.g., Brittingham and Temple 1983, Chasko and Gates 1982, Gates and Gysel 1978, Robinson 1992, Robinson et al. 1993, Wilcove and Robinson 1990). These studies have reported high rates of brood parasitism, and certain vegetation management practices have been implicated in the increase of favorable cowbird habitat. Environmental groups in New England are concerned that vegetation management may enhance cowbird habitat in heavily forested landscapes in New England.

Cowbirds are brood parasites of at least 214 bird species, including 121 species that can rear cowbirds (Friedmann 1963, Morse 1989, DeGraaf and Rudis 1986). Historically, cowbirds roamed with bison, feeding on insects and seeds dislodged and spread by the herds (Mayfield 1965). Because of their feeding habits, cowbirds were once restricted to the open habitats found primarily on the Great Plains. New England forests, however, were largely cleared for agriculture by the mid-1800s, and the newly cleared habitat provided opportunities for an eastward range expansion of cowbirds (Mayfield 1965). Between 1820 and 1840, 75% of arable land in southern and central New England was in farm crops and pastures (DeGraaf et al. 1992). By the mid-20th century, most farms had been abandoned and New England had returned to a primarily forested condition (DeGraaf et al. 1992). Although cowbird populations are decreasing across the northeast (Robbins et al. 1986; Peterjohn et al., Chapter 2, and Wiedenfeld, Chapter 3, this volume), concern still exists over possible localized encroachment of cowbird parasitism through inadvertent creation of cowbird foraging habitat in the process of managing forest vegetation.

The White Mountain National Forest (WMNF) in New Hampshire and Maine is roughly 311,850 ha. Approximately 45% of the vegetation (139,911 ha) is actively managed using even-aged and uneven-aged silvicultural strategies (Figure 35.1). The WMNF Forest Plan limits clearcuts to a maximum of 12 ha (USDA 1986), with an average harvest size less than 8 ha (T. Brady pers. comm.). Stands created using other even-aged practices, such as shelterwood methods, or uneven-aged practices, such as individual selection or group selection methods, are not limited in size. Vegetation management activity is severely restricted on the remaining 55% of the WMNF, which is designated as wilderness or scenic areas. The unmanaged vegetative landscape of the WMNF forms the central core of the forest. Our objectives were to determine the relationships between cowbird occurrence during the breeding season and land-use practices in the White Mountains. We compared the general presence and distribution of cowbirds in relation to land-use classification and vegetation in the actively managed portion of the national forest and surrounding off-forest sites. The WMNF provides an opportunity to contrast an unbroken forest tract managed by a single agency to the adjacent forested landscape in which there are many private owners and where forested and nonforest habitats are more often interspersed.

Methods

All WMNF ranger districts were included in the study, along with the adjacent nonindustrial private forestland and agricultural and residential areas (Figure 35.1). Ten 12.87-km transects were established along secondary or lesser roads penetrating the WMNF from surrounding areas. Each transect consisted of 17 listening stations located 0.8 km apart. Eight of the listening stations were located within the WMNF boundary and comprised a national forest transect segment. Eight listening stations were located outside the boundary and comprised an off-forest transect segment, and one listening station was located at the boundary interface. In all, there were 170 listening stations.

Cowbird Presence

Bird surveys were conducted between 30 April and 1 July 1991 using the point-count method (Ralph et al. 1995) to detect any late migrant activity as well as breeding bird activity in the study area. All birds seen or heard were recorded during a 4-min observation period. The 4-min observation period allowed us to sample all 10 transects and 170 listening stations across this mountainous landscape on a weekly basis. Each transect was surveyed nine times, yielding a total of 36 min of observation per listening station and a grand total of 6,120 min. Surveys were conducted on mornings with low wind and no precipitation, beginning within 15 min of sunrise, and were completed by mid-morning. Three starting points were established on the transects: national forest end, interface, and off-forest end. Starting points were rotated over the survey period to account for bias associated with declining activity levels as morning progressed (Robbins 1981). Four observers rotated assignments to minimize observer bias.

Habitat Assessment

Listening stations were plotted on the most recent aerial photographs available, either 1987 color infrared photo-

graphs (scale, 1 : 12,000) or 1982 black and white photographs (scale, 1 : 1,000). As the recent rate of land-use conversion has been low, photos from 1982 to 1987 were adequate for our purposes. The locations of transect lines and listening stations were transferred from aerial photos to their respective topographic maps using a Bausch and Lomb Zoom transfer scope. A circular vegetation and land-use plot was chosen to characterize the proportion of potential feeding habitat for cowbirds in the general area surrounding a listening station. Land-use and habitat composition within 0.4 km of each listening station were classified into 11 categories: coniferous, deciduous, mixed, and harvested forest; agricultural field; rural, suburban, and town residential; wetlands, developed recreation, and other minor categories. All silvicultural age classes are included in the forest land-use type. These classifications were verified using aerial photographs and ground checks. The dot-grid method was used to estimate total transect acreage, land use, and habitat composition for each transect and in the area surrounding a listening station. In Missouri and Illinois, almost 60% of cowbird movements from breeding to feeding sites were less than 1 km (Robinson et al. 1993).

Statistical Analysis

We examined the weekly occurrence of Brown-headed Cowbirds per station throughout the 1991 breeding season to determine seasonal variation in the degree of flocking seen at stations. We report sums of counts per transect segment (national forest and off-forest) to characterize cowbird occurrence in the breeding season, rather than counts per listening station, because of the lack of independence between listening stations on a transect (Thomas and Taylor 1990). We used SYSTAT (Wilkinson et al. 1992) to analyze most of our data. Log-likelihood ratio tests (Zar 1984) were used to examine the distribution of counts during the breeding season (Table 35.1), and Kruskal-Wallis and Mann-Whitney tests (Zar 1984) were used to examine differences in cowbird occurrence between transects, transect segments, and observation periods (Table 35.2). We used two-way ANOVAs followed by Tukey HSD multiple comparison tests to examine differences in cowbird abundance between transects and transect segments through the spring season (Table 35.3).

Many of the 11 habitat categories generated too few data to analyze habitat use by cowbirds. We therefore collapsed the 11 variables into four grouped categories: forest (coniferous, deciduous, mixed, and harvested), field, residential (rural, suburban, and town), and other (wetlands, developed recreation, and other minor categories). We report mean habitat composition percentages along 20 transect segments, composed of 160 of 170 plots (Table 35.4). We excluded the 10 interface plots. We compared habitat composition on transect segments with and without observations of cowbirds, using a Kruskal-Wallis test (Table 35.5). We examined the relationships between seasonal cowbird

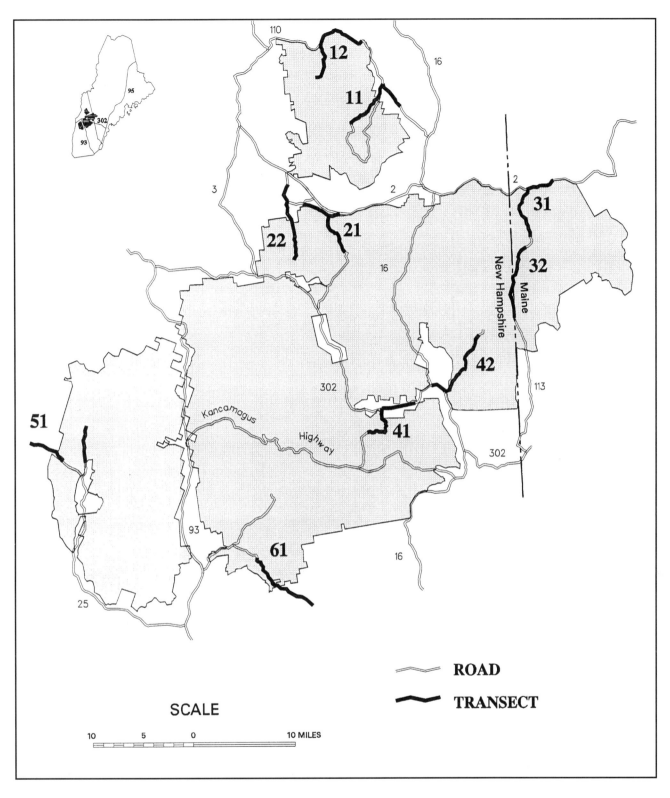

Figure 35.1. Study area in the White Mountains of New Hampshire and Maine, surveyed in 1991. Shaded area denotes the purchase unit authorization boundary of the White Mountain National Forest. Private land inholdings are found within the limits of the purchase unit authorization boundary. Bold numbers indicate the 10 transects used in this study: 11, York Pond; 12, Mill Brook; 21, Jefferson Notch; 22, Cherry Mountain; 31, Evans Notch North; 32, Evans Notch South; 41, Bartlett; 42, Mountain Pond; 51, North-South; 61, Sandwich Notch.

Table 35.1. Brown-headed Cowbirds Counted across 10 Transects in the White Mountains of New Hampshire and Maine, Spring 1991

| Cowbirds/ Transect | Spring Season (Week) | | | | | | | | | Total Transects | Cowbirds/ Season |
| | Early | | | Mid | | | Late | | | | |
	1	2	3	4	5	6	7	8	9		
0	1	4	1	2	2	0	5	1	1	17	0
1	2	1	1	1	1	2	0	3	3	14	14
2	1	3	1	2	1	0	2	0	2	12	24
3	3	0	2	1	3	2	0	1	2	14	42
4	0	0	0	0	0	1	0	0	0	1	4
5	1	0	2	2	0	1	0	0	1	7	35
6	1	1	0	0	0	0	1	1	0	4	24
7	0	0	0	1	2	0	0	1	1	5	35
8	0	0	1	0	0	2	1	0	0	4	32
9	0	0	1	0	0	0	0	1	0	2	18
10	0	0	0	0	0	0	0	2	0	2	20
12	0	0	1	0	0	1	0	0	0	2	24
13	0	0	0	0	1	0	1	0	0	2	26
14	0	0	0	1	0	0	0	0	0	1	14
15	1	1	0	0	0	0	0	0	0	2	30
23	0	0	0	0	0	1	0	0	0	1	23
Transects counted	10	10	10	10	10	10	10	10	10	90	
Cowbird total	39	28	48	39	39	68	31	48	25		365

Note: Counts are weekly sums of 17 listening stations for each transect. The breeding season is also divided into three seasonal periods of three weeks' duration.

abundance and the four habitat composition variables for both transects and transect segments using Spearman rank correlations (Table 35.6).

Results

Cowbird Presence

We counted 365 Brown-headed Cowbirds during the spring of 1991 at 64 of the 170 points. We counted a weekly average of 1.97 ($\pm$ 0.11 SE) cowbirds/station. Weekly counts ranged from 0 to 10 cowbirds/station. Only 4 of the 365 cowbird observations were recorded within the WMNF, all of them during the first three weeks of the survey period (Table 35.2). The first three weeks of observations in late April and early May accounted for 31.5% of all observations. Seasonal variation in the distribution of cowbird counts at stations was not significant (log-likelihood ratio X^2 = 35.244, 30 df, P = .234). Fifteen cowbird observations from interface seg-

ments were excluded from the subsequent analysis. Cowbird occurrences between transects and on transect segments (Table 35.2) were not independent (log-likelihood ratio X^2 = 21.118, 9 df, P = .012). Numbers of cowbird observations were similar between observers (log-likelihood ratio X^2 = 27.984, 24 df, P = .261).

Cowbird abundance did not differ significantly among transects (Table 35.2; Kruskal-Wallis ANOVA, H = 12.616, 9 df, P = .181), nor was it significantly different during the three spring observation periods (H = 0.637, 2 df, P = .727). Cowbird abundance did, however, differ by transect segment (e.g., national forest vs. off-forest, H = 96.357, 1 df, P = .000).

We used a two-way ANOVA to test the hypothesis that cowbird abundance differed across transects and in relation to land use. Land use (national forest vs. off-forest) was a significant source of variation in cowbird abundance (F = 92.119, 1 df, P = .000). Transects also contributed to vari-

Table 35.2. Seasonal Counts of Brown-headed Cowbirds across Transects in the White Mountains of New Hampshire and Maine, Spring 1991

Transect[a]	Segment	Spring Season (Weekly Sums)			Total
		Early	Mid	Late	
11	NF	0	0	0	0
	I	0	0	0	0
	OF	4	20	16	40
12	NF	0	0	0	0
	I	0	0	0	0
	OF	6	6	11	23
21	NF	0	0	0	0
	I	3	2	0	5
	OF	12	3	4	19
22	NF	1	0	0	1
	I	0	0	0	0
	OF	10	3	1	14
31	NF	0	0	0	0
	I	0	0	0	0
	OF	11	10	8	29
32	NF	0	0	0	0
	I	0	0	0	0
	OF	39	44	9	92
41	NF	0	0	0	0
	I	0	1	7	8
	OF	10	12	5	27
42	NF	0	0	0	0
	I	0	0	0	0
	OF	0	1	3	4
51	NF	3	0	0	3
	I	0	2	0	2
	OF	9	28	17	54
61	NF	0	0	0	0
	I	0	1	1	2
	OF	7	14	21	42
Total	NF	4	0	0	4
	I	3	6	8	17
	OF	108	141	95	344

[a]Refer to Figure 35.1 for transect location.

Note: National forest (NF) and off-forest (OF) segments are sums of weekly counts from 8 listening stations per transect segment, and interface (I) segments are sums of weekly counts from 1 listening station per transect. Early spring is the first three-week period from late April through mid-May, mid-spring is the second three-week period from mid-May through early June, and late spring is the third three-week period from early June through the beginning of July.

Table 35.3. Pairwise Probabilities from Tukey HSD Tests Comparing Mean Cowbird Abundances on Different Transects during Spring 1991 in the White Mountains of New Hampshire and Maine

Transect	Mean	11	12	21	22	31	32	41	42	51	61
11	2.22	1.000									
12	1.28	0.998	1.000								
21	1.06	0.991	1.000	1.000							
22	0.83	0.971	1.000	1.000	1.000						
31	1.61	1.000	1.000	1.000	0.999	1.000					
32	5.11	0.229	0.029	0.015	0.007	0.070	1.000				
41	1.50	1.000	1.000	1.000	1.000	1.000	0.053	1.000			
42	0.22	0.771	0.996	0.999	1.000	0.971	0.001	0.983	1.000		
51	3.17	0.998	0.825	0.710	0.575	0.940	0.799	0.914	0.234	1.000	
61	2.33	1.000	0.996	0.983	0.952	1.000	0.313	0.999	0.710	0.999	1.000

Note: N = 180; 10 transects × 2 transect segments × 9 weekly observations.

Table 35.4. Habitat Composition for 20 Transect Segments in the White Mountains of New Hampshire and Maine, 1991

Transect[a]	Segment[b]	Habitat (Mean % Segment ± SE)[c]			
		Forest	Field	Residential	Other
11	NF	97.5 ± 2.0	0	0	2.5 ± 2.0
11	OF	86.3 ± 2.1	5.5 ± 2.6	6.6 ± 2.1	1.6 ± 1.3
12	NF	97.5 ± 2.4	0	2.1 ± 2.1	0.4 ± 0.3
12	OF	78.0 ± 4.2	10.3 ± 3.9	0	11.7 ± 4.4
21	NF	99.7 ± 0.3	0	0	0.3 ± 0.3
21	OF	82.4 ± 5.7	15.5 ± 6.2	0	2.1 ± 1.1
22	NF	97.1 ± 1.5	0	1.3 ± 1.3	1.6 ± 1.1
22	OF	43.4 ± 8.0	50.6 ± 8.1	4.4 ± 2.7	1.6 ± 1.1
31	NF	95.6 ± 1.4	0	0	4.4 ± 1.4
31	OF	73.1 ± 3.4	3.4 ± 2.2	3.4 ± 3.4	20.1 ± 4.1
32	NF	95.1 ± 2.5	0.8 ± 0.8	0	4.1 ± 2.3
32	OF	64.9 ± 5.1	29.1 ± 4.7	1.0 ± 1.0	5.0 ± 3.3
41	NF	99.9 ± 0.1	0	0	0.1 ± 0.1
41	OF	25.5 ± 6.1	9.4 ± 4.2	41.8 ± 12.6	22.3 ± 7.7
42	NF	99.5 ± 0.9	0	0	0.4 ± 0.4
42	OF	77.6 ± 5.9	1.3 ± 1.1	17.2 ± 5.5	3.9 ± 0.6
51	NF	96.4 ± 3.1	0	0	3.6 ± 3.1
51	OF	47.4 ± 5.2	46.2 ± 5.9	1.9 ± 1.3	4.5 ± 2.8
61	NF	92.3 ± 3.7	0	0	7.8 ± 3.7
61	OF	76.1 ± 9.0	4.0 ± 3.1	18.8 ± 9.8	1.1 ± 0.7
Totals	NF	96.9 ± 0.8	0.06 ± 0.61	0.3 ± 0.2	2.5 ± 0.8
	OF	65.2 ± 6.2	17.4 ± 5.7	9.4 ± 4.1	7.3 ± 2.5

[a]Refer to Figure 35.1 for transect locations.

[b]NF, national forest segment; OF, off-forest segment.

[c]Forest habitat includes deciduous, coniferous, mixed, and regenerating forest. Field includes agricultural fields. Residential includes suburban, rural, and town. Other includes wetlands, developed recreation, and other minor categories.

Table 35.5. Habitat Composition for Transect Segments with and without Cowbirds in the White Mountains of New Hampshire and Maine during Spring 1991

Habitat[a]	Mean % Segment ± SE		H^b, P
	Transect Segments with Cowbirds (N = 8)	Transect Segments without Cowbirds (N = 12)	
Forest	70.5 ± 6.21	97.1 ± 0.96	10.500, .001
Field	14.5 ± 5.08	0.08 ± 0.08	9.951, .002
Residential	7.9 ± 3.55	0.3 ± 0.26	6.743, .009
Other	6.5 ± 2.11	2.5 ± 1.01	2.381, .123

[a]Habitats as defined in Table 35.4.
[b]Kruskal-Wallis test statistic H using a χ^2 approximation with 1 degree of freedom.

Table 35.6. Spearman Rank Order Correlation Coefficients for Selected Seasonal Cowbird Abundance, Land-use, and Summary Habitat Variables, White Mountains of New Hampshire and Maine, Spring 1991

	TR[a]	ETR	MTR	LTR	NTR	NFT[b]	OFT	NOFT	FOR[c]	FIE	NFF[d]	NFO	OFF[e]
ETR	0.45	1.00											
MTR	0.97**	0.35	1.00										
LTR	0.73*	−0.12	0.74*	1.0									
NTR	0.99**	0.34	0.95**	0.73*	1.0								
NFT	0.06	0.32	0.06	−0.02	0.02	1.0							
OFT	0.98**	0.39	0.98**	0.73*	0.96**	0.06	1.0						
NOFT	0.95**	0.33	0.98**	0.71*	0.98**	0.02	0.98**	1.0					
FOR	−0.25	−0.38	−0.21	−0.09	−0.10	−0.51	−0.21	−0.08	1.0				
FIE	0.21	0.66*	0.21	−0.09	0.12	0.68*	0.16	0.13	−0.47	1.0			
NFF	−0.54	−0.28	−0.50	−0.31	−0.56	−0.18	−0.66*	−0.59	0.19	−0.07	1.0		
NFO	0.55	0.16	0.52	0.37	0.56	0.10	0.67*	0.63*	−0.05	−0.12	−0.96**	1.0	
OFF	−0.19	−0.39	−0.15	0.03	−0.04	−0.51	−0.15	−0.03	0.98**	−0.39	0.15	−0.04	1.0
OFFI	0.21	0.66*	0.21	−0.09	0.12	0.68*	0.16	0.13	−0.47	1.00**	−0.07	−0.12	−0.39

[a]TR, cowbird abundance by transect; ETR, early spring; MTR, mid-spring; LTR, late spring; NTR, mid- and late spring combined.
[b]NFT, cowbird abundance on national forest transect segments; OFT, off-forest; NOFT, mid- and late spring combined off-forest.
[c]FOR, forest percent by transect; FIE, agricultural field percentage by transect.
[d]NFF, forest percentage by national forest transect segment; NFO, other percentage.
[e]OFF, forest percentage by off-forest transect segment; OFFI, field percent.
*$P < .05$; **$P < .005$.

ation ($F = 5.20$, 9 df, $P = .000$), and there was a significant interaction between transects and land use ($F = 5.254$, 9 df, $P = .000$). Tukey pairwise comparisons suggested that there were more cowbirds on transects with less forested land, and fewer cowbirds on forested transects (Table 35.3; $P < .05$).

Pairwise comparisons of the interaction effects suggested that off-forest transect segments with more cowbirds also had higher proportions of agricultural and residential land use ($P < .05$).

Habitat Assessment

We explored habitat composition on national forest and off-forest segments to explain the differing cowbird abundances (Table 35.4). Analysis of habitat composition by Kruskal-Wallis ANOVAs revealed differences between national forest and off-forest transect segments for both forest and field ($P = .000$) and for residential land-use categories ($P = .005$). No significant difference was detected for the grouped category of wetlands, developed recreation, and other minor habitats, between national forest and off-forest transect segments ($P = .082$).

We found that forest cover was the dominant habitat component on national forest transect segments compared with off-forest transect segments (Table 35.4; mean = 96.9% ± 0.8 vs. 65.2% ± 6.2). Regenerating stands (< 5 years of age) accounted for 5% of national forest transect segments, whereas recently harvested stands covered less than 1% of off-forest transect segments.

Transect segments on which cowbirds were observed had significantly higher percentages of agricultural fields and residential land uses than transect segments without cowbirds (Table 35.5). The mean percentage of managed forest cover was significantly higher on transect segments without cowbirds than on transect segments with cowbirds. The other habitat categories did not differ between transect segments with and without cowbirds ($P = .123$).

Effects of Seasonal Cowbird Abundance and Habitat Composition

We now examine the relationship between seasonal cowbird abundance and habitat composition on both transects and transect segments. Spearman correlations between overall cowbird abundance on transects and later spring observation periods (mid- and late) and off-forest transect segment observations were positive and significant (Table 35.6; $P < .001$). At the transect level, correlations of overall cowbird abundance with transect summaries of habitat composition were not significant ($P > .368$), probably due to the loss of detail from averaging the very different habitat compositions of the national forest and off-forest transect segments into a single variable. However, correlations between early spring cowbird abundance on transects and agricultural field percentages for both transects and off-forest transect segments were positive and significant ($P = .04$). No other significant correlations between seasonal abundance and summary habitat variables were observed. Later in spring, cowbird abundance on transects was weakly negatively correlated with the percentage of forest on national forest transect segments ($P = .10$). Cowbird abundance on transects later in spring was weakly positively correlated with the percentage of other minor habitats on national forest transect segments ($P = .09$).

We also examined the relationship of cowbird abundance

at the transect segment level with habitat composition on both transect and transect segment levels. Correlations between cowbird abundances on national forest transect segments and percentages of agricultural fields for both transects and national forest transect segments were both positive and significant ($P = .03$). The four cowbirds recorded within the WMNF in early spring were most likely transients. Holmes and Sturges (1975) made similar observations at the Hubbard Brook Experimental Forest. Cowbird abundance on off-forest transect segments was negatively correlated with the percentage of forest on national forest transect segments ($P = .04$) and positively correlated with the percentages of other minor habitats on national forest transect segments ($P = .03$).

Discussion and Conclusions

Our results indicate that cowbird abundance in the heavily forested landscapes of the White Mountains of New Hampshire and Maine is generally low. Where cowbirds do occur is strongly related to the occurrence of agricultural and/or residential land use outside the boundary of the WMNF. Cowbirds were essentially absent from the managed forestland within the boundary of the WMNF, as was also found in the New Hampshire Breeding Bird Atlas Project (Foss 1994). In a more intensive long-term study at the Hubbard Brook Experimental Forest in the WMNF, cowbirds have been seen only as occasional transients (Holmes and Sturges 1975, Holmes and Sherry 1988).

Forest covers 86% of the regional landscape of northern New England today (Waddell et al. 1989). At increasingly finer scales of detail, forests continue to be the dominant feature across land management units such as the northern three counties in New Hampshire (92%) and the White Mountain National Forest (96%) (Frieswyk and Malley 1985, USDA 1986). Remaining agricultural and residential land use still occurs in valley bottoms and along some major transportation corridors within the purchase unit boundary of the WMNF. Away from these corridors and along secondary and lesser roads, agricultural and residential land uses within the national forest boundary are minimal. During the breeding season, cowbirds were most often observed on transect segments with agricultural or residential land nearby.

Transect segments dominated by forest cover, especially those within the national forest boundary, showed minimal cowbird activity throughout the breeding season. Cowbird activity was localized and related to nearby foraging opportunities in agricultural land (especially for cereal grains and pasturage) and at feeders near residences. In mostly forested landscapes, recently harvested stands create cowbird foraging areas only if they are converted to cereal grain production or pasturage (i.e., when land use changes). Recently harvested stands that are regenerating to fully stocked forested stands do not encourage cowbird breeding activity, perhaps because foraging opportunities in these stands are minimal. In other regions where the interspersion of agricultural and residential land use is much greater, cowbirds are a more significant concern to managers and owners of both large and small parcels of land (May and Robinson 1985, Robinson 1992, Freemark and Collins 1992, Martin 1992, and Robinson et al. 1993). In the future, land acquisition patterns, habitat composition, and ownership goals on nonindustrial private forest land may need to be addressed to keep cowbird numbers at their current low levels across the White Mountains.

Acknowledgments

We thank L. Rowse for her initial involvement in this study. Her help with designing the study, locating transects, and sampling birds is greatly appreciated. We thank T. Luther and K. Dudzik for the map, T. Brady for timber management information assistance, W. Leak, J. Kanter, and E. Snyder for reviewing this manuscript, and G. Getchell for his editing assistance.

References Cited

Brittingham, M. C., and S. A. Temple. 1983. Have cowbirds caused songbirds to decline? BioScience 33:31–35.

Chasko, G., and J. Gates. 1982. Avian habitat suitability along a transmission-line corridor in an oak-hickory forest region. Wildl. Monogr. 82:1–41.

DeGraaf, R. M., and D. D. Rudis. 1986. New England wildlife: Habitat, natural history, and distribution. USDA Gen. Tech. Rep. NE-108. Broomall, PA. 491 pp.

DeGraaf, R. M., M. Yamasaki, W. B. Leak, and J. W. Lanier. 1992. New England wildlife: Management of forested habitats. USDA Gen. Tech. Rep. NE-144. Radnor, PA. 271 pp.

Foss, C. R. (ed). 1994. Atlas of breeding birds in New Hampshire. Audubon Society of New Hampshire and Arcadia Publishers, Dover, NH. 414 pp.

Freemark, K., and B. Collins. 1992. Landscape ecology of birds breeding in temperate forest fragments. Pp. 443–454 in Ecology and conservation of neotropical migrant landbirds (J. M. Hagan III and D. W. Johnston, eds). Smithsonian Institution Press, Washington, DC.

Friedmann, H. 1963. Host relations of the parasitic cowbirds. U.S. Natl. Mus. Bull. 233.

Frieswyk, T. S., and A. M. Malley. 1985. Forest statistics for New Hampshire, 1973 and 1983. USDA Res. Bull. NE-88. Broomall, PA. 100 pp.

Gates, J. E., and L. W. Gysel. 1978. Avian nest dispersion and fledging success in field-forest ecotones. Ecology 59:871–883.

Holmes, R. T., and T. W. Sherry. 1988. Assessing population trends in New Hampshire forest birds: Local vs. regional patterns. Auk 105:756–768.

Holmes, R. T., and F. W. Sturges. 1975. Avian community dynamics and energetics in a northern hardwoods ecosystem. J. Anim. Ecol. 44:175–200.

Martin, T. E. 1992. Breeding productivity considerations: What are the appropriate habitat features for management? Pp. 455–473 in Ecology and conservation of neotropical migrant landbirds (J. M. Hagan III and D. W. Johnston, eds). Smithsonian Institution Press, Washington, DC.

May, R. M., and S. K. Robinson. 1985. Population dynamics of avian brood parasitism. Amer. Natur. 126:475–494.

Mayfield, H. F. 1965. The Brown-headed Cowbird with old and new hosts. Living Bird 4:13–28.

Morse, D. H. 1989. American warblers. Harvard University Press, Cambridge, MA. 406 pp.

Ralph, C. J., S. Droege, and J. R. Sauer. 1995. Managing and monitoring birds using point counts: Standards and applications. Pp. 161–168 in Monitoring bird populations by point counts (C. J. Ralph, J. R. Sauer, and S. Droege, eds). USDA Gen. Tech. Rep. PSW-149. Albany, CA.

Robbins, C. S. 1981. Effect of time of day on bird activity. Studies in Avian Biol. 6:275–286.

Robbins, C. S., D. Bystrak, and P. H. Geissler. 1986. The breeding bird survey: Its first fifteen years, 1965–1979. Res. Publ. 157. USDI, Fish and Wildlife Service. Washington, DC. 196 pp.

Robinson, S. K. 1992. Population dynamics of breeding neotropical migrants in a fragmented Illinois landscape. Pp. 408–418 in Ecology and conservation of neotropical migrant landbirds (J. M. Hagan III and D. W. Johnston, eds). Smithsonian Institution Press, Washington, DC.

Robinson, S. K., J. A. Grzybowski, S. I. Rothstein, M. C. Brittingham, L. J. Petit, and F. R. Thompson III. 1993. Management implications of cowbird parasitism on neotropical migrant songbirds. Pp. 93–102 in Status and management of neotropical migratory birds (D. M. Finch and P. W. Stangel, eds). USDA Gen. Tech. Rep. RM-229.

Thomas, D. L., and E. J. Taylor. 1990. Study designs and tests for comparing resource use and availability. J. Wildl. Manage. 54: 322–330.

USDA. 1986. White Mountain National Forest Land and Resource Management Plan. U.S. Department of Agriculture Eastern Region, Milwaukee, WI.

Waddell, K. L., D. D. Oswald, and D. S. Powell. 1989. Forest statistics of the United States, 1987. USDA Res. Bull. PNW-RB-168. Portland, OR. 106 pp.

Wilcove, D. S., and S. K. Robinson. 1990. The impact of forest fragmentation on bird communities in eastern North America. Pp. 319–331 in Biogeography and ecology of forest bird communities (A. Keast, ed). SPB Academic Publishers, The Hague, The Netherlands.

Wilkinson, L., M. Hill, J. P. Welna, and G. K. Birkenbeuel. 1992. SYSTAT for Windows: Statistics version 5. SYSTAT Inc., Evanston, IL. 750 pp.

Zar, J. H. 1984. Biostatistical analysis. 2nd ed. Prentice-Hall, Englewood Cliffs, NJ. 718 pp.

PART V

Cowbird Management,
Host Population Limitation,
and Efforts to Save
Endangered Species

36. Introduction

STEPHEN I. ROTHSTEIN AND TERRY L. COOK

Concern over continental declines in the distribution and abundance of passerine birds, many of which are neotropical migrants, has gained increased coverage in the scientific and popular literature (Hagan and Johnston 1992; Terborgh 1989, 1992; Martin and Finch 1995). Some workers (e.g., Robinson et al. 1995b) have attributed declines largely to increasingly fragmented and degraded breeding habitats in North America, which in turn has made populations more vulnerable to predation and parasitism. Other researchers have argued that habitat destruction and degradation in the neotropics is the primary cause of the declines (Morton 1992, Rappole and McDonald 1992).

However, recent studies of passerine population trends have shown that while many species are decreasing, some at an alarming rate, many others are currently increasing (Peterjohn et al. 1995, James et al. 1996). Furthermore, although some species or groups of species are decreasing in some regions, they are increasing elsewhere. For example, trends derived from the Breeding Bird Survey show that most neotropical migrants are decreasing in the central part of North America, but most are increasing in the West (Peterjohn et al. 1995). Ornithologists are coming to realize that even under normal conditions, the abundances of passerines and their distributions are highly dynamic (Johnson 1994). Indeed, it may be normal for many species to be increasing while others are decreasing at any point in time. Nevertheless, some low-profile groups of birds are showing widespread declines. Significantly more than half of all shrubland and grassland birds in eastern North America, most of which are short-distance migrants, are decreasing (Askins 1993).

Analyses of data from the Breeding Bird Survey have failed to demonstrate a link between cowbird population trends and changing abundance in common hosts (Peterjohn et al., Chapter 2, and Wiedenfeld, Chapter 3, this volume). Although brood parasitism can significantly reduce the production of young in local populations of widespread host species, these local populations may not decline because

numbers are maintained by emigrants from more productive populations (Brawn and Robinson 1996). It is likely, however, that parasitism can endanger an entire taxon if habitat destruction and/or highly specific habitat requirements limit it to one or more small and heavily parasitized populations.

Thus, there may be a need for management of cowbirds in some circumstances. The approach of most active population management is to identify key limiting factors and to remove or ameliorate the most important of these (Caughley and Sinclair 1994). In this overview, we discuss active cowbird management programs initiated to minimize the impacts of cowbird parasitism on four endangered species whose numbers have been reduced by extensive degradation and loss of habitat. These four taxa differ in why they are endangered and in their responses to cowbird management. They also differ in the reproductive loss they experience when parasitized; only the Kirtland's Warbler often fledges some of its own young if a cowbird egg hatches in its nest (Mayfield 1960). The other three species are smaller and nearly always lose all of their young if even a single cowbird egg hatches. We first discuss each of these four endangered species briefly and then highlight similarities and differences among them and among the cowbird control programs designed to help their recoveries. We close with a brief discussion of the benefits and costs of cowbird control programs.

Case Studies

Kirtland's Warbler

The first, and perhaps the best known case, of active cowbird management was initiated to protect the Kirtland's Warbler (*Dendroica kirtlandii*; DeCapita Chapter 37). Unlike the other endangered species treated in this book, the warbler has had a limited range and population size throughout recorded history. It nests in several counties of northern lower Michigan and only in jack pine (*Pinus banksiana*) forests

6–24 years after fires. In the last 150 years, it is likely that its numbers peaked at around a few thousand individuals in the late 1800s, which is also the time that it probably became exposed to cowbirds (Mayfield 1960). Warbler numbers were much lower by the 1940s, probably because of fire suppression and cowbird parasitism, and interested parties resolved to conduct a complete census of the species every 10 years. Counts in 1951 and 1961 revealed 432 and 502 singing males, but the 1971 count showed only 201 singing males. This decline, evidence of increased parasitism in the late 1960s, and demographic projections indicating that recruitment rates were critically low with frequent cowbird parasitism led to a cowbird control program initiated in 1972.

DeCapita reports that more than 98,000 cowbirds had been removed from the warbler's nesting area by 1995, with about another 4,000 removed in 1996 (DeCapita pers. comm.). Cowbird trapping reduced nest parasitism from a mean of 70% in 1966–1971 to 5.6% in 1972–1977. During this period, mean host fledgling production increased from 0.8 young per pair per year in 1966–1971 to 3.08 during 1972–1977. Parasitism has remained low in recent years (Bocetti 1994). Despite these impressive reproductive gains, the numbers of Kirtland's Warblers remained fairly constant until they began to increase rapidly in 1990, 10 years after the Mack Lake Burn, a 10,500-ha wildfire, created a massive amount of new habitat. Populations have continued to increase through the 1990s, and 766 singing males were recorded in 1995. As DeCapita shows, much of this dramatic increase is due to birds breeding on the Mack Lake Burn. Thus, the Kirtland's Warbler was limited by the availability of breeding habitat in recent decades (Kepler et al. 1996) after nearly all cowbird parasitism was eliminated. Up till 1996, there was assumed to be plenty of wintering habitat for the species in the Bahamas (Kepler et al. 1996, DeCapita Chapter 37). Recent work, however, provides evidence that winter habitat may also have been limiting (Haney et al. 1998).

Although a shortage of breeding habitat was mentioned as a possible cause for the failure of the numbers of breeding Kirtland's Warblers to increase soon after cowbird control began, it was not identified as the chief limiting factor (Mayfield 1978, 1983; Probst 1986). However, planting, not fires, created nearly all the seemingly suitable habitat that was unoccupied, and planted habitat is less preferred by warblers (DeCapita pers. comm.). It is now recognized that there was a shortage of good breeding habitat prior to the Mack Lake Burn (Weise 1987, Kepler et al. 1996). Indeed, breeding habitat may have had a much greater role in limiting warbler numbers than was assumed in the 1960s and 1970s.

If cowbird parasitism had been the chief or only proximate limiting factor, the warbler population would have increased rapidly after cowbird control nearly eliminated parasitism on the species. Such an immediate increase has happened with most managed populations of the Least Bell's Vireo (Griffith and Griffith Chapter 38).

Given the likely importance of habitat limitation in both summer and winter, was cowbird control in Michigan worthwhile? Cowbird control was clearly prudent in 1971, when it became known that the Kirtland's Warbler population had declined by 60%, and it was also prudent to continue this control while the population hovered around 200 singing males for the next 19 years. Nevertheless, there is no clear evidence that cowbird control was beneficial during that period, although it has often been suggested that cowbird control kept the Kirtland's Warbler from going extinct (Terborgh 1989, Trail 1992, Kepler et al. 1996). Although demographic projections indicated that the 1971 population was not self-sustaining, such projections are heavily dependent on estimates of mortality for young and old birds. Mortality rates are difficult to determine for small birds, and if the true rates are higher than the estimates, then those 200 pairs of Kirtland's Warblers could have been self-sustaining even with cowbird parasitism. In addition, the parasitism rates used in those models were based largely on a relatively small sample of nests, 52, or 8.7 per year found between 1966 and 1971 (DeCapita pers. comm.). These nests may not have been representative of the entire warbler population, especially if they were limited to one or two study sites.

Concluding that the Kirtland's Warbler was headed to extinction before cowbird control began makes it necessary to assume that control just happened to start when the warbler numbers were at the carrying capacity, about 200 pairs, at which they stayed for almost the next 20 years. Such a coincidence seems unlikely to us. Warbler numbers may already have stabilized at about 200 breeding pairs by the early 1970s. In support of this argument, Kirtland's Warblers were not known to be decreasing at the time of the 1971 census; the decrease that occurred between 1961 and 1971 could have occurred anytime in the 1960s and might have been due to losses of wintering habitat (Haney et al. 1998), not to cowbirds. Indeed, Probst (1986) noted that the amount of breeding habitat decreased between 1961 and 1971, and Kepler et al. (1996) suggested that this contributed to the warbler decline during this period. The population size could have stabilized at about 200 pairs because it corresponded to the rate of habitat renewal produced by managed pine plantings and small controlled burns and because that was the carrying capacity of winter habitat at the time. Population stability in the face of cowbird parasitism is also suggested by the steady size of the warbler population between 1971 and 1972 (201 versus 200 singing males, Mayfield 1978), even though there was no cowbird control until 1972. If the population was already declining in response to an excess of mortality over recruitment, as is widely assumed, then there should have been fewer birds in 1972.

Although cowbird control may not have been necessary to save Kirtland's Warbler from extinction in the early 1970s, it is obvious that one cannot do well-controlled experiments to test population-level hypotheses on endangered species. So we again stress that cowbird control was an appropriate management tool for Kirtland's Warbler because it is also reasonable to suggest that the species would have gone extinct without this intervention. But we see no way to distinguish the latter hypothesis from the alternative one that the size of the breeding population was already limited by a shortage of suitable habitat before cowbird control began.

An irony in the recovery of the Kirtland's Warbler is that the Mack Lake Burn was planned as a small controlled burn to benefit the species. It went out of control accidentally, creating the single most beneficial event in this species' recent history. It is even more ironic that because of the destruction of 44 buildings and one human fatality, controlled burns are now rarely used for warbler management (Kepler et al. 1996), although they are clearly the optimal management practice. Agencies create warbler habitat by planting jack pine at suitable densities, but warblers prefer habitat that regenerates naturally after a fire. Mayfield (1993) suggested that Kirtland's Warbler might be more successful on large habitat patches than on small ones, so further large burns may be needed for the species' survival. The outlook for the Kirtland's Warbler is unclear, as the number of singing males on the Mack Lake Burn declined from 300 in 1994 to 276 in 1995 and to 200 in 1996. Numbers on the burn may approach zero in about 5 years (DeCapita pers. comm.). In addition, the total population for the species fell between 1995 and 1996, from 766 to 692 males, for the first time since the warblers began to use the Mack Lake Burn.

Events since 1994, however, also provide cause for optimism. For the first time ever, breeding was documented outside lower Michigan. Although small numbers of singing males have been documented over the years in Michigan's Upper Peninsula, Ontario, and Wisconsin, there was never any evidence of breeding. But at least 2 of 8 singing males in the Upper Peninsula were paired in 1995, and there were 14 males there in 1996. Furthermore, for the first time ever, more than half (57%) of all Kirtland's Warblers bred on planted (artificial) habitat in 1995, and 63% did so in 1996 (DeCapita pers. comm.). This encouraging development may mean that the Mack Lake Burn has generated so many excess birds that more and more have been induced to accept planted habitat as suitable, that agencies have become better at simulating the conditions that prevail after burns, or that the supply of wintering habitat has increased recently (Haney et al. 1998). Unfortunately, severe budget cutbacks have reduced funds for research, so the current breeding success of birds in planted habitat is not known, although past work has shown that males in planted habitat are less likely to attract mates (Bocetti 1994, Probst and Hayes 1987, Kepler et al. 1996).

Least Bell's Vireo

Unlike Kirtland's Warbler, the Least Bell's Vireo (*Vireo bellii pusillus*) experienced a rapid and dramatic increase in its breeding population after cowbird control began in 1983 (Griffith and Griffith Chapter 38). This taxon was originally common over most of California, with 60–80% of its population in the Central Valley (Franzreb 1989). By 1978, only about 140 singing males could be located, and these occurred only on the coast from Santa Barbara County southward (Goldwasser et al. 1980, including addendum p. 745), or in less than 20% of the original range in California. The vireo's past and present ranges in Baja California are known with less certainty. This taxon, like the Kirtland's Warbler, was originally allopatric with respect to cowbirds. But cowbirds colonized its entire range between about 1900 and the late 1930s (Rothstein 1994), and many early records of cowbird parasitism were from vireo nests (Franzreb 1989). A decline in vireo numbers was first noted in the 1930s (Grinnell and Miller 1944). The decline has been attributed to cowbird parasitism and to the destruction or loss of most of this vireo's obligate riparian habitat. For example, the Central Valley lost 95% of its riparian habitat in this century (Smith 1977).

Some of the most extensive riparian habitat left in southern California occurs on the Marine Corps base at Camp Pendleton in San Diego County. About 62 singing males were located there in 1983. These vireos experienced a parasitism rate of about 50% before the start of the cowbird control program described by Griffith and Griffith. After cowbird control, the parasitism rate fell to about 4–20% from 1983 to 1987 and to 1% or less since 1988, when the number of traps and their efficiency was increased. As of 1995, 5,349 cowbirds had been removed from Camp Pendleton, and the vireo population increased more than ten-fold, to 696 between 1983 and 1996. Furthermore, the number of drainages occupied by vireos on Camp Pendleton has increased from 3 to 14, and banded vireos fledged there have been found breeding at numerous other localities in California. One bird dispersed 275 km to Ventura County, one of the northernmost breeding sites for the taxon.

Griffith and Griffith also report briefly on other more recent cowbird control programs in San Diego and adjacent Orange and Riverside counties; these programs have also been followed by large increases in numbers of breeding vireos. The U.S. Fish and Wildlife Service estimates that there are currently more than 1,000 pairs of vireos in southern California (L. Hays pers. comm.), making the increase from about 140 in 1978 one of the most dramatic success stories in bird conservation.

The success is far from complete, however, as the numbers of cowbirds trapped in one year appear to have no effect on the number trapped in the future, as with the Kirtland's Warbler management effort. Furthermore, vireos are not doing as well in their northernmost populations in Ventura and Santa Barbara counties. These populations are especially significant because they are the closest ones to the Central Valley and thus could allow the taxon to recolonize its original center of abundance (Franzreb 1989).

When cowbird control began in Ventura County along the Santa Clara River in 1991, field workers documented about 11 pairs of vireos (M. Holmgren and J. Greaves pers. comm.). By 1996, about 40 pairs could be found, but a larger area was surveyed in that year, and it is difficult to determine how much of the increase in vireos was due to increased sampling effort versus actual population growth. Unlike Camp Pendleton, the Santa Clara River is a management nightmare, as no single entity controls the habitat where vireos are found. The same is true of the nearby Ventura River, which also has a small number of vireos. Vireos along the Santa Clara are scattered along a span of almost 50 km, over which riparian habitat varies from being essentially absent to being present in large patches. Much of the river, and the access to it, is in the hands of numerous private owners, and much of the riparian habitat is under attack from human activities such as gravel mining and agriculture. There is no central authority managing the river, and consistent funding is available only to trap cowbirds. The scarcity of funding for censusing vireos and monitoring their nests means that it is difficult to judge whether cowbird trapping has increased the number of breeding vireos. Even the efficacy of cowbird trapping has been compromised, because human activity and floods have changed the locations of prime riparian habitat.

The northernmost breeding population in Santa Barbara County presents yet other problems (Greaves 1987). Unlike nearly all other populations, it is in a fairly remote area that has low cowbird abundance. This population has experienced a recent parasitism rate of only 15–25% (J. Greaves pers. comm.) and was by far the largest known population in the late 1970s, when its 50 singing males made up over a third of all known individuals in California (Goldwasser et al. 1980). With 57 singing males, it was still the third largest population in 1986. Numbers declined to 25 males by 1987 (Franzreb 1989), and there were only 20–25 males in 1992 and 1994 (J. Greaves pers. comm.), despite the shooting of cowbirds from 1988 and cowbird trapping from about 1991 to 1994. Reasons for the decline are unclear, but it was probably not caused by cowbird parasitism. Unlike the Santa Clara River population, this one could be easily managed. It occupies a small stretch of lush riparian habitat, most of which is within a national forest, and it is not threatened by development. Study of this important vireo population might provide considerable insight into the taxon's biology. However, although there were limited funds for cowbird control and for monitoring vireos in the past, none have been available since 1994 despite the vireo's endangered status.

We have discussed these two northern populations in detail for two reasons: they are geographically important and they demonstrate that full vireo recovery is likely to require more than indefinite cowbird trapping. Complex property rights are involved in protecting the Santa Clara population, and the Santa Barbara one shows that unknown biological factors can have overriding importance. These problems have not occurred on Camp Pendleton.

Griffith and Griffith are more confident than we are in the efficacy of cowbird control. They and others (Gaines 1974, Laymon 1987) argue that there are large tracts of suitable habitat in the Central Valley and elsewhere that are unoccupied by vireos (and other threatened California birds) but that seem suitable except for the abundance of cowbirds there. The Kirtland's Warbler story shows that habitat that seems suitable to humans may not suit the birds, even though that species' biology is known far better than that of the Least Bell's Vireo. In addition, the insight gained from metapopulation theory (Gilpin and Hanski 1991) and source-sink dynamics (Pulliam 1988) shows that small patches of suitable habitat may not be sufficient to forestall extinction. One or more very large and critically located patches may be necessary. Griffith and Griffith also stress the importance of habitat and advocate a vigorous program of habitat protection and augmentation in addition to cowbird control.

The importance of habitat availability is shown by the Arizona Bell's Vireo (*V. b. arizonae*), which breeds from the Colorado River east to central Arizona. It too has declined greatly in this century, but unlike the Least Bell's Vireo, it has a long history of sympatry with cowbirds. Brown (1903) noted that nearly every nest he found along the Colorado River in 1900 was parasitized. Had such parasitism been typical along the entire Colorado River, the vireo would have been extirpated in a few years. It did not decline, however, until the 1950s (Rosenberg et al. 1991), after dam construction reduced the frequency of floods and made it economical to convert extensive tracts of riparian habitat to agricultural uses. Clearly, the nests that Brown found were not typical of the entire population. Perhaps both people and cowbirds are especially likely to find certain nests, such as those on the edges of dense riparian zones. Thus *arizonae* may have been able to survive in the presence of cowbirds because primeval riparian habitat was wide enough to keep parasitism levels in its interior low. Perhaps the Least Bell's Vireo could also survive in the absence of cowbird control if it had extensive and wide tracts of suitable habitat. Although the nominate race of Bell's Vireo is found in the cowbird's past and current center of abundance (Peterjohn et al. Chapter 2, Wiedenfeld Chapter 3) and it is declining (Peterjohn et al. 1995), it is not endangered, presumably because of extensive remaining habitat.

Southwestern Willow Flycatcher

Like the Least Bell's Vireo, the endangered Southwestern Willow Flycatcher discussed by Whitfield (Chapter 40) is also an obligate riparian species of the arid Southwest and has also lost most of its habitat (about 95%). In addition, it also has conspecific populations that have experienced long sympatry with cowbirds but are not endangered, again probably because their habitat is plentiful. Unlike the Least Bell's Vireo, this flycatcher still occupies the extent of its original range from western Texas to southern California, but many previously occupied sites are now vacant. Unitt (1987) estimated the taxon's entire population at 230 to 500 pairs, but more recent estimates are 700–800 pairs, of which about 70 are in California (M. Sogge pers. comm.).

Whitfield reports on a cowbird control program that was designed to aid the largest population in California, which occurs along the South Fork of the Kern River. It decreased from 44 to 27 pairs and experienced a 63.5% parasitism rate from 1989 to 1992 before large numbers of female cowbirds were trapped. Unlike the other control programs discussed in this section, Whitfield's was designed as a controlled experiment, with both a cowbird removal area and a nearby nonremoval area. Cowbird abundance and host breeding success were monitored in both areas. Data from 1993 and 1994 showed declines in both cowbird abundance and parasitism rates on flycatcher nests (to 17.4%) in the removal area. But the proportion of nests producing at least one flycatcher was about the same on the removal and nonremoval areas, in part because the former had a much higher rate of nest predation. However, the number of fledglings per pair of flycatchers was higher in 1993 and 1994 on the removal area than on the nonremoval area (1.87 versus 1.55), and both areas had higher productivity than in years before cowbirds were trapped (0.93 to 1.00). The flycatcher population stopped declining after small numbers of female cowbirds were removed in 1992 and increased slightly to 34 pairs by 1994. However, the population remained at 34 pairs in 1995 and declined to 29 pairs in 1996 (Whitfield pers. comm.).

As Griffith and Griffith describe in Chapter 38, the cowbird removal program to aid the Least Bell's Vireo at Camp Pendleton may have also benefited the Southwestern Willow Flycatcher there. Flycatcher numbers increased from 5 pairs in 1981 to 21–25 pairs in 1989–1991; however, numbers did not increase further between 1991 and 1995. Researchers and managers at the 1996 preseason recovery meeting for the flycatcher and Least Bell's Vireo (sponsored by the USFWS) agreed that cowbird control is not having the same dramatic effect on breeding numbers of flycatchers as it has had on most populations of the Least Bell's Vireo. Nevertheless, cowbird trapping may have stemmed the decline of flycatcher populations in California, and at least one cowbird removal program was undertaken in 1996 in Arizona.

It is unclear why flycatcher populations have failed to increase rapidly in response to cowbird trapping. One possibility is poor reproductive success even in nests that escape predation and parasitism. Rob Marshall (pers. comm.) has found that although nests in Arizona average 2.4 eggs, they average only 1.3 fledglings. This is an extremely low rate of success for a passerine. Usually, about 90% of eggs in successful unparasitized nests result in fledglings (Rothstein 1975). Flycatchers in Arizona commonly nest in riparian vegetation that is nearly a complete monoculture of the exotic tamarisk (*Tamarix* spp.). Although this altered riparian habitat attracts flycatchers, it may be poor habitat in which flycatcher populations cannot grow by local recruitment. In contrast, flycatchers in Whitfield's California study area nest in natural willow-cottonwood woodlands, and most eggs that escape predation and parasitism produce fledglings. So, the failure of the South Fork Kern River population to increase rapidly in response to cowbird removal must be due to some other factor. Perhaps this population is experiencing problems on its wintering grounds.

Black-capped Vireo

The Black-capped Vireo (*Vireo atricapillus*) is unique among the endangered species discussed here in that all of its range is within the cowbird's original center of abundance in the Great Plains. Hayden et al. (Chapter 39) report that it may now be susceptible to extirpation by cowbird parasitism because it has lost extensive amounts of habitat to agriculture, urbanization, and fire suppression. The Black-capped Vireo prefers shrub habitats that exist only 3–25 years after disturbances. Once found as far north as Kansas, this species is now limited to two remnant populations in Oklahoma, larger numbers in Texas (Robinson et al. 1995a), and unknown numbers in Mexico. Hayden et al. describe a cowbird control program initiated at the Fort Hood Army installation in central Texas, after surveys in 1988 indicated that over 90% of all nests were parasitized. Other cowbird control programs to protect this vireo have been conducted elsewhere in Texas and in Oklahoma. As with the previous three endangered species, parasitism decreased significantly, and nest success increased significantly after cowbird control. Research at Fort Hood is complicated by active military operations that prevent researchers from monitoring the entire vireo population with equal survey effort. Hayden et al., however, note an increase from 66 vireo territories in 1991 to 156 in 1994 on part of the base monitored with constant effort. While the active military operations at Fort Hood interfere with research, they benefit the vireo because they destroy habitat that has become too old and promote regeneration of younger, more preferred habitat.

Similarities and Dissimilarities among Cowbird Control Programs

Each of these four cowbird control programs has addressed remarkably similar conservation needs and used similar management strategies. Managers at each location were faced with a declining host population. Losses were largely attributable to fragmentation, degradation, or loss of habitat. The land-use changes that contributed to the decline of these host populations may also have contributed to the range expansion of the cowbird (Mayfield 1978), although the ranges of two of the four endangered species were mostly (Southwestern Willow Flycatcher) or totally (Black-capped Vireo) within the original range of cowbirds. Although other factors may have limited host populations, control of brood parasitism was the most immediate remedy for the drastic declines experienced by these host species.

All four programs show that it is possible to remove most to nearly all cowbirds over large areas with the use of numerous large decoy traps. At Fort Hood, removals covered nearly 400 km^2 around a doughnut-hole area where military activity prevented access. In Michigan, trapping covered about 50 km^2 of patchy warbler habitat, distributed over a total area of more than 19,000 km^2. These programs involved 52 to 67 traps at peak levels of trapping effort. The impressive scale of these two programs raises the question: over how big an area is it practicable to trap and remove breeding cowbirds? The answer probably depends heavily on the landscape setting and on the density of cowbirds. We suspect that the upper limit is in the 500–1,000 km^2 range. Cowbird removal at this scale is probably possible only with great effort and when the distribution of target hosts is patchy, as seen for both Kirtland's Warbler in Michigan and the Black-capped Vireo at Fort Hood. Control programs in mostly agricultural landscapes in the Midwest, where cowbirds are common, may not be practical at even moderate spatial scales of a few square kilometers.

Cowbird removal generally achieves its proximate goals: parasitism rates decrease dramatically and host productivity increases sharply. Cowbird removal thus improves nesting success. However, nesting success is not necessarily the principal limiting factor in songbird populations. Populations may also be limited by variable adult or juvenile survival (McCleery and Perrins 1991, Rogers et al. 1991) or by territorial behavior (Smith et al. 1991). Given this fact, it is perhaps not surprising that the four control programs have had mixed success at increasing the numbers of adult hosts. Only the Least Bell's Vireo can be called a clear success here, as only it has resulted in huge increases in breeding adults over a decade or more. The Black-capped Vireo program seems promising, but the degree of success here is harder to judge given the short period of intensive control. Control of cowbirds has not yet increased the numbers of breeding Southwestern Willow Flycatchers. Although the flycatcher control program lasted only four years, the number of breeding Bell's Vireos had increased by 67% after four years of cowbird removals. Last, it is clear that cowbird control alone did not increase the numbers of breeding Kirtland's Warblers, although it is possible, but by no means certain, that control was necessary to save this species from extinction. Despite uncertainty concerning the effects of some of these control programs, it is clear that cowbird control was an appropriate management tool in each case.

A theme common to all four programs is the need for increased habitat. All workers agree that cowbirds are not the only major problem threatening the survival of these four endangered taxa. Habitat availability is also critical. Because some of these taxa have conspecific populations that have survived long periods of sympatry with cowbirds, it has been suggested that they could have survived in the presence of cowbirds without cowbird management, if large amounts of their original habitat remained (Rothstein 1994). It is one thing, however, to recognize that habitat renewal is needed; it is quite another to actually renew and restore it. The attractiveness of cowbird removal programs is partly that they are cheap and technically straightforward compared to large-scale habitat restoration programs.

All of these programs, except possibly the flycatcher one, show that cowbird control is open-ended in that the numbers removed one year seem to have no effect on the numbers removed in subsequent years. The rapid range expansions of the cowbird (Mayfield 1965, Rothstein 1994) and studies of morphological variation over time and space (Fleischer and Rothstein 1988) indicate that cowbirds have unusually high dispersal rates. This characteristic is undoubtedly responsible for the annual flow of cowbirds into control areas. Management efforts to date have seldom aged cowbirds as yearlings versus older birds (but see DeCapita Chapter 37), which is easily done for males (Selander and Giller 1960, Ortega et al. 1996). It would be interesting to do so to determine whether age ratios are skewed toward yearlings after the first year or two of trapping. The only trap-out study that aged males showed that yearlings and adults had different seasonal capture patterns even in the first year of trapping (Rothstein et al. 1987).

As Whitfield reports, the control program for the Southwestern Willow Flycatcher in the South Fork Kern River Valley may be having long-lasting effects on local cowbird numbers. The number of female cowbirds removed in 1993 (329) was much higher than the numbers removed in 1994 and 1995 (152 and 171, respectively; Whitfield pers. comm.). Removal areas in the three other studies adjoin extensive regions with significant amounts of habitat disturbance by humans and higher densities of cowbirds. In contrast, although the South Fork Kern River Valley has many people, it is surrounded by arid and forested habitats with few people and few cowbirds. Dispersing cowbirds have to come from far away to replace those removed from this valley. This could mean that an intensive control effort that was expanded to the entire valley might not need to be done on an annual basis. It is still too early to confirm that control is indeed having year-to-year effects, since the declines from 1993 to 1995 could reflect region-wide trends. Also, numbers of cowbirds trapped can fluctuate annually by over three-fold (DeCapita Chapter 37), even with roughly constant trapping effort.

One procedural difference among the four control programs is that the Kirtland's Warbler and Least Bell's Vireo efforts have removed all cowbirds while the other two have removed primarily females. In arguing for the value of the latter methodology, Hayden et al. suggest that the resulting extremely skewed sex ratios may disrupt the mating activity of any remaining females. This is a reasonable supposition, as male cowbirds are attracted to females (Dufty 1982a, Rothstein et al. 1988) and harass them with their incessant courtship behavior to the point that females try to drive off courting males (Yokel and Rothstein 1991) even when sex ratios are normal. Even if it has no value in disrupting female breeding activity, removal of only females is preferable on the ethical ground that it is desirable to minimize the killing of any native species.

All three studies that report data on cowbird capture rates as a function of time found that the majority of individuals were caught in the first several weeks of the 2.5- to 4-month trapping periods (DeCapita Chapter 37, Hayden et al. Chapter 39, Griffith and Griffith Chapter 38). DeCapita suggests that some of this trend is due to large numbers of migrants being caught early in the season and that it likely also reflects the fact that cowbirds are extremely social and therefore easily captured in decoy traps. The two studies that report detailed data on numbers of both male and female cowbirds captured (DeCapita, Griffith and Griffith) showed the sex ratio to be skewed in favor of males, as in numerous other studies (Rothstein et al. 1986, Weatherhead 1989). Both studies also show that sex ratios of newly captured birds became even more male biased during the last third to half of the trap-out periods, as also reported by Beezley and Rieger (1987). This strong increase in the male : female ratio may have occurred because males were more likely than females to respond to the local cowbird vacuum due to the trapping by dispersing into the trapped-out area. If birds disperse during the breeding season because they are having low reproductive success, then higher male dispersal would be expected because the male-biased sex ratio and the predominance of monogamy in cowbirds means that many males do not breed (Dufty 1982b, Yokel 1986). A trap-out study in the Sierra Nevada reported the opposite pattern, where the sex ratio shifted toward females near the end of the season (Rothstein et al. 1987). This may have occurred because that study was limited to two heavily used communal feeding sites that some females visited less regularly than most males.

Areas visited by cowbirds on a daily basis can often be separated into sites used primarily for feeding versus areas used for breeding (Rothstein et al. 1984, Thompson 1994, Thompson and Dijak Chapter 10). All of these management studies placed traps in breeding habitat, and all except De-Capita's also placed them at feeding sites. More birds can be caught per trap at communal feeding sites such as horse corrals; however, cowbirds at feeding sites may come from long distances and many may not be affecting sensitive hosts. Furthermore, some of the cowbirds that are affecting hosts may not visit communal feeding sites on a regular basis (Rothstein et al. 1987). So it is clear that trapping should not be limited to feeding sites. Whether traps at such sites benefit the targeted hosts more than an equivalent effort in breeding habitat remains an open question. The answer may vary with the local landscape.

Limitations and Possible Negative Aspects of Cowbird Control

The open-ended nature of cowbird control programs is clearly undesirable from a management standpoint. Funding for control efforts must be continuous for the programs to be successful. Yet funding is far from secure, and sometimes it comes from unlikely sources, such as the Department of Defense for the programs reported by Griffith and Griffith and by Hayden et al. As described above, one control effort for the Least Bell's Vireo has already lost its funding and has ceased. Even the Kirtland's Warbler effort recently suffered a significant decrease in funding (M. DeCapita pers. comm.). We wonder whether the Endangered Species Act even has the potential to deal effectively with species that need yearly management through and after the recovery period. As the act is now written, endangered hosts that recover to the point of being delisted could lose funding for cowbird control along with their endangered status. Stopping cowbird removals could eventually return them to the endangered list. Clearly, we need to do more than just control cowbirds in areas where endangered passerines occur.

The need for sustained and often costly efforts to control local cowbird populations has motivated many to speculate on the efficacy of winter control programs that could kill millions of cowbirds at large roosts (Ortego Chapter 5, Griffith and Griffith Chapter 38). Eliminating millions of cowbirds on winter roosts could reduce the need for local control programs dealing with endangered species and might even reduce the impacts of brood parasitism on numerous host species. Although such actions have some basic appeal, no need for them has been demonstrated, and their potential conservation benefits also are questionable. As discussed by Peterjohn et al. (Chapter 2), Wiedenfeld (Chapter 3), and Rothstein and Robinson (Chapter 1), there is no clear evidence that cowbirds are limiting the numbers of any species other than a few endangered taxa. Because cowbirds that breed in one area winter over a large region (Dolbeer et al. 1982), it is even possible that large-scale control at winter roosts would miss some or all of the cowbirds threatening endangered species locally. Large kills in winter would therefore not affect the need for local control efforts in the breeding season. Rothstein and Robinson (1994) pointed out that it would be extremely difficult to determine whether large-scale winter control affected host breeding success or numbers, because these parameters vary naturally. They also pointed out that the killing of millions of individuals of a native species could attract the attention of animal rights groups, especially if the need for such action is not clear. This attention could endanger the continuation of the local cowbird control programs for endangered species. Finally, we note that a worldwide review of efforts to eradicate large numbers of "pest" birds shows that these have had limited and at best transitory success (Dolbeer 1986).

Despite the controversy surrounding large-scale cowbird control, we suspect that all parties would agree that control is sometimes very effective on a local landscape level even though it is open-ended. But are there potential negative aspects to local cowbird control? Is it possible to overdo cowbird control? We believe that the answer is clearly yes. Any time an active intervention such as cowbird control occurs, it draws attention and resources from other possible management efforts, some of which might be more beneficial. Certainly, control uses funds that might go into the development of other management approaches, and funds are scarce for all four endangered hosts discussed here. For example, there are almost no funds in California for monitoring Least Bell's Vireo breeding biology, and in some places, there are no funds even to monitor the numbers of breeding vireos. If vireos begin to decline, it may be difficult to determine whether the problem lies with breeding success in California or with the availability of wintering habitat in the neotropics.

In light of these research needs, it is perhaps ironic that a recent (October 1997) compilation showed that at least $1 million in state and federal funds is expended annually for cowbird control efforts in California alone (C. Hahn pers. comm.). Much of this funding has limited flexibility; for example, the Department of Defense might be reluctant to expend funds for work away from military bases such as Camp Pendleton. Nevertheless, the lack of funds for research needs such as monitoring vireo populations shows that society's mechanisms for funding endangered species research and management are flawed. It also seems that current funding practices are skewed far too much in favor of applying established management techniques over research that may generate more effective management tools. In other words, cowbird control may have a counterproductive bandwagon effect leading to the view that if it is good in some situations, it is worth doing everywhere. Also, such a view discourages searching for more efficacious approaches.

Another potential downside of cowbird control is that it can be used as mitigation for destruction of the habitat of an endangered species. This is a danger because the long-term continuation of a control program is uncertain, whereas habitat loss can be permanent. We suspect that some land developers consider cowbird control an ally in their attempts to develop habitat of endangered birds and might even push for cowbird control when it is not needed in the particular area they want to develop.

The need for local data on cowbird parasitism raises one last point about cowbird control. As we have stressed, there was good reason to believe that cowbird control was needed in all four of the management programs discussed in this section. High rates of parasitism were documented, and hosts had declined. However, some current control programs are being initiated with little or no prior quantitative information on local parasitism rates. These programs start because of high local cowbird abundance, available manpower (sometimes in the form of volunteers) to service traps, and the presence of species that might decline locally if they are parasitized heavily. One clear message of Part IV in this book and information elsewhere (Friedmann 1963, Robinson et al. 1995a,b) is that parasitism rates vary geographically, even over short distances. Parasitism rates in some control areas may be low, and in such cases, control programs may be attempting to correct nonexistent problems. Even if local parasitism rates are high, managers should consider whether cowbirds are the only problem or whether local habitats are so poor that host populations might not be self-sustaining even in the absence of parasitism. This scenario may be the case for birds nesting in some midwestern woodlots (Robinson et al. 1995a). Furthermore, some of these new control efforts lack monitoring programs to determine whether cowbird control is having an effect on local species. It would be tragic if the bandwagon status of cowbird control diverts the actions of genuinely concerned people from more productive activities and from more profound problems.

We close this overview by pointing out that a species cannot be considered to have recovered as long as permanent and direct human intervention is primarily responsible for its continued existence. Sustainable recovery of a species in the long term will depend on understanding the temporal and spatial factors that are most limiting. Birds such as the Least Bell's Vireo, Southwestern Willow Flycatcher, Kirtland's Warbler, and Black-capped Vireo may be limited by brood parasitism in the short term; however, the long-term solutions to maintaining viable populations of these and other species will be found in the restoration of breeding habitat and the development of compatible and sustainable land management practices. Additional understanding concerning the winter dynamics of these host species and a reduction in factors that threaten wintering grounds will also be critical in maintaining long-term population viability.

Acknowledgments

We thank Mike DeCapita and Mark Holmgren for their comments on part or all of this manuscript. They also shared unpublished data with us, as did James Greaves and Mary Whitfield.

References Cited

Askins, R. A. 1993. Population trends in grassland, shrubland, and forest birds in eastern North America. Curr. Ornithol. 11:1–34.

Beezley, J. A., and J. P. Rieger. 1987. Least Bell's Vireo management by cowbird trapping. Western Birds 18:55–61.

Bocetti, C. I. 1994. Density, demography, and mating success of Kirtland's Warblers in managed habitats. Ph.D. Thesis, Ohio State University, Columbus.

Brawn, J. D., and S. K. Robinson. 1996. Source-sink population dynamics may complicate the interpretation of long-term census data. Ecology 77:3–12.

Brown, H. 1903. Arizona bird notes. Auk 29:43–50.

Caughley, G., and A. R. E. Sinclair. 1994. Wildlife ecology and management. Blackwell Scientific Publications, Oxford, UK.

Dolbeer, R. A. 1986. Current status and potential of lethal means of reducing bird damage in agriculture. Acta 19 Int. Congr. Ornithol., University of Ottawa Press, 1:474–483.

Dolbeer, R. A., P. P. Woronecki, and R. A. Stehn. 1982. Migration patterns for age and sex classes of blackbirds and starlings. J. Field Ornithol. 53:28–46.

Dufty, A. M., Jr. 1982a. Response of Brown-headed Cowbirds to simulated conspecific intruders. Anim. Behav. 30:1043–1052.

———. 1982b. Movements and activities of radio-tracked Brown-headed Cowbirds. Auk 99:316–327.

Fleischer, R. C., and S. I. Rothstein. 1988. Known secondary contact and rapid gene flow among subspecies and dialects in the Brown-headed Cowbird. Evolution 42:1146–1158.

Franzreb, K. E. 1989. Ecology and conservation of the endangered Least Bell's Vireo. USFWS Biol. Rep. 89(1).

Friedmann, H. 1963. Host relations of the parasitic cowbirds. U.S. Natl. Mus. Bull. 233:1–273.

Gaines, D. 1974. A new look at the nesting riparian avifauna of the Sacramento Valley, California. Western Birds 5:61–80.

Greaves, J. M. 1987. Nest-site tenacity of Least Bell's Vireo. Western Birds 18:50–54.

Gilpin, M. E., and I. Hanski. 1991. Metapopulation dynamics: Empirical and theoretical investigations. Academic Press, London, UK.

Goldwasser, S., D. Gaines, and S. R. Wilbur. 1980. The Least Bell's Vireo in California: A de facto endangered race. Amer. Birds 34: 742–745.

Grinnell, J., and A. Miller. 1944. The distribution of the birds of California. Pacific Coast Avifauna 27.

Hagan, J. M. III, and D. W. Johnston. 1992. Ecology and conservation of neotropical migrant landbirds. Smithsonian Institution Press, Washington, DC.

Haney, J. C., D. S. Lee, and M. Walsh-McGehee. 1998. A quantitative analysis of winter distribution and habitats of Kirtland's Warblers in the Bahamas. Condor 100:201–217.

James, F. C., C. E. McCullogh, and D. A. Wiedenfeld. 1996. New approaches to the analysis of population trends in land birds. Ecology 77:13–27.

Johnson, N. K. 1994. Pioneering and natural expansion of breeding distributions in western North American birds. Studies in Avian Biol. 15:27–44.

Kepler, C. B., G. W. Irvine, M. E. DeCapita, and J. Weinrich. 1996. The conservation management of Kirtland's Warbler Dendroica kirtlandii. Bird Conservation International 6:11–22.

Laymon, S. A. 1987. Brown-headed Cowbirds in California: Historical perspectives and management opportunities in riparian habitats. Western Birds 18:63–70.

Martin, T. E., and D. M. Finch. 1995. Ecology and management of neotropical migratory birds. Oxford University Press, New York.

Mayfield, H. F. 1960. The Kirtland's Warbler. Cranbrook Institute, Bloomfield Hills, MI.

———. 1965. The Brown-headed Cowbird with old and new hosts. Living Bird 4:13–28.

———. 1978. Brood parasitism: Reducing interactions between Kirtland's Warblers and Brown-headed Cowbirds. Pp. 85–91 in Endangered birds: Management techniques for preserving threatened species (S. A. Temple, ed). University of Wisconsin Press, Madison.

———. 1983. Kirtland's Warbler, victim of its own rarity? Auk 100: 974–976.

———. 1993. Kirtland's Warblers benefit from large tracts. Wilson Bull. 105:351–353.

McCleery, R. H., and C. M. Perrins. 1991. Effects of predation on the numbers of Great Tits Parus major. Pp. 129–147 in Bird population studies: Relevance to management and conservation (C. M. Perrins, J-D. Lebreton, and G. J. M. Hirons, eds). Oxford University Press, Oxford.

Morton, E. S. 1992. What do we know about the future of migrant landbirds? Pp. 579–589 in Ecology and conservation of neotropical migrant landbirds (J. M. Hagan III and D. W. Johnston, eds). Smithsonian Institution Press, Washington, DC.

Ortega, C. P., J. C. Ortega, S. T. Backensto, and C. A. Rapp. 1996. Improved methods for aging second-year and after-second-year male Brown-headed Cowbirds. J. Field Ornithol. 67:542–548.

Peterjohn, B. G., J. R. Sauer, and C. S. Robbins. 1995. Population trends from the North American breeding bird survey. Pp. 3–39 in Ecology and management of neotropical migratory birds (T. E. Martin and D. M. Finch, eds). Oxford University Press, New York.

Probst, J. 1986. A review of factors limiting the Kirtland's Warbler on its breeding grounds. Amer. Midland Natur. 116:87–100.

Probst, J. R., and J. P. Hayes. 1987. Pairing success of Kirtland's Warblers in marginal versus suitable habitat. Auk 104:234–241.

Pulliam, H. R. 1988. Sources, sinks, and population regulation. Amer. Natur. 132:652–661.

Rappole, J. H., and M. V. McDonald. 1992. Cause and effect in population declines of migratory birds. Auk 111:652–660.

Robinson, S. K., S. I. Rothstein, M. C. Brittingham, L. J. Petit, and J. A. Grzybowski. 1995a. Ecology and behavior of cowbirds and their impact on host populations. Pp. 428–460 in Ecology and management of neotropical migratory birds (T. E. Martin and D. M. Finch, eds). Oxford University Press, New York.

Robinson, S. K., F. R. Thompson III, T. R. Donovan, D. R. Whitehead, and J. Faaborg. 1995b. Regional forest fragmentation and the nesting success of migratory birds. Science 267:1987–1990.

Rogers, C. M., J. N. M. Smith, W. M. Hochachka, A. L. E. V. Cassidy, P. Arcese, and D. Schluter. 1991. Spatial variation in winter survival of Song Sparrows. Ornis Scand. 22:387–395.

Rosenberg, K. V., R. D. Ohmart, W. C. Hunter, and B. W. Anderson. 1991. Birds of the lower Colorado River valley. University of Arizona Press, Tucson.

Rothstein, S. I. 1975. Evolutionary rates and host defenses against avian brood parasitism. Amer. Natur. 109:161–176.

———. 1994. The cowbird's invasion of the far West: History, causes, and consequences experienced by host species. Studies in Avian Biol. 15:310–315.

Rothstein, S. I., and S. K. Robinson. 1994. Conservation and coevolutionary implications of brood parasitism by cowbirds. Trends Ecol. Evol. 9:162–164.

Rothstein, S. I., J. Verner, and E. Stevens. 1984. Radio-tracking confirms a unique diurnal pattern of spatial occurrence in the parasitic Brown-headed Cowbird. Ecology 65:77–88.

Rothstein, S. I., J. Verner, E. Stevens, and L. V. Ritter. 1987. Behavioral differences among sex and age classes of the Brown-headed Cowbird and their relation to the efficacy of a control program. Wilson Bull. 99:322–337.

Rothstein, S. I., D. A. Yokel, and R. C. Fleischer. 1986. Social dominance, mating and spacing systems, female fecundity, and vocal dialects in captive and free-ranging Brown-headed Cowbirds. Curr. Ornithol. 3:127–185.

———. 1988. The agonistic and sexual functions of vocalizations of male Brown-headed Cowbirds, *Molothrus ater*. Anim. Behav. 36: 73–86.

Selander, R. K., and D. R. Giller. 1960. First-year plumages of the Brown-headed Cowbird and Red-winged Blackbird. Condor 62: 202–214.

Smith, F. 1977. A short review of the status of riparian forests in California. Pp. 1–2 *in* Riparian forests in California: Their ecology and conservation (A. Sands, ed). Institute of Ecology Publ. 15.

Smith, J. N. M., P. Arcese, and W. M. Hochachka. 1991. Social behaviour and population regulation in insular bird populations: Implications for conservation. Pp. 148–167 *in* Bird population studies: Relevance to management and conservation (C. M. Perrins, J-D. Lebreton, and G. J. M. Hirons, eds). Oxford University Press, Oxford.

Terborgh, J. 1989. Where have all the birds gone? Princeton University Press, Princeton, NJ.

———. 1992. Why are American songbirds vanishing? Scientific American, May 1992: 98–104.

Thompson, F. R. III. 1994. Temporal and spatial patterns of breeding Brown-headed Cowbirds in the midwestern United States. Auk 111:979–990.

Trail, P. 1992. Nest invaders. Pacific Discovery, Summer 1992: 32–37.

Unitt, P. 1987. *Empidonax trailii extimus:* An endangered subspecies. Western Birds 18:137–162.

Weatherhead, P. J. 1989. Sex ratios, host-specific reproductive success, and impact of Brown-headed Cowbirds. Auk 106:358–366.

Weise, T. F. 1987. Status of the Kirtland's Warbler, 1985. Jack-Pine Warbler 65:17–19.

Yokel, D. A. 1986. Monogamy and brood parasitism: An unlikely pair. Anim. Behav. 34:1348–1358.

Yokel, D. A., and S. I. Rothstein. 1991. The basis for female choice in an avian brood parasite. Behav. Ecol. Sociobiol. 29:39–45.

37. Brown-headed Cowbird Control on Kirtland's Warbler Nesting Areas in Michigan, 1972–1995

MICHAEL E. DECAPITA

Abstract

The Kirtland's Warbler is an endangered species suffering from nest parasitism by the Brown-headed Cowbird and a shortage of its highly specialized habitat. The young jack pine habitat needed by the warbler is normally a result of fires but is now being provided by forest management. The Kirtland's Warbler nests only in northern lower Michigan, where cowbirds were controlled annually by the U.S. Fish and Wildlife Service from 1972 to 1995. The third decennial census of singing male Kirtland's Warblers in 1971 fell 60% below the 1961 count of 502. From 1966 to 1971, 70% of Kirtland's Warbler nests were parasitized by cowbirds, warbler clutch sizes averaged 2.35 eggs, and nesting pairs produced 0.8 fledged young per year. During cowbird removals from 1972 to 1977, parasitism decreased to 6.3% of nests, mean clutch size increased to 4.46, and production of young increased to 3.11 young fledged per pair. Annual censuses of singing male warblers remained relatively stable from 1972 to 1989, averaging about 200, but began increasing in 1990 and reached a record high of 766 in 1995. In 1995, mated pairs were found outside the traditional breeding area for the first time. Cowbirds were captured with 15 to 67 decoy traps, and 98,427 cowbirds were removed in 24 years. More males than females were trapped in most years. The mean proportion of males was 0.549, and the average sex ratio was 1.22 : 1. On average, 63% of all cowbirds were caught by the third week of trapping in early to mid-May. Cowbird control effectively protected Kirtland's Warblers from nest parasitism each year, but had no effect on the cowbird population from one year to the next.

Introduction and Background

The Kirtland's Warbler (*Dendroica kirtlandii*) was first described scientifically in 1852 after a male specimen was collected in May 1851 near Cleveland, Ohio. During the 1880s and 1890s, at least 71 specimens were collected in the Ba-hama Islands (Mayfield 1960). The nesting area was not discovered until 1903, when a nest was located near the Au Sable River in western Oscoda County, Michigan (Holden 1964). All nests of this species ever found have been located within an area of 120 × 160 km (19,200 km²) in northern lower Michigan (Figure 37.1), but strong evidence of breeding outside this area was observed in 1995 (Weinrich 1996). Mayfield (1960) and Middleton (1961) have described the northern lower Michigan area where Kirtland's breeding habitat is located in detail.

Shortly after the discovery of the first nest, the species became the subject of ornithological study, which continues to this day. Studies from the 1920s through the 1950s led to an understanding of its life history and rarity on the nesting ground (Mayfield 1960). Mayfield (1960) and Walkinshaw (1983) provide extensive details of the bird's life history during the nesting and brood rearing period. Many other authors have also written about the species (Huber 1980). The Kirtland's arrives in Michigan from the Bahamas in early to mid-May. Males establish territories near each other in suitable habitat, where their density averages 1.9 per 40 ha and reaches a maximum of 2.8 per 40 ha (Probst and Weinrich 1993). Kirtland's Warblers begin their southward migration in August, with the last birds departing in early October (Sykes et al. 1989). Aside from numerous sightings during migration, little is known about the bird during its migration and wintering in the Bahama Islands. After limited fieldwork in the Bahamas in 1985 and 1986, Sykes (pers. comm.) concluded that adequate wintering habitat existed and that no serious threats were apparent there.

The birds nest only on the ground in young (6 to 24 years old) forests of jack pine (*Pinus banksiana*). The jack pine and the warbler nests are closely associated with a poor sandy soil of glacial origin. Areas suitable for warblers are not continuous but are scattered among other soil and forest types in several counties. The young stands develop naturally after older jack pine forests burn. Most jack pine cones open and release seeds only after being scorched by fire. Once a stand

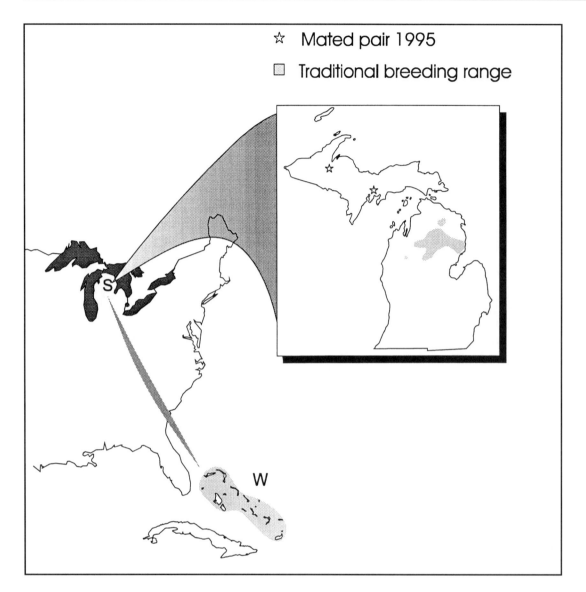

Figure 37.1. Eastern United States showing general location of Kirtland's Warbler summer breeding range (S) in northern lower Michigan, winter range (W) in the Bahama archipelago, and general migration path. On inset map of Michigan, shaded area depicts Kirtland's Warbler breeding range where cowbird control occurred, and stars indicate locations of paired warblers in the Michigan Upper Peninsula in 1995.

of jack pine reaches about 5 m in height, the Kirtland's will no longer nest there. The species must therefore follow its dynamic nesting habitat across the landscape through time as older jack pine stands age and new ones regenerate after forest fires. Such fires once were common, the tree itself is very flammable, and the ecological community it dominates is dependent on fire for its existence. Jack pine in lower Michigan is at the southern edge of its extensive North American range. The jack pine forests, which were lesser components of more extensive mixed pine and hardwood forests of northern Michigan to begin with, began to diminish in the early 20th century as humans suppressed forest fires and planted more valuable tree species.

Brown-headed Cowbirds were not native to heavily forested northern Michigan prior to the late 19th century. Friedmann (1929) believed the cowbird was originally a grassland species not present in the forests of eastern North America. The northern Michigan forests were completely logged or burned off between the 1860s and 1910s. Grasslands, brush, and agricultural lands predominated until the second-growth forest that now dominates this area became established. Cowbirds moved into the area about 1895 and have been parasitizing Kirtland's Warbler nests at least since 1908 (Mayfield 1960). Mayfield (1977) discussed the appearance of cowbirds in northern Michigan and their impact on the Kirtland's Warbler. This warbler may be particularly vulnerable to the cowbird because it had no opportunity to evolve defensive behavior prior to 1895. By 1971, 69% of warbler nests were parasitized, and pairs of warblers were fledging less than 1 young per year (Walkinshaw 1983).

The survival of the Kirtland's Warbler was being jeopardized by a decrease in its very specific nesting habitat and by parasitism of its nests by the cowbird. Beginning in 1957, the state of Michigan and the U.S. Forest Service began setting aside forest management units specifically for the warbler (Mayfield 1963). The Kirtland's Warbler was listed as a federally endangered species in 1967 (USDI 1967) and has had similar state status since 1974. Management for the warbler intensified with the passage of the Endangered Species Act of 1973. There are currently 24 state and federal warbler management units, covering more than 56,000 hectares.

State and federal public forest contains nearly all occupied and protected Kirtland's nesting habitat. Each year the Michigan Department of Natural Resources and the U.S. Forest Service together manage about 700 to 1,100 ha. Management actions consist of commercial harvest of mature jack pine, followed by planting of 2-year-old jack pine seedlings. The goal is to maintain 12,000 to 14,000 ha of suitable warbler habitat (sufficient for 570 to 665 warbler territories, assuming that all habitat is occupied at mean densities). Although fire is important in the ecology of the jack pine forest and the Kirtland's Warbler, forest fires are vigorously suppressed and prescribed fire has been used only occasionally as a management tool, because it is dangerous in an area occupied by humans. Warbler nesting areas are protected and closed to public entry during the nesting season. Free guided tours are provided for persons who wish to see this rare bird.

The first census of singing Kirtland's Warblers in 1951 found 432 males (Figure 37.2). A second census in 1961

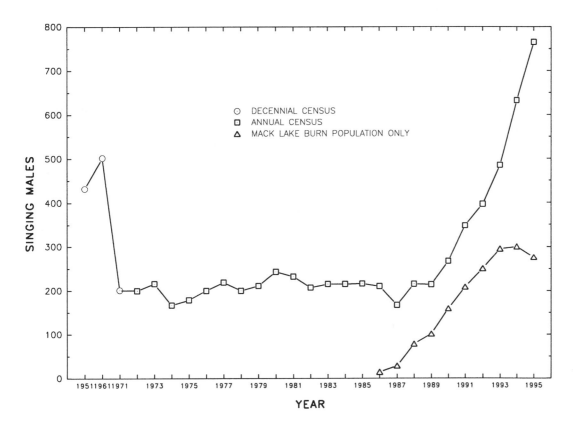

Figure 37.2. Census of singing male Kirtland's Warbler, 1951–1995, with Mack Lake Burn census (Data: Weinrich 1996). Figure adapted from C. B. Kepler, G. W. Irvine, M. E. DeCapita, and J. Weinrich, "The Conservation Management of Kirtland's Warbler *Dendroica kirtlandii*," *Bird Conservation International* 6:11–22, © BirdLife International 1996. Reprinted with the permission of Cambridge University Press.

counted 502 singing males, but the third decennial census of singing male Kirtland's in 1971 counted 201, a 60% decline (Mayfield 1972). Mayfield's (1960) and Walkinshaw's (1983) studies clearly implicated the cowbird in the decline. In May 1972, the U.S. Fish and Wildlife Service began removing Brown-headed Cowbirds from northern lower Michigan nesting areas of the Kirtland's Warbler. Shake and Mattsson (1975) described the first three years of the cowbird control program, Kelly and DeCapita (1982) reported results through 1981, and Kepler et al. (1996) provided results through 1992. This report describes results of the cowbird control program through 1995, including sex ratios, sex-related characteristics, and temporal distribution of captures.

Methods

Cowbirds have been controlled since 1972 on nearly every Kirtland's Warbler nesting area in Alcona, Clare, Crawford, Iosco, Kalkaska, Montmorency, Ogemaw, Oscoda, Otsego and Roscommon counties (total area 19,200 km²). Except for minor changes, methods are as described by Shake and Mattsson (1975). Cowbirds were captured in large (4.88 m square and 1.83 m high) cagelike decoy traps kept stocked with 6 to 24 live decoy cowbirds. Trapping began between 15 April and 1 May each spring and continued for about 11 weeks through to late June. Trapping began about nine days earlier on average in 1983–1995 compared to 1973–1982. Since 1986, increasing numbers of half traps (4.88 m × 2.44 m × 1.83 m high) have been used in addition to the regular 4.88-m-square traps. The number of traps used increased from 15 in 1972 to 67 in 1995 (Table 37.1). Traps made of prefabricated panels were erected in warbler nesting areas. The traps remained in place and were used as long as warblers nested in the area but were deactivated each year at the conclusion of trapping by removing a top panel and wiring open a large door. White proso millet was used as bait beginning in 1984. Cowbirds were banded and released at a small number of banding traps operated outside warbler nesting areas from 1973 through 1979 to gain knowledge about their local movements.

Records kept for each trap included numbers, sex, and age (adult or juvenile) of cowbirds. Male cowbirds were aged (Selander and Giller 1960) beginning in 1994. Capture data are compared by week of trapping rather than by actual calendar dates. Although the onset of trapping varied by up to two weeks (Table 37.1), the temporal patterns in the capture data (see below) were consistent regardless of the starting date. Thus, weekly mean data are presented here. Numbers of cowbirds captured for each week represent the total numbers for a seven-day period.

Results and Discussion

Cowbird control showed immediate results in 1972 with a sharp decrease in parasitism of Kirtland's Warbler nests and increased warbler reproductive success. Nest parasitism dropped from a mean of 70% for 1966–1971 to 6.2% in 1972 and a mean of 5.6% for 1972–1977 (Walkinshaw 1983). Warbler clutch size increased from 2.35 to 4.61 in 1972, and the fledging rate increased from an average of 0.8 in 1966–1971 to 3.35 in 1972 and a 1972–1977 average of 3.08 per pair per year (Walkinshaw 1983).

The Kirtland's Warbler population did not increase for many years (Figure 37.2) but remained stable at about 200 pairs (Ryel 1979, Weise 1987, Weinrich 1996). An increase that began in 1990, however, led to a record high of 766 singing males in 1995 (Weinrich 1996). In 1995, 2 of 8 males found by biologists in the Michigan Upper Peninsula were mated (Figure 37.1). This was the first evidence of Kirtland's breeding outside northern lower Michigan (Weinrich 1996). The increase since 1990 is directly related to the maturation of a large block of nesting habitat resulting from a 10,500-ha fire in 1980, known as the Mack Lake Burn, which the warblers began using in 1986. Figure 37.2 depicts the importance of the Mack Lake Burn to the current population. The Mack Lake Burn habitat has begun to decline in importance, but a high warbler population is being sustained by a large area of managed habitat. The highest recorded proportion of singing males (57%) found on managed habitat occurred in 1995 (Weinrich 1996). This proportion is expected to increase in the next few years, as warblers move out of the maturing Mack Lake Burn.

While habitat quality and availability are critical to the survival of the species (Probst and Weinrich 1993, Probst 1986), it is reasonable to assume that the Kirtland's Warbler would not now exist if cowbird control had not been started in 1972. Mayfield (1977) estimated that the warbler population in the late 1970s would have declined to 20 pairs without cowbird control, instead of the 200 actually observed. Research and management biologists used the VORTEX simulation model (Lacey 1993) to conduct a population viability analysis for the warbler in 1992. These simulations (Seal 1996) indicated that warblers needed to fledge at least 2.3 young per pair per year to maintain the population, well above the fledging rate recorded before the onset of cowbird control (Walkinshaw 1983).

Numbers of Cowbirds

Since 1972, 98,427 cowbirds have been removed from northern Michigan warbler nesting areas (Table 37.1). The mean annual number removed was 4,101, with a range from 2,261 to 7,595. The number of traps used increased from 15 at 7 warbler nesting areas in three counties in 1972 to 67 on 35 nesting areas in nine counties in 1995. The 24-year mean for cowbirds caught per trap is 110.1.

Table 37.1. Summary of Cowbirds Removed from Northern Michigan Kirtland's Warbler Nesting Areas, 1972–1995

| Year | Start Date | Brown-headed Cowbirds | | | | | Traps | Cowbirds per Trap |
		Male	Female	Juvenile	Total	M/F		
1972	unknown	1,621	619	21	2,261	2.62	15	150.7
1973	01 May	1,995	1,195	115	3,305	1.67	18	183.6
1974	29 Apr	2,195	1,717	163	4,075	1.28	22	185.2
1975	28 Apr	2,016	1,431	155	3,602	1.41	28	128.6
1976	27 Apr	2,193	1,994	112	4,299	1.10	38	113.1
1977	25 Apr	1,845	1,405	34	3,284	1.31	38	86.4
1978	24 Apr	1,754	1,639	18	3,411	1.07	40	85.1
1979	17 Apr	1,954	1,721	16	3,691	1.14	35	105.5
1980	29 Apr	1,538	1,429	0	2,967	1.08	37	80.2
1981	26 Apr	1,770	1,085	1	2,856	1.63	36	79.3
1982	26 Apr	1,568	893	38	2,499	1.76	35	71.4
1983	15 Apr	2,128	2,196	0	4,324	0.97	35	123.5
1984	16 Apr	2,183	1,936	0	4,119	1.13	31	132.9
1985	15 Apr	2,644	2,082	14	4,740	1.27	30	158.0
1986	22 Apr	2,328	1,781	75	4,184	1.31	31	135.0
1987	20 Apr	2,291	1,549	60	3,900	1.48	38	102.6
1988	18 Apr	2,932	2,589	19	5,540	1.13	41	135.1
1989	17 Apr	2,907	2,881	2	5,790	1.01	42	137.9
1990	16 Apr	3,818	3,771	6	7,595	1.01	38	199.9
1991	15 Apr	2,576	2,088	6	4,670	1.23	43	106.3
1992	20 Apr	2,003	1,730	4	3,737	1.16	49	76.3
1993	19 Apr	2,361	2,246	7	4,614	1.05	51	90.5
1994	18 Apr	1,862	1,242	5	3,109	1.50	56	55.5
1995	17 Apr	3,070	2,782	3	5,855	1.10	67	87.4
Total		53,552	44,001	874	98,427		894	
Mean		2,231	1,833	36	4,101	1.22	37	110.1

Source: C. B. Kepler, G. W. Irvine, M. E. DeCapita, and J. Weinrich, "The Conservation Management of Kirtland's Warbler *Dendroica kirtlandii*," *Bird Conservation International* 6:11–22, © BirdLife International 1996. Reprinted with the permission of Cambridge University Press.

Catch Distribution

One of several remarkable consistencies in the cowbird catch over 24 years was the distribution of captures during the trapping period. Mean weekly catches from 1973 to 1995 are shown in Table 37.2. Data for 1972 are not included because field records were lost. Calendar dates for initiation of trapping were 17 April to 1 May from 1972 to 1982 and 15 April to 22 April from 1983 to 1995 (Table 37.1). The first and last week of trapping were not full weeks of effort since traps were opened or shut down then. Regardless of when trapping started, the catch distribution each year was very much like the composite shown in Table 37.2, with modest variations in each week's percentage of the total catch. Shake and Mattsson (1975) presented weekly catch distributions for 1973 and 1974, with the 1974 data being most typical.

The catch peaked sharply in the second week, the first full week of trapping, then rapidly declined and stabilized at a lower level. On average, 33% of all cowbirds were caught in the second week and 19% in the third week. Sixty-three percent of the cumulative total was caught by the third week. This type of distribution is expected if cowbirds were being removed from an area without being replaced. A similar though less pronounced pattern of captures occurred in the banding traps (Table 37.3), where most cowbirds were

Table 37.2. Mean Weekly and Cumulative Numbers of Brown-headed Cowbirds Removed from Northern Michigan Kirtland's Warbler Nesting Areas, 1973–1995

Week	Male	Cum. Male	Female	Cum. Female	Juv	Total	Cum. Total	% Total	Cum. % Total
1	195.8	195.6	225.3	225.3	0.0	420.9	420.9	10.2	10.2
2	560.0	755.6	821.4	1046.7	0.0	1381.4	1802.3	33.3	43.5
3	338.8	1094.4	465.9	1512.5	0.0	804.7	2606.9	19.4	62.9
4	217.8	1312.2	159.0	1671.6	0.0	376.9	2983.8	9.1	72.0
5	171.9	1484.1	63.0	1734.5	0.0	234.8	3218.6	5.7	77.7
6	143.2	1627.3	23.1	1757.6	0.0	166.3	3384.9	4.0	81.7
7	133.1	1760.4	20.1	1777.7	0.1	153.3	3538.2	3.7	85.4
8	130.8	1891.2	18.1	1795.9	0.2	149.1	3687.3	3.6	89.0
9	120.1	2011.3	16.9	1812.8	1.0	138.0	3825.4	3.3	92.3
10	130.0	2141.3	29.8	1842.6	7.5	167.4	3992.7	4.0	96.3
11	103.9	2245.2	33.7	1876.3	15.4	153.0	4145.7	3.7	100.0

Table 37.3. Mean Weekly and Cumulative First Captures of Brown-headed Cowbirds in Northern Lower Michigan Banding Traps, 1973–1979

Week	Male	Cum. Male	Female	Cum. Female	Juv	Adult	Total	% Total	Cum. % Total
1	29.7	29.7	32.7	32.7	0.0	62.4	62.4	15.9	15.9
2	35.4	65.1	57.4	90.1	0.0	92.9	92.9	23.6	39.5
3	16.4	81.5	24.7	114.8	0.0	41.1	41.1	10.5	50.0
4	20.7	102.2	14.9	129.7	0.0	35.6	35.6	9.1	59.1
5	14.9	117.1	5.3	135.0	0.0	20.1	20.1	5.1	64.2
6	20.3	137.4	4.0	139.0	0.0	24.3	24.3	6.2	70.4
7	11.0	148.4	4.1	143.1	0.0	15.1	15.1	3.8	74.2
8	7.4	155.8	1.0	144.1	0.0	8.4	8.4	2.1	76.3
9	11.3	167.1	2.3	146.4	0.0	13.6	13.6	3.5	79.8
10	4.1	171.2	1.1	147.5	0.0	5.3	5.3	1.4	81.2
11	8.8	180.0	10.8	158.3	3.0	19.7	22.7	5.8	87.0
12	7.0	187.0	15.2	173.5	29.3	22.2	51.5	13.1	100.0

banded and released at the trap. The numbers reported in Table 37.3 are for first captures only. Dufour and Weatherhead (1991) noted a similar distribution of captures in banding traps operated from April through July in eastern Ontario. Some factor other than removal apparently diminishes cowbirds' susceptibility to capture after the first three weeks.

Data on the movements of banded cowbirds in this project have not been fully analyzed, and it is beyond the scope of this report to do so, but 12.9% of the 3,011 cowbirds banded from 1973 to 1979 were retrapped at least once. Most recaptures were within a day or two at the same trap.

In 1979, 73 of 273 cowbirds were retrapped, 68.5% at the first trap and 31.5% at another trap.

The large number of cowbirds captured in the first three weeks of trapping was probably due in part to migrating cowbirds moving through the area. The temporal distribution of captures may also reflect a change in cowbird behavior and susceptibility to decoy traps. When migrating, cowbirds are in flocks, still mobile, gregarious, sociable, and thus likely to encounter and enter a trap. Once migration ends, flocks break up and cowbirds disperse for breeding, perhaps with females becoming territorial (Darley 1983,

Dufty 1982; Raim, Chapter 9, this volume) or adopting other spacing mechanisms (Teather and Robertson 1985, Raim Chapter 9; see Smith et al., Chapter 8, this volume). Once cowbirds have chosen a breeding site, they may be less mobile and gregarious, distributed less widely, and thus less likely to find a trap.

It must be noted that data presented here were not gathered during research designed to test any hypothesis but were incidental to an operational management program with the single objective of protecting the Kirtland's Warbler by removing as many cowbirds as possible. Resources were not available for inquiry, so conservative assumptions were made to achieve the objective of maximizing cowbird removal. One such assumption, following the above reasoning, was that susceptibility to being trapped is greatest on cowbirds' first arrival in the spring and diminishes thereafter. This assumption led to consistent initiation of trapping about a week earlier beginning in 1983, with the understanding that many of the birds caught would be passing migrants not likely to affect Kirtland's Warblers. No means exist to distinguish between cowbirds passing through from those likely to become local residents.

In spite of a sharp decline in catch after the third week, the traps continued to catch cowbirds throughout the remaining trapping period. Cowbirds remain somewhat gregarious even during the peak breeding period, with a tendency to gather to feed and loaf (Rothstein et al. 1984, Darley 1983, Dufty 1982) in areas away from centers of breeding activity. Although relatively few cowbirds were caught in late May and June, catches still averaged above 100 per week (Table 37.2). Walkinshaw (1983) reported that female Kirtland's Warblers lay their first eggs during the last week in May and the first 10 days of June. Cowbirds removed during this period were likely residents and were thus especially important because of their potential to affect the Kirtland's Warbler adversely. Furthermore, the average weekly numbers of adult cowbirds caught (Table 37.2) decreased to the sixth week, but the decreasing trend did not continue, and numbers caught then stabilized. Thus, cowbirds were never completely removed in any year. Clearly, many cowbirds remained in the region, and despite any behavioral changes during the breeding period, they continued to move about the landscape and may have replaced cowbirds that were removed.

Age

Male cowbirds were not aged until 1994, when 921 were second-year birds, 925 were after second year, and 16 were not aged. There were 1,592 second year, 1,466 after second year, and 11 not aged in 1995.

Sex Ratio

Males consistently outnumbered females in trapped samples (Table 37.1). A statistically equal sex ratio in removal traps happened only three times in 24 years, including the only year in which females outnumbered males (1983). In the other 21 years, males outnumbered females by as much as 2.62 to 1 (1972), but more typically by 1.1 to 1.5 to 1. The overall predominance of males is further indicated by the 24-year sex ratio in removal traps of 1.22 males per female. The proportion of males observed (0.549) differed significantly from a 0.50 expectation ($Z = 30.58$, $P < .001$). In banding traps the proportion of males was 0.52 (1,332 males : 1,221 females), and this proportion was also significantly different from 0.5 ($Z = 2.2$, $P = .022$).

The temporal pattern of males and females in the catch distribution and a seasonal shift in the sex ratio were also very consistent over the years. Each year but one, females predominated for the first two or three weeks (Table 37.2), then their numbers dropped off more sharply than the males', with male captures exceeding female captures thereafter and making up more than half of the total capture (Table 37.1). The same pattern occurred at the banding traps (Table 37.3). The cumulative data in Tables 37.2 and 37.3 illustrate another consistent characteristic of the catch distribution. On average, females outnumbered males until the eighth week for removal traps and the seventh week in banding traps. This point of crossover varied between years. For example, males did not exceed females until the tenth week in 1993, the year with the fourth lowest sex ratio (Table 37.1).

Sex ratio has long been of interest to cowbird observers. Friedmann (1929) reported a ratio of 3 males for every 2 females, and many investigators have subsequently found that sex ratios favor males, with values often close to Friedmann's figure. Dufty (1982), Darley (1971), and Dufty and Wingfield (1986) all reported ratios of males : females of 1.51 : 1 to 1.55 : 1, similar to Friedmann's value. McIlhenny (1940) reported 2.82 : 1, but he felt this was an error. Yokel (1989) reported a higher value of 1.9 : 1, and Rothstein et al. (1987) reported much higher values (from 3.33 : 1 to 7.6 : 1) from both trapping data and field observations. The last two studies were from the Owens Valley and Sierra Nevada of California, which may have extreme sex ratios, while the others were from eastern North America.

Dufour and Weatherhead (1991), who found a sex ratio of 1.35 : 1 for decoy-trapped cowbirds, also noted a sharp decline in catches of female cowbirds in decoy traps after mid-May, although males predominated throughout their entire trapping period from early to mid-April through July. They suggested that females become "trap shy" with the onset of egg laying and that males are over-represented in decoy trap samples because of sex-specific dietary needs. These interpretations fit the data presented here. A preliminary analysis of data from cowbird banding traps in the Kirtland's study area for 1973 and 1975–1978 showed that 70% of retrapped cowbirds were males, suggesting that males have a greater tendency to enter traps. No investigation has been made of cowbird breeding physiology and chronology on or near Kirtland's Warbler nesting areas. However, the decline in female captures recorded in Table 37.2 coincides with the peak egg-laying period of the Kirtland's Warbler (Walkinshaw 1983).

The excess of adult male cowbirds apparently is a result of lower survival of females, since cowbird eggs and fledglings have an equal sex ratio (Weatherhead 1989). Fankhauser (1971), Darley (1971), and Searcy and Yasukawa (1981) all calculated higher survival rates for males, but Arnold and Johnson (1983) found that more females survived at winter roosts. The latter authors suggested that competition for food affected males more, while Darley (1971) and Searcy and Yasukawa (1981) attributed lower female survival to the high demands of egg laying.

Dufour and Weatherhead (1991) found that mist-netted cowbirds exhibited a lower sex ratio than those caught contemporaneously in decoy traps. They cautioned that data from decoy trapping may be biased, because decoy-trapped cowbirds were in poorer condition than mist-netted cowbirds. While a consistent condition bias may exist, the large sample size and consistent sex ratios in the data presented here add to the long list of other reports of excess males. It seems clear that males predominate among adult Brown-headed Cowbirds.

Conclusion

Brown-headed Cowbird control with decoy traps in a large area of Michigan has allowed Kirtland's Warblers to reproduce successfully. There is agreement that the warblers are now limited by habitat, as long as cowbird numbers continue to be controlled. Probst and Weinrich (1993) developed a habitat-based model that accurately predicted Kirtland's populations during the late 1980s. Recent increases in the Kirtland's singing male count due to the Mack Lake Burn and an increase in the area of managed habitat were also predicted. These increases, however, have exceeded maximum projections, perhaps because of higher than expected maximum male densities and/or better than expected habitat quality. Probst (unpubl. reports, pers. comm.) believes that both habitat and numbers of Kirtland's Warblers may decline around the year 2000. Consideration for human safety in the warbler nesting range means that fire suppression must continue and that costly habitat management will be a permanent requirement for this species' continued existence.

This program has shown that Brown-headed Cowbird nest parasitism can be controlled on a limited but still large spatial scale (approximately 19,000 km²). Similar cowbird control programs may be appropriate and successful in other limited situations. The methods used in Michigan are labor intensive and costly and probably would not be practical on a larger scale (see also Hayden et al., Chapter 39, this volume).

Although cowbirds have been removed annually in the same general area for 24 years, large numbers of cowbirds return each year. The overall cowbird population does not seem to have been reduced. An important interest of the participants in the population viability analysis was an examination of the consequences of varying cowbird parasitism rates. The model indicated that with known and predicted habitat and known warbler productivity, survival, and mortality rates, nest parasitism above 30% would lead to extinction of the warbler (Seal 1996). Cowbird control is thus likely to remain a permanent management requirement for the Kirtland's Warbler. Thus, the Kirtland's Warbler faces continued serious jeopardy, not merely of a biological nature but of a political and fiscal nature as well.

References Cited

Arnold, K. A., and D. M. Johnson. 1983. Annual adult survival rates for Brown-headed Cowbirds wintering in southeast Texas. Wilson Bull. 95:150–153.

Darley, J. A. 1971. Sex ratio and mortality in the Brown-headed Cowbird. Auk 88:560–566.

———. 1983. Territorial behavior of the female Brown-headed Cowbird (Molothrus ater). Can. J. Zool. 61:65–69.

Dufour, K. W., and P. J. Weatherhead. 1991. A test of the condition-bias hypothesis using Brown-headed Cowbirds trapped during the breeding season. Can. J. Zool. 69:2686–2692.

Dufty, A. M., Jr. 1982. Movements and activities of radio-tracked Brown-headed Cowbirds. Auk 99:316–327.

Dufty, A. M., and J. C. Wingfield. 1986. Temporal patterns of circulating LH and steroid hormones in a brood parasite, the Brown-headed Cowbird, Molothrus ater, 1: Males. J. Zool., Lond. (A) 208:191–203.

Fankhauser, D. P. 1971. Annual adult survival rates of blackbirds and starlings. Bird Banding 42:36–42.

Friedmann, H. 1929. The cowbirds: A study of the biology of social parasitism. C. C. Thomas, Springfield, IL.

Holden, F. M. 1964. Discovery of the breeding area of the Kirtland's Warbler. Jack-Pine Warbler 42:278–290.

Huber, K. R. 1980. The Kirtland's Warbler (Dendroica kirtlandii): An annotated bibliography, 1852–1980. Museum of Zoology, University of Michigan, Ann Arbor. 99 pp.

Kelly, S. T., and M. E. DeCapita. 1982. Cowbird control and its effect on Kirtland's Warbler reproductive success. Wilson Bull. 94:363–365.

Kepler, C. B., G. W. Irvine, M. E. DeCapita, and J. Weinrich. 1996. The conservation management of Kirtland's Warbler Dendroica kirtlandii. Bird Conservation International 6:11–22.

Lacey, R. C. 1993. VORTEX: A computer simulation model for population viability analysis. Wildl. Res. 20:45–65.

Mayfield, H. F. 1960. The Kirtland's Warbler. Cranbrook Institute of Science, Bloomfield Hills, MI. 242 pp.

———. 1963. Establishment of preserves for the Kirtland's Warbler in the state and national forests of Michigan. Wilson Bull. 75:216–220.

———. 1972. Third decennial census of Kirtland's Warbler. Auk 89:263–268.

————. 1977. Brown-headed Cowbird: Agent of extermination? Amer. Birds 31:107–113.

McIlhenny, E. A. 1940. Sex ratio in wild birds. Auk 57:85–93.

Middleton, D. S. 1961. The summering warblers of Crawford County, Michigan. Jack-Pine Warbler 39:34–49.

Probst, J. R. 1986. A review of factors limiting the Kirtland's Warbler on its breeding grounds. Amer. Midl. Natur. 116:87–100.

Probst, J. R., and J. Weinrich. 1993. Relating Kirtland's Warbler population to changing landscape composition and structure. Landscape Ecology 8:257–271.

Rothstein, S. I., J. Verner, and E. Stevens. 1984. Radio-tracking confirms a unique diurnal pattern of spatial occurrence in the parasitic Brown-headed Cowbird. Ecology 65:77–88.

Rothstein, S. I., J. Verner, E. Stevens, and L. V. Ritter. 1987. Behavioral differences among sex and age classes of the Brown-headed Cowbird and their relation to the efficacy of a control program. Wilson Bull. 99:322–337.

Ryel, L. A. 1979. The tenth Kirtland's Warbler census, 1978. Jack-Pine Warbler 57:141–147.

Seal, U. S. (ed). 1996. Kirtland's Warbler (*Dendroica kirtlandii*) population and habitat viability assessment workshop report. Conservation Breeding Specialist Group (SSC/IUCN), Apple Valley, MN.

Searcy, W. A., and K. Yasukawa. 1981. Sexual size dimorphism and survival of male and female blackbirds (Icteridae). Auk 98:457–465.

Selander, R. K., and D. R. Giller. 1960. First-year plumages of the Brown-headed Cowbird and Red-winged Blackbird. Condor 62:202–212.

Shake, W. F., and J. P. Mattsson. 1975. Three years of cowbird control: An effort to save the Kirtland's Warbler. Jack-Pine Warbler 53:48–53.

Sykes, P. W., Jr., C. B. Kepler, D. A. Jett, and M. E. DeCapita. 1989. Kirtland's Warblers on the nesting grounds during the post-breeding period. Wilson Bull. 101:545–558.

Teather, K. L., and R. J. Robertson. 1985. Female spacing patterns in Brown-headed Cowbirds. Can. J. Zool. 63:218–222.

USDI, Fish and Wildlife Service. 1967. 32 Federal Register 4001.

Walkinshaw, L. H. 1983. Kirtland's Warbler: The natural history of an endangered species. Cranbrook Institute of Science, Bloomfield Hills, MI.

Weatherhead, P. J. 1989. Sex ratios, host-specific reproductive success, and impact of Brown-headed Cowbirds. Auk 106:358–366.

Weinrich, J. 1996. The Kirtland's Warbler in 1995. Michigan Dept. Nat. Res., Wildlife Div. Rep. 3243. 12 pp.

Weise, T. F. 1987. Status of the Kirtland's Warbler, 1985. Jack-Pine Warbler 65:17–19.

Yokel, D. A. 1989. Intrasexual aggression and the mating behavior of Brown-headed Cowbirds: Their relation to population densities and sex ratios. Condor 91:43–51.

38. Cowbird Control and the Endangered Least Bell's Vireo: A Management Success Story

JOHN T. GRIFFITH AND JANE C. GRIFFITH

Abstract

The Least Bell's Vireo is an obligate riparian species. It was recorded as abundant by early ornithologists throughout its range from northwest Baja California, Mexico, to the northern Sacramento Valley, California. It was nearly extinct in the United States by 1978 and restricted, with few exceptions, to San Diego County, California. The decline was caused by the destruction of 95% of the riparian habitat in the historic range and by the effects of Brown-headed Cowbird parasitism on the fragmented vireo population. Cowbirds colonized California about 1890 and soon became abundant. In the early 1980s, between 47% and 100% of all vireo nests contained cowbird eggs. Most cases of successful parasitism result in complete vireo reproductive failure due to the small size of the vireo. Persistent parasitism has caused the extirpation of vireos from thousands of acres of apparently suitable habitat throughout their historic and current range. Cowbird trapping began at Camp Pendleton in 1983 and was extended throughout the current range of the vireo in 1986–1992. In trapped areas, cowbird parasitism of the vireo and other host species was significantly reduced or eliminated. Vireo reproductive success at Camp Pendleton increased by 129% from 1982 to 1988–1991, and populations first stabilized and then grew, expanding into existing habitat. Since cowbird control began, the numbers of singing male Least Bell's Vireos have increased from 60 to 902 at Camp Pendleton, from 19 to 249 at the Santa Ana River, from 9 to 50 at the San Luis Rey River West, and from 13 to 142 at the Tijuana River. No factor other than cowbird control changed significantly during the same period. Similar trapping programs would likely benefit other declining host species throughout the Southwest. Regional cowbird control would possibly yield large-scale benefits.

Introduction

The Least Bell's Vireo (*Vireo bellii pusillus,* hereafter vireo) is a small migratory songbird endemic to California and Baja California, Mexico. Early ornithologists regarded the vireo as common to abundant in riparian and adjacent habitats throughout its historic breeding range (Figure 38.1; Cooper 1874, Grinnell and Miller 1944, Franzreb 1989). Despite the historical abundance and large range of the subspecies, by 1978 only 90 singing males were located in California (Goldwasser et al. 1980). The Least Bell's Vireo was extirpated from the Central Valley, where 60% to 80% of the historic population occurred (Franzreb 1989), and from the coast north from Santa Barbara. The subspecies was subsequently listed as endangered by California in 1980 and by the United States in 1986.

There are two likely causes of this precipitous population decline, neither of which alone might have been sufficient to cause endangerment. First, more than 95% of the vireo's riparian habitat in the Central Valley was destroyed by agriculture, urban development, flood control and water projects, and mining activities (Smith 1977). Similar losses have occurred throughout southern California and Baja California (Federal Register 1986, Wilbur 1981). Much habitat was also so degraded by grazing, invasive exotic plants, and off-road vehicles that vireos stopped using it. Second, beginning about 1900, vireo populations that had already been reduced and fragmented by habitat loss and degradation were subjected to parasitism by the Brown-headed Cowbird (Grinnell and Miller 1944, Franzreb 1989). As early as the 1940s, Grinnell and Miller (1944) recorded "a noticeable decline in numbers . . . apparently coincident with an increase of cowbirds" in some locations. Vireos rarely raise any of their own young when cowbird eggs hatch in their nests, unlike many larger cowbird hosts. Persistent parasitism led to local extirpations and ultimately the endangered status of the Least Bell's Vireo.

After the bird received legal endangered status, a draft vireo management and recovery plan was developed with two objectives: (1) the protection, acquisition, enhancement, and creation of riparian habitat and (2) the reduction or elimination of cowbird parasitism (USFWS 1988). Cowbird trapping began at Marine Corps Base Camp Pendleton, California, in 1983; by 1986–1992 trapping programs were also in place throughout San Diego and Orange counties. With cowbird control, parasitism of the vireo and some other riparian hosts was reduced or eliminated, vireo productivity increased (129% at Camp Pendleton, Table 38.1), and declining or stagnant populations began to grow. By 1996 there were about 1,800 singing male Least Bell's Vireos (USFWS pers. comm.), of which about 85% were likely paired. Many of them were in recently repopulated areas of the historic range south of the Central Valley.

Our objective here is to show that the decline of the Least Bell's Vireo was reversed simply by protecting existing habitat and by reducing the incidence of parasitism by cowbird trapping. Our focus is on the vireo in several drainages at Camp Pendleton, particularly the Santa Margarita River, but we include data from other locations and discuss effects of cowbird removal on some other host species. We have studied the Least Bell's Vireo and conducted cowbird control programs throughout the present range of the vireo since 1986.

Three other Bell's Vireo subspecies occur in the United States and Mexico: the nominate race from the Midwest (*V. b. bellii*), a race from Texas and northern mainland Mexico (*V. b. medius*), and the Arizona Bell's Vireo (*V. b. arizonae*) (AOU 1957). None of these three is federally endangered. The Arizona Bell's Vireo also occurs in extreme eastern California along the Colorado River and, for reasons similar to the Least Bell's Vireo, is listed as endangered by California.

Vireo Breeding Biology

The following general description of the breeding behavior of the vireo is drawn from Franzreb (1989) and from our 12 years of vireo management and research activities in southern California (Griffith and Griffith 1989, 1990, 1991a, 1992).

The riparian habitat preferred by the Least Bell's Vireo is well defined, often linear, and occurs primarily in the lower, flatter sections of streams and rivers. The vegetation in vireo home ranges is dominated in the tree and shrub layers by several willow species: arroyo willow (*Salix lasiolepis*), black willow (*S. gooddingii*), sandbar willow (*S. hindsiana*), and red willow (*S. lasiandra*). Other trees include Fremont cottonwood (*Populus fremontii*), white alder (*Alnus rhombifolia*), California sycamore (*Platanus racemosa*), and coast live oak (*Quercus agrifolia*). Important shrubs include mule fat (*Baccharis glutinosa*), blackberry (*Rubus ursinus*), wild rose (*Rosa californica*), Mexican elderberry (*Sambucus mexicana*), and poison oak (*Toxicodendron diversilobum*). Com-

mon herbaceous species include western ragweed (*Ambrosia psilostachya*), mugwort (*Artemisia douglasiana*), stinging nettle (*Urtica dioica*), and poison hemlock (*Conium maculatum*). Diversity in plant species composition and structure is an important component of vireo home ranges and nest sites; monotypic and senescent willow woodland is generally avoided. Most of the plants listed are used by the vireo for foraging, and all are used as nest substrates. The average height of 1,065 nests located at Camp Pendleton in 1981–1991 was 100 cm; most nests were hung in willow (42%) or mule fat (27%) (Griffith and Griffith 1992). Invasive exotic plants like tamarisk (*Tamarix* spp.) and giant reed (*Arundo donax*) may provide nest substrate but do not provide good forage; vireos are absent from monocultures of these plants.

Least Bell's Vireos arrive in California from their southern Baja California wintering ground from mid- to late March to early April. First-year males generally settle to breed in areas adjacent to other vireos near their natal home range and are subsequently site tenacious, returning each year to the same breeding territory of 0.1–2.0 ha over their life span (up to eight years; B. Kus unpubl. data, Griffith and Griffith unpubl. data). Females may also return to the same

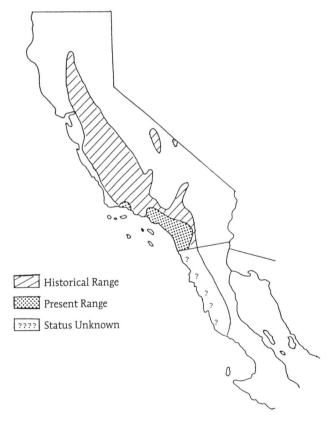

Historical Range

Present Range

Status Unknown

Figure 38.1. Historical range of the Least Bell's Vireo. Source: Franzreb 1989.

Table 38.1. Vireo Fledglings per Nesting Pair at Camp Pendleton, 1982–1991

Year	Number of Nesting Pairs	Number of Fledglings (total)	Number of Manipulated Fledglings[a]	Productivity (total)[b]	Productivity without Manipulations[c]	Parasitism Rate (%)
1982	48	100	36	2.08	1.33	46.7
1983	54	158	11	2.93	2.72	10.5
1984	63	124	4	1.97[d]	1.90	18.0
1985	66	210	2	3.18	3.15	3.9
1986	68	181	1	2.66	2.65	6.3
1987	97	249	13	2.57	2.43	17.1
1988	175	476	3	2.72	2.70	1.1
1989	129	449	0	3.48	3.48	1.1
1990	156	475	2	3.04	3.03	0.5
1991	133	399	0	3.00	3.00	0.0

[a]Number of fledglings from manipulated nests (removal of cowbird egg or young enabled vireos to fledge).
[b]Productivity is the total number of fledglings divided by the number of nesting pairs.
[c]This second productivity measure excludes fledglings from manipulated nests.
[d]Hot dry weather in 1984 is thought to have caused lower productivity (Salata 1984).
Sources: Salata 1983a,b, 1984, 1986, 1987a,b; Griffith field notes 1988; Griffith and Griffith 1990, 1991a, 1992.

breeding territories and males, though some change mates and territories between and among years, usually after breeding failures. Males sing often (10–25 songs per minute for the first 4–6 hours each day from early April through mid-May), making home ranges easy to locate. Breeding begins on arrival and continues through July, though most young are fledged in May and early June. The male and female share nest building, incubation, and parenting duties. Usually 3–4 eggs are laid, although clutches of 5 are known and clutches of 2 are common on third or later nest attempts. There may be up to seven nesting attempts per female per year. Incubation requires 14 days, and the nestling period lasts 10–12 days. Double brooding is common, and triple brooding is rare. About three weeks after fledging, young vireos can forage on their own, but family groups may remain together into September. With few exceptions, the Least Bell's Vireo is absent from breeding areas by the end of September.

Brown-headed Cowbird

Cowbirds first colonized the area west of the Sierra-Cascade axis about 1890 (Rothstein 1994). At that time, the Nevada Cowbird (*M. a. artemisiae*) bred in the Great Basin and the Dwarf Cowbird (*M. a. obscurus*) bred from the Colorado River east to perhaps Texas. The latter invaded the far West from the east and expanded northward beginning around 1900. The first cowbird documented in San Diego County was at Borrego Springs in 1896 (Unitt 1984). By 1930, cow-

birds were "well established" throughout southern California (Willett 1933), and by 1955 they had reached British Columbia (Flahaut and Schultz 1955). It is not clear if cowbirds would have appeared in the far West without the unwitting aid of man. It is likely that large cowbird numbers and consequent impacts on hosts in the region would not have been possible without massive anthropogenic habitat alteration, particularly the provision of year-round cowbird forage by agricultural and livestock operations and the coincident destruction of native habitats. A detailed history of the cowbird's invasion of the far West is given by Rothstein (1994).

Cowbirds are extreme generalists and parasitize nearly every passerine species (at least 220) with which they are sympatric (Friedmann 1963, Friedmann and Kiff 1985). Because this lack of host specificity appears to be true even on an individual basis (Fleischer 1985) and because cowbird productivity is generally proportional to the losses host species experience (Rothstein 1990; but see Robinson et al., Chapter 33, this volume), there are no feedback processes on ecological or evolutionary time scales that lead to the amelioration of parasitism of a particular host. Therefore, unlike parasites whose fate is closely tied to a specific host, cowbirds may drive a rare species (like the Least Bell's Vireo) to extinction with negligible effect on their own population (Rothstein 1990).

Cowbird eggs hatch sooner than eggs of most host species (10–12 days versus 12–16 days), and cowbird young develop faster than most host young. Nestling cowbirds often out-compete their host nest mates. Most small passerine hosts produce only a single cowbird young from parasitized nests in which the cowbird egg hatches. For the vireo and other small hosts, nest parasitism and nest predation have the same end result, except that the host pair renests within 2–14 days after predation, while successful parasitism consumes a substantial part or even all of the breeding season (Grzybowski and Pease, Chapter 16, this volume). In addition, host species in California did not coevolve with cowbirds and may have fewer behavioral defense mechanisms against parasitism than hosts elsewhere.

In contrast to the increase in distribution and abundance of cowbirds throughout California in this century, populations of many native birds in California are in decline, primarily because they depend on increasingly reduced, fragmented, and degraded native habitats in which they are more susceptible to predation and parasitism (Gaines 1974, Goldwasser et al. 1980). Thus there is an inverse relationship between the amount of native habitat and its associated avian populations and the number and subsequent impact of Brown-headed Cowbirds on such populations (see Part IV).

Study Area

Camp Pendleton is a 51,000-ha military reserve in extreme northwestern San Diego County, California (Figure 38.2). Its 30 km of undeveloped Pacific shoreline and interior separate the urban counties of San Diego and Orange. Camp Pendleton has a Mediterranean climate with warm, dry summers and cool, wet winters. Drought conditions prevailed for most of the 1980s. Topography at Camp Pendleton ranges from fairly flat coastal mesas to steep and rugged can-

Figure 38.2. Camp Pendleton, California, showing major drainages. Source: Camp Pendleton GIS database.

yons and hills of up to 817 m in height. Much of the upland area is grassland created and maintained by frequent fires started by military activities, though large areas of native coastal sage scrub, chaparral, and oak-sycamore woodland remain. Additionally, several large drainages (Figure 38.2) and many small ones support extensive and healthy stands of relatively undisturbed mature and successional riparian woodland of the type required by the Least Bell's Vireo. These include (from north to south): Christianitos Creek (CC), San Mateo Creek (SMC), San Onofre Creek (SOC), Piedre de Lumbre Canyon (PDL), Las Flores Creek (LFC), Aliso Creek (AC), French Creek (FC), the Santa Margarita River (SMR), De Luz Creek (DLC), Fallbrook Creek (FBC), Pueblitos Canyon (PBC), Windmill Canyon (WC), Pilgrim Creek (PC), and several miscellaneous habitat patches (MIS); only FBC and PC are dammed. The Santa Margarita River has a watershed of nearly 200,000 ha and is the largest free-flowing river in southern California.

Due to natural resource protection and endangered species management by the Marine Corps and its relatively undeveloped status, Camp Pendleton contains the finest and often largest remaining examples of several habitats and their associated plant and animal populations in southern California, including hundreds of sensitive and endangered species. Threats to these important natural resources include planned housing and facility developments and increased operations by the military (use of Camp Pendleton has increased as other bases have closed), artificially frequent fires, and invasive exotic plants. Natural resource managers at Camp Pendleton are aware of these threats and are working to limit their impacts.

Methods

Vireo Survey and Monitoring

Limited surveys to locate vireos were performed at Camp Pendleton in 1978 and 1980–1981. Comprehensive surveys of all suitable vireo habitat at Camp Pendleton were done in 1982–1991, 1993, and 1995–1996. Nearly all vireos were monitored to document their breeding biology from 1982 to 1991. Adults and nestlings were banded in 1982–1983 (Salata 1983b) and nestlings in 1987–1991 (Griffith and Griffith 1992). Limited nest monitoring was performed in 1992–1996.

Surveys were performed from dawn until heat or wind decreased avian activity or detectability, at 1100–1300 hours. Habitat was traversed at a slow but steady pace. Each occupied location was marked on a USGS 7.5′ quadrangle map or color aerial photograph. Data were collected annually on vireo arrival dates, distribution and abundance, breeding status and chronology, nesting chronology, nest site characteristics, clutch size and nest fate, predation, parasitism, band returns (more than 1,000 vireo young were banded from 1987 to 1991), behavior, home range and nest site

vegetation structure and composition, and human disturbance. A complete description of the methods used to survey, monitor, and band vireos is given in Griffith and Griffith (1989).

Cowbird Control

The removal of cowbirds occurred during implementation of management and recovery programs for a critically endangered species. The removal program was not a planned experiment, as there were no control plots on Camp Pendleton where cowbirds were not removed. However, data from other vireo populations monitored since the early or mid-1980s, but where trapping did not begin until the late 1980s (including some Camp Pendleton drainages), provided reference data for investigating the relationship between cowbird parasitism and vireo decline and between cowbird trapping and vireo recovery.

All cowbird eggs and young discovered in vireo and other nests at Camp Pendleton have been removed since 1981. All known vireo nests were monitored for this purpose until 1989, by which time trapping had essentially eliminated parasitism and the removal of cowbird eggs and nestlings was no longer necessary. Since 1989, samples of 9–195 vireo nests have been monitored each year.

Cowbirds were captured with movable eight-panel modified Australian crow traps measuring 6 × 8 feet on each side and 6 × 6 feet on front and back (5–13 traps were used from 1983 to 1987 and 29–40 traps from 1988 to 1996). Cowbirds entered the traps through a 1 3/8-inch drop-down slot on top, through which they could seldom escape. Traps were placed in sites that maximized the following characteristics: near or within vireo habitat (at the edge or in a clearing), in a cowbird foraging area (dairy, stable, or agricultural field), on a cowbird flight corridor or funnel area (along a river or canyon or in a ridge saddle), near a cowbird roosting area; visible from above (target cowbirds were attracted by the motions and vocalizations of the decoys), under a perch from which a cowbird could inspect the trap before approaching (telephone wire or tree snag); accessible by vehicle; and out of public view or access. We increased trap density and area of coverage as the vireo population grew and spread. Although trap densities varied with habitat quality and vireo densities, intervals of roughly 1 trap per 1 km of narrow (300 m or less) linear habitat eliminated parasitism, while fewer traps did not.

Upon placement, each trap was assembled as follows: the site was leveled, the panels were tightly fastened with carriage bolts and hex nuts, the front mesh floor was covered with sand or dirt (to create a foraging pad), four 1-m perches of giant reed 1.5 cm in diameter were inserted in the trap corners (three high and one low for wing-clipped female and subordinate decoy birds), nylon mesh shade was stapled to the west-facing panel (if the site was unshaded), an informative warning sign was attached to the door, and the trap was labeled with a number. Lastly, the trap was activated by adding a 1-gallon water guzzler, 1.5 pounds of wild bird seed without sunflower seeds (on the slot board and foraging pad), and two male and three female live decoys. The right wing of each female was clipped to prevent parasitism on release by accident or vandals. The trap was secured with a heavy padlock.

A single live male decoy was used in each trap from 1983 to 1987. Beginning in 1988, five live decoys were used, two males and three females, a small flock whose vocal and other social displays were attractive to other cowbirds. Male cowbirds vocalized and displayed most when at least one other male was present, and female cowbirds were more likely to enter a trap containing at least one more female than male, thus the 2 : 3 sex ratio for decoys. That ratio was restored daily beginning in 1992, prior to which the traps were emptied every 3–7 days and the number of decoys fluctuated widely.

Color-marked male "Judas" cowbirds were used beginning in 1989 to increase the effectiveness and range of the traps. One trap-habituated male decoy was released periodically from each trap, a known source of food, water, shelter, and social interaction to which it generally returned within hours or days, often in the company of other cowbirds that may have been otherwise unaware of, or wary of, the trap.

Traps were operated between 15 March and 15 July each year (exact dates varied). The early opening was to catch cowbirds dispersing from wintering flocks to local breeding areas, and the late closure was to capture locally raised cowbird juveniles. The traps were serviced daily in compliance with California live trap regulations and to reduce the otherwise high mortality of generally less hardy nontarget species. Daily visits consisted of adding bait seed, releasing nontarget birds, clipping wings of newly captured females, adding or removing cowbirds to maintain the decoys' 2 : 3 sex ratio, adding water if necessary, repairing any damage from vandals, and verifying that the perches, shade, and sign were intact. All captures and other information were recorded on a daily data sheet. All decoy transfers to or from holding cages were performed inside the trap to preclude accidental releases. In addition to the daily tasks, a complete water change and inspection of trap integrity were performed each week.

All cowbirds captured were euthanized with carbon monoxide by introducing the gas into a sealed holding cage. The birds were anesthetized within 20 sec and expired within 1 minute. Specimens in good condition were donated to local museums, universities, and raptor recovery or reintroduction programs.

No well-managed cowbird control program has failed to reduce or eliminate cowbird parasitism of target Least Bell's Vireo populations. The few failures have been management driven and not due to faulty methods. A complete protocol for trapping cowbirds, including trap construction, place-

ment, activation, daily servicing, disassembly and storage, and operation dates is available in Griffith and Griffith (1994e).

Results and Discussion
Cowbird Control
Egg and Nestling Removal

One or more vireo young fledged from about half of the manipulated vireo nests (nests from which cowbird eggs or young were removed) at Camp Pendleton in 1981–1990 (Table 38.1). This effort increased vireo productivity by 56% in 1982, when 47% of known vireo nests were parasitized. Parasitism rates, and therefore the importance of nest manipulation, dropped dramatically when cowbird trapping began in 1983 and when it was expanded in 1988. Most of the few cases of parasitism since 1983 have occurred on untrapped drainages or far from operational traps, and no cowbirds have fledged from vireo nests.

Mitigation and recovery plans initially involved detailed nest monitoring for cowbird egg and nestling removal. Nest monitoring is labor intensive and invasive, and it does not remove the source of parasitism permanently. It is also less effective in increasing vireo productivity than trapping. Cowbird trapping is now the preferred management strategy because it is effective, nonintrusive, and cheap (less than 25% the cost of nest monitoring).

Trapping

The trapping program removed 5,939 cowbirds from Camp Pendleton in 1983–1996 (Table 38.2). Parasitism of vireos has dropped to 1% or less since 1988, when the number of traps was increased from 12 to 27–40. There have been only two cases of parasitism since 1990.

Free cowbirds were rarely observed in trapped areas at Camp Pendleton (there have been no records on BBS surveys since 1989). We therefore believe that most cowbirds are captured within a few days of arriving at the study area and that capture trends within the year accurately reflect cowbird breeding dispersal (Figure 38.3). About 60% of all male captures and 70% of all female captures occurred by 30 April, and 90% for both sexes by 31 May. The male : female capture ratio began to approach 1.0 when daily maintenance of the 2 : 3 sex ratio for live decoys began at Camp Pendleton in 1992. We also switched to this protocol at about the same time at several other study areas with similar results (Table 38.3).

Trapping was done only at the Santa Margarita River from 1983 to 1987, after which other drainages were added (Table 38.2). Although the number of cowbirds trapped along the river varied from year to year, there was no overall declining trend. Similarly, the total number of cowbirds trapped on all drainages from 1988 to 1996 did not decline, nor did the average number of captures per trap, despite an increase in

Table 38.2. Brown-headed Cowbirds Captured at Camp Pendleton, 1983–1996

Year	Drainage[a]	Number of Traps	Trapping Period	Number of Cowbirds Captured				Number Per Trap	M:F Ratio
				Male	Female	Juvenile	Total		
1983	SMR	5	4/01–7/22	157	79	8	244	48.80	1.99
1984	SMR	13	4/03–7/20	269	215	1	485	37.31	1.25
1985	SMR	12	4/03–7/20	121	80	6	207	17.25	1.51
1986	SMR	12	3/31–6/30	186	134	7	327	27.25	1.39
1987	SMR	12	4/02–6/30	131	76	6	213	17.75	1.72
1988	SMR+2	27	4/01–6/30	252	140	8	400	14.81	1.80
1989	SMR+3	29	4/01–6/30	272	154	9	435	15.00	1.77
1990	SMR+5	32	3/15–6/30	385	268	12	665	20.78	1.44
1991	SMR+5	33	3/15–6/30	277	196	7	480	14.55	1.41
1992	SMR+5	33	3/15–6/15	226	211	0	437	13.24	1.07
1993	SMR+6	32	4/05–6/15	201	198	10	409	12.78	1.02
1994	SMR+6	33	3/15–6/15	307	187	1	495	15.00	1.64
1995	SMR+10	40	3/25–7/15	250	277	25	552	13.80	0.90
1996	SMR+10	40	3/15–6/30	385	201	4	590	14.75	1.92
Mean								17.09	1.42
Total				3419	2416	104	5939		

[a]SMR, Santa Margarita River. Numbers after SMR refer to the number of drainages, in addition to the Santa Margarita River, where cowbird trapping was conducted. 1988, added Las Flores Creek and Pilgrim Creek. 1989, added San Mateo Creek. 1990, added San Onofre Creek and De Luz Creek. 1993, added Aliso Creek. 1995, added Christianitos Creek, Pueblitos Creek, Fallbrook Creek, and Windmill Canyon.
Source: Griffith and Griffith 1996.

traps from 27 to 40. Thus it appears that the removal program has had little or no cumulative effect on the large regional cowbird population (see also DeCapita, Chapter 37, this volume). To protect endangered species, such topical trapping (deployment of multiple traps in targeted habitat) will have to be maintained indefinitely or until other methods of cowbird control are shown to have broader or more permanent effects.

Vireo Recovery

The number of vireos at Camp Pendleton began to increase before trapping began in 1983 because of a 56% increase in productivity via nest manipulation. Known numbers also increased as survey efforts were intensified (1982 was the first comprehensive survey). This was followed by a much greater and sustained increase after cowbird trapping began. Productivity of unmanipulated nests increased by 129% over pretrapping levels (from 1.33 in 1982 to an average of 3.05 in 1988–1991; Table 38.1). From 1983 to 1996, the number of vireos at Camp Pendleton increased from 62 to 902, and the number of occupied drainages increased from 3 to 14 (Table 38.4). The early population growth occurred despite a

prolonged drought. Since the study began, there have been no significant changes in the amount of available habitat or in the predation rate (19–36% annually, and 305 of 1,176 nests or 26% overall). Only the rate of parasitism has changed (Table 38.5).

As the population grew, vireos began to use a larger variety of riparian habitats, similar to historic descriptions of their habitat use (Cooper 1874, Grinnell and Miller 1944). We believe that by the early 1980s, the vireo population had been extirpated by cowbird parasitism from all but large stands of preferred habitat and sites with few cowbirds. Currently, vireos at Camp Pendleton are not excluded from any area by parasitism and breed in suitable willow-dominated riparian woodland and many adjacent habitats, some devoid of willows.

Banded vireos reared on Camp Pendleton have been observed in all major populations (except the desert oases) in numbers roughly inversely proportional to the dispersal distance. Some of the farthest dispersal destinations include the Santa Clara River ($N = 2$; about 200 km north in Ventura County), Santiago Creek ($N = 2$; 69 km north in Orange County), Vail Lake ($N = 1$; about 45 km east in Riverside County), and the Tijuana River and Sweetwater River (N = 10; 80–90 km south in southern San Diego County) (Griffith and Griffith 1992). However, most of the surviving young reared on Camp Pendleton settled there (80% in 1988–1991, N = 166; Griffith and Griffith 1992), an average of 3.45 km from their natal territories (N = 77; Griffith and

Griffith 1990, 1991a, 1992). The few first-year birds that did not settle on Camp Pendleton mostly dispersed immediately to the south (5–10 km) along the San Luis Rey River. Including the latter, 90% of the surviving nestling vireos banded on Camp Pendleton were found in a subsequent year settled in the local area. Because of the short dispersal distances and

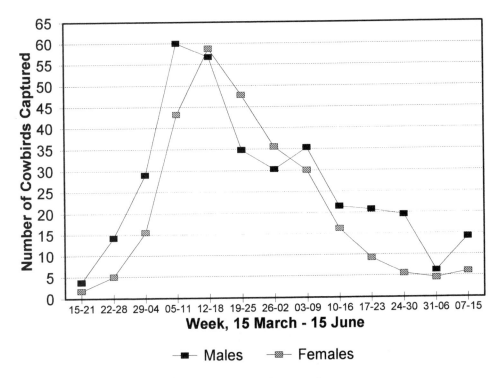

Figure 38.3. Average number of Brown-headed Cowbirds captured per week at Camp Pendleton, 1990–1996. Sources: Griffith and Griffith 1991b,c, 1994a–c, 1995a, 1996.

Table 38.3. Male-to-Female Capture Ratio in Brown-headed Cowbirds Before and After Maintaining a Live Decoy Ratio of 2 Males : 3 Females at Three San Diego County Sites

Location	Years	2M : 3F Ratio Maintained	Number Captured		Capture Ratio
			Male	Female	
Camp Pendleton	1983–1991	No	2050	1342	1.53
	1992–1995	Yes	984	873	1.13
San Luis Rey West	1991	No	161	43	3.74[a]
	1992–1995	Yes	322	318	1.01
Tijuana River	1990–1992	No	207	88	2.35
	1993–1994	Yes	83	86	0.97

[a]Does not include 420 males and 22 females captured at a dairy.

Sources: Griffith and Griffith 1995a,b, 1994d.

Table 38.4. Singing Male Least Bell's Vireos on Drainages at Camp Pendleton, 1978–1996

Year[a]	Field Hours[b]	Drainage[c]														Total	% Growth
		SMR	CC	SMC	SOC	PDL	LFC	AC	FC	DLC	FBC	PBC	WC	PC	MIS		
1978		5		0			0									5	
1980	100	14		1	0		0		0					0		15	200
1981	300	26		0	0		1		0							27	80
1982	750	43		0	0		2		0							45	67
1983	615	60		0	0		1		1							62	38
1984	526	83		0	0		0		0							83	34
1985	511	86					0		1							87	5
1986	539	98		0	0		0		2							100	15
1987	760	142		3	0		4	0	3					4		156	56
1988	1375	200		1	0		3	0	2	0				4		210	35
1989	1184	154		0	0		5		0	2	0			11		172	−18[d]
1990	964	189		1	0		8	0	0	0	0			11		209	22
1991	960	212		1	1		22	2	1	3	0			14		256	22
1993	307	319		4	3	1	59	5	1	3	6	1	1	20		423	65
1995	1243	426	2	23	15	1	125	12	7	24	11	1		44	5	696	65
1996	1120	523	5	48	27	1	148	24	10	26	16	2	2	48	22	902	30

Note: Singing males include transient (present less than 30 days), territorial (present more than 30 days), and paired males. Transients not included for total in 1981–1986.
[a]Comprehensive surveys were not performed in 1979, 1992, and 1994.
[b]Cursory walk-by surveys in 1978 and 1980.
[c]Drainage abbreviations as listed in the text under "Study Area." Blank cells represent no survey performed.
[d]The decrease in vireo numbers from 1988–1989 occurred throughout the subspecies range due to unknown causes.
Sources: Salata 1983a; Griffith and Griffith 1992, 1997; Pavelka 1994; Kus 1996.

subsequent extreme philopatry, we believe that isolated unoccupied or poorly populated drainages are essentially biogeographic islands with few external recruits and, without cowbird trapping, few internal recruits. Vireo populations appear to grow mostly from within; local productivity is thus critical for population growth, and cowbird trapping is essential to increase local productivity.

From 1983 to 1996, parasitism was eliminated at the Santa Margarita River with 5–21 traps, and vireo numbers grew steadily (Figure 38.4). With continued cowbird control, the SMR habitat could soon reach vireo carrying capacity. The vireo population would then fluctuate with habitat availability and stochastic events, and dispersal from the SMR source population could fuel increases in vireo numbers elsewhere.

Several drainages at Camp Pendleton and in the current range served as de facto experimental controls with regard to the effect of cowbird trapping. Where pretrapping survey and monitoring data are available, they uniformly show that before trapping, cowbirds were numerous, parasitism of vireos was high, vireo productivity was low, and vireos were absent or declining. After trapping began, cowbird sightings and parasitism of vireos dropped, and vireo productivity and numbers increased (Table 38.6). Vireo numbers on drainages at Camp Pendleton other than the Santa Margarita River did not begin increasing until cowbird trapping was begun on those drainages in 1988–1990, despite their proximity to the SMR, where trapping began in 1983 (Figure 38.5). For example, the 1–2 vireos present in 1981–1983 at Las Flores Creek were gone by 1984 (at least in part because of parasitism recorded in 1982–1983), and 4 vireos newly present in 1987 were reduced to 3 in 1988. Meanwhile, the number of vireos at the SMR (less than 10 km to the southeast) increased from 26 to 142 from 1981 to 1987 and to 200 in 1988. Cowbird trapping began at LFC in 1988, parasitism was eliminated, and by 1996 LFC supported 148 vireos. Similarly, trapping was begun at Pilgrim Creek in 1988 (0–4 vireos prior to trapping) and San Mateo Creek in 1989 (0–3 vireos prior to trapping); in 1996 each site supported 48 vireos. These and other de facto control areas clearly illustrate the direct relationship between cowbird trapping, reduced parasitism, and vireo recovery. Contingent on cowbird trapping, some other sites at Camp Pendleton and several others in southern California appear poised to begin similar growth, including the Santa Clara River (Ventura County), a potential vireo highway to the Central Valley.

There are now 12 major growing vireo populations and many other locations with 5–20 pairs, though many areas of suitable habitat isolated from the recovering populations and without cowbird control programs remain vacant of vireos. It is likely that the Least Bell's Vireo would currently be at the edge of extinction rather than in the midst of a robust recovery without cowbird control at Camp Pendleton and subsequent control at other southern California sites.

Increases of Other Songbird Species Due to Cowbird Trapping

Many host species at Camp Pendleton in addition to the vireo have increased in distribution and abundance after cowbird removal. This shared beneficial effect is most obvious among other small passerines of riparian and adjacent habitats (those best protected by cowbird traps placed to benefit vireos). Incidental observations of the following species suggest marked population increases from 1987 to 1991, the years of most intensive vireo monitoring and cowbird trapping: Swainson's Thrush (*Catharus ustulatus*), Hutton's Vireo (*Vireo huttoni*), Warbling Vireo (*Vireo gilvus*), Wilson's Warbler (*Wilsonia pusilla*), Yellow Warbler (*Dendroica petechia*), Yellow-breasted Chat (*Icteria virens*), Blue Grosbeak

Table 38.5. Brown-headed Cowbird Parasitism of Known Least Bell's Vireo Nests at Camp Pendleton, 1981–1996

Year	Drainage[a]	Known Nests	Number Parasitized[b]	% Parasitized	Predation Rate (%)
1981	SMR	15	7	46.67	33.33
1982	SMR+1	93	44	47.31	26.88
1983	SMR+2	86	9	10.47	31.40
1984	SMR	78	14	17.95	35.90
1985	SMR+1	26	1	3.85	28.13
1986	SMR+1	32	2	6.25	20.59
1987	SMR	70	12	17.14	28.57
1988	SMR+4	272	3	1.10	24.63
1989	SMR+3	185	2	1.08	25.41
1990	SMR+3	195	1	0.51	18.97
1991	SMR+5	156	0	0.00	21.15
1992	SMR+1	17	0	0.00	
1993	SMR+4	9	0	0.00	
1994	SMR+3	>10	0	0.00	
1995	SMR+1	95	1	1.05	
1996	SMR+1	74	0	0.00	32.40

[a]Drainages monitored in addition to Santa Margarita River (SMR) were: 1982, LFC (some); 1983, LFC, DLC (some); 1985, DLC (some); 1986, DLC (some); 1988, SMC, LFC, DLC, PC (1 of 12); 1989, LFC, DLC, PC (0 of 14); 1990, SMC, LFC, PC (1 of 18); 1991, SMC, LFC, AC, DLC, PC (0 of 32); 1992, FBC (0 of 2); 1993, SMC, AC, FBC, PC (0 of 4); 1994, SMC, AC, PC (none); 1995, LFC (0 of 45); 1996, LFC (0 of 37). See text under "Study Area" for abbreviations.

[b]In 1981–1987 and 1992–1996, monitored nests only; 1988–1991 also includes nests known from observation of family groups.

Sources: Salata 1981, 1983a,b, 1984, 1986, 1987a,b; Griffith field notes 1988; Griffith and Griffith 1990, 1991a, 1992, 1997; Sweetwater Environmental Biologists 1993; Pavelka 1994; Kus 1996.

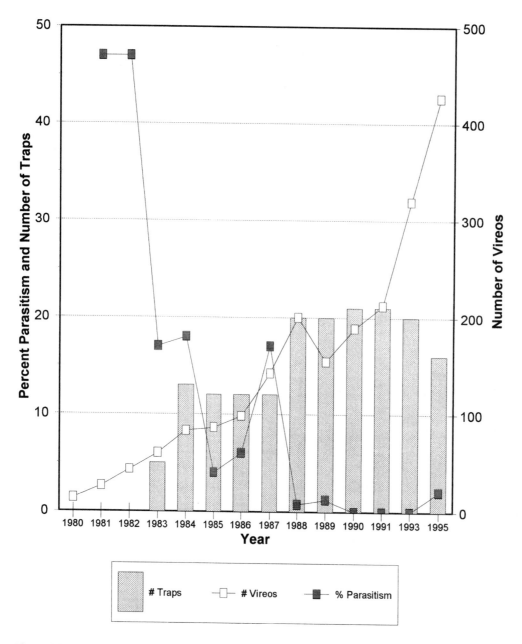

Figure 38.4. Number of vireos, percentage of parasitism of vireos, and number of cowbird traps at the Santa Margarita River, Camp Pendleton, 1980–1996. Sources: Salata 1983b, 1984, 1986, 1987a,b; Jones 1989; Griffith and Griffith 1990, 1991a, 1992, 1995a, 1997; Pavelka 1994; Kus 1996.

(*Guiraca caerulea*), and Lazuli Bunting (*Passerina amoena*) (Griffith field notes). Observers at different locations have also recorded increases in other cowbird hosts and decreased cowbird sightings coincident with cowbird trapping. The number of Yellow Warblers increased from about 8 in 1986 to more than 100 in 1995 at the Santa Ana River (J. Pike pers. comm.) and from fewer than 50 in 1993 to more than 500 in 1997 at the Kern River (S. Laymon pers. comm.).

The federally endangered Southwestern Willow Flycatcher

(*Empidonax traillii extimus*) has also benefited from cowbird trapping. These flycatchers increased at Camp Pendleton from 5 in 1981 (Unitt 1987) to 24–25 in 1989–1991 (Griffith and Griffith 1992). There were, however, no further increases from 1991 to 1995 (S. I. Rothstein pers. comm.). Only three other California populations (the upper San Luis Rey River, Prado Basin, and the South Fork Kern River) have neither decreased nor been extirpated in the last decade. All three have benefited from cowbird trapping begun in 1992

(Griffith and Griffith 1994d, L. Hays pers. comm.; Whitfield, Chapter 40, this volume).

Cowbird parasitism of the federally threatened California Gnatcatcher (*Polioptila californica californica*) has been noted in San Diego, Orange, Riverside, and Los Angeles counties since the 1930s (Atwood 1990). Near Lake Mathews in Riverside County, parasitism rates of these gnatcatchers changed from 0% ($N = 7$) to 71% ($N = 7$) before and after cowbird dispersal in early April (Griffith and Griffith 1993). Data from gnatcatcher surveys in 1989 and 1993–1994 at Camp Pendleton suggest that this upland species has been aided by cowbird trapping done to protect vireos. For example, a ridge covered by sage scrub between San Mateo and San Onofre creeks (where cowbird trapping was begun in 1989 and 1990, respectively) supported 14 gnatcatchers in 1989 before trapping and 79 in 1994 after five years of trapping (USFWS 1991, Griffith and Griffith 1995c). Some of the increase may have been due to increased survey intensity (greater in 1993–1994), habitat changes (sage scrub responds quickly to varying rainfall),

and the dynamic nature of gnatcatcher populations, but we believe a majority of the increase at this site and on all of Camp Pendleton is due to cowbird trapping. In general, more gnatcatchers were located in sage scrub near cowbird traps than in sage scrub far from cowbird traps in the 1993–1994 surveys. Many cowbirds and no gnatcatchers were noted in these surveys in apparently suitable sage scrub farthest from the traps.

Conclusions and Management Implications

We believe the Least Bell's Vireo would not be endangered in the absence of Brown-headed Cowbirds. Persistent cowbird parasitism of vireo populations, already much reduced by habitat loss, fragmentation, and degradation, probably caused the localized extirpation and eventual endangered status of the vireo. By the early 1980s, vireos remained only in large areas of relatively undisturbed habitat and/or in areas with few cowbirds. Many thousands of acres of riparian habitat in southern California and the Central Valley,

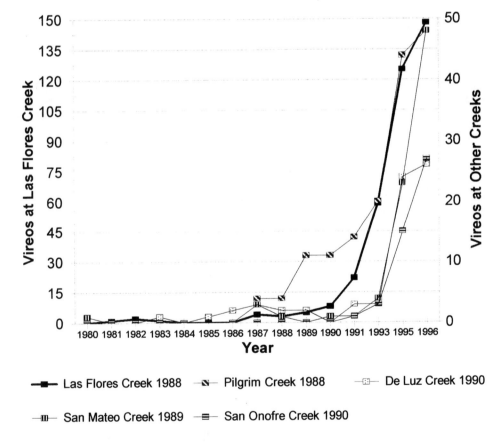

Figure 38.5. Numbers of vireos before and after cowbird trapping at five Camp Pendleton drainages other than the Santa Margarita River. The year when trapping began at each drainage is given after the creek name. Sources: Griffith and Griffith 1992, 1995a, 1997; Pavelka 1994; Kus 1996.

Table 38.6. Change in Cowbird Parasitism Rate and Population Size of the Least Bell's Vireo Before and After Cowbird Trapping at 12 Sites in Southern California

Location[a]	Trapping Begun	% Parasitism		Number of Vireos	
		Before	After	Before	In 1996
Camp Pendleton					
Santa Margarita River	1983	47	0–18	60	523
San Mateo Creek	1989	100	0	0–3	48
San Onofre Creek	1990	NA	0	0	27
Las Flores Creek	1988	some	0	0–4	148
De Luz Creek	1990	some	0	0–3	26
Pilgrim Creek	1988	some	0	0–4	48
Santa Ana River	1986	77–100	16–57[b]	19	249
San Luis Rey River West	1987	63–65[c]	0–40	9	50
San Luis Rey River	1987		20–14	24	89
San Diego River	1987	33–91	0–27	21	32[d]
Sweetwater River	1986	78	0–7	51	61[d]
Tijuana River	1990	some	0–11	13	142

[a]Entire length of drainages not surveyed.

[b]Relatively high posttrapping parasitism rate at Santa Ana River reflects inaccessibility of many areas. For all years, parasitism rate approached zero near traps (L. Hays pers. comm.).

[c]Pretrapping parasitism rate is for the entire study area. High posttrapping rate reflects problems with the program. Since 1991 parasitism has been 0% at the SLR West.

[d]1993 data (most recent data available). Relatively low growth of San Diego and Sweetwater River populations reflects limited habitat in the study areas and does not show real population increases in surrounding areas from increased productivity in the study areas (B. Kus, B. Jones pers. comm.).

including some prime areas, were and are vacant of vireos and other declining host species. Enough habitat exists to support about 5,000 vireo pairs, the number targeted by the draft vireo recovery plan (USFWS 1988). Habitat loss alone did not cause the vireo to become endangered, and habitat protection alone will not provide for its recovery.

Cowbird trapping substantially decreases or eliminates parasitism of vireos and other local hosts, allows subsequent population growth, and leads to a full and diverse complement of native birds in otherwise healthy riparian habitat. Since trapping began, however, the number of cowbirds trapped each year has not declined, suggesting that the regional cowbird population has not been reduced by the seasonal topical trapping. Therefore, we believe that current topical trapping programs must be maintained indefinitely, unless or until the regional cowbird population has been substantially reduced or eliminated. We suggest this might be done by identifying and targeting control efforts toward cowbirds in winter roosts and major foraging areas.

Coordinated topical or regional cowbird control is nec-essary to sustain the present recovery of the Least Bell's Vireo and other co-occurring host species that are vulnerable to the effects of parasitism. In concert with habitat protection, increased topical or intensive regional cowbird control should be implemented throughout the Southwest to aid in the recovery or preclude the further decline, extirpation, or extinction of other cowbird hosts.

Acknowledgments

The Camp Pendleton vireo study was funded and supported by the U.S. Marine Corps. L. Belluomini and L. Salata began the vireo research and cowbird trapping. Subsequent technical representatives were S. Buck, D. Boyer, and R. Griffiths. Several ornithologists contributed to this chapter through their fieldwork and willingness to share data: Santa Ana River, L. Hays and J. Pike; early Camp Pendleton, L. Salata; San Luis Rey, San Diego, Sweetwater, and Tijuana Rivers, B. Jones, B. Kus, and J. Newman. J. Greaves, L. Salata, and J. N. M. Smith reviewed drafts of this chapter, as did S. I. Rothstein, our primary editor. His incisive comments, insight, and perspective greatly improved the final product. Thank you all.

References

American Ornithologists' Union. 1957. Check-list of North American birds. 5th ed. American Ornithologists' Union, Washington, DC.

Atwood, J. L. 1990. Status review of the California Gnatcatcher (*Polioptila californica*). Unpubl. technical report, Manomet Bird Observatory, Manomet, MA.

Cooper, J. G. 1874. Animal life of the Cuyamaca Mountains. Amer. Natur. 8:14–18.

Federal Register. 1986. Determination of endangered status for the Least Bell's Vireo. Final rule. Vol. 51, No. 85: 16474–16483.

Flahaut, M. R., and Z. M. Schultz. 1955. Northern Pacific Coast Region. Audubon Field Notes 9:395–397.

Fleischer, R. C. 1985. A new technique to identify and assess the dispersion of eggs of individual brood parasites. Behav. Ecol. Sociobiol. 17:91–99.

Franzreb, K. E. 1989. Ecology and conservation of the endangered Least Bell's Vireo. USFWS Biol. Rep. 89(1).

Friedmann, H. 1963. Host relations of the parasitic cowbirds. U.S. Natl. Mus. Bull. 233:1–273.

Friedmann, H., and L. F. Kiff. 1985. The parasitic cowbirds and their hosts. Proc. Western Found. Vert. Zool. 2:226–304.

Gaines, D. 1974. A new look at the nesting riparian avifauna of the Sacramento Valley, CA. Western Birds 5:61–80.

Goldwasser, S., D. Gaines, and S. R. Wilbur. 1980. The Least Bell's Vireo in California: A de facto endangered race. Amer. Birds 34: 742–745.

Griffith, J. T., and J. C. Griffith. 1989. Field techniques for surveying, monitoring, and banding the Least Bell's Vireo. Unpubl. manuscript, Griffith Wildlife Biology, Calumet, MI.

———. 1990. The status of the Least Bell's Vireo on Marine Corps Base Camp Pendleton, California, in 1989. Unpubl. report, Natural Resources Office, Camp Pendleton. Griffith Wildlife Biology, Oceanside, CA.

———. 1991a. The status of the Least Bell's Vireo on Marine Corps Base Camp Pendleton, California, in 1990. Unpubl. report, Environmental and Natural Resources Management Office, Camp Pendleton. Griffith Wildlife Biology, Calumet, MI.

———.1991b. Report on the Brown-headed Cowbird removal program on Marine Corps Base Camp Pendleton, California, in 1990. Unpubl. report, Environmental and Natural Resources Management Office, Camp Pendleton. Griffith Wildlife Biology, Oceanside, CA.

———. 1991c. Brown-headed Cowbird removal on Marine Corps Base Camp Pendleton, California, 1991. Unpubl. report, Environmental and Natural Resources Management Office, Camp Pendleton. Griffith Wildlife Biology, Calumet, MI.

———. 1992. The status of the Least Bell's Vireo on Marine Corps Base Camp Pendleton, California, in 1991. Unpubl. report, Environmental and Natural Resources Management Office, Camp Pendleton. Griffith Wildlife Biology, Calumet, MI.

———. 1993. Report on the avian resources observed along the CPA project corridor in 1992, with emphasis on the California Gnatcatcher and the Least Bell's Vireo. Unpubl. report, Metropolitan Water District of Southern California, Los Angeles. Griffith Wildlife Biology, Calumet, MI.

———. 1994a. 1992 Brown-headed Cowbird removal on Marine Corps Base Camp Pendleton, California. Unpubl. report, Environmental and Natural Resources Management Office, Camp Pendleton. Griffith Wildlife Biology, Calumet, MI.

———. 1994b. 1993 Brown-headed Cowbird removal program on Marine Corps Base Camp Pendleton, California. Unpubl. report, Environmental and Natural Resources Management Office, Camp Pendleton. Griffith Wildlife Biology, Calumet, MI.

———. 1994c. 1994 Brown-headed Cowbird removal program on Marine Corps Base Camp Pendleton, California. Unpubl. report, Assistant Chief of Staff, Environmental Security, Camp Pendleton. Griffith Wildlife Biology, Calumet, MI.

———. 1994d. 1994 Western Tijuana River Brown-headed Cowbird removal program. Unpubl. report, International Boundary and Water Commission, El Paso, TX, and Chambers Group, Inc., Irvine, CA. Griffith Wildlife Biology, Calumet, MI.

———. 1994e. Brown-headed Cowbird trapping protocol. Unpubl. manuscript. Griffith Wildlife Biology, Calumet, MI.

———. 1995a. 1995 Brown-headed Cowbird removal program on Marine Corps Base Camp Pendleton, California. Unpubl. report, Assistant Chief of Staff, Environmental Security, Camp Pendleton. Griffith Wildlife Biology, Calumet, MI.

———. 1995b. 1995 U.S. Army Corps of Engineers Western San Luis Rey River Brown-headed Cowbird removal program. Unpubl. report, U.S. Army Corps of Engineers, Los Angeles, and Michael Brandman Associates, Irvine, CA. Griffith Wildlife Biology, Calumet, MI.

———. 1995c. The California gnatcatcher and coastal Cactus Wren at Marine Corps Base Camp Pendleton, California, in 1993 and 1994. Unpubl. report, Assistant Chief of Staff, Environmental Security, Camp Pendleton. Griffith Wildlife Biology, Calumet, MI.

———. 1996. 1996 Brown-headed Cowbird removal program on Marine Corps Base Camp Pendleton, California. Unpubl. report, Assistant Chief of Staff, Environmental Security, Camp Pendleton. Griffith Wildlife Biology, Calumet, MI.

———. 1997. The status of the Least Bell's Vireo and Southwestern Willow Flycatcher at Marine Corps Base Camp Pendleton in 1996. Unpubl. report, Assistant Chief of Staff, Environmental Security, Camp Pendleton. Griffith Wildlife Biology, Calumet, MI.

Grinnell, J., and A. Miller. 1944. The distribution of the birds of California. Pacific Coast Avifauna 27.

Jones, B. L. 1989. Status of the Least Bell's Vireo on Marine Corps Base Camp Pendleton, California, in 1988. Unpubl. report, Natural Resources Office, Camp Pendleton. Sweetwater Environmental Biologists, Spring Valley, CA.

Kus, B. E. 1996. Status of the Least Bell's Vireo and Southwestern Willow Flycatcher at Marine Corps Base Camp Pendleton, California, in 1995. Unpubl. report, Assistant Chief of Staff, Environmental Security, Camp Pendleton, CA.

Pavelka, M. 1994. Status of the Least Bell's Vireo and Southwestern Willow Flycatcher at Camp Pendleton Marine Corps Base, California, in 1993. Unpubl. report, Environmental and Natural Resources Management Office, Camp Pendleton. USFWS Carlsbad Field Office, Carlsbad, CA.

Rothstein, S. I. 1990. A model system for coevolution: Avian brood parasitism. Annu. Rev. Ecol. Syst. 21:481–508.

———. 1994. The cowbird's invasion of the far West: History, causes, and consequences experienced by host species. Studies in Avian Biol. 15:301–315.

Salata, L. 1981. Least Bell's Vireo research, Camp Pendleton Marine Corps Base, San Diego County, California, 1981. Unpubl. report, Natural Resources Office, Camp Pendleton, CA.

———. 1983a. Status of the Least Bell's Vireo on Camp Pendleton, California [in 1982]. Unpubl. report, USFWS, Laguna Niguel Field Office, CA.

———. 1983b. Status of the Least Bell's Vireo on Camp Pendleton, California: Report on research done in 1983. Unpubl. report, USFWS, Laguna Niguel Field Office, CA.

———. 1984. Status of the Least Bell's Vireo on Camp Pendleton, California: Report on research done in 1984. Unpubl. report, Natural Resources Office, Camp Pendleton, CA.

———. 1986. Status of the Least Bell's Vireo on Camp Pendleton, California: Report on research done in 1985. Unpubl. report, Natural Resources Office, Camp Pendleton. Sweetwater Environmental Biologists, Spring Valley, CA.

———. 1987a. Status of the Least Bell's Vireo on Camp Pendleton, California, in 1986. Unpubl. report, Natural Resources Office, Camp Pendleton. Sweetwater Environmental Biologists, Spring Valley, CA.

———. 1987b. Status of the Least Bell's Vireo on Camp Pendleton, California, in 1987. Unpubl. report, Natural Resources Office, Camp Pendleton. Sweetwater Environmental Biologists, Spring Valley, CA.

Smith, F. 1977. A short review of the status of riparian forests in California. Pp. 1–2 in Riparian forests in California: Their ecology and conservation (A. Sands, ed). Institute Ecology Pub. 15.

Sweetwater Environmental Biologists. 1993. 1992 status of the Least Bell's Vireo on Camp Pendleton, California. Unpubl. report, Environmental and Natural Resources Management Office, Camp Pendleton. Sweetwater Environmental Biologists, San Diego, CA.

Unitt, P. 1984. The birds of San Diego County. San Diego Society of Natural History, Memoir 13.

———. 1987. *Empidonax traillii extimus:* An endangered subspecies. Western Birds 18:137–162.

USFWS. 1988. Least Bell's Vireo draft recovery plan. USFWS, Portland, OR.

———. 1991. A survey of the California Gnatcatcher and Cactus Wren on Camp Pendleton, San Diego County, CA [in 1989]. Unpubl. report, Environmental and Natural Resources Management Office, Camp Pendleton. USFWS Southern California Field Station, Laguna Niguel, CA.

Wilbur, S. R. 1981. The Least Bell's Vireo in Baja California, Mexico. Western Birds 11:129–133.

Willett, G. 1933. Revised list of birds of southwestern California. Pacific Coast Avifauna 21:1–203.

39. Cowbird Control Program at Fort Hood, Texas: Lessons for Mitigation of Cowbird Parasitism on a Landscape Scale

TIMOTHY J. HAYDEN, DAVID J. TAZIK,
ROBERT H. MELTON, AND JOHN D. CORNELIUS

Abstract

A significant population of endangered Black-capped Vireos occurs at Fort Hood, an 88,890-ha Army installation in central Texas (279 territorial males documented in 1994). The vireo's endangered status has been attributed partly to frequent parasitism of vireo nests by Brown-headed Cowbirds. In 1987, 90.9% of all vireo nests at Fort Hood were parasitized. We report on a cowbird control program (trapping and shooting) to reduce cowbird parasitism on Black-capped Vireos. During 1988–1994, the number of female cowbirds killed at Fort Hood increased from 8 in 1988 to 3,038 in 1994. Parasitism declined from 90.8% of 33 nests in 1988 to 12.6% of 135 nests in 1994. Nest success increased from 4.7% of 33 nests in 1988 to 51.4% of 214 nests in 1993–1994. Control efforts varied among regions of Fort Hood during 1991–1994, the years of most intensive cowbird control. Regions with cowbird control in 1991–1994 had much lower parasitism levels (13.9% of 230 nests) than regions without cowbird control (44.74% of 103 nests). Numbers of vireo territories more than doubled on parts of Fort Hood from 1991 to 1994 in association with intense cowbird control. We conclude with recommendations to those considering implementing similar cowbird control programs, based on our work at Fort Hood.

Introduction

Cowbird trapping programs have been effective in reducing the incidence of nest parasitism by Brown-headed Cowbirds in small, geographically restricted host populations. Small populations of endangered species that demonstrate reduced productivity due to cowbird parasitism are candidates for intensive cowbird control efforts (Robinson et al. 1993). Populations that have benefited from cowbird control efforts include Least Bell's Vireos (*Vireo bellii pusillus*) in California (Franzreb 1989; Griffith and Griffith, Chapter 38, this volume) and Kirtland's Warblers (*Dendroica kirtlandii*) in Michigan (Walkinshaw 1983, Probst 1986, Mangold et al. 1994; DeCapita, Chapter 37, this volume).

In Texas and Oklahoma, the endangered Black-capped Vireo (*Vireo atricapillus*) is similarly threatened by parasitism by Brown-headed Cowbirds (USFWS 1991). Cowbird control programs to mitigate the impacts of cowbird parasitism of vireo nests have been implemented at several sites throughout the vireo's range (Espey Houston and Associates 1989; Grzybowski 1989, 1990). At the Kerr Wildlife Management Area in Texas, trapping reduced cowbird parasitism of Black-capped Vireo nests from 90% to 9%, and trapping in Oklahoma reduced parasitism from 81% to 24% in the Wichita Mountains Wildlife Refuge (Grzybowski 1989).

Currently, the most intensive cowbird control program to protect Black-capped Vireos is being conducted at Fort Hood, Texas, a large (88,890 ha) U.S. Army installation in central Texas. The Black-capped Vireo population at Fort Hood is the largest known in Texas under a single management authority. In 1994, 279 Black-capped Vireo territories were documented throughout Fort Hood (Weinberg et al. 1994). More than 500 territories of Golden-cheeked Warblers (*Dendroica chrysoparia*), another federally endangered bird species endemic to Texas, also occur on Fort Hood (Hayden and Tazik 1991). Approximately 12,000 ha of suitable warbler habitat occur on the installation. Golden-cheeked Warblers are also susceptible to cowbird parasitism (Pulich 1976, USFWS 1992), but the incidence of parasitism in the Fort Hood population is poorly known because of the difficulty in locating warbler nests.

The large area of Fort Hood and the distribution of endangered vireos throughout the installation present managers with the challenge of mitigating effects of cowbird parasitism at a landscape scale. This chapter evaluates the effectiveness of the cowbird control program in reducing parasitism of Black-capped Vireo nests at Fort Hood and presents recommendations to managers planning large-scale cowbird control programs elsewhere.

The Black-capped Vireo

The U.S. Fish and Wildlife Service listed the Black-capped Vireo as endangered in 1987. Major reasons for listing the species include (1) past population declines, (2) habitat loss due to agricultural practices, urbanization, and fire suppression, and (3) low productivity attributed largely to frequent cowbird parasitism (USFWS 1991). The core U.S. population is currently found in the Edwards Plateau Hill Country region of central Texas. Fort Hood lies at the northern extent of the Edwards Plateau. The current northern limit of the vireo's breeding range is a small, isolated population in the Wichita Mountains Wildlife Refuge in southern Oklahoma (Grzybowski 1990). Populations also exist in northern Mexico, but their extent and size are unknown (Benson and Benson 1990, 1991; Scott and Garton 1991).

The Black-capped Vireo has several characteristics that make it particularly vulnerable to the effects of cowbird parasitism (Friedmann et al. 1977, Friedmann and Kiff 1985). It is a small (9–10 g) neotropical migrant that builds small, open-cupped nests. The incubation period is 14–17 days (Graber 1961), lengthy compared with 11 days for cowbird eggs (Friedmann 1963). The shorter incubation and faster growth rate of cowbird nestlings provide such a strong developmental advantage that vireo young rarely fledge from nests in which cowbird eggs hatch (Graber 1961, Tazik and Cornelius 1993). Black-capped Vireo nesting habitat (shrubland or edge) is often preferred by breeding cowbirds (Brittingham and Temple 1983, Thompson 1994, but see Hahn and Hatfield 1995; Robinson et al., Chapter 33, this volume). Deciduous shrub habitats used by vireos at Fort Hood are maintained by fire or mechanical disturbances in mature mixed oak and ashe juniper (*Juniperus ashei*) woodlands. Fires on Fort Hood are due to lightning strikes, prescribed burns to control juniper encroachment, military pyrotechnics, and human carelessness. Mechanical disturbances include maneuvers by tracked and wheeled military vehicles and brush clearing to maintain military firing ranges.

Vireo habitat at Fort Hood persists for only a finite period without disturbance. Vireos are found primarily in deciduous shrub habitats 3–25 years after disturbance (Tazik et al. 1993b, Weinberg et al. 1994). On a landscape scale, these ephemeral habitats are typically small and patchy, resulting in a clustered distribution of Black-capped Vireo territories in habitats that may also support concentrations of breeding cowbirds.

Black-capped Vireos probably coexisted with fairly high densities of cowbirds prior to European settlement of the area and may exhibit strategies for coping with frequent nest parasitism. For instance, the Black-capped Vireo deserts parasitized nests at a much higher rate than unparasitized nests, which in turn reduces nest success in parasitized nests versus unparasitized nests (Graber 1961, Tazik and Cornelius 1993). Black-capped Vireos compensate for the high de-

sertion rate and low nest success due to cowbird parasitism by renesting, even after successful broods, and by remating during an extended breeding season (Graber 1961, USFWS 1991, Tazik and Cornelius 1993). At Fort Hood, nesting is initiated as early as the first week in April and as late as the second week in July (T. Hayden, unpubl. data). Peak nest initiation for 280 nests at Fort Hood in 1987–1989 was during 21 May–3 June (Tazik and Cornelius 1993). In individual years, peak nest initiation varied between 4–17 June in 1988 and 7–20 May in 1987. Female cowbird trapping success at Fort Hood was greatest during April in all four years from 1991 to 1994 (Figure 39.1).

Management Area

Fort Hood is located in Bell and Coryell counties in central Texas adjacent to the city of Killeen. Grasslands (65%) interspersed with shrub and woodland habitats (31%) dominate the vegetation (Tazik et al. 1993a). This fragmented mosaic of grasslands and woodlands is characteristic of the Lampasas Cutplains physiographic region in which Fort Hood occurs (Raisz 1952). Erosionally dissected uplift plains form narrow, flat-topped mesas with steep slopes and broad intervening valleys. The slopes and mesa tops of Fort Hood are dominated by shrub and mixed oak and ashe juniper woodlands, which provide abundant nesting habitat for host species. The intervening valleys are predominantly grasslands and grassland savannah and provide ample foraging areas for cowbirds. Tazik et al. (1993b) provide a more detailed description of the habitat.

Short grass conditions preferred by foraging cowbirds (Mayfield 1965, Thompson 1994) are maintained by a variety of land use practices on the installation. Fires occur frequently in live-fire areas due to exploding ordnance and prescribed burns to maintain firing ranges. Off-road train-

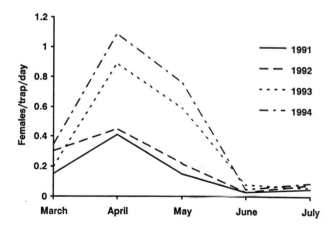

Figure 39.1. Average number of females killed/trap/day by month during March–July 1991–1994 at Fort Hood. Average number of traps operated/day ranged from 43.4 in 1991 to 47.9 in 1994.

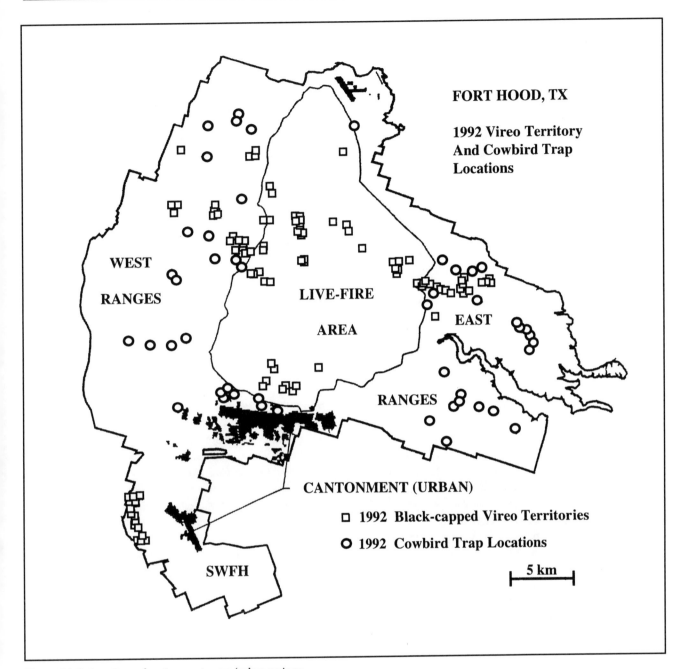

Figure 39.2. Locations of cantonment areas (urban regions within the Fort Hood boundary—filled), cowbird traps, and Black-capped Vireo territories at Fort Hood in 1992. Traps and vireo territories were similarly distributed in 1991–1994. Regions are described in the introduction under Management Area.

ing by tracked and wheeled vehicles maintains short grass conditions in maneuver ranges. Cattle grazing is allowed under lease agreement throughout most of the installation, except for southwest Fort Hood, where the lease was terminated prior to the vireo nesting season in 1992. Burning and chaining of old fields with juniper encroachment is used to maintain pasturelands.

Fort Hood is the only U.S. Army installation currently providing housing and training support for two mechanized armored divisions. In this chapter, five regions of the installation will be treated separately because of differences in cowbird control activities, land-use practices, and topography (Figure 39.2).

The west ranges (26,825 ha) are characterized by rolling hills. The dominant vegetation is grasslands, with scattered woodlands on steeper slopes and along streams. The more open terrain of the west ranges supports the bulk of armored vehicle training on the fort. Forty-two Black-capped Vireo territories (15.0% of all territories known at Fort Hood) were in the west ranges in 1994.

The east ranges (20,256 ha) are more rugged, with more extensive and less fragmented woodlands than the west ranges. The east ranges are not as suitable for off-road vehicle travel, and military training is primarily by small, dismounted units. Vehicular traffic is limited mostly to roads and trails. In 1994, 72 of 279 vireo territories at Fort Hood (25.8%) were in the east ranges.

The live-fire area (25,100 ha) occupies most of the central region between the east and west ranges. It supports small arms, artillery, and tank firing ranges. Much of this area comprises safety buffer zones for impact areas and is mostly unaffected by maneuver training or exploding ordnance. Much of its northern half area is relatively undisturbed woodland. Fires caused by exploding ordnance and pyrotechnics create and maintain the deciduous shrubs preferred by Black-capped Vireos for nesting. Forty-seven percent of the 279 vireo territories at Fort Hood in 1994 were within the live-fire area (Weinberg et al. 1994). Research personnel often had limited access to this area because of scheduling conflicts with live-fire training activities.

Southwest Fort Hood (7,114 ha) is a peninsula of the installation located south of State Highway 190. It is surrounded by urban and ranch lands. The installation's main airfield is in this area. Vireo territories are on a wooded ridgeline running the length of the west boundary of the area. Thirty-four territories (12.2%) were documented in this area in 1994.

The cantonment areas and other developed regions of the installation cover 8,595 ha. No vireo territories were documented in these areas. The main cantonment area is contiguous with the city of Killeen on the south side of the installation. It includes support facilities and housing for military personnel. Large open grassland areas including lawns, parade grounds, golf courses, and airfields provide extensive foraging habitat for cowbirds.

Methods

Black-capped Vireo Monitoring

Black-capped Vireo territories were documented by surveying potential habitats and areas of past occurrence in 1987–1994. Individual adults were identified, when two or more individuals were vocalizing at the same time, by the locations of singing or calling males and by observing marked adults with unique combinations of colored leg bands. Adult vireos were captured for banding using mist nets and taped playbacks of vireo songs and calls. Territory locations and

approximate boundaries were recorded on 1 : 4,800 aerial photographs and 1 : 24,000 USGS topographic maps. Territory locations for 1992 reveal the widespread, but clumped, distribution of vireo territories at Fort Hood (Figure 39.2). Occupied territories were monitored, when possible, to determine mated status and nest success of pairs.

Nests were located by following breeding adult male and female vireos. When access permitted, nests were scored for parasitism, clutch size, hatching success, and fate. Tazik and Cornelius (1993) and Bolsinger and Hayden (1994) give further details of monitoring methods.

Cowbird Trapping

Traps typically were constructed of two-by-four lumber and one-inch poultry mesh based on a U.S. Department of Interior (1973) design. Trap dimensions were 2.5 x 3.7 x1.8 m. Entry of cowbirds to traps was via a wire mesh drop-down basket in the roof of the trap. Modifications to the basic design were made to make traps more predator resistant. Traps were provided with decoy male and female cowbirds, poultry feed grain (millet or cracked corn mixes), and poultry waterers.

Changes in cowbird trapping intensity and approaches during 1987–1994 reflected changing objectives. No trapping was conducted in 1987, but nest monitoring revealed frequent nest parasitism of Black-capped Vireos. In 1988 and 1989, trapping was initiated to evaluate whether cowbird control in occupied vireo nesting habitat could reduce cowbird parasitism. Three traps were operated at one vireo nesting location in 1988. Seven traps were placed in six occupied vireo nesting habitat patches in 1989.

Trapping, at the level of effort conducted in 1988 and 1989, failed to reduce parasitism in vireo nests (see Results). Trapping effort was therefore increased substantially in subsequent years. Twenty-five traps were operated in 1990; traps again were placed primarily in occupied vireo habitats. During 1991–1994, 52 traps were operated annually, though not all traps were operated daily because of repair and maintenance activities. Trap placement during 1992 is representative of the distribution of traps at Fort Hood during 1991–1994 (Figure 39.2). Our approach in 1991–1994 was to increase trap efficiency and to maximize the total number of female cowbirds destroyed annually rather than trapping within vireo nesting habitat. To increase efficiency, trap placement in 1991–1994 was in cowbird foraging areas, and traps were moved frequently from areas with low catches to more productive locations.

Traps were placed in open short-grass areas with nearby perch locations, near corrals, and in areas frequently used by grazing cattle and by flocks of foraging cowbirds. Trap effort in occupied vireo habitats in 1991–1994 was equivalent to or less than the trap effort in occupied habitats in earlier years.

During all years, active traps were visited at least three times weekly. In 1988, all cowbirds captured were killed humanely. In subsequent years, adult females were killed, and adult males and juveniles were usually banded and released. We hoped that the skewing of adult sex ratios toward males might disrupt mating behavior. No studies, however, were done to evaluate this possibility. Juveniles were released because it was believed their dispersal would contribute little to the Fort Hood breeding population in subsequent years. Records were kept of the sex and age of all cowbirds captured and destroyed or banded and released.

Cowbird Shooting

No shooting was conducted at Fort Hood in 1987, 1988, and 1990. In other years, cowbirds were attracted with taped playbacks of chatter calls of female cowbirds, and females were killed with shotguns. In 1989, shooting was conducted in five areas of occupied vireo habitat. Shooting at each site in 1989 was conducted every 4–5 days (Tazik 1991). During 1991–1993, shooting was conducted throughout Fort Hood on an ad hoc basis. Locations of cowbirds killed by shooting in 1991 and 1992 were not recorded; however, most kills were outside the live-fire area, and shooting was conducted in both occupied vireo habitats and cowbird foraging areas (G. Eckrich pers. comm.). Shooting locations were recorded in 1993, and kills were made both within occupied vireo habitats and in cowbird foraging areas. In 1994, early morning shooting was conducted on alternate days in vireo territories in southwest Fort Hood and the west ranges. Vireo monitoring crews reported the locations of cowbirds, and these birds were targeted for shooting. Sexes and ages of shot cowbirds were recorded in all years.

Data Analysis

Data for 1987–1989 are from Tazik and Cornelius (1993), and those for 1990 are from Hunt (1991). Data for 1991–1994 are previously unpublished. Except where noted otherwise, all methods were consistent throughout the study.

Data on cowbird removal are summarized for March–June, the months just prior to and including the peak Black-capped Vireo nesting season. We expected that removal of female cowbirds during this period would have the greatest effect on local parasitism rates. Cowbird parasitism is expressed as the percentage of vireo nests that contained cowbird eggs or nestlings. Intensity of cowbird parasitism is the number of cowbird eggs observed per parasitized nest.

We expressed nest success in two ways. First, we calculated the percentages of nests that fledged at least one vireo young. Although vireos do fledge cowbird young (Tazik and Cornelius 1993), nests that fledged only cowbird young were not considered successful here. Second, we calculated Mayfield estimates (1961, 1975) of nest success to estimate the percentage of nests surviving from the beginning of incubation through fledging. Except for 1990, the probability of nest survival was calculated independently for incubation and nestling periods. The proportion of nests surviving the

total nesting period was calculated as the product of the proportion surviving through incubation and the proportion surviving through the nestling period. For 1990, Hunt (1991) reported a Mayfield estimate for the entire nesting period but did not report separate estimates for incubation and nestling periods. When calculating Mayfield estimates, we assumed a 15-day incubation period and an 11-day nestling period.

Fates of failed nests were assigned to one of three categories: (1) deserted (obviously abandoned with eggs present), (2) destroyed (nest contents removed or destroyed or nest damaged), and (3) failed, cause undetermined (no vireo young fledged but the cause of failure could not be determined reliably). The frequency of nest visits was lower in 1991–1994 than in earlier years to reduce disturbance to vireo nests, which increased the proportion of category (3) failures in 1991–1994. There were no such failures in 1987–1990.

Spearman rank correlation analyses were performed to relate the number of cowbird females removed annually to the incidence of nest parasitism, and to relate parasitism to both estimates of nest success. Linear regression was used to model the relationship of annual percentage of parasitism (dependent variable) to the total number of female cowbirds destroyed annually (independent variable). Proportions were arcsine transformed before regression analysis (Zar 1984).

We could not evaluate how year, region, and cowbird control affected parasitism fully, because cowbird control efforts were not stratified by region. Subsets of 1987–1994 data were used to relate year, region, and cowbird control to patterns of cowbird parasitism using chi-square tests. Four areas were included in regional analyses: east ranges, west ranges, southwest Fort Hood, and live-fire area. All statistical tests were performed using SYSTAT (Wilkinson 1992).

Results

Control Efforts

The number of female cowbirds destroyed during March–June increased substantially from 1988 through 1994 (Table 39.1, Figure 39.3a). The largest increase in control efforts occurred from 1991 to 1994. Trapping effort in 1991–1994 was over four times greater than in 1990 and over 10 times greater than in 1988 and 1989 (Table 39.1). Trap efficiency (mean number of female cowbirds killed/trap/day) also increased from 0.035 in 1988 to 0.510 in 1994. Traps in the cantonment area in 1991–1994 had the highest success, increasing from 0.242 females/trap/day in 1991 to 1.55 in 1994 (Table 39.2).

Adult males were trapped at a higher rate than adult females. Tazik and Cornelius (1993) trapped 326 adult males versus 62 females during 1988 and 1989. Male-biased sex ratios are common in adult cowbirds (Payne 1973, Roth-

Table 39.1. Female Cowbirds Killed at Fort Hood during March–June 1987–1994

	1988	1989	1990	1991	1992	1993	1994
Number of traps	3	6	25	52	52	52	52
Trap days	231	495	1,344	5,295	5,526	5,401	5,849
Total females trapped	8	24	162	1,284	1,046	2,482	2,984
(average killed/trap/day)	(0.035)	(0.048)	(0.121)	(0.242)	(0.189)	(0.460)	(0.510)
Females shot	0	96	0	247	320	149	54
Total females removed	8	120	162	1,531	1,366	2,631	3,038

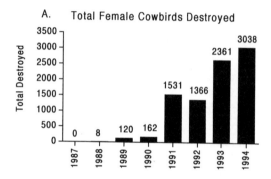

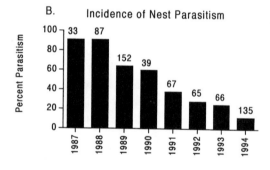

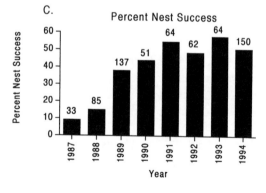

Figure 39.3. A: Total number of female cowbirds killed annually during March–June at Fort Hood, Texas. Total numbers above bars. B: Percentages of Black-capped Vireo nests parasitized by Brown-headed Cowbirds annually. Sample sizes above bars. C: Percentage of Black-capped Vireo nests fledging at least one young. Sample sizes above bars.

stein et al. 1980, Ankney and Scott 1982) and have been found in other trapping programs (Sweetwater Environmental Biologists 1989, Mangold et al. 1994). Sex-related differences in trap wariness or other behaviors may affect capture probabilities and bias adult sex in trapping data.

Greater trapping effort and efficiency increased the total number of adult female cowbirds killed annually (Table 39.1). Totals for 1991 and 1992 were more than eight times higher than in the highest previous year, 1990. Despite a nearly equivalent trapping effort, the number of female cowbirds killed during 1993–1994 was nearly double the number killed during 1991–1992. We believe that the increased trapping efficiency in 1993–1994 was due to improved trap maintenance, to the movement of traps to new locations when capture rates began to decline, and to the more frequent replacement of decoy cowbirds. Changes in cowbird abundance among years could also have affected capture rates, but no census data were available to evaluate this hypothesis. Subjective observations by cowbird control workers did not suggest a doubling of cowbird populations from 1991–1992 to 1993–1994 (G. Eckrich pers. comm.).

During 1991–1994, the percentage of female cowbirds killed by shooting (versus trapping) in March–June ranged from 1.8% in 1994 to 23.4% in 1992. The high shooting total in 1992 is due to 281 female cowbirds killed during March, mostly at a roost site near the main cantonment area (G. Eckrich pers. comm.). Otherwise, the shooting total during 1992 (39 in April–June) was similar to those in 1991 and 1993.

Cowbird Parasitism, Nest Success, and Nest Fate

The incidence of cowbird parasitism of Black-capped Vireo nests declined steadily from 90.9% (N = 33 nests) in 1987 to 12.6% (N = 135) in 1994 (Figure 39.3b). The intensity of parasitism showed a similar decline, from 1.97 cowbird eggs per parasitized nest in 1987 to 1.06 in 1994 (Table 39.3). Intensity of parasitism was consistently higher in 1987–1989 with minimal cowbird control (1.41–1.97 cowbird eggs/parasitized nest) than in 1991–1994 with intensive control

Table 39.2. Cowbird Trapping Effort and Success by Year and Region

	1988	1989	1990[a]	1991	1992	1993	1994
East ranges							
Trap days	231	297	370	2,466	2,364	2,536	2,399
(Number traps)	(3)	(4)	(8)	(25)	(24)	(24)	(24)
Females trapped	8	15	?	729	306	802	1,497
Captures/trap/day	0.035	0.051	?	0.296	0.129	0.316	0.413
Treatment category				Y	Y	Y	Y
West ranges							
Trap days	0	0	126	1,498	2,702	2,165	2,802
(Number traps)			(2)	(15)	(24)	(22)	(21)
Females trapped				321	626	1,429	1,497
Captures/trap/day				0.214	0.232	0.660	0.534
Treatment category				Y	Y	Y	Y
Southwest Fort Hood							
Trap days	0	62	213	614	0	428	342
(Number traps)		(2)	(5)	(5)		(4)	(4)
Females trapped		0		66		47	21
Captures/trap/day		0		0.107		0.110	0.061
Treatment category				Y	N	Y	Y
Live-fire area							
Trap days	0	136	596	507	0	0	0
(Number traps)		(2)	(9)	(5)			
Females trapped		5		24			
Captures/trap/day		0.139		0.047			
Treatment category				N	N	N	N
Cantonment							
Trap days	0	0	39	210	460	272	306
(Number traps)			(1)	(2)	(4)	(2)	(3)
Females trapped				144	114	204	474
Captures/trap/day				0.242	0.248	0.750	1.550
Treatment category				Y	Y	Y	Y

Note: Treatment categories: Y, intensive control in the region; N, limited or no control.

[a] Hunt (1991) reported only total number of females killed (162) in 1990.

(1.06–1.40 eggs/parasitized nest). Hunt (1991) did not report intensity of parasitism for 1990.

In contrast to the declining trend for parasitism, nest success during 1987–1994 increased from 2.7% ($N = 33$) in 1987 to 54.7% ($N = 64$) in 1993 and 50.0% ($N = 150$) in 1994 (Table 39.4, Figure 39.3c). Mayfield estimates of nest success also increased sharply, from 4.2% in 1987 to 40.6% in 1993 and 56.0% in 1994 (Table 39.5).

The annual frequency of nest parasitism was negatively correlated with total number of female cowbirds destroyed annually ($r_s = -0.976$, $P < .001$, $N = 8$). Annual estimates of nest success were significantly negatively correlated with annual percentage of nest parasitism ($r_s = -0.952$, $P < .001$, $N = 8$). Mayfield estimates had a similar negative correlation ($r_s = -0.881$, $P < .001$, $N = 8$). Tazik and Cornelius (1993) considered nest parasitism to be the cause of abandonment in 72.9% of all nests deserted during 1987–1989. Using Mayfield methods, they estimated that 49.1% of parasitized nests were deserted in 1987–1989, versus 7.6% of unparasitized nests.

We expected lower parasitism rates to reduce nest desertion and to increase nest success. Table 39.4 shows the as-

Table 39.3. Intensity of Cowbird Parasitism of Black-capped Vireo Nests in Relation to the Level of Cowbird Control

	No. of Parasitized Nests	Cowbird Eggs/ Parasitized Nest	± SD
Minimal cowbird control			
1987	29	1.966	0.659
1988	77	1.740	0.834
1989	106	1.406	0.731
Annual mean, 1987–1989		1.704	0.282
Intensive cowbird control			
1991	20	1.400	0.681
1992	13	1.077	0.277
1993	8	1.250	0.463
1994	16	1.063	0.250
Annual mean, 1991–1994		1.199	0.160

Table 39.4. Black-capped Vireo Nest Failures during 1987–1994

		Cause of Failure					
		Deserted		Destroyed		Unknown	
	Total Nests	%	(N)	%	(N)	%	(N)
1987	33	63.6	(21)	27.3	(9)		
1988	85	45.9	(39)	40.0	(34)		
1989	137	26.3	(36)	35.8	(49)		
1990	51	13.7	(7)	43.1	(22)		
1991	64	23.4	(15)	15.6	(10)	6.2	(4)
1992	62	16.1	(10)	22.6	(14)	12.9	(8)
1993	64	18.8	(12)	14.1	(9)	9.4	(6)
1994	150	11.3	(17)	23.3	(35)	14.7	(22)

Note: See Methods for definitions of categories of nest failure.

signed causes of nest failure for all nests during 1987–1994. As expected, the percentage of deserted nests was consistently lower during 1991–1994 (mean annual rate = 17.4%) than in 1987–1989 (45.3%). The 6–15% of nests in 1991–1994 where the cause of failure was unknown, however, confound this result. Some of these nests were undoubtedly deserted, which would increase estimates of desertion during 1991–1994. Hunt (1991) attributed the low desertion rate in 1990 (13.7%) to a late start to fieldwork, one month into the vireo breeding season, which may have caused him to miss early desertions. This potential bias may also explain the high nest success in 1990 despite minimal cowbird control.

Effects of Year, Region, and Cowbird Control on Parasitism

Patterns of cowbird parasitism at Fort Hood could be explained by several factors, including cowbird control and year effects due to changing cowbird abundance or habitat use. Two patterns are clear from the data in Table 39.6. First, there was an overall decline in parasitism, regardless of the local region, between 1987–1990 (the years with minimal cowbird control) and 1991–1994 (the years with intensive cowbird control) (Figure 39.4). Second, during 1991–1994 there were significant differences in parasitism between areas with and without intensive cowbird control (Table 39.7). Year and/or regional differences independent of control efforts could explain both of these patterns. The results below, however, suggest that the differences in patterns of parasitism in 1987–1990 and 1991–1994 were largely due to cowbird control.

Year effects are suggested by declines in nest parasitism in all regions between 1987–1990 and 1991–1994 regardless of trapping effort within regions (Table 39.6). Examination of data within periods of equivalent control efforts (i.e., 1987–1990 and 1991–1994) should indicate to what extent annual variation in cowbird parasitism was independent of cowbird control efforts. The frequency of parasitism varied significantly among years during the period of minimal cowbird control, 1987–1990 ($X^2 = 24.122$, df = 3, $P < .001$; Table 39.8). Logistic regression analysis for 1988 and 1989 showed a significant effect of year on parasitism but no effect of the limited cowbird control effort on nest parasitism (Tazik and Cornelius 1993). Tazik and Cornelius hypothesized that the drop in cowbird parasitism in 1989 was due either to lower cattle numbers on the installation that year or to annual variability in cowbird parasitism independent of cowbird control efforts.

During intensive cowbird control in 1991–1994, no significant variation in the frequency of parasitism was detected among years for untreated areas ($X^2 = 5.172$, df = 2, $P = .075$; Table 39.8). Data for 1993 were excluded from this analysis because of a small sample size. Parasitism did, however, vary significantly among years in treated areas during 1991–1994 ($X^2 = 10.983$, df = 3, $P = .012$; Table 39.8). This result was largely due to one region, southwest Fort Hood (Table 39.2). When we excluded the data for southwest Fort Hood, parasitism did not vary significantly among years in treated areas during 1991–1994 ($X^2 = 2.659$, df = 3, $P = .447$; Table 39.8). Declines in parasitism from 1987–1990 to 1991–1994 were greater in the east and west ranges (treated) than in the live-fire area (untreated) and in southwest Fort Hood (mixed treatments; see Methods) (Figure 39.4).

The results for 1987–1990 suggest that annual variation in parasitism, at least during periods of high parasitism rates, could be due to factors other than cowbird control. However, two results suggest that annual variation unrelated to cowbird control does not explain all the difference in parasitism between 1987–1990 and 1991–1994.

First, there was little variation among years during 1991–1994 (except for southwest Fort Hood), and second, the decrease in parasitism between 1987–1990 and 1991–1994 was greater in treated regions than in untreated regions (excluding southwest Fort Hood). Alternatively, the effects of cowbird control may have damped out effects of annual variation in 1991–1994.

The frequency of parasitism was significantly lower in all

Table 39.5. Mayfield Estimates of Nest Success in Black-capped Vireo Nests for Years with Minimal and Intensive Cowbird Control

	Daily Failure Rate (Number of Nests/Exposure Days)		
	Incubation Period	Nestling Period	% Nest Success[a]
Minimal cowbird control			
1987	0.1031 (22n/145.5d)	0.1569 (11n/51d)	4.2
1988	0.0763 (49n/380.5d)	0.0920 (25n/174d)	13.3
1989	0.0447 (105n/939.5d)	0.0438 (82n/66d)	33.7
1990[b]	0.0501 (38n)	0.0501 (38n)	25.5
Intensive cowbird control			
1991	0.0435 (24n/184d)	0.0312 (28n/128d)	36.2
1992	0.0350 (26n/198.5d)	0.0670 (41n/163.5d)	27.1
1993	0.0350 (10n/293.5d)	0.0341 (9n/264d)	40.6
1994	0.0125 (7n/798d)	0.0290 (16n/586d)	56.0

[a]For entire nest cycle.

[b]Data for 1990 (Hunt 1991) were not separated into incubation and nestling periods, and numbers of exposure days were not reported.

Table 39.6. Parasitism of Black-capped Vireo Nests by Cowbirds at Fort Hood by Region

Region[a]	1987	1988	1989	1990	1991	1992	1993	1994
East ranges								
No. parasitized nests (%)	28 (90.3)	27 (84.4)	29 (51.8)	10 (66.7)	3 (13.6)	1 (4.2)	4 (14.8)	5 (10.4)
No. unparasitized nests	3	5	27	5	19	23	23	43
Treatment category					Y	Y	Y	Y
West Ranges								
No. parasitized nests (%)		10 (83.3)	17 (70.8)	1 (25.0)	1 (50.0)	1 (14.3)	2 (13.3)	1 (3.7)
No. unparasitized nests		2	7	3	1	6	13	26
Treatment category					Y	Y	Y	Y
Southwest Fort Hood								
No. parasitized nests (%)	1 (100.0)	19 (100.0)	17 (58.6)	5 (100.0)	2 (25.0)	9 (42.9)	9 (60.0)	4 (11.4)
No. unparasitized nests	0	0	12	0	6	12	6	31
Treatment category					Y	N	Y	Y
Live-fire area								
No. parasitized nests (%)	1 (100.0)	23 (95.8)	36 (83.7)	8 (53.3)	20 (57.1)	8 (61.5)	2 (22.2)	7 (28.0)
No. unparasitized nests	0	1	7	7	15	5	7	18
Treatment category					N	N	N	N

[a] Treatment categories as in Table 39.2.

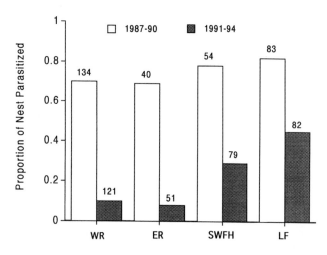

Figure 39.4. Proportions of nests parasitized by region for 1987–1990 (prior to intensive control efforts) and in 1991–1994 (during intensive cowbird control). Total numbers of nests above bars. Proportions of parasitized nests differed significantly (chi-square tests, $P < .001$) between 1987–1990 and 1991–1994 in all regions. Data from Table 39.6.

regions with cowbird control than in regions without control in 1991, 1992, and 1994 (Table 39.7). Nest parasitism was independent of treatment category in 1993. The sample size ($N = 9$), however, is small for areas with no cowbird control because of limited access of researchers to the live-fire area in 1993. Regional differences in habitat use or cowbird abundance independent of cowbird control efforts could explain these results for 1991–1994, but this cannot be addressed because cowbird control efforts were not stratified by region.

We pooled data by region for 1987–1990, the years prior to intensive control efforts, to test for regional differences. In 1987–1990, we found no significant variation among regions in the incidence of nest parasitism ($X^2 = 4.487$, df = 3, $P = .213$; Figure 39.4). There was also no significant difference ($X^2 = 0.556$, df = 1, $P = .456$) in parasitism between the live-fire area (8 of 15 nests parasitized) and the east ranges (10 of 15 nests parasitized) in 1990. Sample sizes for the west ranges and southwest Fort Hood in 1990 were too small for chi-square analysis of regional differences. Logistic regression analysis of data from 1988 and 1989 by Tazik and Cornelius (1993) showed significant regional effects on parasitism, and such effects were marked after 1990 (Figure 39.4). Many fewer vireo nests were parasitized at the east and west ranges than at southwest Fort Hood or on the live-fire area. Although there may have been temporal variability in parasitism among regions, there were no consistent temporal differences in parasitism among regions that could explain the differences between treated and untreated regions during 1991–1994.

The higher levels of parasitism in the live-fire area were expected, given the difficulty of trapping in the midst of active military operations, but the results for southwest Fort Hood are anomalous compared to those for the east and west ranges. Control efforts at southwest Fort Hood during 1990–1994 differed from the east and west ranges both in trap effort and effectiveness (Table 39.2). Despite control efforts in 1991, 1993, and 1994, parasitism in pooled data for 1991–1994 remained frequent compared to other areas with cowbird control (Figure 39.4). Parasitism rates at

southwest Fort Hood also varied widely during 1991–1994, unrelated to control efforts. Cowbird trapping and shooting were suspended temporarily in 1992 to evaluate the effect of cattle removal from southwest Fort Hood. Parasitism subsequently increased from 25.0% ($N = 8$) in 1991 to 42.9% ($N = 21$) in 1992. Parasitism increased further to 60.0% ($N = 15$) in 1993 despite renewed trapping effort and the continued absence of cattle. Parasitism decreased to 11.4% ($N = 35$) in 1994, a level equivalent to that on other regions with cowbird control. The reduced parasitism in 1994 may have been caused by early morning shooting within vireo territories on alternate days.

Table 39.7. Associations between Treatment Category (Cowbird Control versus No Cowbird Control) and Incidence of Parasitism

Year	% Parasitism (N)					
	Treated		Untreated		χ^2	df	P
1991	18.8	(32)	57.1	(35)	10.376	1	0.001
1992	3.2	(31)	50.0	(34)	14.886	1	< 0.001
1993	26.3	(57)	22.2	(9)	0.068	1	0.794
1994	9.1	(110)	28.0	(25)	6.617	1	0.10

Table 39.8. Comparison of Cowbird Parasitism among Years for Periods 1987–1990 and 1991–1994

Minimal Cowbird Control, 1987–1990[a]	1987	1988	1989	1990
No. Parasitized nests (%)	30 (90.9)	79 (90.8)	99 (65.1)	24 (61.6)
No. Unparasitized nests	3	8	53	15
Untreated Areas, 1991–1994[b]	1991	1992	1993	1994
No. Parasitized nests (%)	20 (57.1)	17 (50.0)	2 (22.2)	7 (28.0)
No. Unparasitized nests	15	17	7	18
Treated Areas, 1991–1994[c]	1991	1992	1993	1994
(Including Southwest Fort Hood Data)				
No. Parasitized nests (%)	6 (18.8)	2 (6.5)	15 (26.3)	10 (10.0)
No. Unparasitized nests	26	29	42	100
Treated Areas, 1991–1994[d]	1991	1992	1993	1994
(Excluding Southwest Fort Hood Data)				
No. Parasitized nests (%)	4 (16.7)	2 (6.5)	6 (14.3)	6 (8.0)
No. Unparasitized nests	20	29	36	69

[a]χ^2 (among years 1987–1990) = 24.122, df = 3, $P < .001$
[b]χ^2 (among years 1991–1994, excluding 1993) = 5.172, df = 2, $P = .075$
[c]χ^2 (among years 1991–1994) = 10.983, df = 3, $P = .012$
[d]χ^2 (among years 1991–1994) = 2.659, df = 3, $P = .447$

Discussion

Cowbird control efforts at Fort Hood suggest that reducing female cowbird abundance on a regional scale is necessary to reduce cowbird parasitism of Black-capped Vireo nests effectively. Limited site-specific cowbird control in occupied vireo habitats at Fort Hood in 1988–1989 did not significantly reduce the incidence of nest parasitism (Tazik and Cornelius 1993). Despite equivalent or reduced site-specific control efforts, more efficient cowbird trapping in 1991–1994 increased the total numbers of female cowbirds killed and was more effective in reducing nest parasitism.

Female cowbirds maintain stable home ranges or territories in breeding habitats (Dufty 1982, Darley 1983, Rothstein et al. 1984; Raim, Chapter 9, this volume). The diurnal movement patterns of cowbirds between nesting and foraging habitats (Rothstein et al. 1984; Thompson and Dijak Chapter 10 and Raim, Chapter 9, this volume) and increased social behavior on foraging areas (Dufty 1982, Rothstein et al. 1984) suggest that females are more exposed to traps on foraging areas than in breeding habitats. At the high cowbird densities seen at Fort Hood, the low capture rate of traps placed in breeding habitats in 1987–1990 may not have reduced the pool of replacement cowbirds, and individual females removed from breeding habitats may have been quickly replaced by neighboring females expanding their home range or by female floaters searching for breeding ranges.

Comparison of treated and untreated regions of Fort Hood during 1991–1994 showed that even the highest levels of female removal did not reduce parasitism uniformly throughout all regions of Fort Hood (Figure 39.4, Table 39.6). There are no barriers to diurnal movements of cowbirds on Fort Hood, and some of the females killed in traps on foraging areas in the treated east and west ranges undoubtedly used nesting habitats within the untreated live-fire area. This spillover effect of trapping may explain the lower incidence of parasitism in the live-fire area in 1991–1994 compared with earlier years. Lower parasitism in the live-fire area in 1993–1994 compared with 1991–1992, concurrent with a near doubling of the total number of females destroyed annually, again suggests that trapping in foraging areas and reducing the regional number of cowbirds is an effective management strategy. Alternatively, results in the live-fire area could be explained by decreases in overall cowbird abundance unrelated to cowbird control. The increase in cowbird capture rates from 1991 to 1994, however, argues against a decline in cowbird abundance sufficient to explain the decline in parasitism observed in the live-fire area.

The results for southwest Fort Hood further illustrate the regional dynamics of cowbird control efforts. Southwest Fort Hood is a peninsula of military land in a surrounding landscape of agricultural lands and urban areas. It is thus isolated from regions with more intensive trapping efforts.

Control efforts in this region were less effective in reducing parasitism in 1991–1993 than in other treated regions and may have been inadequate to reduce regional cowbird populations. Regions surrounding southwest Fort Hood probably provided a sufficiently large pool of cowbirds to replace females destroyed by control efforts.

The infrequent parasitism at southwest Fort Hood and the west ranges in 1994 is intriguing, considering the use of shooting on alternate days in occupied vireo habitats in these regions. Whether shooting within occupied habitats on a frequent and regular basis is effective in reducing site-specific parasitism rates awaits more data in future years. If it is successful, this approach may provide an alternative to trapping in areas where it is not feasible to reduce regional cowbird populations. This leaves open the question of whether this shooting strategy can reduce cowbird densities successfully.

Control efforts that target specific nesting habitats may be more effective in physiographic regions with lower cowbird densities and/or where nesting habitat of host species is limited. Under these conditions, cowbirds may be concentrated in their preferred nesting habitats, and replacement of females that are removed may be less frequent than in regions with high cowbird densities.

Removal of cattle from southwest Fort Hood during 1992–1993 had no apparent effect on cowbird parasitism rates. It is unknown whether cattle removal on a larger scale throughout Fort Hood would lower the incidence of parasitism, especially considering that other land-use practices on the installation also maintain short grass conditions. Cattle removal on a spatial scale equivalent to southwest Fort Hood (over 7,000 ha) may have little effect on the incidence of parasitism if there is no concurrent change in surrounding land use. A study of the relationship between cattle and cowbird populations of Fort Hood was initiated in 1994.

The substantial increases in number of females killed annually and trap success (females destroyed/trap/day) during our study were due in part to improved trapping methods. Despite this increased efficiency, our control efforts had little or no effect on cowbird numbers trapped the following year. Griffith and Griffith (Chapter 38) and DeCapita (Chapter 37) have also found this result, but Whitfield noted declines in numbers of cowbirds trapped from year to year (see Chapter 36). Cowbird control efforts at Fort Hood will have to be continued at current levels indefinitely to maintain reduced levels of parasitism of Black-capped Vireos.

Conservation and Maintenance of Endangered Black-capped Vireo Populations

Tazik and Cornelius (1993) calculated that Black-capped Vireos need to produce 2.67 vireo young per female per year to maintain stable populations (assuming 60% adult annual survival and 40% juvenile annual survival). They estimated that this level of production could be sustained at parasitism levels of 16% to 38% (but see Grzybowski and Pease, Chapter 16, this volume). The lower limit (16%) for parasitism in a stable population was based on application of an equation from May and Robinson (1985). The upper limit (38%) was derived from linear regression of observed parasitism and productivity.

Linear regression (Figure 39.5) was used to estimate that 1,712 to 2,735 female cowbirds must be destroyed annually in March–June to maintain cowbird parasitism levels within 16 to 38%. The current level of trapping effort appears to be sufficient to keep parasitism levels low enough to sustain stable vireo populations.

Reduced parasitism and increased nest success in recent years may be enhancing vireo populations at Fort Hood. Although survey effort has varied among years and regions, three areas (Training Area 2, Training Area 44B, and southwest Fort Hood) had equivalent survey effort during the years 1991–1994. In these areas, numbers of territories increased from 66 in 1991 to 156 in 1994. Vireos are also establishing territories in previously unoccupied habitats at Fort Hood (Weinberg et al. 1994). Population sizes of vireos, however, are probably influenced by a variety of factors other than reduced parasitism, including habitat availability, immigration and emigration, and adult and juvenile survival rates.

Recommendations for Managers

1. Managers should evaluate regional dynamics of cowbird populations and host nesting habitats when selecting approaches to cowbird control. Managers should try to maximize the number of females killed in regions with high cowbird densities and abundant host nesting habitat.

2. Trapping success in landscapes with abundant breeding and foraging habitat for cowbirds is maximized by trapping in cowbird foraging areas rather than in the nesting habitat of hosts.

3. Once initiated, cowbird control efforts must be continued indefinitely to maintain protection for host populations.

4. Efforts to reduce parasitism in small, isolated host populations may not succeed without control efforts sufficient to reduce cowbird densities in the surrounding region (see Chapter 22). Potential benefits to small, isolated populations of species of concern must, however, be weighed against the substantial effort and cost of reducing regional cowbird abundance.

5. Even control efforts as intensive as those currently conducted at Fort Hood have a limited area of effect in reducing cowbird populations and parasitism. The success of cowbird controls at Fort Hood cannot be extrapolated to the control of cowbird populations on a national or statewide scale or at winter roosts.

6. Cowbird control efforts such as those implemented at Fort Hood are costly in personnel and other resources. Before initiating such a program, managers should consider five questions. (a) Is the viability of host populations of concern in jeopardy from cowbird parasitism? (b) Are target host populations viable in the absence of nest parasitism? (E.g., a very small population with a low probability of long-term persistence may not be a suitable candidate for intensive control efforts.) (c) Are resources available to continue the control effort indefinitely? (d) Have management goals been established based on measurable population parameters that reflect associations between cowbird parasitism and host productivity (Robinson et al. 1993)? (e) Will monitoring be implemented to evaluate the effectiveness of control efforts, with the ultimate measure of success being a marked and sustained increase in population size of the target hosts?

References Cited

Ankney, C. D., and D. M. Scott. 1982. On the mating system of Brown-headed Cowbirds. Wilson Bull. 94:260–268.

Benson, R. H., and K. L. P. Benson. 1990. Estimated size of Black-capped Vireo population in northern Coahuila, Mexico. Condor 92:777–779.

———. 1991. Reply to Scott and Garton. Condor 93:470–472.

Bolsinger, J. S., and T. J. Hayden. 1994. Project status report: 1993 field studies of two endangered species (the Black-capped Vireo and the Golden-cheeked Warbler) and the cowbird control program on Fort Hood, Texas. Unpubl. report, HQ III Corps and Fort Hood, DEH, Fort Hood. 54 pp.

Brittingham, M. C., and S. A. Temple. 1983. Have cowbirds caused forest songbirds to decline? BioScience 33:31–35.

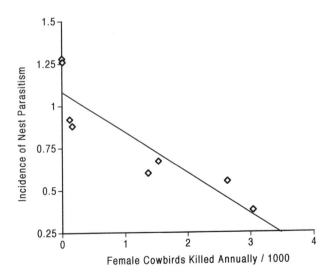

Figure 39.5. Annual percentage of cowbird parasitism (arcsine transformed data) relative to total number of female cowbirds killed annually at Fort Hood, Texas. Regression equation: $y = 1.087 - 0.247x$, $r^2 = 0.821$, $N = 8$.

Darley, J. A. 1983. Territorial behaviour of the female Brown-headed Cowbird (*Molothrus ater*). Can. J. Zool. 61:65–69.

Dufty, A. M., Jr. 1982. Movements and activities of radio-tracked Brown-headed Cowbirds. Auk 99:316–327.

Espey Houston and Associates. 1989. Results of the 1989 Brown-headed Cowbird trapping program of the Austin Regional Habitat Conservation Plan. Unpubl. report, Nature Conservancy of Texas and Executive Committee of Austin Regional Habitat Conservation Plan, Austin, TX.

Franzreb, K. E. 1989. Ecology and conservation of the endangered Least Bell's Vireo. USFWS, Biol. Rep. 89(1). 17 pp.

Friedmann, H. 1963. Host relations of the parasitic cowbirds. U.S. Natl. Mus. Bull. 233. 273 pp.

Friedmann, H., and L. F. Kiff. 1985. The parasitic cowbirds and their hosts. Proc. Western Found. Vert. Zool. 2:225–302.

Friedmann, H., L. F. Kiff, and S. I. Rothstein. 1977. A further contribution to knowledge of the host relations of the parasitic cowbirds. Smithsonian Contrib. Zool. 235.

Graber, J. W. 1961. Distribution, habitat requirements, and life history of the Black-capped Vireo (*Vireo atricapillus*). Ecol. Monogr. 31:313–336.

Grzybowski, J. A. 1989. Black-capped Vireo investigations: Population and nesting ecology. Unpubl. report, USFWS Office of Endangered Species, Albuquerque, NM.

———. 1990. Ecology and management of the Black-capped Vireo (*Vireo atricapillus*) in the Wichita Mountains, Oklahoma, 1990. Wichita Mountains Wildlife Refuge, USFWS, Indiahoma, OK. 28 pp.

Hahn, C. M., and J. S. Hatfield. 1995. Parasitism at the landscape scale: Do cowbirds prefer forests? Conserv. Biol. 9:1414–1424.

Hayden, T. J., and D. J. Tazik. 1991. Project status report: 1991 field studies of two endangered species (the Black-capped Vireo and the Golden-cheeked Warbler) and the cowbird control program on Fort Hood, Texas. Unpubl. report, HQ III Corps and Fort Hood, DEH, Fort Hood. 81 pp.

Hunt, J. 1991. Status of Black-capped Vireos on Fort Hood, Texas. Unpubl. report, U.S. Army Construction Engineering Laboratories, Champaign, IL.

Mangold, L. K., J. M. Richter, and M. E. DeCapita. 1994. Control of Brown-headed Cowbirds on Kirtland's Warbler nesting areas in northern Michigan. Unpubl. report, USFWS, East Lansing, MI. 11 pp.

May, R. M., and S. K. Robinson. 1985. Population dynamics of avian brood parasitism. Amer. Natur. 126:475–494.

Mayfield, H. F. 1961. Nesting success calculated from exposure. Wilson Bull. 73:255–261.

———. 1965. The Brown-headed Cowbird with old and new hosts. Living Bird 4:12–28.

———. 1975. Suggestions for calculating nest success. Wilson Bull. 87:456–466.

Payne, R. B. 1973. The breeding season of a parasitic bird, the Brown-headed Cowbird in central California. Condor 75:80–99.

Probst, J. R. 1986. A review of factors limiting the Kirtland's Warbler on its breeding grounds. Amer. Midland Natur. 116:87–100.

Pulich, W. M. 1976. The Golden-cheeked Warbler: A bioecological study. Texas Parks and Wildlife Department, Austin.

Raisz, E. 1952. Land forms of the United States. International Geological Congress, Washington, DC.

Robinson, S. K., J. A. Grzybowski, S. I. Rothstein, M. C. Brittingham, L. J. Petit, and F. R. Thompson. 1993. Management implications of cowbird parasitism on neotropical migrant songbirds. Pp. 93–102 in Status and management of neotropical migratory birds (D. M. Finch and P. W. Stangel, eds). USDA Forest Serv. Gen. Tech. Rep. RM-229.

Rothstein, S. I., J. Verner, and E. Stevens. 1980. Range expansion and diurnal changes in dispersion of the Brown-headed Cowbird in the Sierra Nevada. Auk 97:253–267.

———. 1984. Radio-tracking confirms a unique diurnal pattern of spatial occurrence in the parasitic Brown-headed Cowbird. Ecology 65:77–88.

Scott, J. M., and E. O. Garton. 1991. Population estimates of the Black-capped Vireo. Condor 93:469–470.

Sweetwater Environmental Biologists. 1989. Cowbird removal program during 1988 at Marine Corps base Camp Pendleton, California. Unpubl. report, U.S. Marine Corps, Natural Resources Office, Camp Pendleton. 17 pp.

Tazik, D. J. 1991. Proactive management of an endangered species on Army lands: The Black-capped Vireo on the lands of Fort Hood, Texas. Ph.D. Thesis, University of Illinois, Urbana-Champaign. 247 pp.

Tazik, D. J., and J. D. Cornelius. 1993. Status of the Black-capped Vireo at Fort Hood, Texas, vol. 3: Population and nesting ecology. USACERL Tech. Rep. EN-94/01.

Tazik, D. J., J. D. Cornelius, and C. A. Abrahamson. 1993b. Status of the Black-capped Vireo at Fort Hood, Texas, vol. 1: Distribution and abundance. USACERL Tech. Rep. EN-94/01.

Tazik, D. J., J. A. Grzybowski, and J. D. Cornelius. 1993a. Status of the Black-capped Vireo at Fort Hood, Texas, vol. 2: Habitat. USACERL Tech. Rep. EN-94/01.

Thompson, F. R. III. 1994. Spatial and temporal patterns of breeding Brown-headed Cowbirds in the midwest United States. Auk 111:979–990.

U.S. Department of Interior. 1973. Building and operating a decoy trap for live-capturing cowbirds and other birds. USFWS Rep. AC-211, Twin Cities, MN.

USFWS. 1991. Black-capped Vireo (*Vireo atricapillus*) recovery plan. Austin, TX.

———. 1992. Golden-cheeked Warbler (*Dendroica chrysoparia*) recovery plan. Albuquerque, NM. 88 pp.

Walkinshaw, L. H. 1983. Kirtland's Warbler: The natural history of an endangered species. Cranbrook Institute of Science, Bloomfield Hills, MI.

Weinberg, H. J., J. S. Bolsinger, and T. J. Hayden. 1994. Project status report: 1994 field studies of two endangered species (the Black-capped Vireo and the Golden-cheeked Warbler) and the cowbird control program on Fort Hood, Texas. Unpubl. report, HQ III Corps and Fort Hood, DEH, Fort Hood. 71 pp.

Wilkinson, L. 1992. SYSTAT for Windows, version 5.03. SYSTAT, Inc., Evanston, IL.

Zar, J. H. 1984. Biostatistical analysis. Prentice Hall, Englewood Cliffs, NJ.

40. Results of a Brown-headed Cowbird Control Program for the Southwestern Willow Flycatcher

MARY J. WHITFIELD

Abstract

In 1992, I initiated a Brown-headed Cowbird control program in the South Fork Kern River Valley, California. My goal was to reduce cowbird parasitism and to establish and maintain a self-sustaining population of the Southwestern Willow Flycatcher. Flycatcher numbers had declined from 1989 to 1992, when 63.5% of flycatcher nests were parasitized. Low average nest success (20%) during this period was mainly due to cowbird parasitism. In the first year of cowbird control, I shook cowbird eggs and removed cowbird nestlings from Willow Flycatcher nests. In addition, 30 female cowbirds found near flycatcher nests were shot. Nest success of flycatchers in 1992 increased to 31%. In 1993 and 1994, I monitored flycatcher nests and trapped cowbirds in half of a 500-ha study area, using four and seven traps respectively. Point-count surveys suggested that trapping reduced cowbird abundance by about 50% on the trapped area, while cowbird numbers in the control area remained stable. Cowbird numbers declined from May 1993 to May 1994, suggesting that trapping in 1993 reduced cowbird abundance in 1994. Parasitism was reduced to 17.4% in the trapped area compared to 56.3% in the control area. Parasitism in the trapped area declined from 60.4% before trapping to 17.4% during trapping. In addition, nest success and numbers of flycatcher young fledged per pair were higher in trapping years than in years without cowbird control. Cowbird control in 1992–1994 resulted in a self-sustaining flycatcher population in those three years.

Introduction

The southwestern subspecies of Willow Flycatcher (*Empidonax traillii extimus*) was once a common breeder in riparian willow thickets in southern California, southern Nevada, southern Utah, Arizona, New Mexico, western Texas, and extreme northwestern Mexico (Unitt 1987). Today it is very rare throughout most of its former breeding range, and it has probably been extirpated from Nevada and Utah. The U.S. Fish and Wildlife Service estimates that there are between 230 and 500 pairs remaining in the rest of its range (Federal Register 1993). Because of this dramatic decline in distribution and abundance, California, Arizona, and New Mexico have listed the Willow Flycatcher as an endangered species and the U.S. Forest Service has listed it as a sensitive species. Furthermore, the U.S. Fish and Wildlife Service has listed the Southwestern Willow Flycatcher as endangered (Federal Register 1995).

Unitt (1987) designated the South Fork Kern River, Kern County, California, as the northern extent of the Southwestern Willow Flycatcher's range in the Sierra Nevada Mountains in California. Unitt (1987) based his decision on a specimen collected in the area in 1911 by an expedition led by Joseph Grinnell of the Museum of Vertebrate Zoology, University of California, Berkeley.

The decline of the Willow Flycatcher has largely been attributed to the widespread destruction of riparian habitat essential for nesting and breeding (Harris et al. 1987, Unitt 1987). Johnson and Haight (1984) estimate that only 5% of the Southwest's original lowland riparian habitat remains. Other factors such as livestock grazing and cowbird parasitism have also played a role in this species' decline (Gaines 1974, Garrett and Dunn 1981, Harris 1991, Unitt 1987). Cowbird parasitism was of particular concern because Willow Flycatchers had a high rate of parasitism in several areas in the West, and their nests usually fail when a cowbird egg hatches in them (Walkinshaw 1961, Sedgewick and Knopf 1988, Harris 1991).

Release from cowbird parasitism appeared to increase numbers of Southwestern Willow Flycatchers on the Santa Margarita River, San Diego County, California. Cowbirds have been trapped there since 1983 to help an endangered Least Bell's Vireo (*Vireo bellii pusillus*) population (Unitt 1987, S. Buck pers. comm.; Griffith and Griffith, Chapter 38, this volume). The Willow Flycatcher population increased from 5 pairs in 1981 (prior to cowbird trapping) to

19 pairs in 1991, eight years after trapping (Unitt 1987, Pavelka 1994). The population estimate in 1993, however, was only 6 pairs. Because of reduced survey intensity (302 hours compared to 960 hours in 1991), it is unknown whether this constituted an actual decline (Pavelka 1994).

Cowbird parasitism is clearly affecting Willow Flycatchers on the South Fork Kern River. In 1987, Harris (1991) found 19 Willow Flycatcher nests on eight territories; 13 of them (68.4%) were parasitized by cowbirds. In addition, 3 nests were abandoned for unknown reasons, possibly related to cowbird parasitism. From 1989 to 1992, cowbird parasitism averaged 63.5%, while only 20.0% of the nests successfully raised at least one Willow Flycatcher young; 51.3% of unparasitized nests and 7.8% of parasitized nests raised at least one Willow Flycatcher young (Whitfield 1990, Whitfield et al. in press). After the 1991 breeding season, I hypothesized that the flycatcher population was not reproducing at a self-sustainable rate. To test this, I calculated the net reproductive rate (R_0) for the flycatchers. R_0 is the sum, over time, of the survivorship rate (l_x) multiplied by the fecundity rate (m_x), where fecundity is defined as the number of females produced per year assuming a 1 : 1 ratio of males to females (Pianka 1983). A R_0 value of less than 1.0 indicates that the population is not self-sustaining. For the years 1989 to 1991, I determined that the flycatchers were producing an average of 1.0 young per pair, which approximately equals a fecundity rate of 0.5 females per year. The adult survival rate of 0.58 was taken from McCabe (1991). Using these estimates, $R_0 = 0.66$, which indicates that the population is not self-sustaining. As expected with such a low R_0, the number of Willow Flycatchers breeding in the area dropped from 44 pairs in 1989 to 27 pairs in 1992.

In 1992, after observing the Willow Flycatcher population decrease for three years, I initiated a cowbird control program designed to reduce the impacts of brood parasitism on Willow Flycatchers. The primary goal of the control program was to increase the reproductive success of the Willow Flycatcher to a level where $R_0 \geq 1$.

Study Area

The study area is located on the Nature Conservancy's Kern River Preserve (KRP) and the adjoining USDA Forest Service South Fork Wildlife Area (SFWA), Kern County, California. It encompasses approximately 500 ha of riparian habitat along the South Fork of the Kern River from Isabella Lake upstream to the eastern boundary of the Kern River Preserve (Figure 40.1). The area is approximately 762 to 805 m above sea level. The riparian woodland is dominated by three tree species: red willow (*Salix laevigata*), Goodding's black willow (*Salix gooddingii*), and Fremont cottonwood (*Populus fremontii*). The forest is interspersed with open areas that are often dominated by mule fat (*Baccharis salicifolia*) and hoary nettle (*Urtica dioica holosericea*) and flooded areas

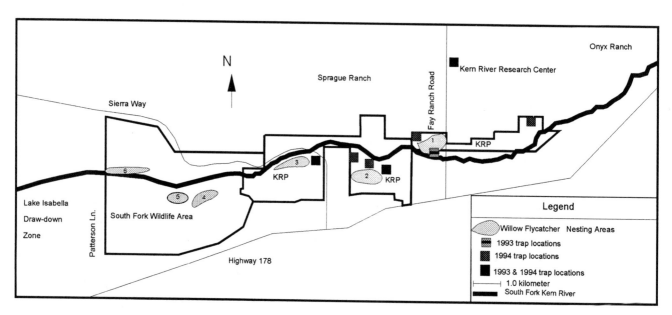

Figure 40.1. Willow Flycatcher nesting areas and Brown-headed Cowbird trap locations, South Fork Kern River. (1) Mariposa Marsh, (2) slough channel, (3) Prince Pond area, (4) South Fork Wildlife Area pond, (5) South Fork Wildlife Area Skunk Forest, and (6) South Fork Wildlife Area river channel.

that support freshwater marsh dominated by cattail (*Typha* spp.) and bulrush (*Scirpus* spp.). Hoary nettle is also a common understory plant.

Within the study area, Willow Flycatchers tend to nest in clusters in distinct areas of the forest. I studied six such clusters, which constituted 65% to 75% of the population. They were located in the following areas (as indicated on Figure 40.1): (1) Mariposa Marsh, (2) slough channel, (3) Prince Pond area, (4) South Fork Wildlife Area pond, (5) South Fork Wildlife Area Skunk Forest, and (6) South Fork Wildlife Area river channel.

Methods

Cowbird Population Monitoring

I surveyed relative densities of Brown-headed Cowbirds using point-count surveys at 60 points. Three other observers conducted approximately 20% of these surveys. The points were at least 200 m apart along forest edges near where Willow Flycatchers were nesting. I counted cowbirds at each point for 10 minutes and visited each point three times between 1 May and 10 July. We ended the point counts by mid-July, since cowbirds in the Sierra Nevada and along the South Fork Kern River show a noticeable decline in detectability by late July (Rothstein et al. 1980).

Cowbird Control

There was no cowbird control from 1989 through 1991. In 1992, between 10 June and 10 July, 30 female cowbirds found near Willow Flycatcher nesting areas throughout the study area were shot. I also addled all cowbird eggs by shaking them and removed all cowbird nestlings in Willow Flycatcher nests.

In 1993, I set up four cowbird traps (2 x 2 x 2.5-m modified Australian crow traps, Zanjac and Cummings 1965) in May and early June. Three of the traps were located near Willow Flycatcher nesting areas, while one was located at a Brown-headed Cowbird feeding area (Figure 40.1). The area that received traps (KRP areas 1, 2, and 3) usually had 16 to 24 pairs of Willow Flycatchers per year from 1989 to 1993, whereas the monitored areas without traps (SFWA areas 4, 5, and 6) had 4 to 7 pairs per year. I baited each trap with wild bird seed, water, and live wing-clipped cowbirds (four females and two males). I checked the traps twice daily to release nontarget birds and male cowbirds and to remove female cowbirds. In 1994, I opened seven cowbird traps on 1 May, baited them with three females and two males, and checked them as in 1993. Three traps were in locations used in 1993, while four were in new locations in the study area. In 1993 and 1994 I addled all cowbird eggs found in Willow Flycatcher nests by shaking them and removed all cowbird nestlings in both the trapped and nontrapped areas.

Willow Flycatcher Monitoring

I monitored Willow Flycatchers from 1989 to 1994 to determine population trends, reproductive success, and parasitism rates. I started surveying for Willow Flycatchers and searching for their nests during the last week in May, when they first begin to breed (Harris 1991, Whitfield pers. obs.). I surveyed portions of the study area that contained suitable habitat using playbacks of a recording of a singing male Willow Flycatcher (Harris et al. 1987). When a nest was located, I checked it daily using a small mirror during the egg-laying stage and then every other day during incubation and nestling stages. When the Willow Flycatcher nestlings were eight days old, I banded them with a USFWS band and one color band.

The following definitions are used in this chapter. (1) Nest success: the number of nests successfully fledging at least one Willow Flycatcher divided by the total number of nests found that had at least one egg (flycatcher or cowbird). (2) Parasitism rate: the number of nests with at least one cowbird egg or nestling divided by the total number of nests found that had at least one egg or nestling (flycatcher or cowbird). (3) Number of young fledged: the summed total of Willow Flycatcher young successfully leaving all nests. (4) Number of young fledged per pair: the total number of young fledged per year divided by the number of pairs.

Statistical Analysis

The Mann-Whitney test was used to test for differences between KRP and the SFWA for numbers of cowbirds counted and for the number of young fledged per pair (Zar 1984). I also used this test to compare the number of young fledged per pair before and after trapping. In addition, I used the Wilcoxon paired sample test when comparing cowbird count data between years (Zar 1984).

Since the data were categorical, I used chi-square tests to test for differences between the nest success and brood parasitism rates of Willow Flycatchers at different levels of cowbird control (Zar 1984). When I found a significant difference, I used multiple comparisons for tests of homogeneity to find what factor(s) was responsible for the difference (Marascuilo and Serlin 1988).

Results

Cowbird Population Monitoring

In 1992, prior to trapping, there were no significant differences in female cowbird numbers between KRP, the trapped area in 1993 and 1994, and SFWA, the nontrapped area in 1993 and 1994 (May count, $Z = 0.5$, $P > .60$; July count, $Z = 1.45$, $P > .10$; Table 40.1). In addition, there were no significant differences in female cowbird numbers between the first and last counts for all 60 points ($Z = 0.68$, $P > .50$).

Table 40.1. Brown-headed Cowbird Point-Count Results

Area	N	1992 May	1992 July	1993 May	1993 July	1994 May[b]	1994 July
		Number of Females/Point ± 2SE[a]					
KRP[c]	41	1.59 ± 0.14	1.37 ± 0.17	1.51 ± 0.16	0.71 ± 0.13	0.93 ± 0.14	0.39 ± 0.09
SFWA	19	1.58 ± 0.17	1.74 ± 0.18	2.00 ± 0.22	2.22 ± 0.18[d]	2.05 ± 0.30	1.89 ± 0.22
Total	60	1.57 ± 0.11	1.48 ± 0.13	1.67 ± 0.13	1.17 ± 0.14[e]	1.28 ± 0.15	0.87 ± 0.13

[a]June survey results are not shown.
[b]In 1994, the first cowbird count started April 28 and ended May 4.
[c]1993 and 1994 trap area.
[d]$N = 18$, due to flooding.
[e]$N = 59$.

However, in 1993 and 1994, when I trapped and removed 329 and 152 female cowbirds, respectively, the number of female cowbird detections decreased by 50% or more in the trapped area but remained stable in the nontrapped area. The decline in the trapped area between May and July was significant (1993, $Z = 2.40$, $P < .02$; 1994, $Z = 2.92$, $P < .005$, Table 40.1). Furthermore, in May 1993, before trapping began, there were no significant differences in female cowbird numbers between the trapped and nontrapped areas ($Z = 1.74$, $P > .08$). Yet, by July there were significantly fewer female cowbirds in the trapped area than in the nontrapped area ($Z = 4.95$, $P < .001$).

In 1994, however, there were significant differences between the trapped and nontrapped areas in May and July (May count, $Z = 3.29$, $P < .001$; July count, $Z = 5.15$, $P < .001$). In addition, there were significant differences in female cowbird numbers in the trapped area between May 1993 and May 1994 ($Z = 2.70$, $P < .008$) but not in the nontrapped area ($T = 38.5$, $P > .5$).

Willow Flycatcher Monitoring

The parasitism rates of Willow Flycatchers in the trapped area differed for the different levels of cowbird control between 1989 and 1994 ($X^2 = 26.56$, df = 3, $P < .005$; Figure 40.2). This difference was caused by significantly lower parasitism in years with major cowbird control (1993 and 1994) compared either to no cowbird control (prior to 1992, $Z = 5.61$, df = 2, $P < .005$) or to minor cowbird control (1992, $Z = 4.80$, df = 2, $P < .005$). In contrast, the parasitism rates were similar for the different levels of cowbird control in the nontrapped area ($X^2 = 5.44$, df = 3, $P > .10$).

Parasitism was not reduced significantly by the removal of 30 female cowbirds in 1992. In contrast, cowbird trapping did reduce parasitism rates significantly. Moreover, parasitism did not decline significantly in the nontrapped area for any level of cowbird control.

In contrast to parasitism rates, nesting success from 1989 to 1994 was not significantly different for the different levels of cowbird control in the trapped area ($X^2 = 3.70$, df = 3, $P > .10$) but it was significantly different in the nontrapped area ($X^2 = 8.11$, df = 3, $P < .05$; Figure 40.3). This difference was caused by significantly higher nest success in years with major cowbird control ($Z = 2.67$, df = 2, $P < .005$).

To evaluate the success of the trapping program, I compared five measures of breeding performance in the trapped and nontrapped areas before and during cowbird trapping (Table 40.2). These measures were (1) predation rate, (2) parasitism rate, (3) nest success, (4) number of young

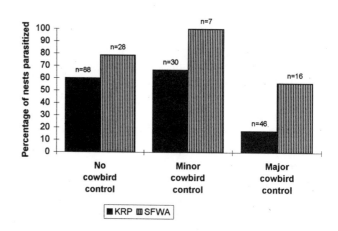

Figure 40.2. Effects of two levels of cowbird control on parasitism rates of Willow Flycatchers. Minor cowbird control was the removal of 30 female cowbirds by shooting in both areas in 1992. Major cowbird control involved cowbird trapping in KRP only in 1993–1994. Cowbird eggs were addled by shaking and cowbird nestlings were removed from Willow Flycatcher nests, in both study areas in 1992–1994. In the no-control period (1989–1991) there was no egg or nestling removal and no trapping and shooting.

fledged, and (5) number of fledged young per pair. The pre-dation rate was higher in the trapped area than the non-trapped area in 1993 and 1994, but this difference was not significant ($X^2 = 2.07$, df = 1, $P > .10$). The parasitism rate was significantly lower in the trapped area than in the non-trapped area in 1993–1994 ($X^2 = 8.35$, df = 1, $P < .005$) but not in 1989–1991 ($X^2 = 3.13$, df = 1, $P > .08$). Nest success was not significantly different in the trapped area when compared to the nontrapped area both before and during trapping (1989–1991, $X^2 = 0.80$, df = 1, $P > .30$; 1993 and 1994, $X^2 = 0.02$, df = 1, $P > .90$). The number of Willow Fly-catcher young fledged, however, was higher in the trapped area than in the nontrapped area both before and during trapping. The number of young fledged per pair was not dif-ferent in the trapped and nontrapped areas before or during trapping (1989–1991, $Z = 0.31$, $P > .50$; 1993 and 1994, $Z = 0.41$, $P > .50$). However, the number of young fledged per pair was significantly higher in the trapped area during trap-ping than before trapping ($Z = 2.18$, $P < .03$). This was not the case in the nontrapped area ($Z = 1.34$, $P > .18$).

Discussion

The low-level control efforts of 1992 had little effect on par-asitism rates and nest success of Willow Flycatchers at KRP, although cowbird removal was beneficial to the birds in the SFWA. However, the production of young in both areas in 1992 increased significantly to 1.94 fledged young per pair compared to an average of 1.0 young per pair from 1989 to 1991 ($Z = 2.13$, $P < .03$). This increase was mostly due to a lower than usual predation rate for 1992, 17.9%, compared to the average of 41.3% between 1989–1991.

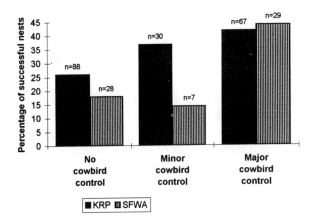

Figure 40.3. Effects of two levels of cowbird control on nesting success in Willow Flycatchers. Categories as in Figure 40.2.

In contrast, the significant decline of parasitism in 1993 and 1994 can be attributed to the cowbird trapping. In the trapped area, the parasitism rates in Willow Flycatcher nests were significantly lower than in years with no cowbird con-trol and minor cowbird control, and the flycatcher nest suc-cess was the highest recorded during the previous five years. There was, however, no significant difference in nest success between the trapped area and nontrapped areas. On the other hand, it is worth noting that the predation rate was markedly higher in the trapped area (50%) than the non-trapped area (26.7%) for 1993 and 1994, even though the difference was not significant. In addition, predation may explain the significant difference in nest success in the non-trapped area with different levels of cowbird control. The

Table 40.2. Reproduction by Willow Flycatchers in Kern River Preserve versus South Fork Wildlife Area before and during Cowbird Control

| | Before Cowbird Control 1989–1991 | | During Cowbird Control 1993–1994 | |
	KRP	SFWA	KRP[a]	SFWA
% predation (no. nests predated)	41.5 (34)[b]	41.0 (9)	50.0 (22)[c]	26.7 (4)[d]
% parasitism (no. nests parasitized)	60.2 (53)	78.6 (22)	17.4 (8)[e]	56.3 (9)
% successful (no. nests fledged)	26.1 (23)	17.9 (5)	41.6 (20)	43.8 (7)
No. young fledged	56	7	60	14
Fledglings per pair	1.00	0.93	1.87	1.55
No. nests	88	28	48	16

[a] 1993 and 1994 trap area.

[b] $N = 82$.

[c] $N = 44$.

[d] $N = 15$.

[e] $N = 46$ because 2 nests were too high to check.

predation rate averaged 41.0% with no cowbird control and 26.7% during major cowbird control. Because sample sizes were small in the nontrapped area, it is difficult to draw firm conclusions here.

During cowbird trapping, nest success was higher (41.6%) in the trapped area than before cowbird trapping (26.1%), but this difference was not statistically significant. While cowbird trapping did not increase Willow Flycatcher nest success greatly, it did increase the mean number of fledglings produced per pair per year from 1.0 before trapping to 1.87 after trapping.

One explanation for this increase in breeding success is that 1993 and 1994 were especially good years for flycatcher reproduction. Evidence from other studies along the South Fork Kern River contradicts this idea. Surveys on the preserve in early June 1993 indicated that insect numbers were below normal (Lorquin's Entomology Society unpubl. data). This decrease in insects may have caused below-normal production of young Yellow-billed Cuckoos (*Coccyzus americanus*, S. Laymon pers. comm.) and Song Sparrows (*Melospiza melodia*, B. Larison pers. comm.) in 1993. The productivity of these two species was even lower in 1994 (S. Laymon pers. comm., B. Larison pers. comm.). Also, the mean number of Willow Flycatcher young produced per successful nest in 1993 and 1994 (2.62) was almost equal to the mean production in 1989 to 1992 (2.64).

The cowbird control efforts of 1992–1994 increased Willow Flycatcher reproduction to self-sustaining levels. The average number of young produced per pair was 1.88, which yields a R_0 of 1.24, indicating that the population was increasing slightly and thus was self-sustaining for those two years. Indeed, numbers increased from 27 to 34 pairs of Willow Flycatchers from 1992 to 1994.

Cowbird point counts were conducted in 1991, but the results are not included here for two reasons: (1) They were not significantly different from the 1992 data (Wilcoxon paired sample test, $Z = 0.35$, $P > .50$), and (2) they were conducted for 15 minutes without being broken up into 5-minute intervals; therefore I could not compare the results with 1994 directly.

The decline in numbers of female cowbirds in the trapped area, together with stable cowbird numbers in the nontrapped area, suggests that female cowbirds do not move readily into vacated areas (trapped area) after they have established breeding ranges. This implies that only female cowbirds in the vicinity of Willow Flycatcher territories need to be removed to reduce parasitism rates sharply. Therefore, in some areas, a small-scale trapping program targeted on a single host may be effective.

On the other hand, the reduced numbers of female cowbirds in the trapped area for the first count in 1994, along with stable numbers in the nontrapped area and the reduced number of females caught in 1994, indicate a carryover effect from trapping in 1993. Thus, a larger trapping program may be better than a small trapping program for this area, because the more intense initial effort would reduce cowbird numbers faster. Trapping might then be required only every other year or even less often. This would save time and resources that could be better spent on alternative conservation actions such as habitat restoration and reduced cattle grazing in the riparian zone.

One should interpret the data from the cowbird counts cautiously. First, cowbirds have large home ranges (Rothstein et al. 1986). Thus, I may have counted the same cowbirds at different points, and the points may not have been independent. However, because these data were collected in the same manner each year and in both areas and were used only to estimate relative numbers, the question of independence may not be important.

Another problem with the data is that the counts were started on different dates. The first counts began on the following dates: 21 May 1992, 3 May 1993, and 28 April 1994. The last counts, however, had similar dates: 5 July 1992, 29 June 1993, and 29 June 1994. This could have affected the year-to-year comparisons of cowbird numbers for May. Last, other trapping programs have not reported a carryover effect (DeCapita, Chapter 37, Griffith and Griffith, Chapter 38, and Hayden et al., Chapter 39, this volume; M. Holmgren pers. comm.). Thus, management plans for cowbird control need to be made on a site-by-site basis because cowbird populations may react differently depending on topography and neighboring land use.

I believe that it is important to monitor the cowbird population as well as the host population when trapping cowbirds. Monitoring cowbird populations can reveal the cowbird population levels that a host species can tolerate before its reproduction falls to an unsustainable level. There may be long-term solutions, such as habitat manipulation, that keep cowbird levels below this threshold and eliminate or reduce the need for cowbird trapping.

Acknowledgments

The California Department of Fish and Game (CDFG), El Dorado Audubon, Tulare County Audubon, Monterey Audubon, Golden Gate Audubon, Santa Clara Valley Audubon, North Kern Water Interests, and private donors funded this study. Many other people contributed to this study: Ron Schlorff of CDFG provided advice and support. S. Laymon and P. Williams provided advice and encouragement and reviewed the manuscript. E. Gorney, Z. Labinger, T. Gallion, and J. Uyehara helped to improve the manuscript with their insightful comments. Tulare County Audubon built and donated one cowbird trap. B. Bergthold and R. Friedman built two cowbird traps. N. Multach built and donated two cowbird traps. I am grateful to H. Green, J. Humphrey, T. Leukering, and J. Uyehara for volunteering their time doing fieldwork. J. Placer was an invaluable field assistant for two years, and A. Storey also provided much needed help in the field. R. Tollefson of the Nature Conservancy's Kern River Preserve provided invaluable support, assistance in the field, housing for my field assistant, and three rebuilt cowbird traps.

References Cited

Federal Register. 1993. Endangered and threatened wildlife and plants: Proposed rule to list the Southwestern Willow Flycatcher as endangered with critical habitat. 58:39495–39522.

———. 1995. Endangered and threatened wildlife and plants: Final rule determining endangered status for the Southwestern Willow Flycatcher.

Gaines, D. 1974. A new look at the nesting riparian avifauna of the Sacramento Valley, California. Western Birds 5:61–80.

Garrett, K., and J. Dunn. 1981. Birds of southern California: Status and distribution. Artesian Press, Los Angeles, CA.

Harris, J. H. 1991. Effects of brood parasitism by Brown-headed Cowbirds on Willow Flycatcher nesting success along the Kern River, California. Western Birds 22:13–26.

Harris, J. H., S. D. Sanders, and M. A. Flett. 1987. The status and distribution of the Willow Flycatcher in the Sierra Nevada: Results of the 1986 survey. California Dept. Fish and Game, Wildlife Management Div., Admin. Rep. 87-2.

Johnson, R. R., and L. T. Haight. 1984. Riparian problems and initiatives in the American southwest: A regional perspective. In California riparian systems: Ecology, conservation, and productive management (R. E. Warner and K. M. Hendrix, eds). University of California Press, Berkeley.

Marascuilo, L. A., and R. C. Serlin. 1988. Statistical methods for the social and behavioral sciences. W. H. Freeman and Co., New York.

McCabe, R. A. 1991. The little green bird: Ecology of the Willow Flycatcher. Rusty Rock Press, Madison, WI. 171 pp.

Pavelka, M. A. 1994. Status of the Least Bell's Vireo and Southwestern Willow Flycatcher at Camp Pendleton Marine Corps Base, California, in 1993. USFWS, Carlsbad, CA.

Pianka, E. R. 1983. Evolutionary ecology. Harper and Row, New York.

Rothstein, S. I., J. Verner, and E. Stevens. 1980. Range expansion and diurnal changes in dispersion of the Brown-headed Cowbird in the Sierra Nevada. Auk 97:253–267.

Rothstein, S. I., D. A. Yokel, and R. C. Fleischer. 1986. Social dominance, mating and spacing systems, female fecundity, and vocal dialects in captive and free-ranging Brown-headed Cowbirds. Curr. Ornithol. 3:127–185.

Sedgewick, J. A., and F. L. Knopf. 1988. A high incidence of Brown-headed Cowbird parasitism of Willow Flycatchers. Condor 90:253–256.

Unitt, P. 1987. *Empidonax traillii extimus:* An endangered subspecies. Western Birds 18:137–162.

Walkinshaw, L. H. 1961. The effect of parasitism by the Brown-headed Cowbird on *Empidonax* flycatchers in Michigan. Auk 78:266–268.

Whitfield, M. J. 1990. Willow Flycatcher reproductive response to Brown-headed Cowbird parasitism. M.S. Thesis. California State University, Chico.

Whitfield, M. J., K. M. Enos, and S. P. Howe. In press. Is Brown-headed Cowbird trapping effective for managing populations of the Southwestern Willow Flycatcher? Studies in Avian Biology.

Zanjac, A., and M. W. Cummings. 1965. A cage trap for starlings. University of California Agricultural Extension Service OSA 129, Davis.

Zar, J. H. 1984. Biostatistical analysis. Prentice Hall, Englewood Cliffs, NJ.

Contributors

The institutions listed below are the affiliations of the contributors at the time the work published here was done. They do not represent the current affiliations for all contributors.

Bollinger, Eric K., Department of Zoology, Eastern Illinois University, Charleston

Bowen, Bonnie S., National Biological Survey, Northern Prairie Science Center, Jamestown, North Dakota

Buehler, David A., Department of Forestry, Wildlife, and Fisheries, University of Tennessee, Knoxville

Burhans, Dirk E., Division of Biological Sciences, University of Missouri, Columbia

Capen, David E., School of Natural Resources, University of Vermont, Burlington

Carello, Christy A., Department of Environmental, Population, and Organismic Biology, University of Colorado, Boulder

Chace, Jameson F., Department of Environmental, Population, and Organismic Biology, University of Colorado, Boulder

Coker, Daniel R., School of Natural Resources, University of Vermont, Burlington

Cook, Terry L., The Nature Conservancy, Austin, Texas

Cornelius, John D., U.S. Army Headquarters III Corps, Fort Hood, Texas

Costello, Christine A., Northeastern Forest Experiment Station, Durham, New Hampshire

Cruz, Alexander, Department of Environmental, Population, and Organismic Biology, University of Colorado, Boulder

Davis, Stephen K., Department of Zoology, University of Manitoba, Winnipeg

Dawson, Deanna K., National Biological Service, Patuxent Wildlife Research Center, Laurel, Maryland

DeCapita, Michael E., U.S. Fish and Wildlife Service, East Lansing, Michigan

DeGraaf, Richard M., Northeastern Forest Experiment Station, University of Massachusetts, Amherst

Dijak, William D., USDA Forest Service, North Central Forest Experiment Station, University of Missouri, Columbia

Donovan, Therese M., Division of Biological Sciences, University of Missouri, Columbia

Dowell, Barbara A., National Biological Service, Patuxent Wildlife Research Center, Laurel, Maryland

Dufty, Alfred M., Jr., Biology Department, Boise State University, Idaho

Ehrlich, Paul R., Center for Conservation Biology, Department of Biology, Stanford University

Faaborg, John R., Division of Biological Sciences, University of Missouri, Columbia

Fallon, Frederick W., National Biological Service, Patuxent Wildlife Research Center, Laurel, Maryland

Fallon, Jane E., National Biological Service, Patuxent Wildlife Research Center, Laurel, Maryland

Farmer, Chris, Department of Biological Sciences, University of California, Santa Barbara

Ford, Thomas B., Department of Biology, Indiana University, Bloomington

Gill, Sharon A., Department of Zoology, University of Manitoba, Winnipeg

Greenberg, Grant M., Department of Biology, Indiana University, Bloomington

Griffith, John T., Griffith Wildlife Biology, Calumet, Michigan

Griffith, Jane C., Griffith Wildlife Biology, Calumet, Michigan

Grzybowski, Joseph A., College of Mathematics and Science, University of Central Oklahoma, Edmond

Hahn, D. Caldwell, National Biological Survey, Patuxent Wildlife Research Center, Laurel, Maryland

Hatfield, Jeff S., National Biological Survey, Patuxent Wildlife Research Center, Laurel, Maryland

Hayden, Timothy J., Natural Resources Division, U.S. Army Corps of Engineers, Champaign, Illinois

Herkert, James R., Illinois Endangered Species Protection Board, Springfield

Hoover, Jeffrey P., Illinois Natural History Survey, Champaign

Howe, William H., National Biological Survey Bird Banding Laboratory, Laurel, Maryland

Knopf, Fritz L., National Biological Survey, National Ecology Research Center, Fort Collins, Colorado

Koford, Rolf R., National Biological Survey, Northern Prairie Science Center, Jamestown, North Dakota

Koukal, Matthew A., Department of Biology, Indiana University, Bloomington

Kruse, Arnold D., National Biological Survey, Northern Prairie Science Center, Jamestown, North Dakota

Larsen, David R., School of Natural Resources, University of Missouri, Columbia

Lokemoen, John T., National Biological Survey, Northern Prairie Science Center, Jamestown, North Dakota

Marvil, Rebecca E., Department of Environmental, Population, and Organismic Biology, University of Colorado, Boulder

McLellan, Toni M., Northeastern Forest Experiment Station, Durham, New Hampshire

McMaster, D. Glen, Department of Zoology, University of Manitoba, Winnipeg

Melton, Robert H., Natural Resources Division, U.S. Army Corps of Engineers, Champaign, Illinois

Miles, R. Kirk, Department of Forestry, Wildlife, and Fisheries, University of Tennessee, Knoxville

Nakamura, Tammie K., Department of Environmental, Population, and Organismic Biology, University of Colorado, Boulder

Neudorf, Diane L., Department of Biology, York University, North York, Ontario

Ortego, Brent, Texas Parks and Wildlife Department, Victoria

Pease, Craig M., Department of Zoology, University of Texas, Austin

Peer, Brian D., Department of Zoology, Eastern Illinois University, Charleston

Peterjohn, Bruce G., USGS, Biological Resources Division, Patuxent Wildlife Research Center, Laurel, Maryland

Petit, Daniel R., Smithsonian Environmental Research Center, Edgewater, Maryland

Petit, Lisa J., Smithsonian Migratory Bird Center, National Zoological Park, Washington, D.C.

Post, William, Charleston Museum, Charleston, South Carolina

Prather, John W., Department of Environmental, Population, and Organismic Biology, University of Colorado, Boulder

Raim, Arlo, Illinois Natural History Survey, Champaign

Robbins, Chandler S., National Biological Service, Patuxent Wildlife Research Center, Laurel, Maryland

Robinson, Scott K., Illinois Natural History Survey, Champaign

Rothstein, Stephen I., Department of Biological Sciences, University of California, Santa Barbara

Sauer, John R., USGS, Biological Resources Division, Patuxent Wildlife Research Center, Laurel, Maryland

Schwarz, Sandra, USGS, Biological Resources Division, Patuxent Wildlife Research Center, Laurel, Maryland

Sealy, Spencer G., Department of Zoology, University of Manitoba, Winnipeg

Smith, James N. M., Department of Zoology and Centre for Biodiversity Research, University of British Columbia, Vancouver

Snyder, Gregory K., Department of Environmental, Population, and Organismic Biology, University of Colorado, Boulder

Tazik, David J., Natural Resources Division, U.S. Army Corps of Engineers, Champaign, Illinois

Thompson, Frank R. III, USDA Forest Service, North Central Forest Experiment Station, University of Missouri, Columbia

Trine, Cheryl L., Department of Ecology, Ethology, and Evolution, University of Illinois, Champaign

Uyehara, Jamie C., Department of Biology, University of California, Los Angeles

Verner, Jared, Pacific Southwest Research Station, USDA Forest Service, Fresno, California

Ward, David, Mitrani Centre for Desert Research, Ben-Gurion University of the Negev, Israel

Whitehead, Donald R., Department of Biology, Indiana University, Bloomington

Whitfield, Mary J., Kern River Research Center, Weldon, California

Whyte, Carolyn Frazer, Department of Biology, Indiana University, Bloomington

Wiedenfeld, David A., George M. Sutton Avian Research Center, Bartlesville, Oklahoma

Wiley, James W., Grambling Cooperative Wildlife Project, Grambling State University, Grambling, Louisiana

Winslow, Donald E., Department of Biology, Indiana University, Bloomington

Yamasaki, Mariko, Northeastern Forest Experiment Station, Durham, New Hampshire

Index